CW01276702

Edited by Steve Chilton and Alexander J. Kent
Layout and formatting by Martin Lubikowski of ML Design, London
Printed by MC Print Services, Stoke-on-Trent
ISBN 978-0-9930089-0-0

Published by the Society of Cartographers, September 2014.

Cartography: *A Reader*

Celebrating 50 years of the Society of Cartographers

Edited by
Steve Chilton and Alexander J. Kent

Cartography: A Reader

Celebrating 50 years of the Society of Cartographers

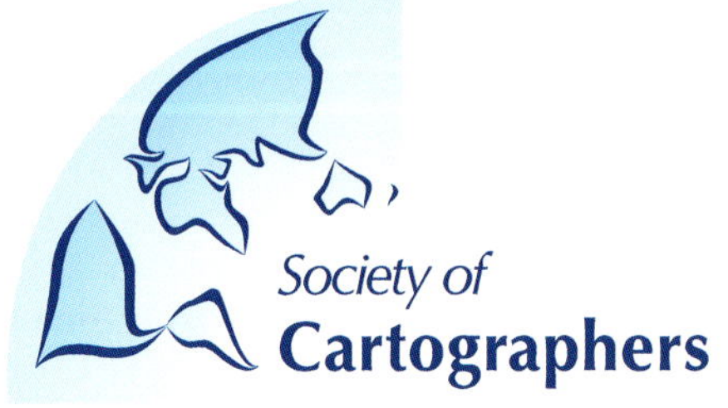

CONTENTS

FOREWORD

Over the decades (what started as) the Society of University Cartographers has widened it reach and its remit. Today it is a society for all cartographers. Tomorrow it may have an even broader appeal. This beautifully produced and carefully chosen selection of articles, cherry-picked from fifty years of the work of its members and supporters, shows not only the diligence and vivacity of UK cartography, through pieces most written by cartographers based in the United Kingdom, but also how the profession has evolved through time.

Just as a tiny number of US cartographers (those with the first access to computers in the 1960s) were beginning to explore the possibilities of automated cartography, UK cartographers were organizing themselves to better define what good practice in mapping might be and how to better think about cartography in the round. The Society's contribution pushed the practical and professional practice of map-making to the fore. It is interesting to speculate how much knowledge might have been lost, or have never been brought together and disseminated, and with what effect, had the Society not been formed in 1964.

Thanks largely to the Society of Cartographers, those who made most of the maps which appeared in UK University and other textbooks over the course of the last half century also contributed to the debate of how those maps should look. From Jane Thake's and H.A. Sandford's work on mapping for children, through to debates about the most complex of cartographic techniques – many included in this volume – the contribution of those making maps and being trained to make maps became part of the theory and education in cartography that typified the UK experience.

Writing on the first 25 years of the history of the Society (1964–1989) and also included in this volume, Carson Clark and colleagues described how, before the advent of the Society, it was common for University Cartographers to be said to have to '…possess a high degree of "manual dexterity" – a derogatory term used frequently by academics to emphasize their view of the inferiority of all non-academic staff' (see chapter 'The SUC: 1964–1989'). The Society, and its *Bulletin*, was one of the ways in which such views were greatly – if not yet entirely – changed for the better.

At times, and especially at the height of the popularity of Geographic Information Systems, articles appeared in the *Bulletin* lamenting whether cartography was becoming a dying art. The recent political history of cartography and cartographers in the UK appears to contain episodes of rebellion, despair, and anger over the incursion of machines, as well as celebration. Gary Brannon's chapter in this volume, written in 1998 suggested that: 'it is quite possible that in a few decades from now, we will look back at cartography, not in terms of a viable salary-paying profession, but as a service provided by a machine, looking perhaps much like the ubiquitous bank machines found in every high street, where a multicoloured map sheet will be custom produced at a kiosk while you wait.'

We are now, almost, a few decades from Gary's 'now'. I would suggest that if the images in this volume produced after 1998 are compared to those before that date then perhaps there is less need to despair – unless the works published by the Society in its *Bulletin* have become less and less representative of the wider world. Flick through this book, from start to end and you see a kind of animation of the evolution of contemporary UK cartography. Fifty years in fifty seconds if you can flick the pages slowly enough. The final maps, of London's subterranea by Stephen Walter, could not contrast more with Elwyn Edwards' initial images of 'ideal', 'satisfactory' and 'poor' lettering.

How might the next fifty years of cartographic evolution play out? What objects that are currently not considered maps might become part of the remit of cartography and what things that cartographers once did will be taken for granted more and more – will be automated? I suspect we'll be going off the page, into the third and fourth dimensions more routinely, and perhaps not knowing we are doing so if that becomes more ubiquitous.

I suspect the old will continue to lament the lack of knowledge of the young. In my younger days you had to learn to programme a computer to use one! I suspect there will continue to be a need for solidarity in the face of derogatory views, perhaps new forms of not very nice older views. And I suspect another fifty years will shows just as much, if not an even faster, rate of change. Hopefully this change will be based on a knowledge of the past and then the change that will be all the better for that.

Steve Chilton and Alex Kent deserve to be warmly congratulated for bringing together a volume which is much more than the sum of its parts, and which, I think, can be looked at as a whole to learn something many of us had not seen before about the evolution of map-making, especially in what now looks, in hindsight, to have been a particularly vibrant period in UK cartographic history.

Danny Dorling

President of the Society of Cartographers

INTRODUCTION

This special book is being published to celebrate the first 50 years of the Society of Cartographers. Founded at the University of Glasgow in 1964 as the Society of University Cartographers, the Society continues to serve its object to foster and encourage the study of cartography in all aspects, and in particular, to promote and maintain high standards of cartographic illustration. First published in November that year as *The Bulletin of the Society of University Cartographers*, the *Bulletin* has since earned a reputation that extends well beyond its initial readership of cartographers working in university departments around the UK and enjoys a wide international following. Since its first Issue, the scope of the *Bulletin* has encompassed the whole breadth of cartography, from critical interpretations of early maps to the use of volunteered geographical information for disaster relief. The *Bulletin* has always been oriented towards the practising cartographer, with an emphasis on showcasing the latest innovations in cartographic methods and techniques.

The idea for this book arose from a discussion between the two most recent editors of the Bulletin whilst returning from the 26th International Cartographic Conference at Dresden. Our idea was to bring together some of the best articles from the vast range that have been published in the Bulletin since the Society's foundation (a complete online archive is being made available for this special anniversary year via the Society's website *www.soc.org.uk*). Going through the archives, we each drew-up a list of papers which we felt had lasting significance and/or offered a particularly poignant reflection of the times. It was a struggle to leave out any from our lists, but we had to draw the line somewhere and eventually decided on a 'final cut'.

All five decades are represented and cover as wide a scope of themes and applications as possible within this ever-changing field. Topics include map design, planetary cartography, historical cartography, children's map reading skills, propaganda mapping, early methods of computer mapping, typographic design, census mapping, maps and the news media, topographical mapping, citizen science, crowd-sourced mapping, and map art. These are complemented by a number of reflective pieces that offer contemporary views on the status and future direction of cartography, from fear of the de-humanization of the mapmaking process to confidence in the enduring relevance of cartographic design. The sequence is laid out chronologically, rather than thematically, allowing the reader to appreciate the technological and theoretical developments within each decade and over the period as a whole. Each decade of articles is preceded by a short essay from one of the editors that reflects upon the selected contributions and their influence, and sets these within the wider context of the development of cartography in the UK.

Aside from their layout, the papers themselves are reproduced as originally published, with any identified spelling or typing errors corrected. To assist those wishing to use these articles as a starting point for their own research, we have endeavoured to identify any missing sources and existing source material has been verified and standardized. Regarding illustrations, we have aimed to reproduce original maps and diagrams as faithfully as possible, but occasionally these have been digitally restored to enhance their clarity. Where colour versions of images exist, we have attempted to locate these with the help of authors and have replaced the original black and white images where appropriate.

The book therefore brings together some of the *Bulletin's* best and most fascinating articles in a single volume, which have been contributed by many leading authors who have made major contributions to the development of cartography and of other fields, including, among others, Alan Hodgkiss, Elwyn Edwards, Mike Wood, Roger Anson, Michael Blakemore, Martin Kemp, Peter Vujakovic, Alan Collinson, Henry Castner, Ifan Shepherd, Peter Haggett, Muki Haklay, and the Society's current President, Danny Dorling. It is, however, one of the strengths of the *Bulletin* that it has encouraged many practising mapmakers and early career researchers, such as Carson Clark, Gary Brannon, Andrew Smithers, Bernhard Jenny, Nathaniel Kelso, Steve Coast, Richard Fairhurst, and Michalis Vitos to publish equally meaningful and influential contributions in its pages and we are delighted to have their work represented here.

The preparation and production of this book to celebrate the Society's 50th year has relied on the dedication and goodwill of many. We thank the founder members of the Society and those who edited the *Bulletin* before our successive turns, namely Alan Hodgkiss, Jack Render, Andrew Tatham, and Bob Parry. Without the wisdom of those forming the Society in 1964 and those who sought to publish the best work of interest to the practising cartographer thereafter, the idea of creating such a book could not have been conceived. For scanning the *Bulletin's* many pages used in the production of this book and for the compilation of the online archive, we are grateful to Graham Allsopp. We also thank John Hills for undertaking a number of higher-resolution scans of the figures which illustrate the papers that are included here. The splendid map adorning the cover is by long-standing SoC Secretary Mike Shand and we appreciate him giving permission for us to use it for this book. Jenny Kynaston kindly researched and compiled the table of the *Bulletin's* Volumes and Issues, while Heather Browning dealt with the intricacies of registering the Society as a publisher with Nielsen. For the professional services that we have relied upon to create and publish this book within such a short timeframe we owe our special thanks to Martin Lubikowski for the page layout, and to Alan Cuell for printing the work. Lastly, and certainly not least, we thank the members and committee of the Society of Cartographers, without whose generous support the production of this book could not have been possible.

We hope you join us in thanking those above and enjoy reading some of the best of the work published in the *Bulletin* as we celebrate the first 50 years of the Society's history.

Steve and Alex

THE NEW SOCIETY

Alan G. Hodgkiss

Originally published in Bulletin No 1, pp.1–3

The years 1963 and 1964 have been auspicious ones for University Cartographers. September 1963 saw the formation, at Leicester, of the British Cartographic Society, a body designed to interest all those connected with cartography in any way, whether as mapmaker or map user. Whilst the B.C.S. has much to offer those engaged in cartographic illustration, it was felt that another more specialized organization was needed to cater for the particular interests of those people engaged in producing maps for the purpose of book illustration. Consequently at the Summer School held in the University of Glasgow this autumn, a meeting was held at which it was decided to form a new society, to be called the Society of University Cartographers. The question of the title of the society provoked some discussion, it being felt by some of those present that the Society of University Cartographic Illustrators might be more suitable. However, 'Cartographer' has been recently defined as 'One whose art or business is the making of maps and charts from an accurate compilation based on expert assessment of all relevant material'. This definition suggests that the title chosen for the new society is a valid one.

Cartographic illustration, or the art of compiling and drawing maps for geographical textbooks and journals has developed considerably in the post-war years. Until then no Geography Department had a drawing office with full-time staff fully engaged in illustrative work, though one or two such as Nottingham had personnel doing a little of this kind of work among many other tasks. In the University of Liverpool, Department of Geography, in 1946, Professor H.C. Darby inaugurated a drawing office with a staff of two full-time cartographic illustrators. It is probably true to say, then, that as well as having the first Honours School in Geography in the British Isles, Liverpool was well to the forefront in establishing a cartographic office. At the present every Department has staff engaged on cartographic illustration, the larger offices such as that of University College, London having as many as six draughtsmen.

No longer do University lecturers have to spend time drawing their own maps for publication, they can rely on being able to find professional assistance to do the job for them. The work is of a specialized nature, yet the illustrator must be a man of many parts. Primarily he should be something of an artist with a flair for layout and design but able to adapt himself to the fairly rigid disciplines of cartographic drawing. He should be a neat and accurate draughtsman with the ability to produce first class lettering by any suitable method. A knowledge of the printing trade is a great advantage as much of his work is designed to be printed as book illustrations. He should have a good knowledge of the theory of cartography, know something about geography and geology, have the ability to make filed sketches and sketches from photographs, have some knowledge of photography as many of his drawings will be made into slides – and lastly and importantly, he should be able to spell. Where is such a person to be found? So far there has been no training grounds for this kind of work and entrants to the profession have come in via the Ordnance Survey, art school, straight from school or from a variety of occupations completely unconcerned with any form of cartography. A one-year full-time course in cartography has recently been announced at the Oxford Technical College in conjunction with the Clarendon Press. This is encouraging but it is doubtful whether the University authorities would permit junior cartographers to attend such a course for a year, and it is certainly very unlikely that the present level of salaries for junior technicians would attract new personnel to the profession from such a course.

The present system of training is for juniors to be trained in the different offices by senior members of staff as and when time permits – learning as they work – but the Department of Geography, University of Southampton have inaugurated a systematic training scheme in which the chief cartographer devotes one full day each week solely to training his juniors and works with them to a carefully prepared syllabus with an examination and certificate at the end of one year. This certificate is recognized by the University authorities in Southampton as a suitable qualification for which a bonus is paid on top of the normal annual salary. It would seem that one of the major tasks of the new society should be to try to formulate a national scheme of training with a recognized syllabus and examination approved by the University authorities. Difficulties might be raised in a Department where there was only one illustrator and no one to supervise the training scheme. This could perhaps be overcome by a regional system of training where the junior could attend the nearest Department having senior members of staff, perhaps one day per week or fortnight. It should be possible by cooperation between Departments to establish a workable scheme. To have a set training scheme with a qualification at the end of it would certainly be an incentive to new entrants to the profession and should also do much to improve the general level of work being carried out throughout the country.

A Summer School for University Cartographers, Glasgow, 1964

The Department of Geography, University of Glasgow, organized a most successful week's course for University cartographers from September 21st to 28th, 1964. Twenty people attended the course and fourteen different University departments were represented. All the participants were greatly appreciative of the enormous amount of work put into the course by Mr. J.S. Keates, Mr. G. Petrie and Mr M. Wood. The very interesting programme of lectures and discussions ranged over methods of changing scale, relief depiction, photomechanical methods, photogrammetry, scribing, airbrush work and the care and cataloguing of maps. It was especially valuable to be able to try out new skills such as scribing and airbrush work which are not normally encountered in the daily work of the cartographic illustrator, but which should certainly be part of his stock in trade. A full day was devoted to a visit to the Edinburgh establishment of John Bartholomew & Sons Limited. This was a fascinating and valuable experience and it was particularly refreshing to encounter the craftsmen such as the engravers on the copper sheets, with their very real and obvious pride in their work. The brothers, John and Robert Bartholomew were most generous in giving a great deal of time to explaining what goes on in the different departments and to escorting the party round the works. Everyone was impressed by the skill of engravers, draughtsmen and printers and will no doubt study Bartholomew maps with a new knowledge and interest. The final event of the week was a most enjoyable excursion into the Highlands visiting Stirling, St. Fillans, Killin and Loch Lomond. Professor Miller made this a most rewarding tour by his fascinating commentary on the region. All in all a splendid week and something of an occasion, as never before have so many University Cartographers had the opportunity to meet each other.

Editorial Note

This article was the very first one in the inaugural issue of *The Bulletin of the Society of University Cartographers*, which was published in November 1964. It had no volume number at that time. It is unsigned, but was written by Alan Hodgkiss, who first took on the role of Editor. It provides a fascinating insight in to cartography at the time of writing and the background to the Society's formation.

BULLETIN EDITORS AND VOLUME INDEX

Editors	
1964–1973	Alan Hodgkiss
1974–1979	Jack Render
1980–1983	Andrew F. Tatham
1983–1987	Robert B. Parry
1988–2006	Steve Chilton
2007–present	Alexander J. Kent

* According to date specified on Issue; actual publication date occasionally later.
† The first three Issues were not published in a Volume.
‡ Published together from 2003.

Index to Volumes*			
1†	November, 1964	21, No.1	1987
2†	February, 1965	21, No.2	1988
3†	June, 1965	22, No.1	1988
1, No.1	June, 1966	22, No.2	1988
1, No.2	December, 1966	23, No.1	1989
1, No.3	June, 1967	23, No.2	1989
2, No.1	December, 1967	24, No.1	1990
2, No.2	June, 1968	24, No.2	1990
3, No.1	December, 1968	25, No.1	1991
3, No.2	June, 1969	25, No.2	1991
4, No.1	December, 1969	26, No.1	1992
4, No.2	July, 1970	26, No.2	1992
5, No.1	December, 1970	27, No.1	1993
6, No.1	September, 1971	27, No.2	1993
6, No.2	March, 1972	28, No.1	1994
7, No.1	September, 1972	28, No.2	1994
7, No.2	March, 1973	29, No.1	1995
8, No.1	Autumn, 1973	29, No.2	1995
8, No.2	Spring, 1974	30, No.1	1996
9, No.1	1975	30, No.2	1996
9, No.2	1975	31, No.1	1997
10, No.1	1976	31, No.2	1997
10, No.2	1976	32, No.1	1998
11, No.1	1977	32, No.2	1998
11, No.2	1977	33, No.1	1999
12, No.1	1978	33, No.2	1999
12, No.2	1978	34, No.1	2000
13, No.1	1979	34, No.2	2000
13, No.2	1979	35, No.1	2001
14, No.1	1980	35, No.2	2001
14, No.2	1980	36, No.1	2002
15, No.1	1981	36, No.2	2002
15, No.2	1982	37, Nos.1 & 2‡	2003
16, No.1	1983	38, Nos.1 & 2	2004
16, No.2	1983	39, Nos.1 & 2	2005
17, No.1	1984	40, Nos.1 & 2	2006
17, No.2	1984	41, Nos.1 & 2	2007
18, No.1	1984	42, Nos.1 & 2	2008
18, No.2	December ,1984	43, Nos.1 & 2	2009
19, No.1	June, 1985	44, Nos.1 & 2	2010
19, No.2	December ,1985	45, Nos.1 & 2	2011
20, No.1	June, 1986	46, Nos.1 & 2	2012
20, No.2	December ,1986	47, Nos.1 & 2	2013

The 1960s

Comedian Robin Williams has been credited with saying that 'if you remember the '60s, you weren't there'. I was there, going through secondary school education, playing a lot of football and thinking about career opportunities. However, while I wasn't there when the Society of University Cartographers (SUC) was founded in 1964, I know enough history to now see what the significance – initially locally in the UK, and later internationally – of the formation of the SUC has been. Although the British Cartographic Society had been founded the previous September, the SUC was originally set up because 'a more specialized organization was needed to cater for the particular interests of those people engaged in producing maps for the purpose of book illustration' (Hodgkiss, 1964).

One of the main aims of this new organization was to coordinate appropriate training and offer support to these 'practical' cartographers, who often worked in small units or individually. The UK's first, full-time diploma course in cartography was established at Oxford College of Technology in the same year of the Society's foundation, with a diploma course in Geographical Techniques at Luton College of Technology being offered from 1967. Textbooks for students of the art and science of cartography were already available in English; Erwin Raisz's standard text, *General Cartography*, first published in 1938 and revised in 1948, was soon accompanied by the first edition of Monkhouse and Wilkinson's *Maps and Diagrams* (1952), while the first of Arthur H. Robinson's *Elements of Cartography* was published in 1953 (and was to go through five further editions until its last in 1995, gathering co-authors as it went). The cover blurb to the first edition stated: 'One of the important innovations in *Elements of Cartography* is the inclusion of a chapter on map design, a phase rarely covered in other books'.

So, what was happening in cartography in this decade? 'Commercial cartographers' were peeling their scribecoats to produce the artwork for multi-colour lithographic printing to take place. Meanwhile, 'university cartographers' were generally using rOtring® pens, Letraset® and Letratone® to produce their map illustrations. However, a revolution was about to happen, initially in a very small way. There were the first signs of computers being used for mapping tasks, admittedly in an experimental fashion. One of the first programs was SYMAP, which was created in 1964 at Harvard University. A little later on this was the first computer mapping package I used and I well remember the tedium of setting up the card input and the rather poor resolution of the grid-based lineprinter output.

This presaged a period of very strong influence of the Harvard Laboratory for Computer Graphics and Spatial Analysis, which was founded in 1965. They soon released a second program in 1967, SYMVU, 'a computer graphics program written for the purpose of generating three dimensional line-drawing displays of data'.

Shortly, there was the first known use of the term 'Geographic Information System' by Roger Tomlinson, in his 1968 paper, 'A Geographic Information System for Regional Planning'. While Tomlinson is considered by many as the father of GIS, the influence of this new technology was soon to take an upturn that has eventually changed the whole industry. Jack and Laura Dangermond founded the Environmental Systems Research Institute (Esri) in 1969 as a privately held consulting group. The business began with $1,100 from their personal savings and operated out of their home in Redlands, California. Esri now a global supplier of GIS software, web GIS and geodatabase management applications, and is very prevalent in some sectors of the cartography field.

The content of the *Bulletin* reflected cartography at the time. Although it had started life with discussions of some quite narrow practical issues, by Volume 1, No.3 (published in June 1967), its content had widened significantly. The first article we have included in this section of the book is by Carson Clark, a founder member of the Society, and was the first of several by different authors on aspects of map design – a topic not often considered in the present era of getting the data mapped and available as soon as possible. Carson was a Cartographer in the Department of Geography at Southampton University at the time of writing. In 1969 he established the Carson Clark Gallery on Edinburgh's Royal Mile which was labelled as 'Scotland's Map Heritage Centre', and is still run as a family business.

Secondly, we present an early example of the Society encouraging interaction between cartography and other disciplines. Published in Volume 2, No.1 (December, 1967), Elwyn Edwards' views on the psychological aspects of map design were possibly ahead of his time. An ergonomist and aviation psychologist, he was a Lecturer and Reader in Psychology at Loughborough University from 1960 to 1976, Professor of Applied Psychology at Aston University from 1976 to 1984, then Director of Human Technology until 1993. His obituary in *The Independent* (25th November 1993) noted: 'In Britain we have a serious and, for the future, damaging attitude of almost total neglect of our technologists. Our way of life is entirely dependent on their efforts and achievements, but so long as they do their work well, we can ignore them and they remain unknown outside their particular sphere. Edwards and his circle of aviation technologists are in this category'.

The influence of the *Bulletin* can be judged by the fact that many of the authors of articles went on to have a significant influence in the wider cartographic community. Volume 3, No.1 of the *Bulletin* carried the first of a series of contributions by Mike Wood. He was one of the first 'academic cartographers' to be involved in the Society, sharing his serious academic rigour and yet being very much in touch with the practical aspects of the cartographer's field of work. Later he became President of the Society of Cartographers, of the British Cartographic Society and of the International Cartographic Association (ICA).

The range and depth of articles the Society was publishing in the *Bulletin* was now increasing, and by Volume 4, No.1, it was starting to carry some very substantive papers. The final 1960s article we reproduce here is by Alan Hodgkiss, the first editor of the *Bulletin* and author of one of the classic cartography texts for the next decade, *Maps for Books and Theses* (1970), which I studied during my Cartography course at Oxford Polytechnic. Here Alan charts the evolution of the Ordnance Survey's one-inch maps in a comprehensive paper that helped set the pattern for several other authors to establish their reputations by pursuing in-depth studies of the work of the Ordnance Survey.

Steve Chilton

References

Hodgkiss, A.G. (1964) "The New Society" *The Bulletin of the Society of Cartographers* Issue No.1 pp.1–3.
Hodgkiss, A.G. (1970) *Maps for Books and Theses* Newton Abbot: David & Charles.
Monkhouse, F.J. and Wilkinson, H.R. (1952) *Maps and Diagrams: Their Compilation and Construction* London: Methuen.
Raisz, E. (1948) *General Cartography* (2nd ed.) New York: McGraw-Hill.
Robinson, A.H. (1953) *Elements of Cartography* New York: John Wiley.
Tomlinson, R. (1968) "A Geographic Information System for Regional Planning" In Stewart, G.A. (Ed.) *Symposium on Land Evaluation, Commonwealth Scientific and Industrial Research Organization* Melbourne: MacMillan of Australia.

MAP DESIGN

A. Carson Clark

Originally published in Volume 1, No 3, pp.26–28

Of all the aspects of cartography, map design is perhaps the most complex. Whilst the main purpose of the map may be to show essentially accurate information, if the map has not been properly designed it must be a cartographic failure. It is not necessary to be an artist to learn to design effectively. The basic elements of good design lend themselves to systematic analysis and their principles can be learned by the average person, provided the person has a willingness to exercise imagination. Having said that, it must not be forgotten that the cartographic field, like many others, has developed with established traditions and conventions and these must be considered carefully before any are discarded, particularly conventional symbols, which, if disregarded, could cause the user considerable inconvenience and frustration and would be evidence in itself of poor design.

The aim of cartographic design is to present the map data in such a manner that the map appears as a completely integrated unit, and each component part must be clear and legible, but with no part receiving more or less prominence than it should.

We should remember that everything we see has visual significance and this applies no less to a map. The eye quite naturally looks first to the centre of a map, though on a piece of typescript the eye looks first to the top. An item that is visually significant is usually something that is different from the information surrounding it. A simple location map in black and white should have its town or city centrally placed on the map with roads, railways, etc., automatically drawing the eye to the visually significant and most important part.

Just as a well designed house has a plan, so must also the well designed map. Each item to be included on the map must be considered, its importance related to how visually significant it should be, and all considered together with its possible impact on the author or reader. My remarks are of course restricted to the design of the specialized maps we are generally called upon to prepare in university departments and not atlas maps or complex reference maps which come into another category. Usually our maps are required to show just one or two specialized items rather than complex dictionary type maps.

For the majority of our maps then we need a plan, an exact layout of all the information to be portrayed on the map. This plan, usually referred to as the 'mock-up', should have every item that is to appear on the finished map exactly in its position, together with all names that are to be shown. This mock-up is not only absolutely essential to the good design by the cartographer, but is of great assistance when called upon to work with a cartographically untrained author, since this may save time and patience otherwise required for the production of one or more trial maps for the benefit of the author. Unfortunately, though the author may be a perfectly competent author, and have a clear concept of the intellectual content of the map, he may be quite unable to see or understand the problems of good design. In fact, it appears to be quite impossible for those inexperienced in design problems to show any understanding whatever of this aspect.

Summing up this first part of design then, we must remember that clarity and design, not losing sight of the actual content of the map, are the primary items to be considered.

Legibility and clarity is achieved by the proper choice of symbol, line and lettering. These must be clearly defined, lines must be dense and sharp shading patterns, uniform coastlines and their generalization must be carefully compared from map to map. An important element in design is legibility, bearing in mind the size at which the map is to be published and read. Nothing looks worse than a map in which the lettering and/or symbols are too small to be read comfortably, or in some cases read at all. At the design and drafting stage the map may look pleasant and well designed, but may appear as an abysmal failure if due planning and consideration has not been made for its reduction to page size.

Perhaps no element of cartographic design is as important as contrast, particularly in black and white maps. It should be remembered that the degree to which a map appears precise and clearly drawn is dependent on the contrast structure of the varying elements of the linework, symbols and lettering. For example, in linework the weight of one line, or the curve of a line of the same weight, may give it the appearance of something quite unexpected. The shape of letters may blend into the background pattern of lines, or the opposite may occur and the lettering appear to stand out in relief. If one element of design is changed as far as weight, density or shape, then the relationship of all the other parts may likewise appear changed. Careful consideration and judgement is required in order to arrive at the right combination of all the component parts.

It has already been stated that contrast is the most important element of good cartographic design, and this may be seen either as tonal contrast or pattern contrast. For example, any grouping of symbols or shadings have a tonal value, achieved by widths of lines, shading patterns, names, titles, key, etc. and their arrangement within the map frame is a basic part of good design. It is estimated that the human eye can perceive differences in tonal value of about eight shades of grey between black and white and beyond this it becomes necessary to add a different shading such as dots or some other symbol. It will be found that a dark area next to a light area will make the dark appear darker and the light appear lighter. This can make a map difficult for a reader to understand, since a given value may appear differently in different parts of the map. This effect can be lessened by leaving a white line around a shaded area.

Shading patterns are usually those composed of lines or of dots. Any line anywhere on a map has, to the eye of the viewer, a direction. He tends to move his eyes in the direction of the line. If irregular areas are thus shaded by lines which do not vary much in value the reader's eyes will be forced to change direction frequently, and in consequence he will experience considerable difficulty in noting the positions of boundaries. Replace the line patterns with dot patterns, and the map immediately becomes more stable. The eye no longer jumps, the boundaries become clearer and of course lettering is easier to read against a background of dots rather than lines. If lines must be used, generally they should be fine and closely spaced.

Next we must consider balance and layout. The aim of the cartographer is to balance everything so that the map looks right, or I should say right for the purpose for which the map is required. Here we are required to fit all the visual components in such a way that their relationship seems logical or at least appears ordered to the reader's mind. In short, in a well balanced design nothing appears too light or too dark, too long or too short, or too small or too large. Here again we must refer back to the layout or mock-up sheet, and of course if in doubt a thumb-nail sketch is a useful and quick way to clear up the situation.

Finally, a word about boxes for the key or legend. Usually it is recommended that they should bear some relationship to the overall size and shape of the map. When this is impossible it may be necessary to divide the key into separate boxes, though this is not recommended. Alternatively, a key in a box across the full width of the map may be necessary, and provide a happier design than that of separate boxes within the frame.

The mock-up, work sheet, or whatever name you care to give to it, is surely the key to success in good map design and here everything can be laid out and planned in order that the finished product will be accurate and well pleasing to the eye.

PSYCHOLOGICAL ASPECTS OF MAP DESIGN

Elwyn Edwards

Originally published in Volume 2, No 1, pp.10–13

To the applied psychologist, the problem of map design stands as one example of the general class of problems concerning the representation, in two dimensions, of three-dimensional space. Other examples of analogous projects in the same area are engineering drawings, exploded assembly diagrams and architectural representations. Obviously, the cartographer is well aware of the basic issues involved in his art, and has developed numerous techniques to provide design solutions to the many difficulties encountered in the production of adequate maps. The contribution of the psychologist, it is suggested, derives from three sources:

1. He is aware of the basic mechanisms of human perception, and can therefore bring to the problem some fundamental knowledge of the principles relating to representations.
2. He is concerned to design and conduct experiments to extend the scope of available knowledge concerning human performance in any particular area.
3. In view of his wider interest in the problems of perception and representation, he may well be able to bring to cartography the ideas and solutions which have been generated in superficially different, but nonetheless related, problem areas.

It is convenient, in the systematic consideration of a design problem, to view the human factor aspects in two broad groups. The first is concerned with the overall basic features which determine the general form and nature of the completed product. The second group, which inevitably succeeds the first in time order, is concerned with the details of design. Thus, in the context of cartography, the basic features of design would include such decisions as map size and type of projection employed. Following upon these principal characteristics the aspects of detail design would include the colouring of different map features and the design of letters and numerals.

The principal characteristics of a good map, or indeed of any piece of equipment designed for human use, can only be properly determined in relation to the exact use to which the device is to be put. We may list some typical questions which would need attention in the case of, say, an aeronautical map:

1. Should shortest distance routes be represented by straight lines?
2. Should constant direction routes be represented by straight lines?
3. Should any topological features be represented? If so, which, and how?
4. How much ground surface is to be represented?
5. What size of sheet is most convenient?
6. Should any features of the airspace be represented? If so, which, and how?
7. Under what conditions (e.g. of space restriction, vibration and lighting) will the map be used?
8. By whom (e.g. highly skilled personnel or relatively unskilled personnel) will the map be used?
9. In conjunction with what other equipment (e.g. aircraft instruments, drawing instruments) will the map be used?
10. Is the map for temporary or semi-permanent use?

The answers to the first three questions are associated with the type of projection upon which the map will be based, and probably determine the most fundamental decision of all with regard to map design. In practice it may well be impossible to achieve simultaneous satisfaction of all the desired map properties. We cannot, for example, preserve true representation of the shape of ground features, and at the same time have straight rhumb lines. Compromise is necessary; and this must take into account the performance difficulties which result from providing characteristics which are less than ideal.

Questions 4 and 5 are concerned with the scale of the map which must be determined in relation to the manner of its use. The scale of aeronautical charts is dictated largely by the speed and height of the aircraft in which they are to be used, and is, of course

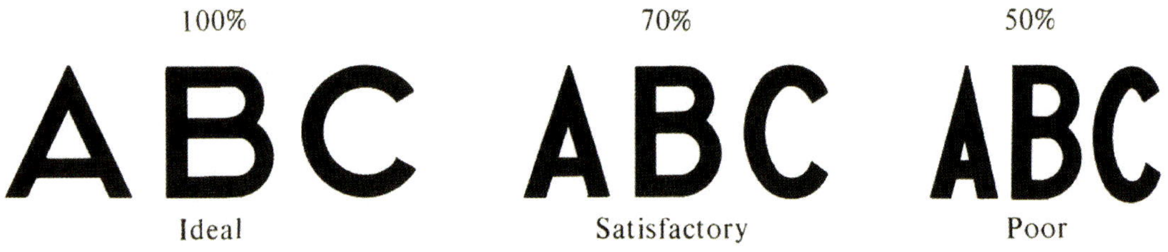

Figure 1 Loss of legibility when width of letters is reduced from 100 per cent of height to 70 per cent and 50 per cent

intimately bound up with the amount of detail which requires representation.

The environmental conditions under which equipment is used require careful study early in the design process. Vibration, for example, severely hampers visual acuity, and loud noises tend to make judgments difficult.

Compatibility between items of equipment which must be used con-jointly is essential if efficient performance is to result. Thus, for example, bearings in relation to magnetic north, rather than to true north are helpful on a map which is to be used in conjunction with an aircraft compass. Here is an example of the importance of an integrated conception of a total system (in this case that of air navigation), as opposed to a fractionated collection of separate and unreliable designs.

Hence, by a systematic study of the use to which a map is to be put, and by due consideration of the relevant aspects of human skills and their limitations, it is possible to reach rational decisions regarding the general form which the map should take. Let us now turn our attention to some of the details of design to which human psychology might contribute.

A good deal of knowledge is available concerning the eye, and its sensory capabilities. From this knowledge it is possible to derive certain recommendations regarding the visual presentation of information.

Lettering on maps should be clearly legible, and this legibility is dependent upon size, shape and contrast, in addition to such factors as viewing distance and lighting which are not inherent in the map itself. It has been found that lettering should be about 1/8" high at a viewing distance of 28" to ensure good legibility. For a different viewing distance of x", the size may be multiplied by the factor x/28.

Block capitals should ideally be as wide as they are high, but only a small amount of legibility is lost if the width is reduced to 70 per cent of the height. Further reduction of width, although possibly advantageous when there are space restrictions, produces considerable decrements in legibility (see Figure 1).

For dark letters on a light background, the width of the stroke should be about 1/6 or1/8 of the letter height. Due to the phenomenon of irradiation, the stroke width should be reduced to about 1/12 for light letters on a dark background (see Figure 2).

Figure 2 Black-on-White, stroke width is one-sixth of letter height; White-on-Black, stroke width is one-twelfth of letter height

Special designs for letters and numerals have been produced which have advantages over the conventional alphabets. Details of these may be found in the references quoted below.

The cartographer should be familiar with some of the optical illusions which have attracted a good deal of attention in the psychological study of visual perception. These geometrical illusions can often produce very misleading effects when they occur unintentionally. Some of the better known ones are illustrated in Figure 3.

There is scope for a good deal of psychological research into the problems of cartography. The few points mentioned in this paper merely indicate some of the ways in which the experimental psychologist might assist the cartographer in his task. But very little has yet been done; it is hoped that this whole area may soon attract the attention which it demands.

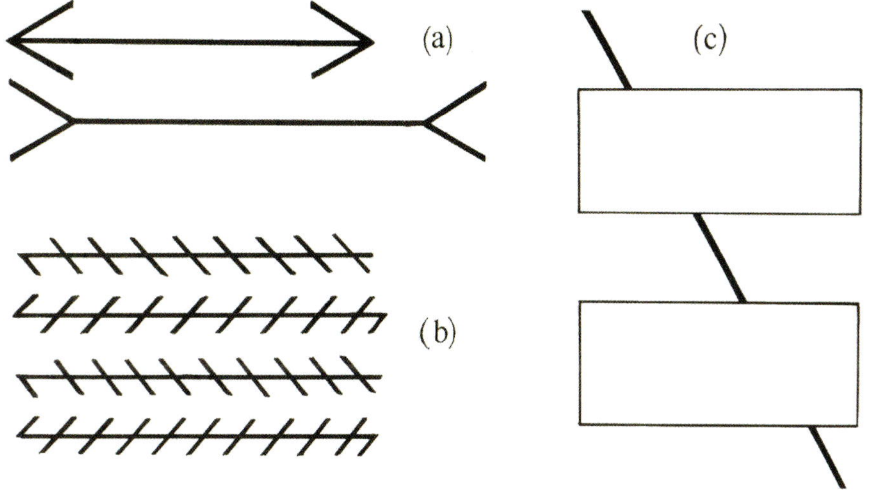

Figure 3 Some well-known geometric illusions. In (a) the long lines are the same length. In (b) the four lines are parallel. In (c) the diagonal is a straight line.

References

McCormick, E.J. (1957) *Human Engineering* New York: McGraw-Hill.

Morgan, C.T., Cook, J.S., Chapanis, A. and Lund, M.W. (Eds) (1963) *Human Engineering Guide to Equipment Design* New York: McGraw-Hill.

SOME THOUGHTS ON VISUAL PERCEPTION AND MAP DESIGN

Michael Wood

Originally published in Volume 3, No 1, pp13–16

The psychology of perception often brings to mind certain strange illusions of shape and colour. Although far from the core of the subject this aspect of the science will now be used as a point of entry into it. To begin with, consider examples of one of the phenomena of visual perception namely simultaneous contrast i.e. how adjacent colours affect one another.

(i) The affect of contrast on brightness alone where a grey ring is light on a dark background and dark on a light background.

(ii) The effect of contrast on hue, the appearance of hue being changed by its surroundings.

(iii) Other examples of this show how surrounding colour influences a central colour.

Having seen a few of these illustrations (which form part of the study of the visual perception of colour), one might say 'Our maps are seldom as simple as these and thus what value can be drawn from such studies?' The maps we draw are complex images and they contain many variable elements which must communicate certain information to the user. Can study of the above examples of colour contrast be of any benefit to the man designer? Can they help him to see his problems more clearly or even provide some answers?

There have been a number of experiments into the perception of map symbols and these have been reported in various journals over the past ten years. Names such as Williams, Jenks, Knos, Saunders and of course Professor Robinson of Wisconsin come to mind. Some research workers have taken a general approach while others have made more specific measurements of how and what people see and have attempted to formulate laws to govern certain aspects of the design of symbols. For instance, we have seen analysed the judgment of the sizes of isolated point symbols, also the grading of shaded tones into equal steps and, of course, the legibility of letters and figures. All these are very necessary, as it is on such basic research that new ideas and theories can be founded. So far, however, there are few specific guides and it is not unusual for the closing paragraphs of the experiment reports to contain the statement, 'Cartographers should be aware of the special circumstances described'.

As in other fields, an approach to the design of the functioning object must be preceded by a clear statement of its intended uses and users. This is difficult. For example, has the formula been devised for the perfect chair? Not at all, and yet all one has to do with a chair is sit on it! How much more difficult, then it is to define all the uses of a map.

The problems which normally face those who draw maps are:

a) a certain quantity of information must be shown on

b) a map of a certain size and

c) with limited colours, often only one.

We soon discover that for most of our work, the results of existing research are still of little practical value and are difficult to relate to the map in hand. We may feel that the subtle modifications suggested to improve the judgement of proportional symbols will be annulled in the complex environment of the map. None of this is the fault of the workers, but merely of the stage of investigations. Nevertheless it means that we must return to our empirical rules which do not provide all the answers, but on which we can rely. Rules such as 'boldly striped areas disturb the balance and attract too much attention', 'the map should have complexity and interest but also possess unity', 'do not juxtapose large areas of bright colour' or 'dots are more stable than line patterns'. At the same time we must have a healthy respect for the conventional uses of colour and pattern in maps, e.g. blue sea, red/orange deserts, white snow and ice.

However to return again to perception psychology, are there any other alleys to be explored or shortcuts to be found? We have read that colour and pattern affect our basic feelings. The former can have meanings and associations and provide for harmony or discord. But what of the more tangible aspects?

The eye sees colour and shape. Colour is seen in several qualities commonly known as:

Hue: the colour

Saturation: the dullness or richness of a hue

Brightness: the darkness or lightness of a hue

As was described at the outset of the paper, however complex environments can cause false impressions of the real colour. Consider some more examples and examine their relevance to cartography:

- The smaller the size of an object the duller will be its colour.
- Dull greens and blues at small sizes are difficult to distinguish.
- Thematic map makers sometimes employ blue and green dots for distributions.

- Yellow is very difficult to see at small scales.
- The 'Spreading Effect' of black lines makes other colours look darker. This can be observed when red roads are cased by black lines. In some maps, narrow orange roads look darker than the wider orange roads, because of black.
- Identical grey areas, if influenced, one by black the other by yellow, are given either a darkish or a yellowish appearance.
- White inclusions in the yellow areas make them weaker than areas of solid, continuous yellow.

The eye perceives shape only if there is adequate contrast with the background and the size of the object is sufficient. Hence size and contrast are important. Some shapes are easier to detect than others. One of the main concerns of the map reader is to see groups of associated objects within the map, as well as being able to pick out individual points and names. This user requirement leads us into what is known as the 'Gestalt' psychology, this being the German word for 'organization'. This theory states that we see things in groups and various laws are as follows:

Area: the smaller a closed region, the more it tends to be seen as a figure.

Proximity: objects close together tend to be grouped.

Closedness: areas with closed contours are seen sooner than those with open ones.

Good condition: that which makes fewest changes from straight or gently curving lines.

Some maps can be seen to comply with or contravene these 'laws' which may thus provide an approach towards a set of guiding principles for map design.

What is essential in the image for easy detection?

First of all, clarity and contrast. We can, however, have both clarity and contrast without comprehension. The second requirement, therefore, is 'meaning'; to reinforce the object, and we can add this by drawing the symbol in a familiar form. But even when we have contrast and meaning there may still be confusion, as one area or symbol may stand out at one moment and another at the next. This is called the 'figure-ground' effect and is illustrated by the well-known diagram in which we may see a vase in the foreground, or, alternatively, two faces in the foreground. This results when the ideas associated with the alternative shapes, inside and out, are either both clearly understood, or both vague. Hence if we had a map showing two regions separated by a line only, the eye would have no preference, but if one side could be recognized as the boundary of a familiar town, then this might take preference and become figural. Generally, however, colours or patterns are added to the areas to enhance visibility. This secondary stimulus nevertheless, could make clear recognition more difficult, as one side of the line could be dull but familiar, while the other is bright but irrelevant.

This figure-ground is one of the most significant phenomena to affect our reading of maps, and again, some examples can soon be found.

One has experienced the frustration of not being able to pick out a pattern of lines etc., from a map, and also the satisfaction of achieving this goal. We must ask what it is which aids good figural effect. The problem of perceiving patterns or distribution is greatest when the map is visually flat, and there is inadequate contrast of line thickness, tonal gradations and name sizes. One appreciates this and generally tries to introduce the required contrast of colour or tone, but only according to rules of experience. Is there no theory of perception, or clue in nature which can draw all these rules together, like the law of gravity which is a constant force acting upon a body?

Next to landscape paintings and photographs, there are few graphic images which contain so much complex, overlapping and superimposed information as a map, but at least the landscape has depth. If a map has to hold much information, it cannot be flat or shallow, it must have depth. This idea of having visual planes of depth or distance in maps is not a new one. Psychologists, also, have studied depth perception, and have indicated clues which increase the illusion of depth, e.g. the size of a familiar object decreasing in size with distance. What was lacking, however, was a simple, logical idea to link all the clues together.

American psychologist, G.G. Gibson, published some ideas in 1950 which may provide this link.[1] He analysed the normal field of view of the human eye, be it a landscape or a room, and isolated various qualities which give us clues to depth (apart from the binocular one.) These he called 'gradients' i.e. a change from large to small, coarse to fine, etc. His gradients were texture, size, and all the effects associated with aerial perspective, which is the influence of haze on the colours and tones of a landscape. For example the natural textures of grass, stones and walls grow finer with distance.

He was concerned with continuous gradient, but perhaps we can modify this to a series of receding planes. In this way we can stack data according to importance, or effect, and introduce adequate contrast to make the study of one of a series of superimposed distributions much easier.

What are the rules?

The foreground has coarsest texture, richest hues and darkest shadows, but, with increasing distance, saturation falls off, and colours get lighter and weaker, while brightness also falls off and shadows get duller. Objects and lines become smaller and thinner.

Can we see any of these manifested in maps?

If colours are strong they appear to be in the foreground, and image is flat. Clear separation is required. By subduing the background we can add more symbols. Crisp black and white circles, for instance, can be placed on a background which is of dull and impure hues, or the background can have paler tints, while the foreground has richer hues. A pale background with grey lines is surmounted by thick black distribution lines in the foreground.

These rules can be applied in black and white maps as well as those in full colour.

The thickening of one set of lines to raise them above another is the most obvious example. If lines are too similar and the texture too coarse, the whole map is 'too close' to the observer, and overpowering. Finer textures, thicker lines and better contrast improve this situation. This method is effective, even over dark images.

One could examine the thorny problem of the choice of shading tones under this theory of planes.

If a purely mathematical basis is used for the choice of graded tones e.g. spacing of parallel lines, it is not graphically effective. The more common solution is to associate darkness with importance, using various textures, Recall, however, the textural gradients, where near represented coarse and distant represented fine grain. The top four darker steps of a graded series may seem to work, but, although the fifth has the correct percentage of ink a coarse pattern or grain makes it come out towards the viewer. This helps to explain how 'correct-percentage-of-ink' steps may get out of place in the visual scale if texture is not considered.

Having studied examples one can now make the hypothesis that in the map, as in the landscape, the foreground could have bright, purely coloured objects, strong contrast and very dark, crisp shadow colours; thick lines, large objects and large and bold letters. The background has pale colours, dull shadow colours, poorer contrast, thinner lines, smaller symbols and small or light letters, and, between the two, a gradient of visual planes.

This is as far as one can advance with an inadequately tested theory. It may seem to be a very obvious idea for the linking of the rules of map design, but perhaps it may help to clarify things in certain, if not all situations. It can be analysed and explained through the study of visual perception, but the secret or clues may well lie in something as simple as a landscape painting.

Editorial Note

1. See, for example, Gibson, J.J. (1950) *The Perception of the Visual World* Boston: Houghton Mifflin.

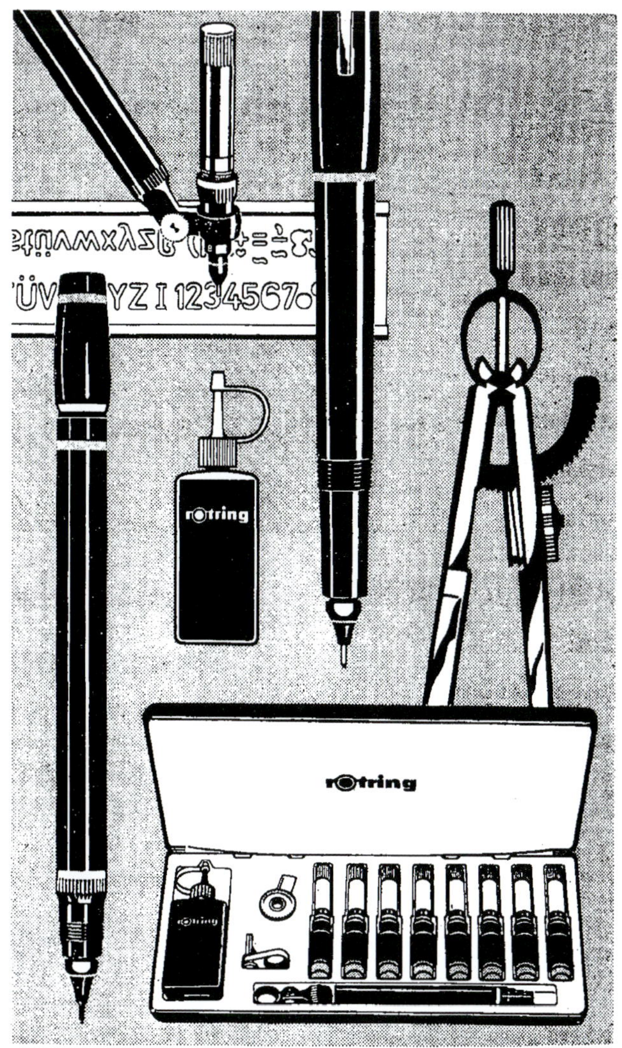

rotring

makes drawing easier

For the Expert the name rotring means quality drawing pens. These are the Radiograph, Varlant, Varioscript, Foliograph, etc. Comprehensive programme of drawing materials for all who want precision work. Practical, comprehensive and reliable, that is the rotring programme. Designed by technicians for technicians. Available from all Drawing Materials Dealers.

THE ORDNANCE SURVEY ONE-INCH TO ONE MILE MAP: AN OUTLINE HISTORY, WITH SPECIAL REFERENCE TO THE EARLY EDITIONS

A.G. Hodgkiss

Originally published in Volume 4, No 1, pp19–34

The Origins of the Ordnance Survey

It is a regrettable fact, though one which can hardly be disputed, that on many occasions throughout cartographic history a threat of war has been necessary in order to provide the stimulus for mapmaking. Our own Ordnance Survey is no exception for, although the official beginnings of the Survey date from the year 1791, the impetus which pushed forward its foundation can be traced further back to the last Jacobite rising when the Young Pretender, Prince Charles Edward, landed in Scotland and thence marched over the Border at the head of his largely Highland army, duly penetrating as far south as Derby and by so doing causing so great a panic in London that the Bank of England resorted to paying its customers in sixpences in order to delay those who wished to fly with their money or to hide it. Charles Edward, however, finding that support for his cause was not forthcoming in England was forced to retreat back over the border into Scotland where he was defeated by the Duke of Cumberland's forces at the Battle of Culloden in April 1746. The subsequent pacifying operations in the Highlands by Cumberland were considerably hindered by the lack of good maps. Those maps which were available were simply military sketches executed with the magnetic compass and plane table. In 1747, therefore, a dual programme of mapping the Highlands and opening them up with military roads was instituted by the Government of the day. The roads were to be built under the supervision of General Wade and by their construction troops would be enabled to move rapidly about the country in the event of more Jacobite excursions. The concurrent task of mapping originated with Major-General Watson and this survey may be fairly regarded as the real birth of the Ordnance Survey. Work commenced in 1747 from a headquarters which was established at Fort Augustus in Invernessshire and the principal name associated with it is that of William Roy, a young engineer then only 21 years of age, and an assistant to the Quartermaster-General. Roy, a Scot born in Linlithgow, and Watson carried out their survey by compass traverses filling in the detail by means of field sketching at a scale of 1,000 yards to one inch. These historic maps are now housed in the British Museum and they were the first to be made by the military authorities for the Government.

The work of survey and mapmaking was interrupted by the Seven Years' War but in Roy's words 'on the conclusion of the peace of 1763, it came for the first time under the consideration of Government, to make a general survey of the whole island at the public cost. Towards the execution of this work whereof the direction was to have been committed to my charge, the map of Scotland was to be made subservient'.[1]

In 1765 Roy was established in a new post – Surveyor-General of Coasts and Engineer for Making and Directing Military Surveys in Great Britain under the Honourable Board of Ordnance. Thus was established the first connection between the Board of Ordnance and mapmaking. Roy's duties included the inspection and surveying of the coasts and districts of the kingdom and the islands belonging to it and also the submission of periodic reports. He was paid the sum of twenty shillings per day.

As we have seen the threat of war has on many occasions provided the necessary stimulus for cartographic development. Wars themselves, however, have proved to be a hindrance and the pursuance of the War of American Independence meant that a tight rein was kept on the country's economy and also meant that the military was employed on more arduous and dangerous tasks than mapmaking. No more progress was made in fact until peace came in 1783. The initiative was to come from across the English Channel and was in the form of a suggestion made by the prominent French astronomer, M. Cassini de Thury, that it would be advantageous to all parties if, for the purpose of checking certain astronomical calculations, the two observatories of Paris and Greenwich could be linked by precisely surveyed triangulation and their precise positions determined. Correspondence on Cassini's proposal passed to and fro between the French and English Governments and as the observatories were of considerable importance both to navigation and to time-keeping, the proposal was approved and work commenced. At the English side of the Channel Roy was placed in charge of operations, his task being to measure a new baseline on Hounslow Heath from which a triangulation system would reach out, linking Greenwich with Dover and subsequently joining up with the French triangulation. Hounslow Heath was selected as the site of the baseline for a number of reasons. It was flat, close to London and Greenwich, and it was admirably sited as base for subsequent triangulation. The general direction of the baseline having been established, engineers cleared a wide path in preparation for the actual work of measurement. For this task, Jesse Ramsden, an instrument maker of renown, was instructed to prepare a 100 ft steel chain, the measurements from which were intended to be supplemented by measurements carried out with deal measuring rods, precisely graduated. These deal rods were, however, found to be unsatisfactory for the purpose due to the way in which they expanded with any increase in humidity. As an alternative it was decided after considerable experiment to use accurately calibrated glass rods. The carrying out of such experiments as this meant that the techniques and the instruments used for linear measurement were now at a high degree of perfection, and the accuracy obtainable was such that when a check base was measured on Romney Marsh and its length then calculated by triangulation from Hounslow Heath, the error was only 1 in 28,000.

The measurement of the base line was completed in 1784 but a further three years were to elapse before work re-commenced to unite the French and English triangulations. The delay was in large part due to the slow progress made by Jesse Ramsden in constructing the great 3 ft theodolite which was to be used in the triangulation. Ramsden was an instrument maker of world standing, the son of a Halifax innkeeper, and at one time, in spite of employing sixty workmen his business was so extensive and his orders so great that he was unable to keep pace with the volume of work. The theodolite used in the survey was his greatest achievement and, together with a second great theodolite and his two standard 100 ft chains, did quite remarkable service for the survey until 1862. During the principal triangulation of Great Britain sights were read with these theodolites at distances of over one hundred miles. Triangulation in south-eastern England was completed in 1784 and paved the way for the subsequent triangulation of the whole of Great Britain.

Roy died in 1790 after being responsible, by his enthusiasm and skill, for the successful culmination of the surveying operations. He had been a ceaseless campaigner for a systematic national survey and in 1791 the third Duke of Richmond, Master General of Ordnance and a keen admirer of Roy, decided to bring surveying under the wing of the Ordnance and established the Trigonometrical Survey. In so doing he was no doubt influenced by the possibility of invasion from across the Channel.

The offices of this new body were set up in the Tower of London together with those of the Ordnance and the main object of the Survey was to produce a map for the whole United Kingdom at a scale of one-inch to one mile. This was to be their principal task until 1824. The Trigonometrical Survey consisted of a small group of surveyor¬ draughtsmen together with a few military personnel from the Engineers and Artillery and they commenced work on both the triangulation of the whole country and on the one-inch to one mile survey. The work was under the direction of Colonel William Mudge, one of the most significant names in the development of the Ordnance Survey. Some of the early survey work carried out by temporary civilian surveyors did not always measure up to the exacting standards demanded by Mudge and consequently he established in 1805 the Corps of Royal Military Surveyors and Draughtsmen. This new organization improved the level of work and existed until 1817 when it was disbanded, to be reborn in 1823 at the instigation of Mudge's successor, General Colby when it became the 13th (Survey) Company of the Royal Engineers.

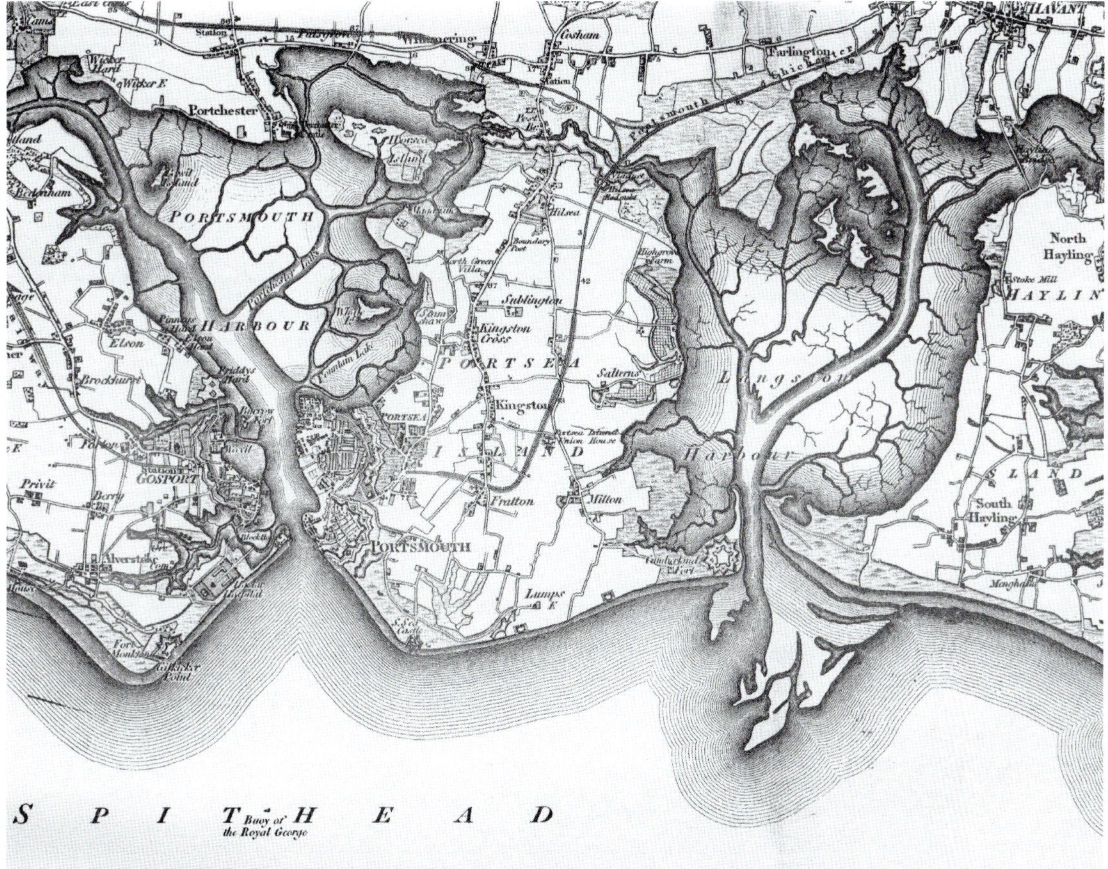

Figure 1: *Part of Sheet 10, First Edition, Old Series, 1810*

The First Sheets of the One-inch Series

It is not surprising that work commenced in the London area and Southern England for, following the outbreak of the French Revolution in 1789, there was a real fear that England and France would soon be engaged in hostilities with the consequent possibility of invasion. The very first sheet was published on the 1st January 1801, a memorable date for cartography indeed, and was entitled 'General Survey of England & Wales: An entirely new and accurate survey of the county of Kent, with part of the county of Essex' and it provided a healthy and promising beginning to what was to prove a century of continuing cartographic achievement in Britain. The next sheet to follow was of the county of Essex and, like their successors for some considerable time to come, the maps were engraved on copper, an extremely slow method of printing; twenty copies an hour being considered to be a good rate of production. Copper engraving, however, had considerable advantages; it was possible to reproduce thick and thin lines with the same outstanding clarity and the very finest of lines would print clearly on the final map. These attributes are readily apparent if a study is made of the hachuring on these early maps. A secondary advantage of the copper plate was the ease with which alterations could be made, simply by scraping the part to be amended away, grinding and burnishing the scraped section and finally re-engraving as necessary. These early copper plates measured 36" x 24" and they weighed about 35 lbs each.

Before any engraving actually took place the draughtsman would prepare a drawing on fine cardboard in readiness for the engraver, reducing to the one-inch scale the survey work which was carried out at a basic scale of two inches to one mile, though this was not invariably so. For drawing purposes he would use Indian ink and a camel hair pencil, and having prepared a rough draft from the survey information he would decide from a thorough study which information was to be omitted, which exaggerated and so on. He would indicate roads and lettering in their correct conventions and add hachuring with broad strokes. Progress on this work would be far from speedy, a single sheet taking some months to prepare, but the results would provide a clear and comprehensive study at the reduced and correct scale from which the engraver could work.

A Rural Landscape

These early maps depict a rural landscape crossed by a network of turnpike and secondary roads with prominent areas of heath, common and woodland and little sign of industry. Although they represented a considerable improvement over some of the late eighteenth century county maps there was, notwithstanding, quite a fair amount of criticism as to their accuracy. Mudge's successor, Colby, became the butt of such criticisms and to the Select Committee on the Survey and Valuation of Ireland he was provoked into excusing inaccuracies by declaring that 'the employment of civilian surveyors on the topographical survey of England had resulted in considerable inconvenience; they had performed their work without a sufficient degree of accuracy'.[2] It might be fair to say that a further excuse could have been that the Survey was hardly lavishly treated as regards its financial budget; from 1791 to 1811 Close tells us that expenditure was a mere £52,000. This financial stringency may have influenced the Survey officers in their attitude to the relative importance of the trigonometrical and topographical stages of the survey for, certainly, their overriding interest was in the basic trigonometrical constructions rather than in the detailed topographical survey. It is apparent that there was a lack of an overall policy in the design of the maps for in the early stages the Board of Ordnance was still thinking in terms of the familiar county map such as had been produced by private surveyors, rather than of a single comprehensive series of sheet lines independent of administrative boundaries. Some maps, in fact, have the standard border omitted from their inner edges so that it would be possible to have them mounted as county maps.

The normal scale of field survey for the maps was at a scale of two inches to one mile but parts of Somerset, Devon, Hampshire, Kent, Essex and Sussex were surveyed at three inches or six inches to one mile. This meant that certain one-inch sheets would be made up from surveys at three different scales, made at different dates and of varying degrees of accuracy. There was, in fact, a lack of overall guidance and planning in the early conduct of the survey; the only document which clearly laid down instructions for procedure being written before 1790 by General Roy. These, however, were merely general instructions based on accepted surveying practice and were not specifically designed to apply to a new one-inch map of England. As a starting point the topographical surveyors were given a diagram of points at the two-inch scale established by the interior triangulation and from this they carried out traverses with the small theodolite or compass. Hills were sketched in and distances established partly by calculation and partly by pacing or the use of a perambulator or waywiser. A further reason for some inaccuracies in the survey was the system of payment used for reimbursing surveyors employed by the Board of Ordnance. Surveyors received a basic set pay but were given a further payment for every square mile surveyed. This meant, naturally, that the extra payment was easily earned where the terrain was easy but became an underpayment where surveying was difficult. The logical consequence was that surveying which had to be carried out in awkward country tended to be skimped.

The Irish Survey under Colby

In the meantime a new development had ensued which caused an interruption to the Survey. A Select Committee of the House of Commons had made a recommendation for a survey of Ireland to be made at a scale of 1:10,560 or six inches to one mile; the survey to be carried out with all possible despatch for purposes of valuation. At the same time the Committee expressed its appreciation of the work

Figure 2 Part of Sheet 3, First Edition, Old Series, 1801

which had taken place so far in England while deploring the fact that it had been going on for no less than thirty-three years and was still only two-thirds complete. Colby was placed in charge of the Irish Survey which was completed in 1846, the year of his retirement. Those methods of linear measurement such as pacing or perambulating which had been used by Roy in Scotland made way for accurate measurement; the triangulation system of Ireland being built on a great triangle which had its base at Lough Foyle. Two thousand men were employed to carry out the Irish Survey and the total cost was £800,000. In addition to the work of the survey itself Colby compiled 'Memoirs' of various types of information about the country. Most of these were never printed but are still kept in manuscript in Dublin.

The Influence of Colby on the English Survey

Back in England work proceeded at a much reduced pace during the 1820s and after 1830 there was growing pressure brought to bear on the Board of Ordnance both to speed up and improve the quality of the work. Colby had learnt a number of lessons from both his failures and his successes in Ireland and he was prompted by them to take a new look at the obvious weaknesses of the English survey. Thoroughly dissatisfied with the shortcomings he found he began to institute measures designed to improve matters. For instance we find him in 1834, writing that to prevent imperfect plans being made by surveyors, he had obtained an order that those who produced plans which were grossly inaccurate would be charged for their correction. He further decided that all those surveyors who, by virtue of their age or infirmity, were unable to match the quality of work required, should be removed. Colby's third step was to make provision for a supply of qualified assistants to carry on the work and this was done by instituting a training school at which a Mr Dawson was required to instruct assistants for the survey. Colby remarks that 'several able assistants were reared under his instruction, and the present state of the survey, is greatly due to his exertions'. In 1834 a further important step was taken, that of removing the iniquitous system of payments for every square mile surveyed. It was decided, furthermore, to separate the two functions of outline surveying and hill sketching which had formerly been carried out by the same surveyor. For the future the outline surveying would be carried out at the two-inch scale while the hill sketching would be dealt with by specialists in this particular field. Surveyors under a Lieutenant Bailey of the Royal Engineers were responsible for laying down roads, rivers, buildings, etc. which needed to be laid out from actual measurements and those assistants who had been trained under Mr Dawson were employed in delineating the hills, woods and other ornamental matters on the maps. These latter assistants were regarded as being higher qualified and received a higher rate of pay. This rather suggests that more artistic work of hill sketching and filling in other ornamentation was of a more skilled nature than the actual work of measurement, or at least it would appear that this was how the authorities of the time regarded it.

During the twenty-six years of Colby's directorship the one-inch map took on something of a new appearance – it had a wholly lighter and more delicate look with the finest of hachuring and small, neat lettering, some of it indeed almost too small to be legible with the naked eye and in some places too delicate to stand up against the background detail on which it was placed. Colby was something of a perfectionist, always striving to the utmost to improve the imperfections of the survey as he saw them and to him much credit is due.

Conventions on the Early Maps

By 1840 only Scotland and the six northern counties of England remained unmapped at the one-inch scale. Sheets 1 to 90 of the Old Series covered England and Wales south of the Hull-Preston line, mainly reduced from the two-inch to one mile survey but with some areas of southern England as we have seen reduced from three-inch and six-inch to one mile survey. The size of the maps was mainly 36" x 24" though occasionally the maps are found as quarter sheets. Though numerous symbols were used on the maps no table of conventional signs was printed. Turnpike roads are clearly shown by means of parallel lines, one thick and one thin, with various annotations throughout the series – T.P. or T.G., T. Pike, T. Pike Gate, Toll Gate and so on. From 1852 onwards, however, the standard abbreviation T.P. is used. From 1822 distances were indicated in Roman figures along the roads in one direction only, but from 1833 this was amplified to show distances in each direction. Secondary roads are indicated by lighter lines engraved closer together, a practice continued throughout the series as was the indication of fenced and unfenced roads by either solid or broken lines. Tracks were shown throughout by a double dotted line but the use of a special symbol for footpaths, a single broken line, is intermittent – being used initially on sheet 25 but not being employed with any consistency until after 1824. Bridges over streams are named occasionally but without the use of a symbol, and fords are shown as at Crayford in Kent where the River Cray crosses the turnpike in the middle of the village. On the earlier sheets Roman roads are named in Roman serif lettering but later sans-serif lettering similar to that used for Roman remains is generally employed.

Buildings in villages and hamlets or those representing isolated settlements are shown in solid black but in the towns the built-up area is depicted with either a fine regular dot tint or with close, diagonal hatching. In the London area, for instance, the buildings in London itself have a regular fine dot tint. Romford has buildings which are in-filled with close hatching while the village of Woodford has buildings in solid black.

Churches are symbolized by a plain, small cross with no indication being provided of whether they have a tower or spire. Occasionally they are surrounded by a rectangle to indicate the churchyard. In 1852 the name of the church is given followed by lettering to say whether the status of the incumbent was Rector, Vicar or Curate.

A particularly attractive feature of the earlier maps, as indeed it must have been of the countryside itself, was the remarkable profusion of windmills, particularly of course in eastern England. The mills were depicted by two symbols, charmingly drawn so as to indicate whether the mill was a post or a tower mill. The remarkable number of mills in Fenland is well displayed on sheet 65 where a count over only an eighth section of the map reveals over a hundred mills. Watermills, surprisingly, are not allocated a distinctive symbol but are simply labelled as 'mill' unlike some of the earlier county surveys such as that by Yates in his Lancashire map where watermills are depicted by a small wheel.

Gardens are shown clearly around the houses on the sheets of southern England by a very delicate stipple but are omitted on the later maps of the six northern counties. Individual farms and the larger houses are indicated by name as are also the occasional public house where the thirst of the weary traveller could be appeased – so that we find the hostelry of The Three Pigeons indicated near Thame in Oxfordshire and the Old Lamb Inn at Kingston Bagpuize, names redolent of rural England. On certain later sheets, that of Snowdonia for instance, hotels which would be of importance to travellers are indicated such as the Victoria Hotel at Llanberis, a hotel still offering a welcome to the tourist today. On sheet 95 we find 'tea gardens' where holidaymakers could take their ease in the developing watering place of Scarborough.

There is no lack of fascinating and evocative names on the maps. If sheet 3 of the county of Kent is examined we find some especially attractive names – Muscles Farm, Etchings Hill, Hearts Delight, Beaver Green, Cuckolds, Godcheap, Petty France, Bigberry Wood, Fishpond Wood – to list but a few. This Kent sheet is remarkable also for the number of parks to great houses; for the piers shown at the watering places of Ramsgate, Deal and Broadstairs; for the military implications of the coastal fortifications – castles at Walmer, Deal and Dover, batteries, signal houses and beacons, Martello towers and so on – all evidence of the military purpose of these early maps of southern England. Many of these maps have most attractive cliff and rock drawing, others carry a wealth of symbols for nautical features – sheet 2 of the county of Essex has a symbol for the Nore light, a symbol for beacons, separate symbols for black and red marker buoys, signal staffs and so on. Soundings are given in italic figures and sandbanks are shown by form lines with a light stipple. Salt marsh is differentiated from other forms of marsh by a distinctive symbol.

Sheet 25 of part of south western England is a heavily hachured sheet but one full of interesting titbits of information. At St. Giles in the Heath we find an inclined plane; a steam ferry at Saltash; an aqueduct at Tavistock. A postbridge is unusually given a symbol in its own right and the Dartmoor railway is indicated by a double line which clashes with the symbol used for roads – as yet no individual symbol was needed for this new form of transport. The Tors up on Dartmoor are delightfully drawn.

Unlike the more impersonal age in which we now live it was the usual practice to credit the original workers on these earlier maps and we therefore find the name of Ebenezer Bourne constantly recurring as the person responsible for the lettering. Benjamin Baker and assistants are credited with the depiction of the hills. These names continue up to 1834 when numerous new names begin to appear – lettering by J.A. Harrison, hills by L. D'Elboux, hills by J. Peake, outline by G. Baker and so on. Perhaps these were evidence of the weeding out and introduction of new blood by Colby in 1834.

The type of lettering used on the early maps varied with the feature being depicted so that county names are shown in Roman capitals about an inch in height. This was later reduced to about nine-sixteenths of an inch. Major towns were in Roman characters about a sixteenth of an inch in height and other towns in smaller italic capitals. Villages were in Roman upper and lower case and hamlets in italic upper and lower. Roman camps and other Roman sites were indicated by sans-serif and up to 1830 antiquities were in italic. After 1830 an Early English type similar to that still employed on O.S. maps was used.

The maps were surrounded by a piano key type of frame whose treatment varied slightly on different maps, while the mile scale and imprint were usually below.

The Battle of the Scales

Six different meridians were used during the mapping of southern England, i.e. south of the Hull-Preston line, those of Dunnose, Clifton Beacon, Burleigh Moor, Delamere, Moel Rhyddled and Greenwich. This of course meant that there was no exact correspondence between any one map and its neighbours and the maps could not be fitted exactly together. For the northern maps, however, one meridian only was used, that of Delamere Forest and the sheets were published in quarter sheet form, 18" x 12". Of the earlier maps, those for Kent, Surrey, Essex and Sussex had been first issued in bound folios for each county and the remainder of the maps in sheets all measuring 23" from north to south but varying from 29" to 34" from east to west. The northern maps were to be reduced from the 6" to one mile survey as the success of the Irish Survey had resulted in what came to be known as the Battle of the Scales. A suggestion that Britain should be mapped at a scale of 6" to one mile was not greeted with any enthusiasm by the private land surveyors who had been called in whenever a canal, a railway or a new road was proposed or perhaps an alteration to someone's estate. As we are told in an early Ordnance Survey report 'enormous sums of money have been expended upon surveys for railways and on plotting fields [...] in the employment of surveyors at five and ten guineas per day. This competition for surveyors and draughtsmen raises the price far beyond that of any national survey'.[3] The Battle of the Scales, however, resulted in victory coming to the Government and the 6" to one mile was continued northwards to the Border with one inch reductions made from it. This latter survey was known for a time as the 'cadastral' one inch from the Latin word 'capitastrum'. This indicated a connection with taxation and probably derived from the original survey of Ireland at the six-inch scale which had been made for valuation purposes. The six-inch scale thus became associated with taxation.

The 1:2500 Survey

During the mid part of the nineteenth century the great industrial developments coupled to the rapid expansion of towns and communications led to an insistent demand for large scale plans and, after sounding out the feelings of various types of map user; a Departmental Committee recommended that the whole country, with the exception of uncultivated areas such as mountains and moorland, should be surveyed at a scale of 1:2500. This was a momentous decision to take for as yet no other country had attempted to map a sizable part of its territory at such a large scale. Despite a certain amount of opposition the 25" to one mile survey scale was adopted in 1858 and began with a pilot survey being made in Durham. In 1863 the survey was extended to the southern counties and was completed in 1895. The twenty-five inch was now the base from which all small scale maps were to be derived.

Problems of Dating the Old Series

By 1870, then, the whole country had a complete control system which was based on the primary triangulation and a series of maps which encompassed the entire land and which surpassed in quality any to be found elsewhere. This Old Series of maps numbered from 1 to 110 covered a period of three-quarters of a century, those made in the early days were of a rural England; the later sheets saw England firmly in the grip of the Industrial Revolution. This is a point which should be remembered when making comparisons between individual sheets of the series. Though lacking in consistency and without any real overall uniformity of style due to the absence of long term planning from the outset, the Old Series was a remarkable achievement. It is a complex series to study because the various printings and revisions which incorporated cumulative additions to each sheet make the dating of any particular sheet extremely difficult. Dr J.B. Harley, who has carried out considerable research into this problem of dating the Old Series Sheets, has prepared short monographs on the history of the sheets and these are incorporated into a facsimile reprint of the later printings of the series. This new edition is printed by the Devonshire publishers, David and Charles, and it is on slightly different, re-numbered sheet lines – some sheets which were mainly areas of sea have been incorporated into others for convenience and economy. The publisher has chosen to make his facsimiles from the later printings so that the original edition may be presented with its detail unimpaired but with the cumulative revision and particularly the railways added. This is a valuable project which makes the very rare sheets of the Old Series now freely available.

Revisions of the Old Series

The Ordnance Survey officially records the following revisions of the Old Series: seven sheets for which new copper-plates were engraved; eleven sheets with part of their surface only re-engraved. To this must be added the various local amendments which occur on almost all sheets. In the 1830s some geological symbols – dip arrows, etc. – were added to the maps of southwest England for the new Geological Survey was making use of the topographical series as base maps, just as it continues to do at the present day. Dr Harley informs us that revision took place as circumstances demanded and not simultaneously over the whole country – areas where there was rapid industrial and urban expansion would consequently have more printings in the Old Series than would a more static rural area. The building of a railway would provide an obvious incentive for the revision of a sheet. There was also a special printing which was known as the Index to the Tithe Survey on which the names and boundaries of those parishes which were not tithe-free, or in which tithes had not been commuted on enclosure, were indicated.

The Old Series then represents not just a set of maps but a momentous era of mapmaking, an era of change and development not only in mapmaking techniques but also in the nature of the countryside which was being committed to paper in map form. The Old Series had taken almost 80 years to complete – the ensuing hundred years from 1870 to the present day were to see no less than nine further editions of the one-inch to one mile map, an average of eleven years per series.

In 1870, control of the Ordnance Survey passed from the War Office to the Ministry of Works, evidence that the civil work of the Survey now far exceeded in importance its military significance.

The New Series or Second Edition

Those sheets of the Old Series north of the Hull-Preston line which had been reduced from large scale survey were, for purposes of the Old Series, issued with numbers which followed on from those of southern England, namely 91 to 110. Each sheet number was divided into quarters as the maps were issued as quarter sheets and so we have 91NW, 91NE, 91SW and 91SE, etc. This large sheet numbering was abandoned in 1872 when it was decided to carry the 6" survey southwards and the original quarter sheets were withdrawn from circulation, then reissued with new numbers to form sheets 1 to 73 of the New Series or Second Edition. This series was to consist of 360 small sheets of dimensions 12" x 18" and was completed in a relatively short space of time as compared with the Old Series. South of the Hull-Preston line the sheets took only twenty-five years to complete and as these sheets were now based on a six-inch survey the New Series represents a considerable overall improvement in accuracy over the original Old Series. The New Series also presents less difficulty in dating than had the Old Series where the main problem had been that the date of original publication was always the one which appeared on the map whatever the date of printing and revision might have been. With the new series the date of field survey is printed on the bottom margin of each sheet together with the date of publication and when revision or re-issue of a sheet occurred this was also normally added to the imprint. The New Series was issued in different version – primarily an Outline Edition with contours indicated in black; secondly a printing in colour with contours in red, roads sienna, water blue and so on; thirdly a printing with hachures in black; and fourthly a version with hills shown in brown. The reason for showing the hachuring in brown on the last named version is that, because of the alterations which were constantly being made to the copperplates, the delicate hachures became damaged and it was therefore decided to put the hachures on a different plate from the rest of the map and print them in brown. Although the maps were generally in black and white only – the basic outline edition anyway – the New Series has a more modern look than the early maps of the Old Series, hardly surprising considering the time which had elapsed. The appearance is generally much more delicate, a table of conventional signs has appeared, types have become more standardized and the whole layout bears more resemblance to that of the maps of today. Parish boundaries are now included and contours, indicated by a dot-dash line, are at 100 ft intervals. Being in black the contours do tend to be lost among the mass of other detail, particularly as the line used is very fine. Railways are now strongly shown by a double line in-filled with black and white sections while mineral lines and tramways have a light line with cross ticks. The familiar symbols distinguishing those churches which have towers, those with spires and those with neither have now appeared; metalled roads are shown as double lines of varying strengths and distance apart to symbolize first class, second class and third class roads. The light pecked line used for footpaths is retained and so too is the symbol for windmills. Parks to great houses are still charmingly drawn with the open ground finely stippled and delicate trees. A new and useful feature on these maps was the indication of letter boxes by the letters L.B. On the hachured edition the hachures themselves are finely drawn for this was a period when hachuring had reached its peak. It, nevertheless, still leaves something to be desired as a method of representing relief forms. Well-defined valleys are certainly well brought out but often there is little convincing overall impression of the topography.

The variants of the outline edition which have been mentioned were based on a full-scale revision of the New Series between 1893 and 1898 and the experiments carried out in the use of colour mark the transition from black and white of the Old Series to the coloured maps as we know them today. Authorities differ as to whether this revision should be regarded as part of the Second Edition or as the Third Edition of the one-inch series but general opinion relates it to the Second Edition.

The Third Edition

Work on the actual Third Edition commenced in 1901 and survey was complete in 1912. Like the New Series it consisted of 360 sheets each measuring 12" x 18" and the system of sheet numbering and the sheet lines were, in fact, identical. This edition represents the second national revision of the New Series, the first being that of 1893–98. In the margin of some of the sheets is given the dates when certain railways were added or when alterations were made. The Third Edition was issued in a black and white outline edition; the appearance of the sheets being very similar to that of the Second Edition or New Series. Contours are shown at 100 ft intervals up to 1,000 ft and thence at 250 ft intervals. Trig' points and spot heights are shown and there is some very fine rock drawing on certain sheets, the Eglwyseg crags, for instance, on sheet 121 of the Wrexham area. In addition to the outline edition there were also coloured and hachured editions and also what was known as The Large Sheet Series. This last name consisted of 152 sheets mostly at 18" x 27" formed by amalgamating the smaller sheets. The large sheets had hachuring in brown, with contours as in the outline edition at 100 ft interval up to 1,000 ft and thence at 250 ft intervals. The contours are now in red and, together with the delicate brown hachures, give a good impression of relief. Woodland is now coloured green with streams and lakes in blue – the lakes being shown by blue form lines. Sea areas are left white except for form lines round the coast and some marine contours. Altitudes are given in feet from a datum at Liverpool, determined by taking the level of the tide at Victoria Dock every five minutes for about an hour around high and low water, every tide from the 7th to the 16th of March. This datum had been in use since 1840 and continued until well on into the 20th century when it was replaced by the Newlyn datum.

The Outline Edition of the Third edition was printed from copper-plates but the coloured editions were printed by lithography; the transfers being made from the copper plates to the stone, a different stone being used for each colour, six in all, meaning six different printings.

The Fourth Edition

In 1913 survey for the Fourth Edition of the one-inch map commenced. It was completed in 1923, the sheets being actually published between 1918 and 1926. This edition maintained the larger sheet lines of the Large Sheet Series of the Third Edition, there being 146 sheets 18" x 27" in size, and it represented the third national revision of the New Series. Production was hampered considerably by the 1914–18 War and the Government's economy drive which followed it put an abrupt end to some lavish hopes which had been fostered for the elaborate use of colour. An experimental sheet of the Killarney area had been beautifully printed in no less than thirteen colours but there was now insufficient money available for such lavish colour processes. The old copper plates had to be used but they were cut and re-joined to make more convenient sheet lines. Hachures were abandoned and more surprisingly were parish boundaries – a retrograde step indeed. To offset the loss of hachures to some extent the contour interval was stepped up to 50 ft using interpolated contours put in from sketches made on the ground. This Fourth or Popular Edition as it is alternatively called used seven colours: black, two shades of blue, green for woods, orange for contours, and red and brown to differentiate two classes of road. A black and white edition was also published for those who had special requirements and wished to apply their own colours. Thirty¬-three sheets of the Tourist Edition were issued between 1919 and 1938. They were based on revisions of the Fourth Edition together with a number of special District sheets. Some of the latter were particularly attractive and in mountainous areas such as Glencoe, Snowdonia or the Lake District a splendid impression of relief was given by the hill shading and the twelve-colour printing. It is most unfortunate that one sheet only of this Tourist Edition, the Lake District sheet, survived the Second World War but this was a really beautiful map and one which must have been a splendid introduction to the Ordnance Survey for many people. The Fourth Edition represents the end of an era for the one-inch map. Up to and including it the maps were based on engraved copper plates; material for each subsequent revision being amended on the plates and the maps themselves being made by transferring the engraved outline from the copper to stone or zinc. In subsequent editions this engraving tradition was dispensed with and the maps were produced from original drawings directly by photo-lithography.

The Fifth Series

Work commenced on the Fifth Series in 1928 and publication of the first sheets began in 1931. This edition appeared in two distinct guises both in colour but one, known as the Fifth (Relief) Edition, with layer colouring and shadow. The second set had contours only. A new projection which was a modified Transverse Mercator was employed and, all in all, this was the most ambitious map attempted so far. On the maps of the Relief Edition, contours at 50 ft intervals drawn in orange are combined with a shadow effect produced by delicate hachuring in grey to give a pleasing impression of relief. Woods are green and it is enlightening to study the drawing of the woodland closely for the craftsmanship does not compare with the delicacy and charm of the engraved trees in the early editions. A number of new conventional signs appear on this edition – electricity transmission lines, pipe lines, National Trust properties, road numbering, telephone call boxes, wireless masts, post offices with telegraph and so on. Gradients of 1 in 5 and steeper are symbolized by a double arrow-head while those between 1 in 5 and 1 in 7 have a single arrow-head. The map is covered by a grid system of lines 5,000 yards apart. Sheet lines were generous and so was the price – one shilling and ninepence. Sadly this splendid series of maps was to reach no further north than Birmingham when publication ceased with the outbreak of the Second World War in 1939. This marks the end of the history of the one-inch map before the introduction of the National Grid – it is a period full of interest and much complexity with plenty of meat for the cartographic historian.

The Davidson Committee

In 1935 an inter-departmental committee set up under the chairmanship of Lord Davidson investigated the future policy of the Ordnance Survey. This Davidson Committee made four main recommendations:

1. That a National Grid calibrated to the metre should be employed as a standard system of reference for the whole country and superimposed on all maps and plans.
2. That urban areas should be surveyed at a scale of 1:1250 and a method of continuous revision be devised.
3. That the 1:2500 survey should be overhauled and planned on a single national projection.
4. That the six-inch maps should continue as a national series.

These proposals made a fresh triangulation of the whole country necessary and the Ordnance Survey began to increase its recruitment of civilians to meet the requirements of the Davidson Committee. Unfortunately, the outbreak of war once again brought mapping to a standstill and the resources of the Ordnance Survey were largely placed at the disposal of the War Office.

War Revisions

During the war period two editions of the one-inch map were prepared for military purposes. These were termed the War Revision and the Second War Revision and the sheets were produced in great numbers. As a matter of wartime expediency they were printed on poor paper and were generally rather unattractive, partly as the result of a prominent military grid printed in purple. The headquarters of the Ordnance Survey was bombed in 1940 and a considerable proportion of the plates, records and stocks of maps were lost. A temporary establishment was organized at Chessington in Surrey which sufficed until the completion of the new headquarters in Southampton in 1968.

The Sixth (New Popular) Edition

After the war, work commenced on the Sixth or New Popular Edition of the one-inch series, a series which was based on the Fourth Edition in Northern England and Wales and on the Fifth Edition for those sheets south of Birmingham. This series covered England and Wales only, the Popular Edition continuing to serve for Scotland. It incorporated the new National Grid System recommended by the Davidson Committee and had grid lines in black at 1 km apart to facilitate the reading of National Grid references. Each kilometre square covered the same area as a sheet of the new 1:2500 (two-and-a-half-inches to one mile) series. The final sheets were produced by photolithography and incorporated various improvements and additions; Ministry of Transport road numbering was shown on the maps, parish boundaries appear once again, contours are printed in brown at an interval of 50 ft, water is blue, woods green, 'A' classified roads red and 'B' roads orange. An outline edition of the New Popular Edition was also produced for those who required a black and white map only on which they could add their own detail or colourings.

The Seventh Series

Although the New Popular Edition was an undoubted improvement on the two makeshift War Revisions its shortcomings become apparent when any sheet is compared with its fellow in the Seventh Series which superseded it. The Seventh was an entirely new series and was unique in that it was the first set of one-inch maps to cover the whole of Great Britain (not Northern Ireland) on a single system of sheet lines which numbered right through from north to south. Several new touches were incorporated; a better method of indicating National Trust property with a red bounding line and a red N.T., National Park boundaries prominently shown by a bold yellow line, and bus stations and coach stations and Youth Hostels were included. Relief depiction is still by contours only at 50 ft intervals without layer tinting and this does not always convey an immediately striking visual impression of relief. One of the biggest improvements compared with the Sixth Edition was to replace the black built-up areas by a grey tint. This gives a much lighter and more pleasing appearance to the sheets particularly in densely populated areas. The Seventh Series covers the whole of Great Britain in 189 sheets which are revised at the rate of about twelve per year. Particularly important changes, however, such as new motorways are specially surveyed and added when a sheet is reprinted.

One-inch Tourist and Special Maps

A limited number of sheets have been produced to cover areas of particular interest to the tourist. They are based on the standard one-inch series but are on different and generous sheet lines with some additional information included. The areas so far covered are Cairngorms, Lorn and Lochaber, Lake District, Cambridge, North York Moors, Dartmoor, Exmoor, New Forest, Peak District, Loch Lomond and the Trossachs, and Wye Valley and Lower Severn. The maps vary in treatment, mostly fairly conventional with relief indicated by contours and hill shading but the Dartmoor sheet, for instance, incorporates several breaks with tradition, the most obvious being in the layout of the legend and in the colours used for highland areas. The very detailed legend includes in addition to the normal information a small inset map of the Dartmoor National Park, a similar map showing the extent of information available on Public Rights of Way, a list of selected places of interest (Archaeological, Beauty Spots, Ecclesiastical, Museums, Stately Homes and Gardens, View Points and Others). Symbols used on the map of direct use to tourists and holiday makers are camp and caravan sites, the National Park Information Office, Parking Facilities and View Points. Public footpaths, which give a right of way on foot, are distinguished from Bridleways, which offer right of way both on foot and on horseback, and roads used as public paths are also shown.

Three linear scales in miles, yards and kilometres are included. The separate headings in the layout of the legend are lettered in black superimposed on a panel in carmine and whether this is a good move or not is very much a matter of taste. It is certainly a departure from convention. Again, instead of having the title of the sheet centred in the top margin in serif lettering, the Dartmoor sheet has its title in sans-serif type placed in an inset of the frame at top right with 'Ordnance Survey of Great Britain' also inset in sans-serif at top left. A much more striking green is used for woodland on the Dartmoor sheet; 'A' roads are in a rather unpleasant carmine instead of the earlier vermillion and blues and purples are now introduced at the higher end of the layer tinting scale. The arrangement of this colour system for the relief goes from pale green through pale pink and orange to a reddish brown below 1,000 ft with a marked transition at 1,000 ft into grey, blue and purple followed by white for land over 2,000 ft. This new colour scheme may take a little getting used to as it gives quite a new and different appearance to the one-inch map. There is no doubt, however, that these Tourist Maps with their large sheet lines – 33.5" x 40" in the case of the Dartmoor sheet – their clarity and the wealth of information provided for the traveller are excellent value at 6s.0d. for the paper flat copy in most cases.

What future developments will be and how the change over to the metric system will affect the one-inch series remains to be seen. We can say, however, that as far as mapmaking is concerned in this modern age Great Britain has something of which it can be proud in the Seventh Edition of the one-inch to one mile map.

Editorial Notes

1. From William Roy (1785) "An Account of the Measurement of a Base on Hounslow-Heath" *Philosophical Transactions* LXXV (Part II) pp.385–480 (quote on p.387).
2. From Major Thomas Colby's (1824) evidence given in *Report from the Select Committee on the Survey and Valuation of Ireland* British Parliamentary Papers (House of Commons series) (445) VIII, 79 (quote on p.28).
3. From an unidentified report.

We gratefully acknowledge the assistance given by Richard Oliver in identifying the source material for these quotations.

Selected Bibliography

Close, C. (1969) *The Early Years of the Ordnance Survey* (Reprint of the 1926 edition with an introduction by Dr J.B. Harley) Newton Abbott: David & Charles.
Davidson, J.C.C. (1938) *Final Report of the Departmental Committee on the Ordnance Survey* London: H.M.S.O.
Harley, J.B. (1964) *The Historian's Guide to Ordnance Survey Maps* London: National Council of Social Service for the Standing Conference for Local History.

The 1970s

The 1970s were transformative years for cartography. Perhaps more so than in any other decade, the identity of the discipline reoriented towards the sciences. This had several consequences, from the ways in which maps were produced and evaluated to the theoretical avenues of research that were pursued and published, as reflected in the Society's *Bulletin*.

As Steve has already mentioned, *Maps for Books and Theses*, the handbook written by founder member Alan Hodgkiss for university cartographers, was published in 1970 and became a standard text for the increasing numbers of students enrolling on cartography courses in the UK, along with the third edition of Monkhouse and Wilkinson's *Maps and Diagrams*, published in 1971, and Keates' *Cartographic Design and Production* in 1973. The latter presented a stronger theoretical basis for map design that more closely echoed the scope of *Elements of Cartography*, which had reached its third edition by 1969.

Yet, for their encouragement of good draughtsmanship and coverage of practical techniques, these texts did not foresee the major technological advances that were to influence mapmaking during the new decade. The gradual incorporation of computers within university geography departments, especially with the combined use of the SYMAP program and graph plotter, brought a whole new realm of possibilities to how maps could be used as tools for visualizing geographical data.

Changes were also afoot on a deeper, theoretical level. Chorley and Haggett's (1967) *Models in Geography* had offered fresh theoretical perspectives to the wider discipline of geography that extended towards cartography, particularly through Board's influential chapter, "Maps as Models", which offered a new rationalization of the cartographic process. Theories of cartographic communication that had emerged at the end of the 1960s were taking hold in the 1970s and maps were now judged according to their efficiency of communication. Following Robinson and Petchenik's treatise *The Nature of Maps*, published in 1976, 'noise' was to be eliminated in the path towards the creation of the optimum map.

The papers we have chosen to represent this decade reflect these changes and the response within the cartographic community at the time. We begin with Carson Clark's reminder of the historical roots of mapmaking, sustaining the view that art was the original keynote of cartography, and perhaps offering readers a moment's solace from the march of technology that is all too apparent in the following paper by Guy Lewis, Senior Cartographer in the Geography Department of Swansea University. Lewis' survey provides an insightful snapshot of the new technology of the time, with the warning that 'university cartographers can no longer ignore the use that is being made of computers; it is a case of "if you can't beat it, join it"'.

The following paper is an example of the theoretical work which students undertaking the Diploma in Cartography course at what was then Oxford Polytechnic were encouraged to pursue. It is easy to forget that Jane Thake's early contribution towards understanding children's comprehension of maps was published at a time when the dominant paradigm in cartography tended to ignore the diversity of map user groups and their various needs and abilities. Another theoretical contribution from Oxford Polytechnic, John Ager's substantial and well-illustrated paper on maps and propaganda won the Society's Wallis Award in 1977 as a written entry and has often been cited since. His perceptive and systematic analysis of how propaganda maps utilize various aspects of cartographic design to achieve their goal provides a rare contribution to the field that was ahead of its time.

The following paper by Tomlinson Amachree and his supervisors from Ohio University demonstrates the maturity of the *Bulletin* in the type of article that was being published towards the end of the 1970s, characterized by a growing international dimension and reflecting the increasing scientific identity of cartographic research. Based on Amachree's Master's thesis, which involved a series of experiments that required users to compare the legibility of fonts, the paper is an example of how scientific rigour was being applied to optimize certain aspects of map design. It is interesting to note the confidence put in the communication paradigm as a sound theoretical framework and the impact of *Cartographic Design and Production* (Keates, 1973), upon which the authors heavily rely for their approach to map design.

The developing stature of cartography as an academic discipline, as reflected in the *Bulletin* towards the end of the 1970s, is demonstrated in our return to the history of cartography and Roger Kain's essay on tithe maps. Based on substantial new research, the paper is one of the author's earliest contributions to the topic and provides a concise synthesis of tremendous value to the local historian. The paper concludes by setting an agenda for further avenues of study which were subsequently pursued and materialized with the publication of several key reference works on tithe maps and the history of mapping in England.

The re-orientation of cartography towards its new identity as a communication science is embraced in our final paper of the decade, which advocates a full 'strip down and rebuild' of the mapmaking process to optimize each 'unit'. By this time, various authors had put forward successive cartographic communication models which bore lesser similarity to each other. Patrick Sorrell's paper provides a helpful account and offers an evaluation of the effectiveness of the leading models of the time. It ends by posing some challenging questions for the development of cartographic communication theory that few were able to address.

Alexander J. Kent

References

Board, C. (1967) "Maps as Models" In Chorley, R.J. and Haggett, P. (Eds) *Models in Geography* (pp.671–725) London: Methuen.

Keates, J.S. (1973) *Cartographic Design and Production* London: Longman.

Monkhouse, F.J. and Wilkinson, H.R. (1971) *Maps and Diagrams: Their Compilation and Construction* London: Methuen.

Robinson, A.H. and Petchenik, B.B. (1976) *The Nature of Maps: Essays Toward Understanding Maps and Mapping* Chicago: University of Chicago Press.

SOME EARLY CARTOGRAPHERS AND THEIR MAPS

A. Carson Clark

Originally published in Volume 6, No 1, pp.20–23

From the records that are available to us one of the earliest written references to a map was made by a Greek historian, Herodotus, who said '[…] I cannot help but laugh when I see numbers of persons drawing maps of the world without having any reason to guide them, making as they do, the ocean-stream to run all round the earth and the earth itself to be an exact circle, as if described by a pair of compasses, with Europe and Asia just of the same size'.[1] The best known example was probably the Babylonian idea of the world as a disc surrounded by water.

Some of the materials on which early maps were made are surprising – clay, iron, wood, bone, papyrus or skins – so cartographers today who may think they have problems with plastic film or paper should spare a thought for earlier mapmakers who struggled to draw or cut a line on much more difficult surfaces.

Imagination played a great part in the life of the early cartographer; he saw animal shapes in almost every map he produced. Even today we can, in fancy, imagine all sorts of shapes as we look at maps of different parts of the world, e.g., William Blake's description: '[…] his left foot near London covers the shades of Tyburn, his instep from Windsor, to Primrose Hill stretching to Highgate and Holloway. London is between his knees, its basement fourfold. His right foot stretches to the sea on Dover cliffs, his heel on Canterbury's ruins, his right hand covers lofty Wales. His left, old Scotland; his bosom girt with gold involves York, Edinburgh, Durham and Carlisle and on the front, Bath, Oxford, Cambridge, Norwich; his right elbow leans on the rocks of Erin's Land, Ireland ancient nation. His head bends over London […]'.[2] We can readily see how the imagination of a more leisurely age gave rise to such amusing descriptions.

It has been customary to begin a talk about early cartographers by speaking of Ptolemy, Laudios Ptolemaios of Alexandria, who lived AD 87–150. From my study of this subject, I begin to wonder if Ptolemy was really a cartographer at all, for I find references to him as an astronomer and a geographer but not as a cartographer. He was undoubtedly a great astronomer and a follower of Hipparchus, the founder of scientific astronomy, and also an eminent geographer as he was a follower of Marinos of Tyre.

The earliest existing manuscript map based on Ptolemy's writings is no earlier than the twelfth century, a thousand years after his death. Certainly in the Arab world the work attributed to him was the accepted standard amongst Arab geographers for three centuries, but there is probably insufficient evidence to allow us to say categorically that Ptolemy was a cartographer. I find myself coming to the same conclusion as Raymond Lister in his book on early maps when he says 'It may merely be that the maps have been put to his credit because of his fine reputation as a geographer'. [3]

I would move on now to an undisputedly fine cartographer who was born in Rupelmonde in East Flanders in 1512. The famous cartographer, Gerhard Mercator (actual name Kramer), was first heard of as an engraver of the gores for Gemma's globe in 1536 when he was 24 years old; his background is somewhat difficult to trace, though it would appear that he started his adult life as a land surveyor, later learning the crafts of a mathematical and astronomical instrument maker, and finally becoming the highly skilled cartographer by which craft he is so well remembered. His reputation as a cartographer became well known throughout Europe, and he certainly was greatly influenced by cartographers from Spain and Portugal. Some of his contemporaries described him as '*ingenio dexter, dexter et ipse manu*' (clever in mind and hand). Pedro Nunes, the famous Portuguese mathematician and navigator, may have been one who influenced Mercator, due to the close contacts between Portugal and Flanders (Mercator's home).

Mercator's map of Europe published in 1554, gave him international acclaim. The map, published in fifteen sheets, had many improved features over any previous map of Europe. Until this time most maps had been slavishly prepared from the writings of Claudius Ptolemy, but Mercator, with his mathematically enquiring mind, could not accept all Ptolemy's measurements and in effect the new map of Europe was really a completely new map which reduced the previous length of the Mediterranean and established the position of several offshore islands – including the Canaries correctly. Mercator's maps introduced a fine clear style of italic lettering for descriptive names. Today Mercator's name is probably best remembered for the map projection which bears his name and is the basis on which many world maps are produced and on which sea charts are based. He was a hardworking conscientious cartographer, always revising and correcting his work. Undoubtedly Mercator loved his maps; their preparation was no task which he took lightly but one to which he gave his best. Even today his maps

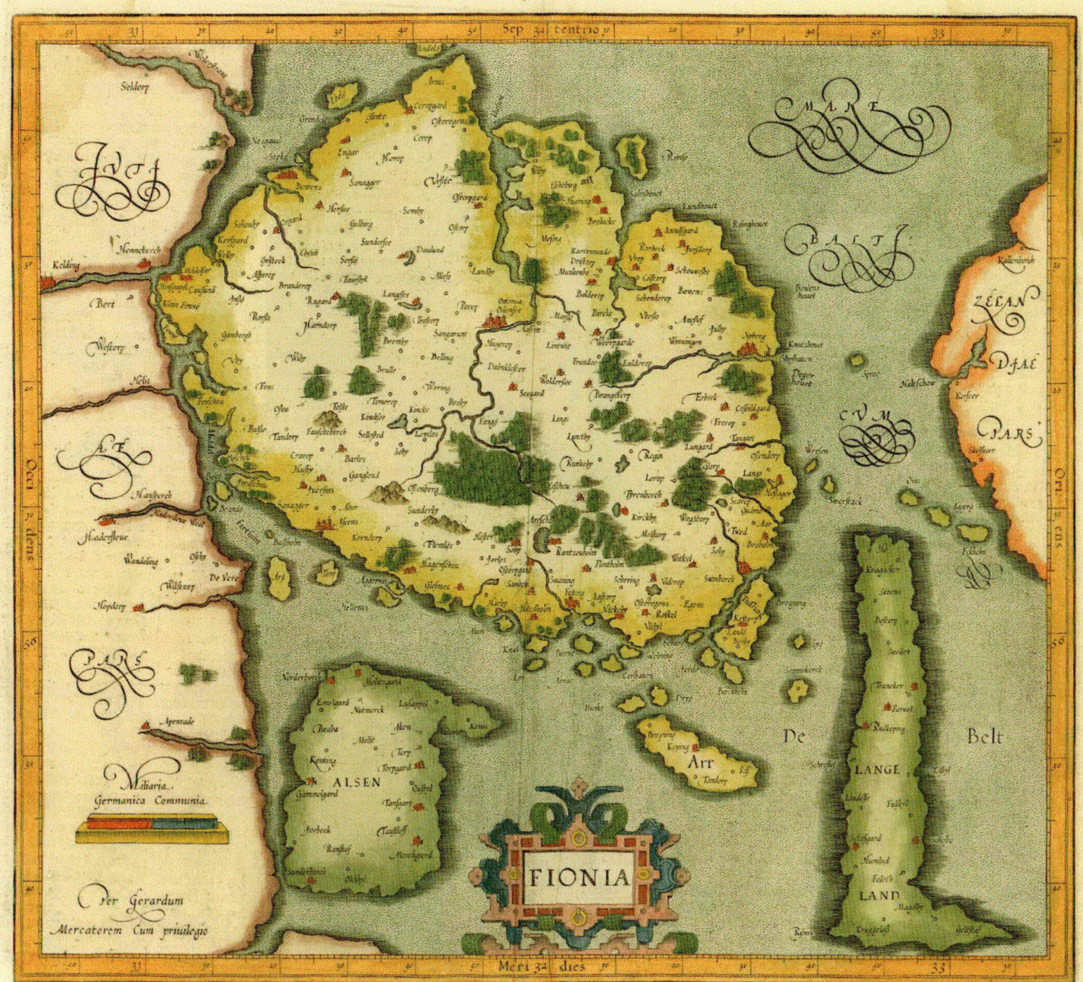

Figure 1
Fionia by Gerhardus Mercator, 1636 (reproduced courtesy of Sanderus Antiquariaat)

with the vast source of information they provide and the delightful, clear italic writing, can only draw from us the greatest admiration.

The second cartographer whose maps I wish to mention is that of Ortelius, or Abraham Ortelius to use the Latinized form of name which he seemed to prefer. He was born at Antwerp in 1527, the son of Leonard Wortels who appears to have been a collector of fine works of art, if not an antique dealer. Abraham's father died while Abraham was still a child and he inherited his father's valuable collection of objects of art; however, he had to support two younger sisters and in order to do this set up in business as a map colourist and map seller. He was registered in this manner in 1547 by the Guild of St. Luke. The business flourished, he made many friends and travelled through Europe and England. Amongst his friends in England were William Camden, who later produced his Britannia and Richard Hakluyt, the famous geographer. Ortelius was essentially a compiler-geographer who collected much information which he portrayed on his maps; in his world map of 1564 he used a projection based on one already used by the German cartographer, Peter Apian, in 1530 and in his map of Asia (1567) he used much of the earlier work of Venetian cartographer, Jacopo Gastaldi. Though the individual maps were important, Ortelius is remembered first of all for his monumentally successful work *Theatrum Orbis Terrarum* published in 1570. This could truly be described as the first modern Atlas, though it was not known by that name. It was undoubtedly an unprecedented achievement and in 1573, three years after publication, King Philip the Second bestowed on him the title of Geographer to the King. What a volume it was, containing over seventy individual maps, a world map, four maps of continents and sixty-five regional maps, three of Africa, six of Asia, and the remainder of Europe. There were black and white editions and hand-coloured editions, the coloured ones being usually done to special order. Ortelius, as cartographer and editor-in-chief, gives generous acknowledgement to 87 cartographer-compilers, who had helped to collect the data and produce the individual plates. Ortelius was not a geographer like Mercator, but rather a scholarly craftsman-cartographer and, of course, a business man whose maps were produced by gathering information from others. Sometime between 1560 and 1569 Mercator is recorded as having discussed his plan for a world atlas with his friend, Ortelius, who was also planning his own atlas, and looking back today it seems strange that Mercator's work was not published as a complete atlas until after his death in 1595, though it should be remembered that the Mercator Atlas ran to 57 editions between 1595–1642.

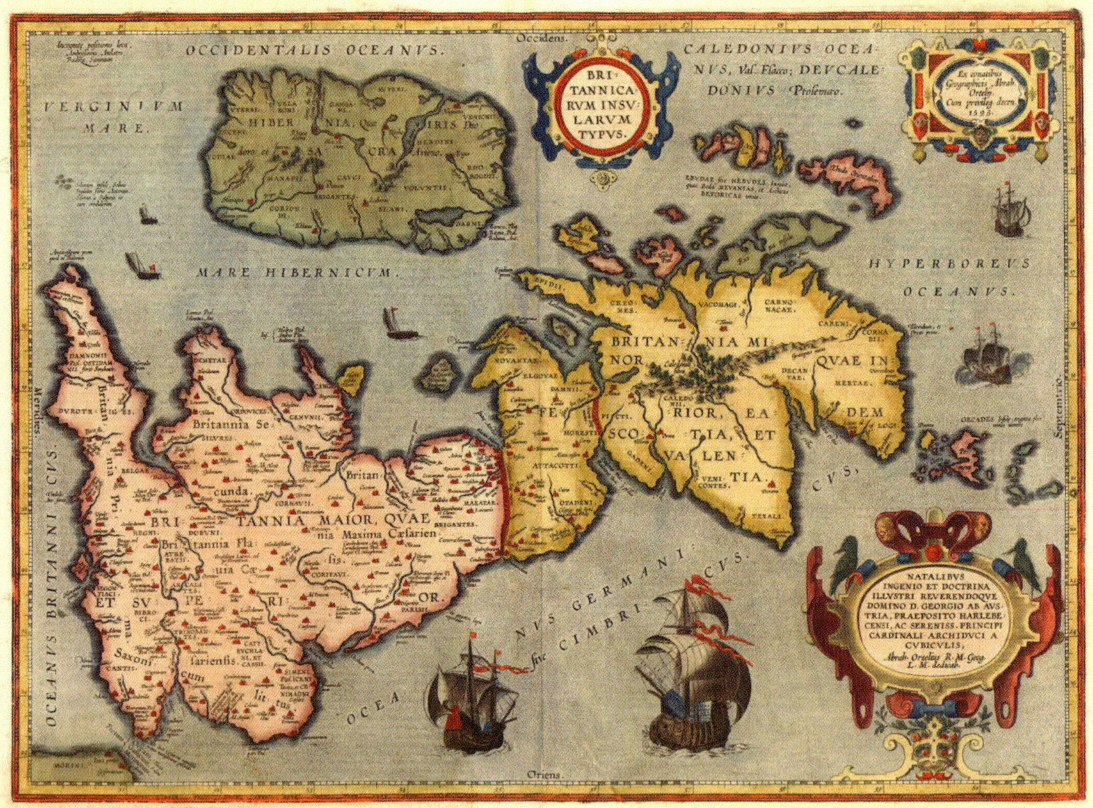

Figure 2
Britannicarum Insularum Typus by Abraham Ortelius, 1609 (reproduced courtesy of Sanderus Antiquariaat)

For a considerable time the fame and popularity of Ortelius's *Theatrum* in a sense overshadowed the achievements of Mercator, but the cartographic achievements of both men must be remembered for all time. Theirs was the basis of cartographic expertise which was to last for more than a century in the cartographic world.

The third cartographer whose maps I would like to consider is Hondius, from the same part of the world as Mercator and Ortelius. Jodocus Hondius, born in 1563, was a prolific map engraver and cartographer, who settled in London in 1583 and engraved plates for Waghenaer's *Mariners' Mirror*, published in 1588. He also produced globes whilst in London, but later we find him established in Amsterdam where he built up a business as map engraver and publisher. Though Hondius is better known as an engraver, he was undoubtedly a cartographer of considerable ability. It was in 1604 that he acquired the plates of Mercator's Atlas, and in 1606 after just two years he published his first edition of Mercator's Atlas, which included no less than 37 completely new maps in addition to many revised and altered plates from the original Mercator Atlas (some measure of his ability).

One can hardly speak of Hondius without some reference to his brother-in-law, Peter van den Keere, born in 1571 and son of the famous typefounder, Hendrik van den Keere. I made reference to Hondius in England and in Holland; the friendship of these two engraver-cartographers started while they were both in England and developed into a real partnership, with Keere producing an enormous number of maps for the Hondius Atlases, as well as for people like John Norden, John Speed and many others.

Do we begin to see here an inter-relationship, a sharing and working together in the common bond of being great map producers?

Undoubtedly it was the acquisition of the Mercator plates to which I have already referred which played an enormous part in the making of Hondius. He saw the advantages of Mercator's projection and used it as a basis for his large world map, first published in 1608. In detail, the Hondius engraved maps from the Mercator plates are even finer in calligraphy than those of Mercator, especially the maps in late editions of the atlas which, incidentally, ran to some fifty editions, the last appearing in 1642. Hondius never dropped Mercator's name from his atlases though it was sometimes hyphenated with his own as the Mercator-Hondius world atlas. Here again we see something of the character and wisdom of Hondius who might well have dropped the Mercator name altogether when one remembers that he expanded the number of Mercator maps from 107 to over 300 in three separate volumes.

Jodocus's son, Henry, continued in the tradition of his father, though he appears to have been the one who introduced the rich decor to the neat-line, a practice which the meticulous Mercator would have frowned upon.

The fourth cartographer is another Dutchman, Jan Jansson, born in 1596. We know very little of his early life, but in 1633, at the age of 37, Jansson joined Henry Hondius in the map publishing business. This partnership again reflects the family co-operation and relationship which seems to have played a prominent part in the early development of cartography. Though first of all a publisher, Jansson was a fine engraver-cartographer,

Figure 3
Guineae Nova Descriptio by Jocodus Hondius, 1606 (Wikipedia)

though it is considered that his maps lack something of the originality of those by Ortelius and Mercator. Again, like Ortelius, he was a great compiler and collector of data; he may, however, be criticized for his failure to acknowledge sources and to give credits to earlier cartographers whose work he undoubtedly copied. Jansson's most famous work was his *Atlas Novus*, originally in three volumes, but gradually extended until in 1661 eleven volumes had been published. This was a monumental cartographic achievement equalled only, perhaps, by his contemporary

Figure 4
Siciliae Veteris Typus by Joan Janssonius, c.1650–1660 (reproduced courtesy of www.bergbook.com)

Figure 5 *Africæ Nova Descriptio by Willem Blaeu, 1635 (Wikipedia)*

and rival, Blaeu. Jansson gained in popularity as his atlases became accepted, and on the death of his partner, Henry Hondius, he inherited a vast collection of earlier plates going right back to Mercator's time. Without such an inheritance many of these earlier cartographers would not have achieved such eminence or made such a name for themselves. So consider now some of the work of Blaeu, born 1571. William Jansson Blaeu, whose name of Jansson caused some earlier confusion with his rival of the same name, was the founder of a family publishing business and father of the famous Joan Blaeu. The father is associated with the first documented discovery of Australia; he was a hydrographer and commissioned by the Dutch East India Company. This discovery was in 1606 though at that time

Figure 6 *Lothian and Linlitquo by Timothy Pont, published by Joan Blaeu, 1664 (reproduced courtesy of Sanderus Antiquariaat)*

it was not realized that this was Australia. He is reported to have returned to his masters, saying he could only find man-eaters and felt constrained to return to the comparative civilization of Java.

It is little wonder that with a family tradition of discovery and mapmaking that Joan Blaeu was destined to become so famous; for over a century the Blaeu maps and atlases were something of a legend. The English diarist, John Evelyn, visited Blaeu's establishment in 1641 and wrote 'I went to Hondius's shop to buy some maps, greatly pleased with the designs of this indefatigable person, to Blaeu the setter forth of the Atlases and other works of that kind is worthy of seeing',[4] and today we can substantiate that statement for they are still well worth seeing.

Blaeu, together with his brother, produced a world map in 1606, and as far as is known today only one copy survives. Their most important work, however, was the completion of the *Atlas Novus* in 1662, a monumental work which extended to 11 volumes and was the first to contain a national Atlas of Scotland. It was in the mapping of the Scottish counties that Timothy Pont did so much excellent work; if one is looking for a Blaeu map of Scotland, it will invariably bear his name.

The maps and atlases which were the products of this great publishing house are even today generally accepted, certainly as far as the county maps of England and Scotland are concerned, as the finest ever produced. The original coloured copies, hand coloured in the Blaeu workshops are most frequently seen, uncoloured examples being very rare. Furthermore, as a result of a disastrous fire, many late editions of the maps are today extremely scarce.

In conclusion may I say how conscious I am that I have only touched the edge of the subject, but I trust that I have conveyed to you some of the spirit and purpose of these early cartographers, and shown the relationship not only between them as cartographers but between the cartographer and his maps, and in addition, their undoubted sense of dedication to their life's work and the value of their tradition and fine work in the field of mapmaking.

Editorial Notes

1. From Book IV of *The History of Herodotus*, written by Herodotus in 440 BC (Rawlinson's translation).
2. From the second book of *Milton* (1804–1810) by William Blake.
3. From Lister, R. (1970) *Antique Maps & Their Cartographers* London: G. Bell & Sons.
4. From the entry for 24th August 1641. See, for example, Bray, W. (Ed.) (1870) *Memoirs Illustrative of the Life and Writings of John Evelyn* New York: G. Putnam & Sons.

The figures which originally accompanied this paper have been re-ordered and substituted by full-colour high-resolution versions for this book.

THE UNIVERSITY CARTOGRAPHER AND COMPUTER CARTOGRAPHY

G.B. Lewis

Originally published in Volume 8, No 1, pp.13–17

The move to process everything on the computer has not bypassed cartography and it is time that university cartographers took a long hard look at the advancements that have been made in that part of computer cartography that touches upon their own particular field. It is necessary to look at the types of maps that have been produced by computers and the various programs that are available.

Computer cartography must not be confused with automated cartography which is really the work of the Experimental Cartography Unit of the Royal College of Art,[1] the O.S. and other large organizations. Computer cartography is really an analytical tool and involves the use of a simple base map, which needs to be only a sketchy outline, together with a representation of geographical data in a form decided upon by the program used, and by the computer equipment available to the user.

The proliferation of data produced today means that a quick graphic representation is necessary to the understanding of the significance of the data, 'rapid and cheap listings of data are virtually meaningless without some cartographic display' (Davies, 1970). Computer mapping is the method for producing the rapid graphic, representation and it will produce numerous maps enabling the geographer to interpret and test hypotheses.

The computer is now an integral part of the university set-up and the university cartographer has the opportunity to learn to use the computer as an aid, but often a knowledge of one of the computer languages is necessary before one can feed information to the university computer and obtain maps. The alternative is to use a package already devised by someone else and then only a knowledge of how to punch cards is required. However, if the university cartographer is to 'succeed it is inevitable that he will have to know a minimum about the working of a generalized system' (McGullagh and Sampson, 1972).

The speed with which computers produce maps is the essential factor of this type of cartography, the maps are not works of art themselves, the output often being very rough, too rough for use as an illustration. When a set of data has been processed the computer can produce many versions of the map, sometimes at different scales, and it will be the task of the university cartographer to produce a presentable map from the computer map that he selects.

Figure 1
A drum plotter

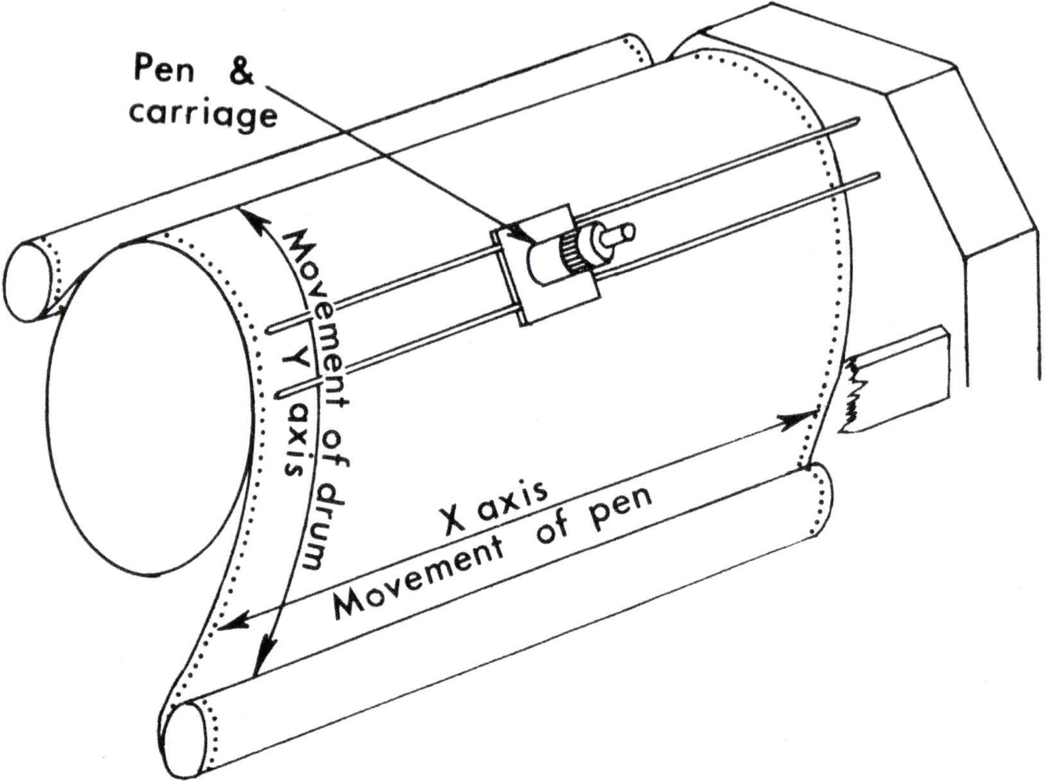

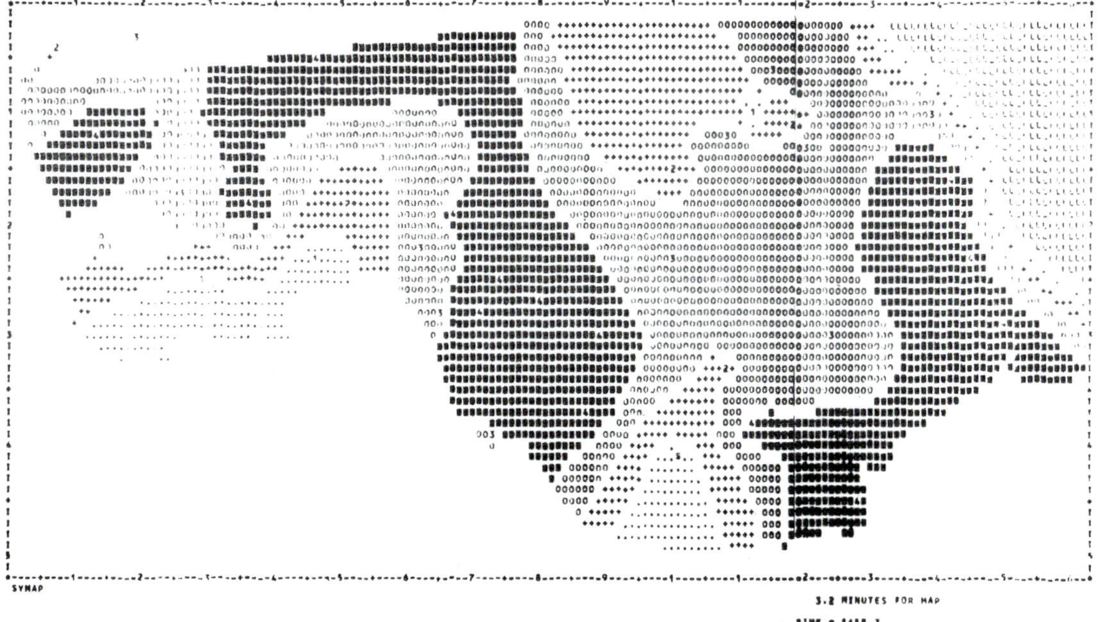

Figure 2
An example of a contour map of South Wales drawn using SYMAP

In producing a computer map other devices are required such as a digitizer. This consists of a measuring table, on which an outline map of the area is mounted, and a cursor with cross hair for positioning over desired points on the map. Attached to the digitizer there should be an output device such as a paper tape punch. The digitizer is operated by placing the cursor over a point on the map, the operation of a switch causes the (x,y) coordinates of the point, relative to the table axes, to be punched out on the paper tape. When the paper tape is fed into the computer it will

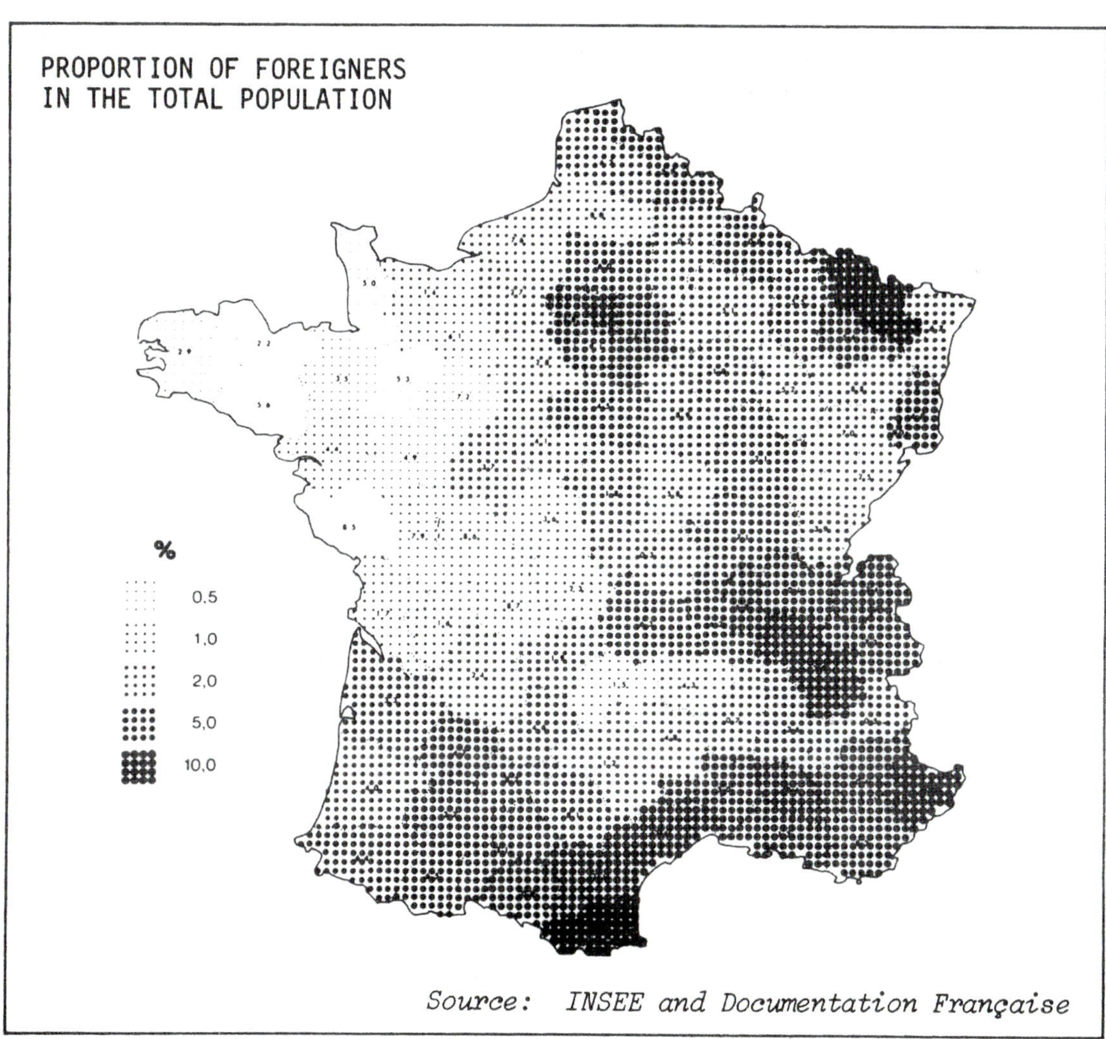

Figure 3
An example of a choropleth map created using a photosetter

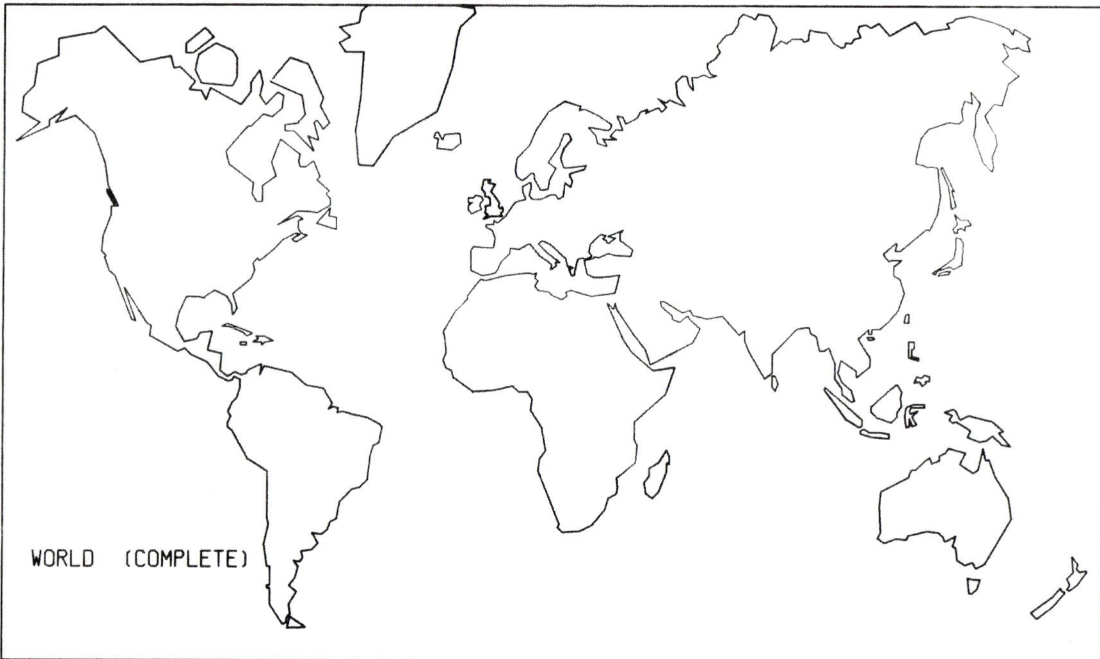

Figure 4 *World map drawn using a basic computer program and a drum plotter*

draw a copy of the original outline, though the outline may be more generalized than the original according to the number of points selected. Not all universities have this device but a way to surmount this obstacle is suggested later in this paper.

Another piece of equipment is a plotter, which may be of a flat bed type but is more usually a drum plotter as in Figure 1. The plotter, often called a Graph Plotter because it has been used in the past to produce graphs, is the device which finally draws the maps. On the plotter the pen moves in one direction only, across the paper, while movement of the drum causes rotation of the paper perpendicular to the direction of the pen movement, i.e. the plot (map) is produced by rotary action of the drum (x-axis) and lateral motion of the pen carriage (y-axis). Electronic plotters utilizing cathode ray tubes tend to be much more expensive and will not be found in many universities. 'This is a field in which technology is advancing rapidly, and cathode ray tubes or their successors are likely to be important cartographic output devices' (Experimental Cartography Unit, 1971: 92).

Maps may also be produced using the line printer which is standard in any computer centre. Using an ordinary typewriter keyboard and allowing it to overprint symbols, shadings of different intensity are produced (Figure 2). By altering the keyboard, another series of maps may be produced. In this case by using a photosetter which uses a series of cartographic symbols instead of letters, as in Figure 3. The maps are automatically produced by punched cards directing the operation of the specially keyed electric typewriter.

The best known computer mapping program is SYMAP, an acronym for Synagraphic Mapping, and developed by Professor H.T. Fisher of the Laboratory of Computer Graphics at Harvard University, who also coined the term synagraphic which means 'acting together' and 'visual method of presentation'.

SYMAP has been developed over almost ten years and is now a highly sophisticated system. It is one of the packages mentioned earlier in this paper and can be purchased by a university, then set up in their computer. Three types of maps are produced by this program:

a) The contour or isoline map which 'should be restricted to the representation of continuous information' (Fisher, 1968: 7). Figure 2 is an example of a contour map of South Wales drawn using SYMAP.

b) The second type of map is, what Fisher calls, the conformant map, but which is more commonly called the choropleth map and 'is best suited for data, either qualitative or quantitative, whose areal limits are of significance, and whose representation as a continuous surface is inappropriate' (*ibid.*).

c) The final type of map is called the proximal map but geographers would know it as 'the nearest neighbour map'. 'Each character location on the output map is assigned the value of the data point nearest to it. Boundaries are assumed along the line where the values change' (*ibid.*), and the resulting map is similar in appearance to the conformant map.

A continuous series of maps may be produced and though the size of these maps is limited to the width of the paper there is an automatic system of fragmentation which allows for the production of larger maps. The map in Figure 2 was too large to be produced on one sheet and the computer automatically produced the eastern section of the map on a second sheet. Normally, a map is made up of a very much larger number of sheets that are joined together and reduced photographically. One of the disadvantages is the unevenness in the printing produced by the line printer, as in Figure 2 along the longitude of Cardiff. The isolines between the shaded areas are white consisting of spaces left by the line printer (a complete account of SYMAP is

Figure 5
Reduced photograph of a map drawn by the O.S. using a plotter

given in "The SYMAP Programme for Computer Mapping" by J.C. Robertson, *The Cartographic Journal*, Volume 4, Number 2, December 1967, pp.108–113).

A second type of computer map is that produced by a program called LINMAP (Lineprinter Mapping), and developed by the Department of Environment with an output very similar to that of SYMAP. The system has been devised for use by planners and intended for the mapping of census data (originally, the maps were 'restricted to thirty-six inches wide and one hundred inches long'3) and was developed using the ten per cent sample census in 1966. A redesigned version of LINMAP is now in use, called LINMAP 2 and it produces maps with up to ten classes of shading. 'The most exciting part of LINMAP 2 is the facility by which it is possible to obtain printed thematic maps of graphic arts quality in colour' (Gaits, 1969: 58). This part of the system is called COLMAP and is 'designed to become a full blown atlas production system' (ibid.).

CAMAP is a choropleth mapping program, similar to part of SYMAP, and has been used principally to produce parish-based maps for the Agricultural Atlas of Scotland.

Another large group of computer maps is produced by programs that utilize a drum plotter. The drum plotter draws its maps in a series of straight lines and can only move in eight directions, a curve is produced in short incremental steps based on these eight directions formed by the combination of the two movements shown in Figure 1.

Figure 4 shows a map of the world drawn using a simple basic program and a drum plotter. In this case the map consists of approximately one thousand coordinates only. Figure 5 is a reduced photograph of a map drawn by the O.S. using a plotter and it can be seen that the quality is much better than in the map of the world owing to better spacing of coordinates and to the use of a digitizer to produce the coordinates.

The university cartographer must decide on the type of computer program which best suits his/her purpose. Most universities possess a computer which he/she could use through the Geography Department. Further the drum plotter may also be part of the equipment housed in the Computer Science Department. It is the digitizer that is least likely to be found and here the university cartographer must resort to a form of hand digitizing. A map is drawn on graph paper or a transparent grid placed over an existing map. Points along the coast are taken and the (x,y) coordinates are read off from an origin, usually the south west corner of the frame. The coordinates are then punched on cards or on paper tape. This task is tedious but the coordinates so produced can provide the university cartographer with a rapid succession of outline maps.

The computer can be further programmed to place symbols or to produce a quantitative map using a program such as MAPIT or KOMPLOT.

Figure 6 is the computer program used to produce the map of the world in Figure 4. The program, written in Fortran, is used together with the cards of coordinates, which are themselves punched according to the format in line 31, i.e. 2F6.1,I1, this means that there are two coordinates per card and that the coordinate can consist of six figures with one decimal point. A typical coordinate card would be punched as bbb543bbb2671, where b represents a blank, and the coordinate would be 54.3 cm, 26.7 cm. The final figure indicates whether the pen on the drum plotter is in the up or down position enabling the pen to draw a line (down position) or when an island has been completed the pen can move on to draw the mainland without drawing a connecting line between the island and the mainland (up position).

Two further programs which deserve mention are MAPIT and KOMPLOT. In the case of MAPIT several types of thematic maps may be drawn, including Dots, Graduated Symbols and Flow Maps, though geographers would know the type of flow lines on the last maps, as desire lines. The development of this type of program could alter methods of geographical research. Kern and Rushton (1969: 137) recognize that 'situations can be anticipated where geographical research will consist to a large extent of the researcher sitting at a computer console and calling data files for the area that interests him'. If this is true then university cartographers will have to draw maps from compilations done by the computer.

KOMPLOT (see Kadmon, 1971) is a program that goes further than MAPIT, for not only is it able to produce an outline map together with the map types drawn by MAPIT, but it can also subdivide the graduated symbols to include

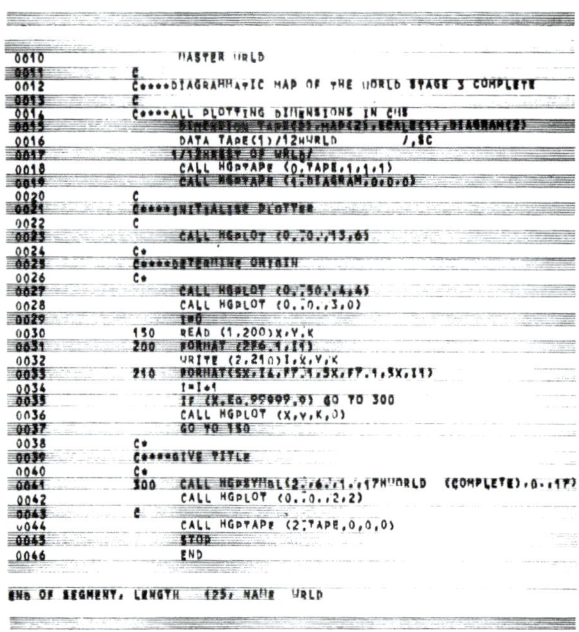

```
      MASTER WRLD
C
C****DIAGRAMMATIC MAP OF THE WORLD STAGE 3 COMPLETE
C
C****ALL PLOTTING DIMENSIONS IN CMS
      DIMENSION TAPE(2),MAP(2),SCALE(1),DIAGRAM(2)
      DATA TAPE(1)/12HWRLD         /,SC
     1/12HREST OF WRLD/
      CALL HGPTAPE (0,TAPE,1,1,1)
      CALL HGPTAPE (1,DIAGRAM,0,0,0)
C
C****INITIALISE PLOTTER
C
      CALL HGPLOT (0.,0.,13,6)
C*
C****DETERMINE ORIGIN
C*
      CALL HGPLOT (0.,50.,4,4)
      CALL HGPLOT (0.,0.,3,0)
      I=0
150   READ (1,200)X,Y,K
200   FORMAT (2F6.1,I1)
      WRITE (2,210)I,X,Y,K
210   FORMAT(5X,I4,F7.1,5X,F7.1,5X,I1)
      I=I+1
      IF (X.EQ.99999.0) GO TO 300
      CALL HGPLOT (X,Y,K,0)
      GO TO 150
C*
C****GIVE TITLE
C*
300   CALL HGPSYMBL(2.,6.,1.,17HWORLD  (COMPLETE),0.,17)
      CALL HGPLOT (0.,0.,2,2)
C
      CALL HGPTAPE (2,TAPE,0,0,0)
      STOP
      END

END OF SEGMENT, LENGTH   125, NAME  WRLD
```

Figure 6
A computer program, written in Fortran, that was used to produce the world map in Figure 4

histograms, pie diagrams and flowgrams, in the sense that geographers think of flow maps, and within the limitations of the drawing capabilities of the plotter, i.e. the flow from one town to another is shown as a band of calculated thickness, though straight rather than curved. A bonus with this program is that the line printer produces a complete listing of the results of the calculations performed by the computer, so that for a map using pie diagrams it would list all the radii of the proportional circles and the percentages used for the subdivision of each circle into segments.

This survey has no pretensions of being complete but it is sufficient to show that university cartographers can no longer ignore the use that is being made of computers; it is a case of 'if you can't beat it, join it'. However, to 'join it' one must attempt to learn the jargon of what has been described as 'computerese'; the use of words such as hardware, software, do loops, and core tend to develop a mystique that often frightens off the outsider. Many of the programs are designed for use by persons who have no knowledge of program languages, but some of the jargon will have to be understood. Bertin (1968: 12) says 'the cartographer should no longer be only a skilful and experienced draughtsman, but above all the man who deals with a "language" which shares its logic and its aims with mathematics but differs from the latter by its means and by its laws'.

Notes

1. Experimental Cartography Unit (1971) *Automatic Cartography and Planning* London: Royal College of Art.
2. Several examples of maps produced in this way can be seen in Rosing, K.E. and Wood, P.A. (1971) *Character of a Conurbation: A Computer Atlas of Birmingham and the Black Country London*: University of London Press.
3. Department of the Environment (1968) *Mapping by Computer* Urban Planning Directorate, 3rd August, p.1.

References

Bertin, J. (1968) "Cartography in the Computer Age" *Paper presented at the ICA Technical Symposium (S 40), New Delhi, December 1968.*

Davies, R. (1970) "Computer Graphic Techniques" *Planning Outlook* 8 pp.29–34.

Fisher, H.T. (1968) *Reference Manual for SYMAP Version V* Cambridge, Massachusetts: Laboratory for Computer Graphics, Harvard University.

Gaits, G.M. (1969) "Thematic Mapping by Computer" *The Cartographic Journal* 6 (1) pp.50–68.

Kadmon, N. (1971) "KOMPLOT: 'Do-it-yourself' Computer Cartography" *The Cartographic Journal* 8 (2) pp.139–144.

Kern, R. and Rushton, G. (1969) "MAPIT: A Computer Program for Production of Flow Maps, Dot Maps and Graduated Symbol Maps" *The Cartographic Journal* 6 (2) pp.131–137.

McGullagh, M.J. and Sampson, R.J. (1972) "User Desires and Graphics Capability in the Academic Environment" *The Cartographic Journal* 9 (2) pp.109–122.

Acknowledgements

I am indebted to Mr. T. Fearnside, University College of Swansea, for permission to use the map of South Wales drawn by SYMAP.

specialised drawing services ltd

Technical Fountain Pens

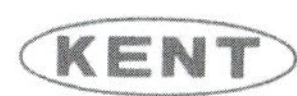

Introducing the new "Kent" technical fountain pens.
*3 different set combinations
*Can be obtained singly
*12 different line combinations
These pens are used in most industries including engineering, construction publishing, plastics, oil, cartography, etc.

This technical fountain pen is non-clogging and easily cleaned when used with drawing ink for drafting, ruling, guide and freehand lettering, tracing, writing and commercial artwork.

Designed to meet the needs of the professional who requires frequent changes of the five, seven or twelve colour-coded line widths.

These pens can also be used as reservoir pens with the Kent Lettering Set.

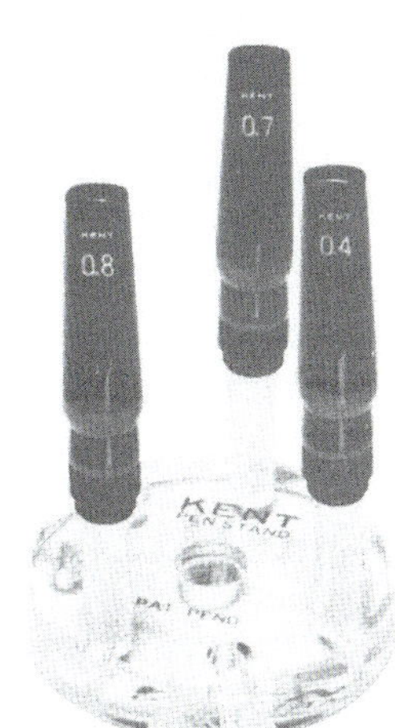

pen stand

0.1mm
0.2mm
0.3mm
0.4mm
0.5mm
0.7mm
0.8mm
1.0mm
1.4mm
1.7mm
2.0mm
2.4mm

various line widths

Prices

Single Pens

0.1mm, 0.2mm, 0.3mm, 0.4mm, 0.5mm, 0.7mm, 0.8mm, 1.0mm, 1.4mm, 1.7mm, 2.0mm, 2.4mm
83p each (16/7d)

Pen holder
14p each (2/9d)

Sets

850–0005 (5 pens and case)
£5.73 (£5.14.7d)
850–0007 (7 pens and case)
£7.40 (£7.8.0d)
850–0012 (12 pens and case)
£11.80 (£11.16.0d)

Sets include compass attachment, pen stand, pen holder, pelican ink.

set of seven pens

pen complete

pen pin body ink reservoir holder

Please send remittance with order to:
Specialised Drawing Services Ltd., Equipment Division, 4, West Street, Ware, Hertfordshire. Telephone: Ware 4816

1971

CHILDS PLAY: MAPPING CONCEPTS FOR 7–9 YR. OLDS

Jane Thake

Originally published in Volume 10, No 2, pp.29–37

'Children need to learn how to read maps before they read maps to learn' (Kohn, 1953). In an increasingly sophisticated educational system, it does not seem unrealistic that the basic skills of mapping should be further developed and understood at Primary School level. Maps for this age range have been produced in recent years by many publishers in the form of atlases, map work books and various cartographic concepts introduced in a wide range of mathematical series. But the question must arise as to whether the end product is really acceptable; initially to the teacher, but above all to the child.

Before attempting consumer research into this problem, their attitudes and level of comprehension must first be studied. To teach a subject one must also understand the pupil. It is often this side of the problem, the sometimes strange and erratic workings of a child's mind, that is so wrongly put aside and forgotten when products such as maps and atlases are designed. Throughout the time devoted to this study, attempts were made to relate all research, results and conclusions directly towards maps, their understanding and their success as media of communication. Initially products already available were used as test-pieces for the children, whereupon it was found that other methods of representing and describing cartographic concepts might be more effective for the child, questionnaires were devised – practical tasks were set for the children directly relating to the basic concepts of mapping, while both the teachers and student teachers questionnaires were aimed at finding out the type of material they required for successful cartographic communication. Tables and charts were drawn up of the results, and while certain conclusions could be made on paper, it became obvious, after many hours spent with the children, that nothing in the child's mind is as 'clear-cut' as we adults believe it to be. At school, subjects have become integrated and topic teaching and learning have emerged. The position of the map within the field of' educational progress has to be recognized in that it can now adapt itself under many headings, and relate to almost every topic.

It has long been customary to introduce the map to a child by first teaching him to understand, draw and use localized plans and scale drawings. However, investigations with children were not aimed solely at the question of how one should introduce maps, for it is certain there is no demonstrably correct way, but rather at the way the child actually appreciates map concepts, data and colour, and his general ability to perceive, read and interpret them. As Bartz (1970) says 'if we truly believe that development of skills in using spatial logic and representation of the earth-space effectively is an important part of being educated, then by and large, results of research are discouraging'. Why then, with an adequate amount of material, is the child still not understanding the mapping technique?

Map communication is possible only when an appreciation of concepts such as scale, direction and symbols has been fully developed. Like a child who fails to understand the text of a book because he cannot read, one who possesses no map skills will see no more than shapes and lines on a map (Harris, 1972). While it does appear that a child has an ability to use maps in some way or another, and that to some extent this is a natural development, the problem to be overcome is the nurturing of an ability to interpret a graphic language which is more sophisticated than he personally is capable of imitating. This appears only common sense, but have the publishers recognised this basic factor? Some atlases include plans of class-rooms and photographs of towns, in the front of the book, but then subsequently introduce highly technical maps of areas unknown to the child, with complicated symbols and no explanation as to what is being shown, how it is being shown, or what can be learnt from it.

Primary atlases have in the past often tended to be mere simplifications of the works produced for secondary level, and even today the publishers' research is very limited. It has proved time-consuming, expensive and difficult to obtain information from children themselves, so design and content research is derived from the teachers' opinion on the needs of his/her pupils. Moreover, it has been suggested that the publisher must finally decide upon the format and content of an atlas or map, due to the vastly varying opinions received from educationalists.[1] It is no wonder then that the content of these products is not always what the child requires, or even understands.

If the conclusions reached from an analysis of questionnaires completed by teachers and student teachers were used in atlas planning, the included maps would mainly be of local areas (Figure 1) (the problems of which are readily apparent) and the British Isles. They would represent details in a topic form, with inter-relating text, general photographic illustrations and include aerial photos. Teachers are aware of children's needs and, above all, should know the type of content and presentation required. But there can surely be no substitute for direct research involving the child, where true ability can be

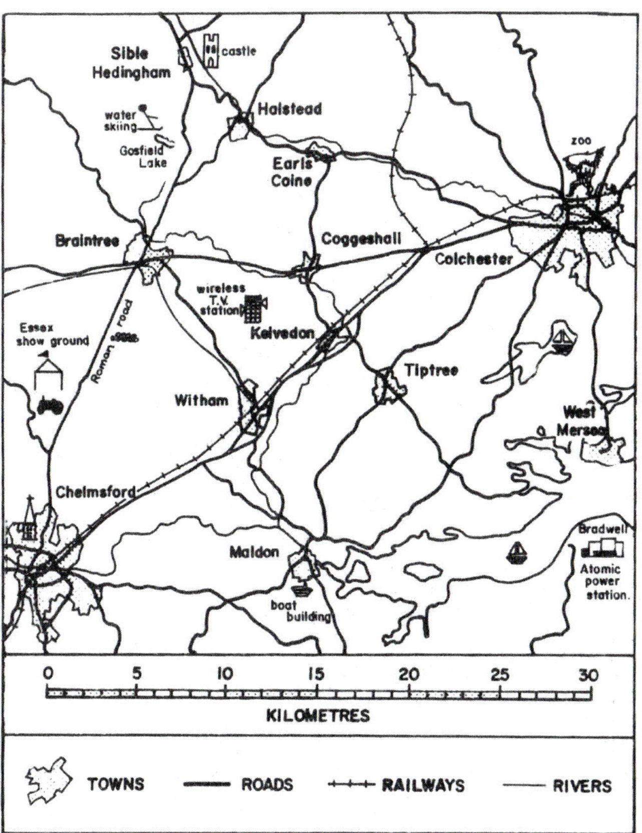

Figure 1
A suggestion for a Local Area map for children

How to find the treasure : up the road 5 k turn left 5 k North 5 K East 10k North 5 K turn left stop.

Figure 2
Description of a route – confusing East and West with left and right

tested, immediate reactions observed and methods of approach recorded.

When dealing with this young age group, research must begin with basic concepts, without which a child cannot be expected to interpret detail, despite the fact that, before scale is understood, a child is able to draw a house and garden in a reduced form in near proportion to each other.

The first problem to be encountered by the child is that of the 'map view', pre-supposing that the viewer is directly above every part of the map. Spatial orientation and direction must next be considered, followed by area and linear scale, grid reference systems and symbolization. Other concepts, such as colour and type must also be examined if map design is to be researched in depth.

Direction and Spatial Orientation

The realization of the existence of direction on maps is an important step. At first children will naturally orientate any plan with that which lies in front of them, at the top of the page. As smaller scale plans are drawn, so the appreciation of the cardinal points needs to be developed. Whether the north point should actually be placed on all maps the child draws is debatable. Some claim that the premature introduction of cardinal directions perpetuates such symbolic equivalents as left equals west, and up equals north (Figure 2), and that as long as the child can orientate himself, the use of North, South, East and West could be left until a later stage (Meyer, 1973). Others state that the setting of every map in relationship to the cardinal directions enables the child to 'set' himself when referring to local and world maps (Harris, 1972).

Whichever idea is accepted, the concept of direction must be understood, and its introduction could easily be incorporated within an atlas at the appropriate stage. Research indicates that children between seven and nine years of age were already aware of the cardinal directions on a four point system. Their ability to describe a simple route on a large scale map proved that they could also apply their knowledge. Lord's (1941) study led him to believe that although children may have a type of understanding of cardinal directions based on maps, this is not extended over nearby places of which they have direct knowledge. Here they form a personal direction system (Meyer, 1973). Many atlases include a north point on maps but if Lord's findings are correct, which is likely, it is during practical work, such as mapping the school grounds, that the north point and its implications should be explained.

Scale

(a) Area scale: The area scale concept is one that is used by every child as soon as he can draw, but is seldom specifically explained although repeatedly used in atlases. Area scale is directly related to spatial relationships in that if the areas of objects are wrongly represented often their spacings are also incorrect. This was well illustrated during my researches when children were unable to place the correct scaled shapes onto a reduced version of a large square, bearing similar shapes. Results showed that only about half the children tested could relate shapes after a change of scale had taken place, or make a correct spatial arrangement.

This poses the question as to whether countries shown on a world map can be related to maps of individual countries at larger scales, especially when these are subsequently sub-divided, i.e. regional maps of the British Isles often included in Primary Atlases. This theory is certainly not conclusive. There does not seem to be an immediate solution to the problem but maps on the same page of an atlas should be of the same scale, and shapes of countries should always remain consistent, at whatever scale. There is then perhaps a need for even greater accuracy in Primary map work.

(b) Linear scale: Recognition of scale should be seen

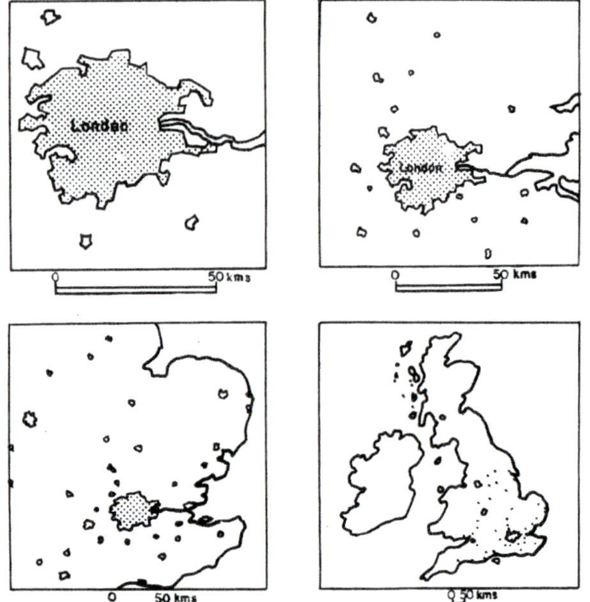

Figure 3 An introduction to the linear and area scale concept

as having a progression from simply 'drawn smaller' or 'larger' to showing distances and relative positions of features (Harris, 1972). It is still apparent to many teachers that accurate scale ideas are best developed from plans of the school or surrounding areas, when a certain amount of measuring is involved. Even if this is only done in relation to a child's pace, a scale bar can be made upon the completion of the map.

Exact scale conceptions often cause great difficulty to children. This was proved when a selection of scale bars was chosen from atlases, and then used by the children to measure a certain distance on a small scale map (Figure 4). Less than half of those tested were able to give an accurate answer, and many did not even attempt to gain a result. Figure g.4(c) shows the traditional scale bar which was found easiest to use, although two points must be made clear. Firstly, all the scale bars were movable and could be placed directly upon the length to be measured, which admitted direct use of a ruler in many instances – a skill few of the children obtained. The other point, which perhaps affected the results most of all, is that the bar which was simplest to use was the only one whose length extended further than the distance to be measured. Indirectly these results proved extremely valuable to the design of maps. The relatively new idea of writing the scale of the map in words, e.g. 2 cms on the map equals 30 miles on the ground, involves a certain amount of mathematical calculation which, although quite straightforward, could only be done by about one-fifth of the children tested. Where scale bars fell short of the length to be measured, a variety of methods of obtaining a result followed. Some children noted the length of the scale bar itself, and then guessed the rest of the distance. Others measured the distance on a ruler and tried to relate it to the scale bar – in the main unsuccessfully. The rest either gave an answer which corresponded only to the length of the scale bar, or gave up completely!

Research by others has shown that the growth of this concept is 'positively and significantly correlated with chronological age and intelligence' (Meyer, 1973). In consequence, as long as the child is aware of scale in the form of size and shape, no detailed knowledge is needed until he is well-acquainted with mathematical figures and has carried out practical exercises involving measuring. Some have stated that it is not until a mental age of 10 or 11 years is reached, that scale can be really understood (*ibid.*).

Symbolization

When children draw their first maps, they frequently record a feature in terms of its side elevation. As the scale of plans becomes smaller, and the understanding of mapping is developed so it will be realised that not all features cover a large enough area in proportion to the map scale to be shown. Thus symbolic representations of features develop. At first, the child should be allowed to devise his own symbols although later on conventional signs will have to be introduced (Harris, 1972).

The use of symbols on Primary maps seems to have a special significance, possibly due to the belief that children can relate symbols to fact more easily than understanding text – especially as there are still children of this age who cannot read. Symbols are designed to resemble the actual object and as Piaget suggests, at this stage the child still depends on concrete reality for the working out of mental operations (Stones, 1966). This simplified method of expressing matters related to specific localities seems also to be a good intermediate step between becoming accustomed to symbols, and understanding conventional signs. Work has been carried out into whether concrete objects should be used for symbols, or whether a more abstract level of symbolization could be adopted at age seven. Rogers and Layton (1966) concluded that

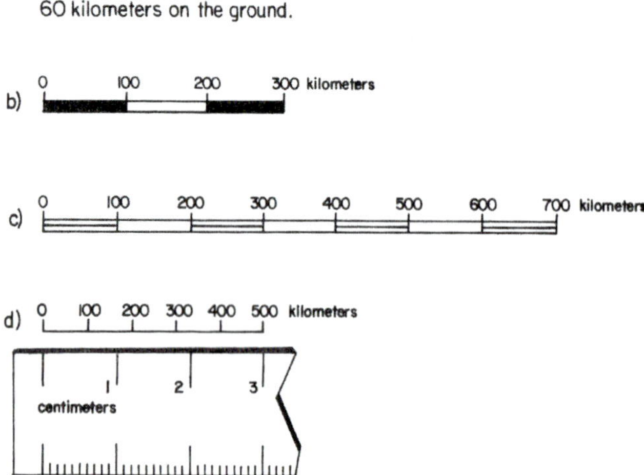

Figure 4 Scales used in research

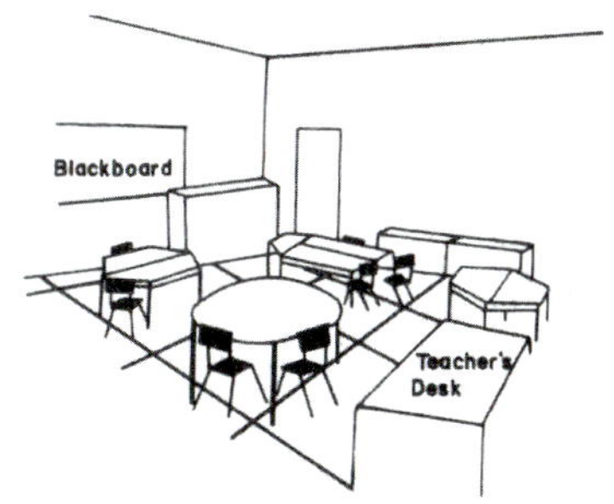

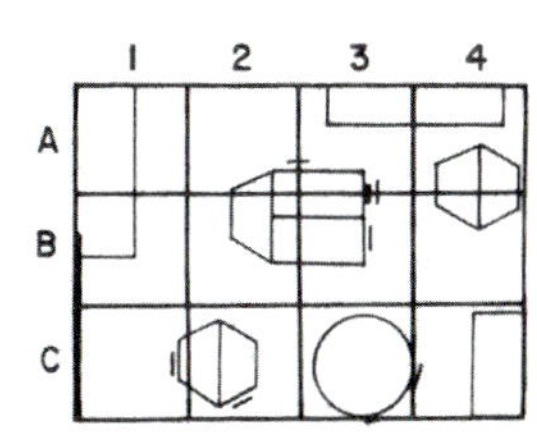

The reference :

Cupboard Al and Bl

Teacher's Desk C4

Figure 5 *Introducing the Grid Reference System*

children will, and can, learn on a more abstract level than is being incorporated on Primary maps.

Thought must also be given to the fact that in a great many schools Ordnance Survey maps are being used for local excursions. Nearly half the teachers questioned used Ordnance Survey sheet maps, and nearly all the student teachers had encountered them during relevant lectures or during teaching practice. It might then be suggested that conventional symbols should be introduced earlier, possibly even on Primary atlas maps. If children of this age can understand and use conventional signs on maps then perhaps an interim stage is unnecessary.

In conclusion it must be stated that generally symbols (whether conventional or otherwise) representing locational information on the map are the easiest for the child to interpret. Some children who find map concepts relatively simple to understand may be able to use conventional signs immediately but, for the sake of the majority who still cannot relate the abstract to real life, the intermediate stage of concrete symbolization must be adopted.

Grid Reference System

The child will often initially describe particular points on large scale maps in terms of their relationship to other recognizable features. This process may often originate from real life situations – when describing locations of things or places, one naturally relates them to known objects. Children encounter grid systems from an early age in their puzzle books and in games such as 'Battleships'. It is often a good idea to suggest drawing plans on squared paper so that from the beginning pupils become accustomed to having squares and lines over their maps – even if initially they are only used as a guide to drawing straight lines. Reductions of drawings on squared paper forms a valuable exercise and shows the child the relevance of this system regarding location and reference.

The majority of Primary atlases tend to omit any type of grid; even though a world graticule may seem a somewhat abstract concept to grasp, its presence need not confuse a map and may lead to a heightening geographical awareness, Conversely, a simple grid could be incorporated on wall maps where the location of features might be a regular occurrence in order that a set of one letter, one figure references could be built up. Likewise, on atlas maps, a straightforward grid could be included with perhaps a simple index (Figure 5) to enable the child to find information in a manner which he will soon have to use frequently when researching alone. During experimentation, the children were reluctant to use a grid to explain the position of a route, but their acquaintance with the system was much in evidence, although they were perhaps unable to apply it.

Colour

Most contemporary map designs seem to be based more on colour conventions than on investigations designed to determine the factors that make maps interesting, attractive and readable to children.

Keates (1962) points out that gradient brightness value colour schemes are easier to read than spectral schemes. He also recognizes that variation in brightness is more significant than variation in hue. This latter point could be extremely valuable if more research were undertaken and perhaps the publisher could produce a cheaper atlas while better satisfying the child's requirements.

Investigations into child colour preferences were carried out independently of any map research. Subjects were simply asked to choose the three colours they 'liked' the best. Results showed red to be the most popular, followed by a light, bright green and dark purple (Figure 6). Few chose colours at the end of the spectral range; the least attractive appearing to be yellow, black and brown, the last two being chosen almost solely by boys. This type of result is valuable only in as far as it provides a guideline as to colours which attract a child's eye, the colours which appear most dominant, and those which are recessive.

There seems to be a particular interest in the relationship of colour to altitude, even though atlases and maps still lean towards conventional yellows, greens and browns. Portrayal of 3-D phenomena on a 2-D surface poses one of the most difficult problems both of comprehension and depiction, especially if relatively sophisticated contour lines are used. There is some controversy amongst teachers upon this subject (Harris, 1972), but publishers still adhere to convention regarding map appearance.

Research has suggested that the child relates change of colour to varying heights, rather than tonal gradation to differences of altitude. As far as could be determined only about half the children also attempted to grade their choice of colours (whether consciously or not), some with the darkest at the top and others with it at the lowest level. Overall, the colours used to represent altitudes were extremely conventional – being light and dark green, and light and dark brown (Figure 7). Most important of all,

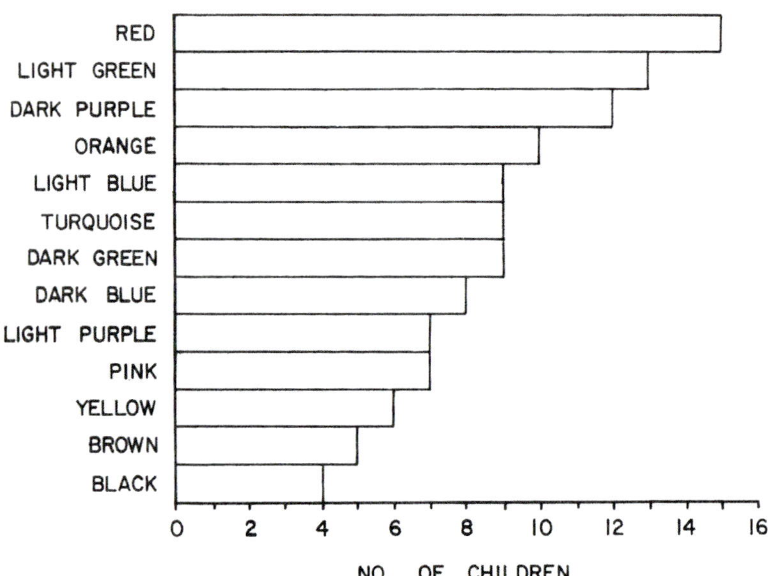

Figure 6 Children's choice of colours

being more attractive than the non-serif, although perhaps less legible. Those type faces are consistently advised by the educationalists, but a child will have to encounter many differing styles throughout his early days and even at eight will probably read newspapers, where type faces are not specifically designed for the child. The question of coloured type to aid comprehension and attraction is a relatively new idea amongst publishers. It could prove extremely confusing on some coloured maps, but associated text could certainly be made more cheerful using this method.

Bartz's (1970) research into type legibility showed that all faces were equally easy to read, there being no appreciable difference between serif or non-serif, weight and point size. The only confusion encountered was during experiments with mixed faces when interpretation became

this proved that the children were old enough to discriminate between the colours they 'like' and those they really thought related to different heights.

These findings could result from the children having already come into contact with maps representing relief in these colours, but the percentage who chose these was quite decisive. Even if the aforementioned reason was true, then these particular colours must have impressed the child

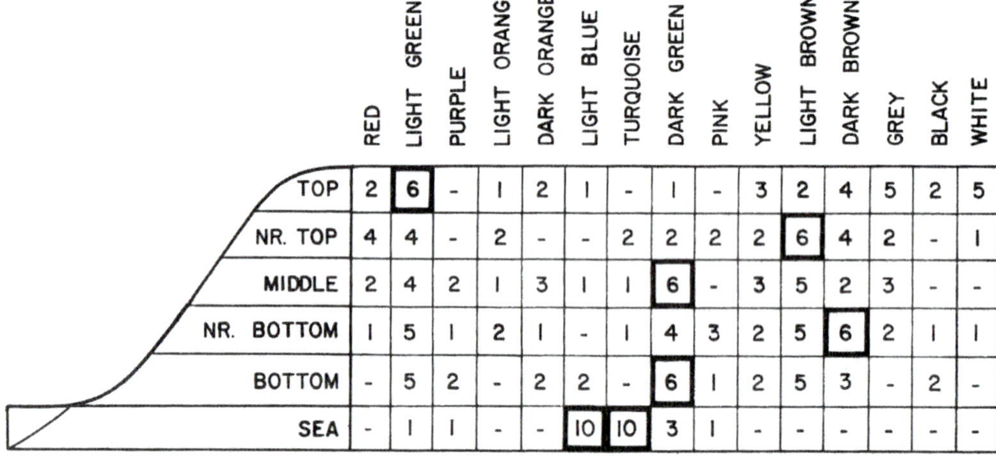

	RED	LIGHT GREEN	PURPLE	LIGHT ORANGE	DARK ORANGE	LIGHT BLUE	TURQUOISE	DARK GREEN	PINK	YELLOW	LIGHT BROWN	DARK BROWN	GREY	BLACK	WHITE
TOP	2	6	-	1	2	1	-	1	-	3	2	4	5	2	5
NR. TOP	4	4	-	2	-	-	2	2	2	2	6	4	2	-	1
MIDDLE	2	4	2	1	3	1	1	6	-	3	5	2	3	-	-
NR. BOTTOM	1	5	1	2	1	-	1	4	3	2	5	6	2	1	1
BOTTOM	-	5	2	-	2	2	-	6	1	2	5	3	-	2	-
SEA	-	1	1	-	-	10	10	3	1	-	-	-	-	-	-

Figure 7 Children's choice of colours to represent altitudinal layers

enough for them to be remembered. Which is surely more than one might hope to achieve!

Like colour brightness, the type size employed on children's maps is a widely discussed topic. Most people tend to regard large type faces as more suitable for a child than those normally used in adult books and this generally accepted idea has many logical foundations. To begin with, an amount of text in small type tends to look monotonous and uninviting, and is more difficult to read than larger lettering. Collins Publishers have suggested that although print size differed greatly in their books, normally type for the seven to nine year olds would be either 12 on 14 pt or 14 on 16 pt.[2] The use of a serif type face is most popular

difficult. Children who were given six small scale maps, identical except for their type faces, took a considerable time before recognizing the differences. Even then their choice of those easiest to read was varied and indecisive.

The relevance of differently sized and spaced map type is very apparent. England, London and Dover were three names written on a map of the British Isles by children, after discussing their relative numerical size and importance (Figure 8). Although a few systematically spaced 'England' over its whole extent and wrote London larger than Dover, the majority insisted upon writing the names in a normal manner, with no differentiation. They were definite about some minor points such as ensuring

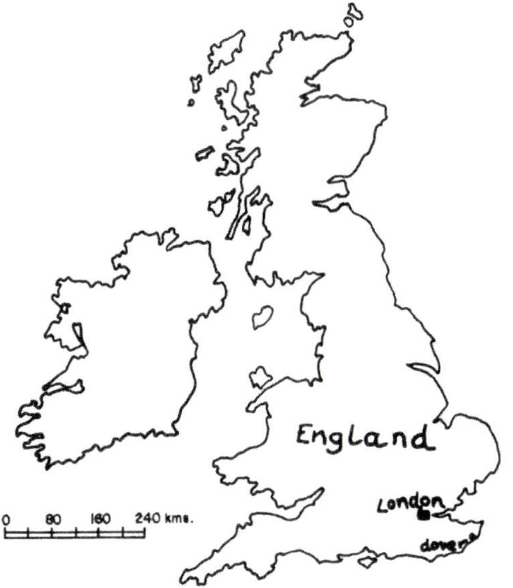

Figure 8
Testing the relevance of type size

that names didn't go into the sea and always printed horizontally. It appears from this and from Bartz's results, that differing types size neither aids in representing dimensions or importance of an area, nor assists in its legibility. The concept must, therefore, be represented by another method such as coloured symbols.

Until such time as a child is capable of understanding all the mapping concepts, a sequence of stages tends to develop whereby maps can be drawn corresponding to the nature and extent of the concepts appreciated at that particular time. Area scale and direct on are two concepts a child appears to understand at an early age. He will draw the garden all round the house, if that is how it is, and objects will be proportional to each other. The first kind of map drawn will probably only involve the area concept – plans of tables, cups, etc. When the child is required to draw a classroom and its contents, direction will become involved. Linear scale will only be used in an extremely generalised fashion, and representation of objects will be purely in a 'concrete' manner.

The next stage will probably be mapping the school, and will involve such factors as linear scale and compass directions. Finally, when a complete local area (perhaps a street or small village) is mapped, more symbols may be introduced – in either conventional form or of the child's own design. A grid may also be incorporated with a more accurate scale bar and a north point.

From this it can be seen that a sequence of stages may emerge, relating the practical side of mapping where experience is gained, to the conceptual drawings of the 'map world' (Figure 9). The map in its own right must be taught, but its use in the widening horizons of education

CONCEPTS	AGE INTRODUCED	SUGGESTED TOPICS	REMARKS
Size and shape	Immediately	Drawing of objects from all angles.	
Direction	7-9 yrs. orientated on a 4 point system.	Routes. North point on floor. Treasure maps.	Especially when on outside exercises.
Area scale and spacial organisation	Only half the children tested (8yrs) understood.	3.D. models convert-ed to plans.	Keep all map scales comparative.
Linear scale	Often encountered in maths work at about 7yrs.	Use measuring in-struments on all scales of maps and plans.	Always make scale bars easy and simple to manipulate.
Symbolisation	On plans of schools & streets Certainly at 8yrs.	Aerial photographs - add symbols to them. Make 3.D. symbols.	Should devise own symbols and then O.S.
Grid Reference System	Often much later but should be between 6 and 8yrs.	Reduction of drawing on squared paper. "Battleships"	Keep grids a per-sonal thing. e.g. co-ordinates for each child's seat position.

Figure 9
A child's ability to receive map concepts

must also be recognized. With so much emphasis on the study of a locality, maps tend frequently to become incorporated within the child's work. Their existence outside this region seems to be as important, in that topics can often be related back to a map of the World, Europe or the British Isles. The introduction of map concepts is covered by atlases, maths and geography books but may well also occur in a new type of pre-atlas workbooks. There is abundant map material – if the teacher is aware of its existence – but if the child does not understand the mapping techniques employed it might be asked if the content and approach is suitable.

But where does the fault lie? Publishers are really in the hands of teachers, although certain decisions are made solely by the Cartographic Editors, who are thus partly to blame. Teachers can assess the needs of the children only if they themselves have both enough interest and time to research and experiment. On the other hand, there are probably many sound and viable ideas on the improvement of maps as communication media which might be exploited if the publishers could only research them.

It is now certain that within this specific and specialized area of Cartography there is a great need to present material which is relevant to the child's perspective on life, material which affords the child some insight into what the educational task is about, and which initiates the investigatory techniques commonly used by the age group for which products are designed. Whether this involves the teacher, the publisher, the researcher or, more likely, all three, above all else the outcome will affect the child.

Notes

1. Discussion with H. Fullard, George Philip & Sons Ltd., London.
2. Discussion with A. Davis, Collins Publishers, London.

References

Bartz, B.S. (1970) "Maps in the Classroom" *Journal of Geography* 69 (1) pp.18–24.

Harris, M. (1972) *Starting from Maps* (Schools Council Environmental Studies Project) London: Rupert Hart-Davis Educational Publications.

Keates, J.S. (1962) "The Perception of Colour in Cartography" *Proceedings of the Cartographic Symposium*, Edinburgh.

Kohn, C.F. (1953) "Interpreting Maps and Globes" (Skills in Social Studies) *Twenty-fourth Yearbook of the National Council for the Social Studies* Washington, D.C.: National Council for the Social Studies.

Lord, F.E. (1941) "A Study of Spatial Orientation of Children" *Journal of Educational Research* 34 pp.481–505, quoted in Meyer (*op.cit.*).

Meyer, J.M.W. (1973) "Map Skills Instruction and the Child's Developing Cognitive Abilities" *Journal of Geography* 72 (6) pp.27–35.

Rogers, V.R. and Layton, D.E. (1966) "An Exploratory Study of Primary Grade Children's Ability to Conceptualize based upon Content Drawn from Selected Social Studies Topics" *Journal of Educational Research* 54 pp.481–505, quoted in Savage and Bacon (*op.cit.*).

Savage, T.V. and Bacon, P. (1969) "Teaching Symbolic Map Skills with Primary Grade Children" *Journal of Geography* 68 (8) pp.491–497.

Stones, E. (1966) *An Introduction to Educational Psychology* London: Methuen.

Further Reading

Cole, J.P. and Beynon, N.J. (1969) *New Ways in Geography* (Series) Oxford: Basil Blackwell.

Department of Education and Science (1972) *New Thinking in School Geography* (Pamphlet No.59) London: H.M.S.O.

Evans, H. (1970) *The Young Geographer* (Series) Exeter: Wheaton.

Fahy, E.M. (1972) *Basic Map Work* Dublin: Gill and Macmillan.

Ferriday, A. (1971) *A Picture Map Book of the British Isles* London: Macmillan.

Houghton, D.M. and Morgan, V. (1974) "Children's Reasoning about their Environment" *Journal of Geography* 73 (5) pp.5–10.

Myatt, J. and Payne, H.C. (1972) *Mapping out Geography* (Series) Edinburgh: Oliver & Boyd.

O'Kelley, M.W. and Napp, J.L. (1973) "Teaching Geography Today: The View from Above: Selective Primary Level Geographic Concepts and Map Skills" *Journal of Geography* 72 (9) pp.53–57.

Pemberton, P.H. (1970) *Geography in Primary Schools* Sheffield: Geographical Association.

Rushdoony, H.A. (1968) "A Child's Ability to Read Maps: Summary of the Research" *Journal of Geography* 67 (4) pp.213–222.

Rushdoony, H.A. (1971) "The Geographer, the Teacher, and a Child's Perception of Maps and Mapping" *Journal of Geography* 70 (7) pp.429–433.

Towler, J.O. and Nelson, L.D. (1968) "The Elementary School Child's Concept of Scale" *Journal of Geography* 67 (1) pp.24–28.

Walker, E.A., Walker, M.J. and Wilson, T. (1972–3) *Location and Links* (Series) Oxford: Basil Blackwell.

MAPS & PROPOGANDA

John Ager

Originally published in Volume 11, No 1, pp.1–15

Some Misconceptions

A subject that should be briefly touched upon is misconception in cartography as this could be and has been exploited by the propagandist. Since the beginning of cartography many misconceptions have occurred about the relative importance of countries, due to the size of the map. The average school atlas depicts a student's own country and his own continent at a larger scale than other countries and continents. Individual countries are often given more prominence than even larger constituent parts of still larger states. Many people have the impression that France is a very large country, when actually it is only about the size of Minnesota, Iowa and Wisconsin combined. On most maps showing individual continents, Europe is presented on a larger scale than Asia, although non-Soviet Europe is nearly as large as India.

The power of the map or the wrong map may be exemplified by one case, amusing whether true or not, discussed by K.E. Boulding (1961): 'It has seriously been suggested that the history of World War One was profoundly affected by the fact that in school atlases of the old German Empire, the United States and Germany each occupied a single page. This has led to a serious underestimation on the part of the German people of the size and capacity of the United States' (Figure 1). If the U.S. had realized this could they not have used maps, in their war of propaganda, to show their immense size and power compared with the Germans, thus helping to break morale?

The Main Variables in Propaganda Maps

When producing a map, a cartographer should aim to show the information he wishes to communicate accurately, comprehensibly, and with a balanced design. However, if a cartographer wishes to produce a map to propagate an idea or favour one side of an argument, this is not so. It could be said that the propaganda cartographer's main aims are to produce a map which has visual impact and is not only believable, but goes a stage further – is convincing.

Both the impartial and the propaganda cartographer have certain techniques or variables that they can use when producing a map (Figure 2). These variables can be adjusted and manipulated to suit their own purpose and fulfil their separate aims.

Selection

This perhaps is the most important variable used by the cartographer – the actual selection of information that will appear on the final map. When a cartographer wishes to draw a map on a particular subject, often on a client's instructions, his first action is to collect information on that subject, i.e. the geographical area in question and the purpose for which the map is intended. From this mass of information a primary selection process takes place, where some information is excluded for the sake of:

1. Clarity
2. Limitations of the final product, e.g. scale, monochrome or colour reproduction
3. Information not lending itself to mapping.

However, in some cases a secondary selection process takes place (often alongside the primary process) in which the cartographer selects material that supports the client's argument or viewpoint, and excludes material which would otherwise diminish his case. The flow diagram in Figure 3 shows this procedure where, after the primary selection process, the route divides up into the 'Propaganda Cartographer' and the 'Perfect Cartographer', who produces ideally 'perfect' maps, in which the selection and presentation processes have in no way unbalanced the situation. An example of selection affecting the balance of the map is illustrated thus: A map appeared recently, produced by the London design firm 'Diagram' showing German and Russian advances into Poland at the beginning of the Second World War. This map later appeared in a Chinese publication; the only alteration, apart from translation, being the omission of arrows showing the direction of German advances.

However, it must be remembered that selection is just one of the variables the propaganda cartographer uses to mislead the reader. In combination they become more effective. Figure 4 is an example, quoted by Prestwich (1973) where selection, as well as projection and colour, is used to show communism as a threatening force during the Cold War. By using the Mercator projection but selecting only part of it, the distortion of the land areas is not obvious.

Symbols

In *Die suggestive Karte* (1928) Karl Haushofer states that in creating the 'suggesting' map the greatest stress is placed on the use of the proper symbols. Maps designed to illustrate the speed of modern war, in battles and campaigns, must suggest movement. He states that parallel lines illustrating the position of front lines of action are not sufficient. No propagandist would wish to suggest that a

Figure 1: Misconceptions of size

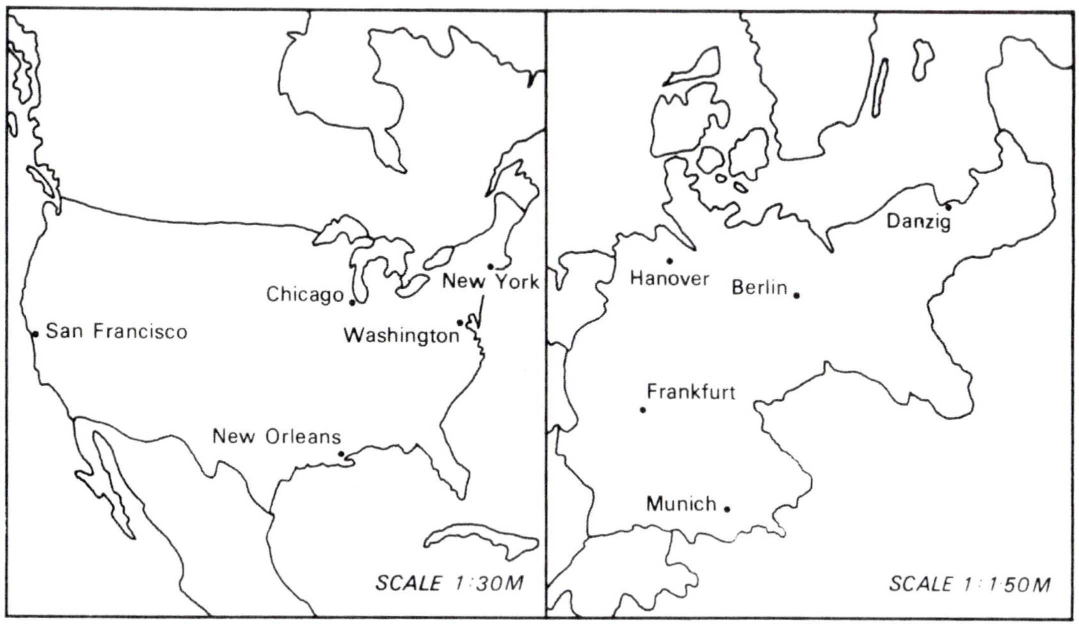

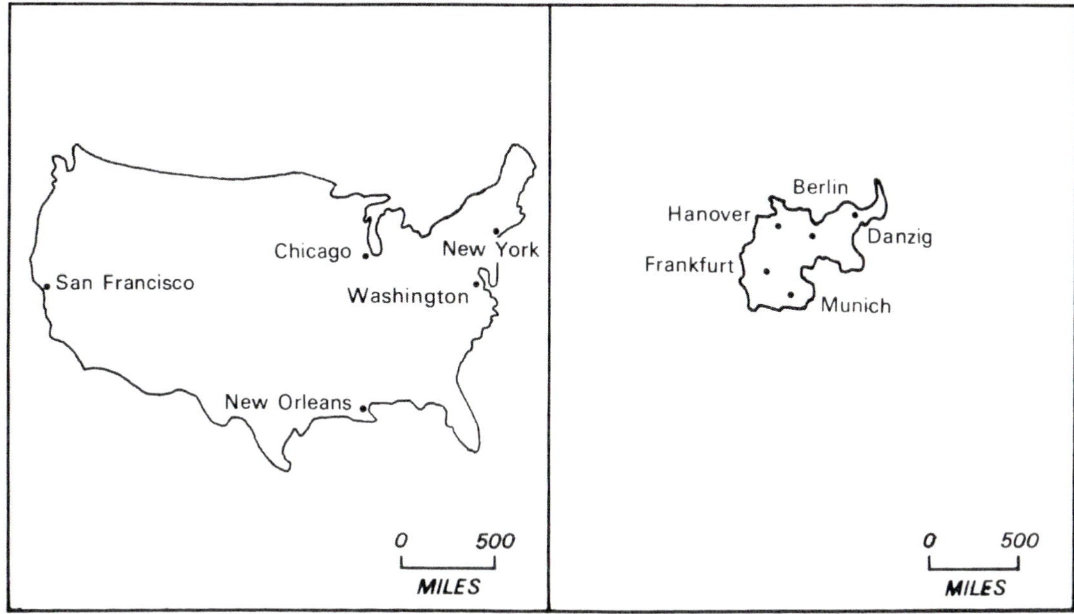

stalemate prevails or that both sides have an even chance. He must indicate that the enemy is being crushed.

As action itself cannot be shown on a map, the cartographer must develop a system of symbols which suggest movement, the most common of which is the arrow. A strong thick straight arrow is a sign of aggression, an arrow curved back upon itself a sign of frustration, and so on. By altering the size, tone and design of the symbol, the visual effect of 'who has the upper hand' can vary considerably (Figure 5). Louis B. Thomas (1949) suggested that this symbolism to show movement in this 'dynamic cartography' seemed to derive in part from military maps, which have long used arrows and similar devices to show troop movements, and in part from the political cartoon. Every cartographer realizes the difficulty of size of symbols in relation to the scale of the map. Rivers, roads, towns and pictorial symbols cannot, except in the case of very large scale maps and plans, be drawn true to scale. Geographers and cartographers try to overcome these difficulties but the propagandist manipulates them so as to focus attention on the idea he wants to stress.

Hans Speier (1941) mentions an example of a map that appeared in *Facts in Review* (November 1939), a magazine published in the U.S.A. by the German Library of Information. 'On a map to illustrate the repatriation of

Figure 2
The main variables in propaganda maps

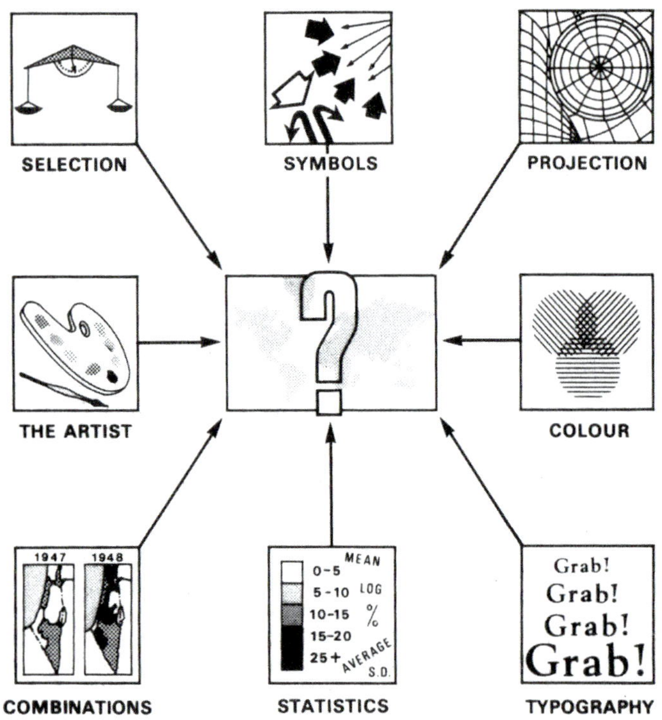

Figure 3:
The 'Propaganda' versus the 'Perfect' map

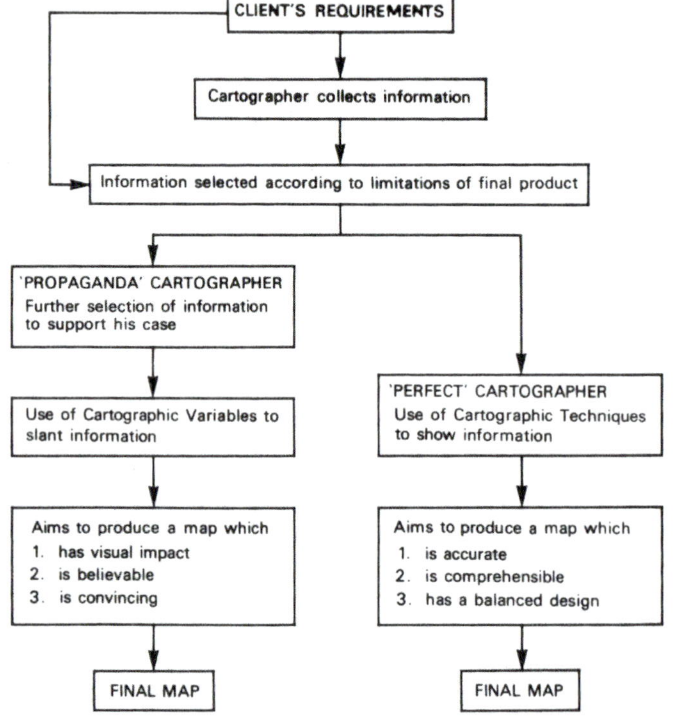

NOTE:- It should be pointed out that in reality there is not a clear division between 'Propaganda' cartographers and 'Perfect' cartographers, but both are at opposite ends of a spectrum in which all cartographers fall, and their positions vary in accordance with the production of each map.

Germans who had been living in Latvia, the German minority about to return to the Fatherland is represented by a row of thirteen identical symbols, each standing for five thousand men. The symbols extend over the whole area of Latvia where it is widest, from Libau in the west to the eastern border. The size of the symbol is so selected that the country seems to be largely populated by Germans, whereas, in point of fact, the German minority amounted to 3.7 per cent of the total Latvian population'. A map entitled 'Karte von Englands Landeraub und Weltknechtung' printed in Germany at the start of the First World War shows a map of the world on the Mercator projection, with U.K. territories and Empire in bright red, similar to maps found in Britain, but Germany turned this pro-British map into an anti-British map by the symbols that were added. All countries and territories in the Empire were linked together with heavy black chains lying in the oceans. Shackles appeared round the major cities of the Empire, and the chains all connected to Britain with a large ring around it.

Projection

To develop the concern regarding misconceptions of size and relationship that could be exploited by the propaganda cartographer, it can be stated that many of these misconceptions are the result of the choice of projection. On any map it is impossible to show the world without some distortion. Over the years cartographers, mathematicians and many in other fields have developed dozens of projections, each with certain advantages, but all have some inherent disadvantage, the main factors in question being correct shape and relative areas. Conscientious cartographers choose the projection for their map depending upon the information to be put on it, weighing up the properties of the projections. For example, it would be wrong to show a dot distribution of population on a non-equal area projection such as Mercator.

When the question of projection falls into the hands of the propagandist, the decision is taken differently. N.J.G. Pounds (1972) in his book *Political Geography* mentions a prominent German geographer during the inter-war years: 'Haushofer made his points with the aid of maps, which were simple, striking and misleading. Arrows were used to suggest action, expansion or attack, shading was used to suggest the associations which Haushofer desired to convey, and his maps were drawn not upon the projection which gave him the least distortion but upon that which emphasized the relationships which he wished to exaggerate. He was a master of cartography as a tool of propaganda used for nationalistic purposes'.

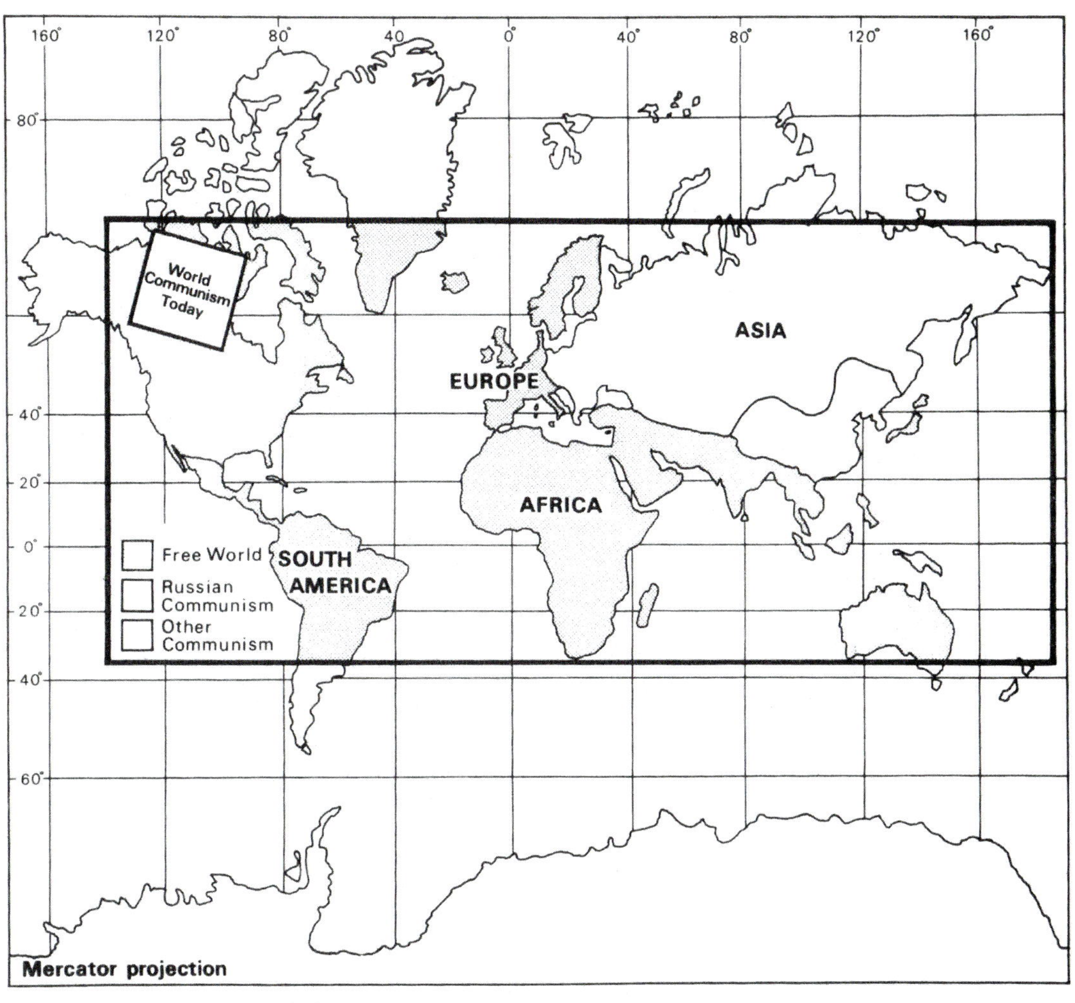

Figure 4
Use of projection, selection and colour (after Prestwich, 1973)

AIM:-
To show Communism as a threatening force.

METHOD:-
1. Use the Mercator projection, thus exaggerating U.S.S.R land area.
2. Delete area outside box, thus exaggerated projection is not apparent.
3. Delete Alaska, and place title box over Canada, thus reducing area of Non-Communist countries.
4. Use bright colours for Communist areas, so they appear larger than the 'recessive' grey-toned areas.

THIS MAP IS SIMILAR TO ONE IN AN AMERICAN SCHOOL ATLAS PRODUCED IN THE 1950s.

Probably the most well-known of all projections has been used widely in propaganda maps – Mercator. Although it shows true compass directions and, therefore, it is still the ideal projection for navigators, the Mercator projection has serious shortcomings. Except in the vicinity of the equator, it does not even pretend to show the correct relative size of the land areas of the globe. For, as you go further away from the equator, so E-W distances are stretched as well as N-S. So, if you want to show on a map how far the British Empire extended, you use the Mercator projection, as Canada and Australia are grossly exaggerated in area (up to 250 per cent). Combine this with the traditional bright red or rosy pink (symbolic of courage?) which helps the areas to stand out from the map, and you achieve a map showing almost world domination. Alternatively, if you wish to show the British Empire appearing less dominantly, you would use a projection such as the Eckert IV or Mollweide, which reduce area towards the poles. Mercator's projection has also been blamed for giving the impression of 'the Russian menace looming so large in people's minds'. Compare the effect of projection in Figures 4 and 6 when used with other variables. Both maps are about Russia and the world, but communicate two entirely different messages.

In recent years, if there was any one man who tried to get people away from thinking of the world in terms of the Mercator projection it is Richard E. Harrison. Harrison worked as a graphic artist/cartographer on the American news magazine *Fortune* in the 1930s, and for over twenty years used a unique form of map. He developed perspective maps showing views from one country to another, in order that the American people should understand what it was like for the country concerned in a particular situation. In 1937 he drew a map called 'Hitler's

Figure 5
One situation – two maps (size, tone and symbol design)

ANGOLA ON NOVEMBER 15 1975

1. Stressing Encirclement.

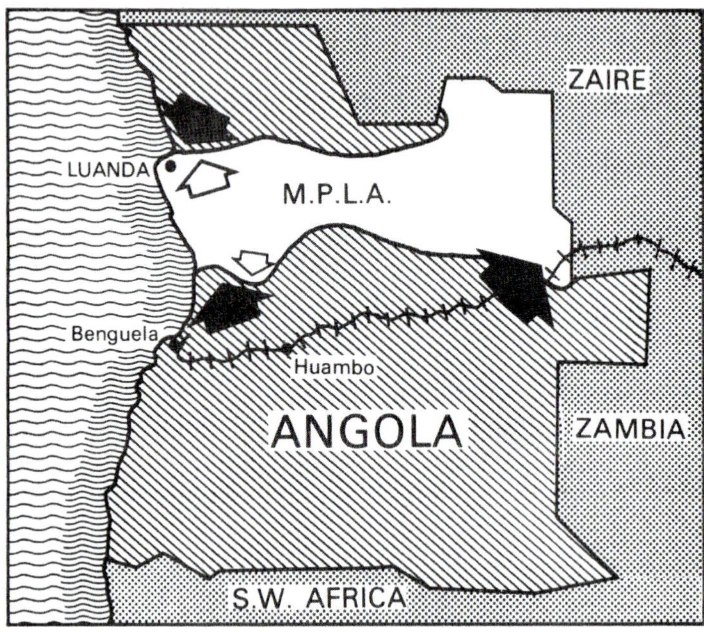

2. Underplaying Encirclement.

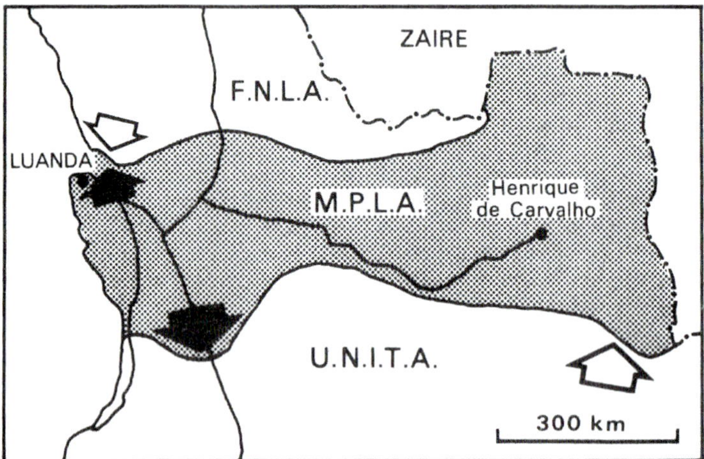

Eye View of Britain'. Orientated with west at the top of the page, the map gave the reader the 'view' of the British Isles as observed by the Fuhrer. He drew similar views during the war, and in the 1950s during the Cold War, drew 'Europe's Oil in Danger' – a double-page perspective map which pictured the Middle East as viewed from the U.S.S.R. At this time it was a novel approach and an interesting effect was achieved by departing from the orthodox practice of placing north at the top of the map. The perspective maps are based on the visual illusion that the shapes of countries change when the viewpoint changes. Harrison's skill has left an indelible mark on cartography. Even today occasional 'bird's-eye views' are found in newspapers and periodicals and are used to influence opinion. Figure 7 shows a map that appeared in the *Daily Express* (27th January 1976). It shows Russia's view of Africa, in the style of Harrison's maps. Its headline and title is 'So this is why the Russians are trying to gag Maggie – Satisfying view as Moscow looks out on the world today'. It marks countries that have major, moderate and potential Soviet influences and shows sweeping arrows in the Mediterranean, supposedly Soviet Navies, and the oilfields appear as if about to be grabbed by Russia.

It can be said that the choice of an unusual or conventional projection can be crucial to the case the cartographer wishes to present. A country can appear to be on the defensive or offensive, expanding or surrounded. Projection can be used quite adequately on its own but when combined with the other variables it becomes most effective.

Colour and Shading

It should be pointed out from the start that not all propaganda maps utilize colour. This is due to the fact that many are found in newspapers and cheaply produced pamphlets which are only printed in black and white. However, other outlets do use colour in the attempt to sway opinion. In using colour, makers of propaganda maps during the first half of the century portrayed the territory to which they were favourably disposed in bright red, while enemies were often coloured a sickly yellow. An example of this is found by looking at British atlases where the British Empire is always found to be rosy pink, whereas in pre-war German atlases this empire is found to be a sickly yellow colour which is perhaps symbolic of cowardice. One of the best known examples of a propaganda map, which uses colour with great effect, was published in a German atlas *Der Krieg 1939–41 in Karten* (Munich, 1941). This atlas was also published in the U.S.A. as part of German propaganda. It aimed at persuading Americans against becoming involved in the European war, by discrediting the Allied powers. The technique of using maps in the propaganda war was perfected in Nazi Germany by members of a school of 'Geopolitik' – professional manipulators of political geographical information.

The map's aim demonstrates the constriction of Germany by its enemies during the First World War. The text, and a second map, mention that this time it will be different, as Germany has occupied Norway, Denmark, the Netherlands, Belgium and France, and has an alliance with Italy, which substantially reduces the extent of encirclement. The projection used exaggerates area away from the centre of

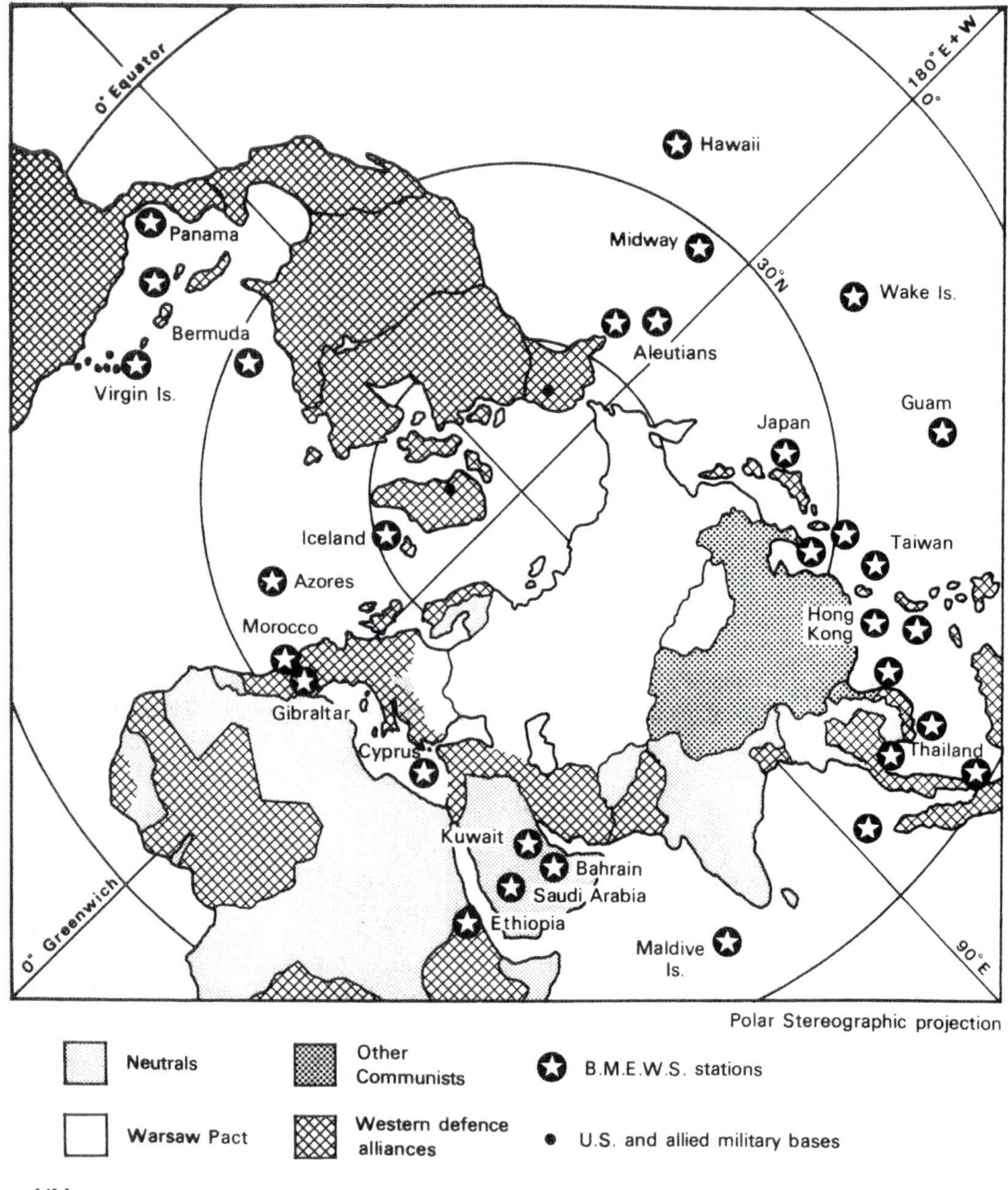

Figure 6
Use of projection, symbols and shading (after Prestwich, 1973)

AIM:-

To show the encirclement of the U.S.S.R.

METHOD:-

1. Use Polar Stereographic Projection, so Russia is near the centre.
2. Dominant symbols to give prominence to strategic bases.
3. Dark shading for Western alliances as well as other Non-Russian countries.
4. Latitude lines add to visual effect of encirclement

the map. Concentric circles are used to shade enemy countries and produce what is known as 'Gestalt perception', in that a circular shape is perceived, although there are gaps in the arcs. Again, colour is used here; red for Germany and its allies and yellow for its enemies.

Today the colour red has obvious psychological connotations with Communism, so that a cartographer can indirectly say a country or area has left wing sympathies or communist connections by colouring it red. This is seen in Figure 8, the emotive title 'Militant Minorities', and the actual areas on the map are coloured in a bright red and stand out from the pale grey base of Europe. To the reader the analogy with communist activity may be immediate, even though some of the regions have no connections with communism, e.g. Northern Ireland. The region stands out as a whole yet only a minority in the region may be militant and in agreement with the particular political activity, i.e. a wish for independence. The shading giving the areas thickness, also increases size and, therefore, importance. Alsace-Lorraine on this map appears to be larger than the Netherlands.

Colour can also be an important factor on dot distribution maps, e.g. illustrating the ethnic composition of the population. Where a small circle is coloured a bright

Figure 7
"Satisfying view as Moscow looks out on the world today" (reproduced by permission of the Daily Express)

Satisfying view as Moscow looks out on the world today

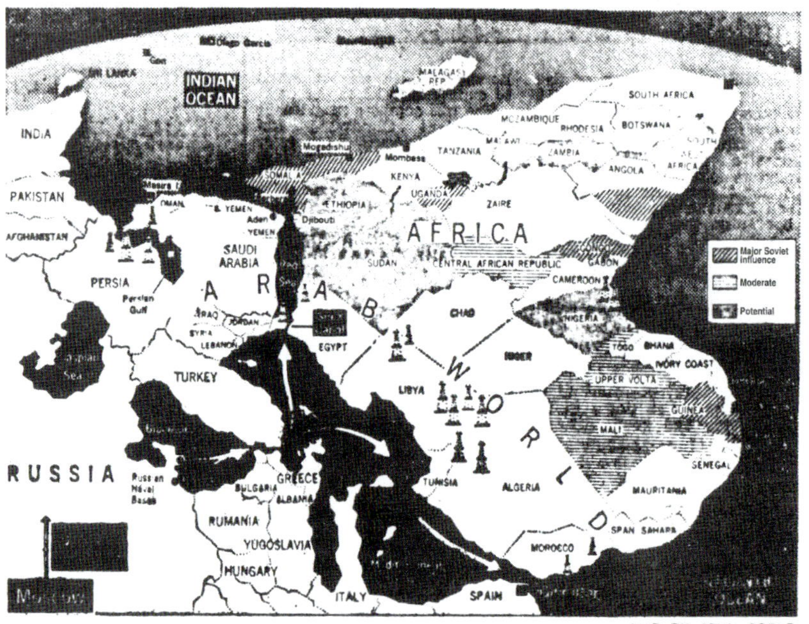

THE Soviet leaders have good reason to congratulate themselves as they gaze out from Moscow on this map of the world that lies beyond their southern borders.

Whether they turn to Africa, the Middle East, or South-East Asia, the story is the same—a steady increase in Russian influence.

And especially so in Africa. This is how Russia has been extending its colonial hold throughout the continent since the Arab-Israeli Six-Day War of 1967 :—

GUINEA-BISSAU : Russian aircraft have been there since 1973 and warships since 1970.

SOMALIA : This is now the main logistics base for Russia's Indian Ocean fleet, with communication stations, barracks, a 15,000ft. runway, and missile storage and handling facilities.

UGANDA : Russia gives military aid (42 MIG jets with Czech instructors recently arrived), as she also does to Mali, Mauritana, Nigeria, Guinea and Uganda.

ANGOLA : Under increasing Soviet domination with the recent successes of the Communist-backed M. P. L. A.

solid red it has the effect of appearing larger than a circle of equal size tinted in a lighter colour, such as yellow or grey. Thus, without falsifying the statistical values, a minority can be made to appear dominant or nearly equal in number to the actual majority. Louis Thomas (1949) states: 'Where a map propagandist wishes to exaggerate importance of the rural population he can avoid dots and graphs and use solid colours. When these are spread over wide, sparsely populated rural areas they do not emphasize the urban concentrations which often make up the greater percentage of the total population'. A further example of colour in propaganda maps is illustrated in the *War in Maps Atlas*, it is called 'Divide and Rule'. This is a map of India on which the outer boundaries are represented by smooth conventional lines, all bordering countries being shown in one solid colour. Political subdivisions within India, however, are shown in a different colour and each small area is separated from the others by broken pecked lines. The impression is one of utter disunity and disintegration of India.

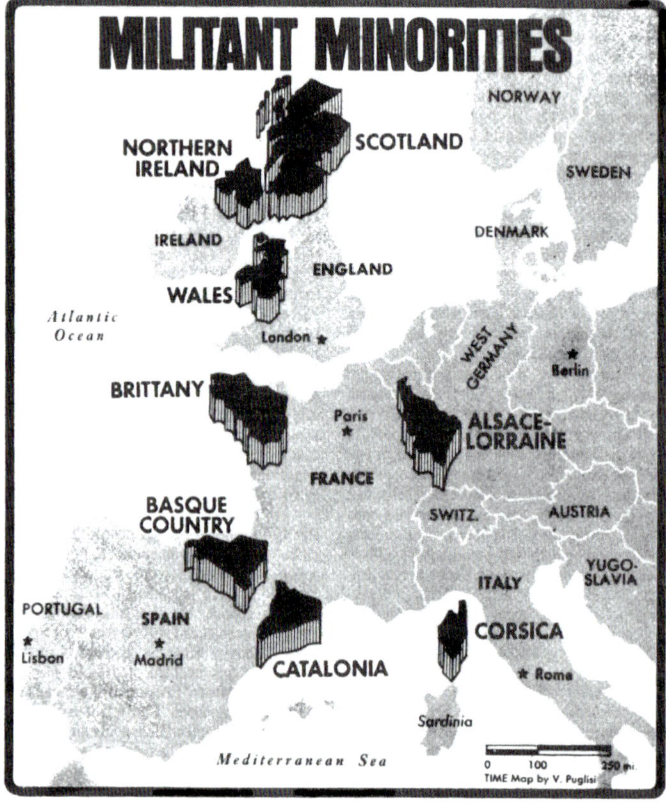

Figure 8
"Militant Minorities" (by permission of Time, The Weekly News Magazine © Time Inc. 1977)

During the Second World War, one of the methods used in propaganda was to drop leaflets from aircraft in order to break morale. Civilians were usually the target but occasionally leaflets were dropped on the enemy army. Figure 9 is a photocopy of a map which was dropped in large quantities on the beaches of Dunkirk from German aircraft in order to demoralize the retreating allies, with its impression of encirclement. It should be noticed that the propaganda value of this map does not lie in its cartographic accuracy, but in its presentation. Since the sea is shaded, and the large area overrun by the Germans is dotted, the white space showing the allied forces would seem completely surrounded. Seven arrows pointing from all directions to the allied area reinforce this impression. This 'encirclement' method is frequently used today. Figure 10 appeared in the *Daily Express* (13th February 1976, p.4) and uses similar devices as the Dunkirk map. In this case the sea is solid black and the attacking forces are shaded, with a thick black

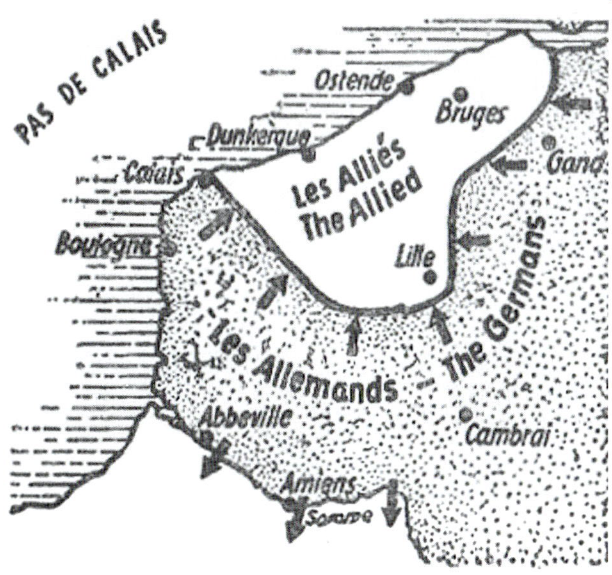

Camarades!

Telle est la situation!
En tout cas, la guerre est finie pour vous!
Vos chefs vont s'enfuir par avion.
A bas les armes!

British Soldiers!

Look at this map: it gives your true situation!
Your troops are entirely surrounded —
stop fighting!
Put down your arms!

Figure 9
Map dropped by German aircraft on the beaches of Dunkirk to demoralize the retreating allies

'fighting front' reinforced with twelve arrows aimed towards the white centre of the map. The headline above the map reads 'South Africans gear up for red siege'.

Typography

Another of the variables that the propaganda cartographer uses in producing his map is typography. This includes all lettering that appears on or around the map, e.g. place-names, titles or captions. With regard to place-names, conventions such as using a larger type size, bolder lettering or upper case letters to imply more important towns are misused to attract attention and mislead. The arrangement of place-names in a cluster can imply a high density of settlement, whereas with a smaller point size, an impression of emptiness can be given.

Maps, particularly those found in newspapers and leaflets, often contain a propaganda element in the use of title, calculated to arouse attention, anger or sympathy. Figure 11 poses a question to the reader who is asked to make a decision on the basis of the evidence presented on the two maps.

As with the Arab-League maps, the Martin Gilbert atlas, *The Arab-Israeli Conflict in Maps* uses quotations from statesmen. One map in this atlas, called 'Lebanon – a base for terror!' shows areas inside Lebanon that are designated 'Palestinian terrorist bases' and 13 border settlements in Israel that are frequently attacked, along with a sample Israeli casualty list. A quotation from the Lebanese Prime-Minister says 'Lebanon will continue to honour its commitments and will service in solving the Palestine problem'. Taken out of context quotations are often used to discredit parties.

Figure 12 was produced as part of a leaflet promoting Governor George Wallace in the 1968 U.S. presidential election. Here typography totally dominates the crudely drawn map. This, with the large flowing arrows and the emotive language of a 'Treason Cycle' resulting in DEAD AMERICANS, puts the case very effectively for stopping U.S. trade with communist countries (in which England appears to be included).

Statistics

Statistics are often used in conjunction with the written word to prove a case or justify an argument, so it is only logical that the propaganda cartographer applies this 'variable' to the map. Occasionally, maps appear with startling figures around them to further suggest an opinion, but to use statistics more effectively the cartographer often shows statistics pictorially. These are usually in the form of graduated circles, bar graphs, spheres or squares. As one of the aims of the propaganda cartographer is to produce a map which has visual impact, often symbols, such as the 'Scotsman' and the 'German' (mentioned in the Combinations section) are drawn to compare situations, such as the number of people under the rule of the British and German Empires. In this particular example the area symbols were worked out on a linear symbol scale, thus exaggerating the difference. Although these methods of showing statistics are used in conventional cartography, by misusing the methods of working out the size of the symbol, the cartographer can misrepresent, either intentionally or unintentionally, the true situation.

The choice of class interval and shading scales, particularly in choropleth maps and isopleth maps, can also be vital to the message the map conveys to the user. By using an unevenly balanced series of class intervals or grey scales, spatial patterns can be over- or under-emphasized.

Combinations

In *Magic Geography* Hans Speier refers to 'combination maps', another of the variables at the disposal of the propaganda cartographer. Combination maps consist of two or more maps placed alongside each other in order to compare events or unexpected parallels. A very good example of combination maps was published in Germany in 1940, entitled 'A Study of Empires'. The maps and the message are most effective because of the simplicity of their design (Fig. 11). One map shows Britain with outlines of all its empire and possessions scattered like pieces of a jigsaw. The other map shows just a very small Germany

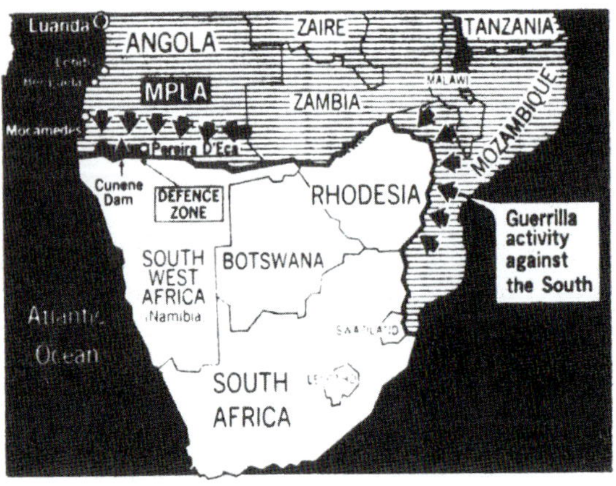

Figure 10
"South Africans gear up for red siege" (reproduced by permission of the Daily Express)

with no territories, with the caption 'The Aggressor Nation?'. The German 'record' is empty and clean, the British cluttered and greedy. This map contributes nothing to the question of war guilt, but plays on the theme of haves and have-nots.

Another excellent example of a combination map is in the Imperial War Museum, again from Germany, but earlier than the previous example. Published as a poster at the start of the First World War, it contains three maps of the world, one under another; they are on Mercator's projection and compare the British and German Empires at their different stages of growth in 1800, 1871 and 1914. They endeavour to show how Britain has added more and more of the land areas of the world to her empire and has gradually moved towards world domination. After 1870, the race for colonies began in earnest among the European powers. Germany entered the race rather late, as Bismark had been reluctant to embark on colonial adventures. Although colonial rivalries were not in themselves a cause of the First World War, they helped to aggravate the tensions already existing between European nations. The poster is entitled 'England and Deutschland' and on the left hand side of each map there is a proportional German and Scotsman to show the population of conquered people in each empire in that year. Needless to say, by 1914 the giant Scotsman is over five times the size of the small German. On the right of each map there is a comparison of land areas of each empire in that year in the form of a two-bar histogram. Again, Britain dominates Germany and, as with the rest of the poster, the impression is given of Britain as a nation striving for world dominance.

The previous example was produced in Germany, primarily for home consumption. But as with 'A Study in Empires', the first example, this is not always the case. Often a country may publish maps in foreign journals or distribute them through societies or embassies abroad in order to win support and sympathy in a particular situation. This is further illustrated by two maps published in America in about 1930 (Figure 13). These were prepared to encourage revision of the Treaty of Trianon which divided up much of Hungary's territory and apportioned it to neighbouring countries after the First World War. If a map had been drawn showing only the division of Hungary it would have had little meaning for Americans who knew

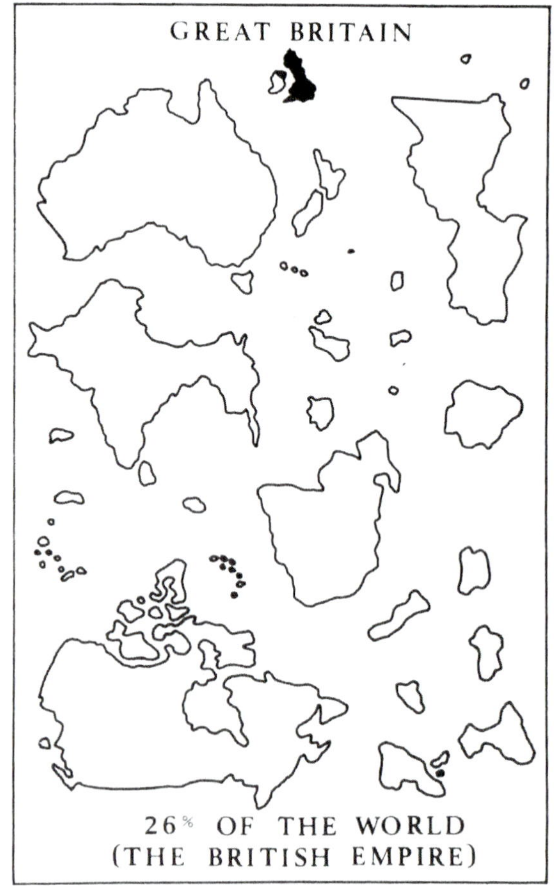

Figure 11
"A Study in Empires"

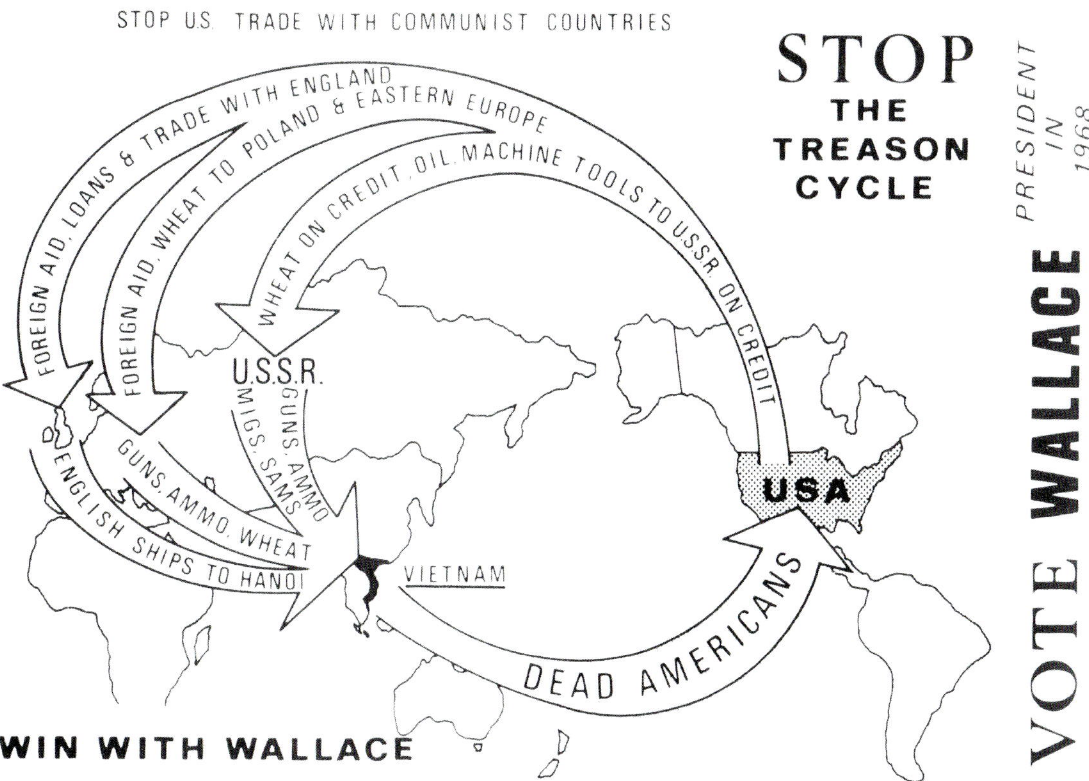

Figure 12
Part of a leaflet promoting George Wallace in the 1968 US presidential election

little about this relatively small state in Europe. By illustrating Hungary's losses in terms Americans could understand, this map did succeed to some extent in arousing sympathy, although the effectiveness of these maps is difficult to evaluate. As well as in the U.S.A., similar maps were published in France, Italy and the U.K., with maps drawn as if each country had been subject to the treaty. For example, in the British case, Scotland was shown as belonging to Norway, Northern England to Germany, Ireland to the U.S.A., and the south coast to France. Many of the examples of propaganda used in this project have been German, and many first appeared before or during the last war.

The Artist

Whilst researching for this project, I frequently came across propaganda material which contained maps or illustrations with roughly drawn geographic outlines, but they did not contain enough information or were of sufficient accuracy to be termed cartographic. This type of graphic artwork, which includes posters, advertisements, cartoons and motifs, containing a geographical element, I have termed 'non-cartographic Maps'. A 'non-cartographic map' can be defined as a map that appears as part of a graphic design, e.g. poster or cartoon, in which the map is of secondary importance to the artwork. It is usually a highly stylized

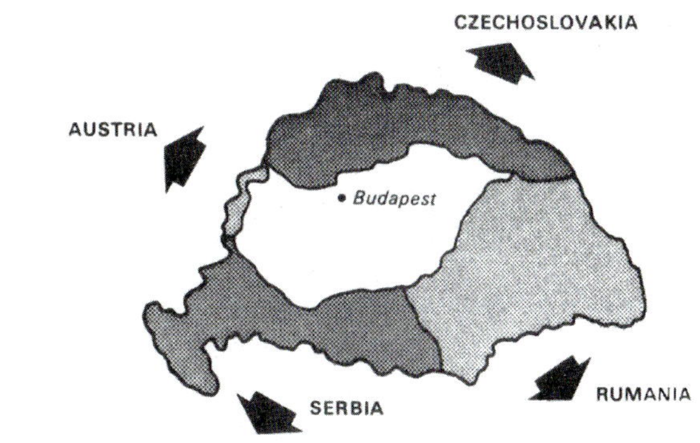

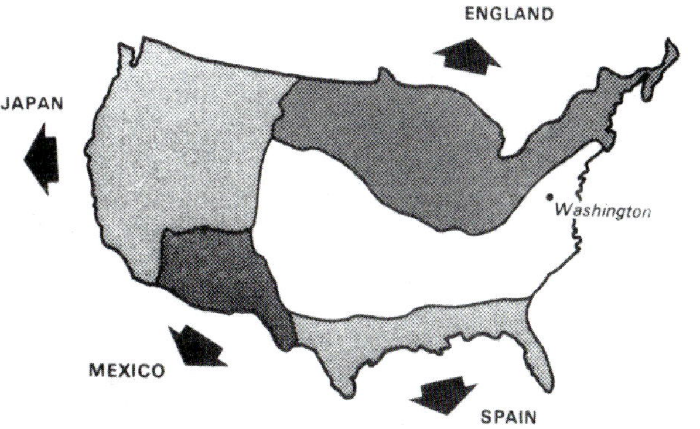

Figure 13
Two Hungarian maps published about 1930 (after Thomas, 1949). These were prepared to encourage revision of the Treaty of Trianon, by illustrating Hungary's losses in terms Americans could understand.

outline of a geographic area, and relates to the primary artwork. For this reason, most non-cartographic maps are drawn by the graphic artist or cartoonist rather than the cartographer. This type of map is not restricted to the field of political propaganda but is also found in commercial advertising.

Often the outline of the map is used in a variety of ways. It can be used as a backdrop or stage for the artwork, it can be filled or dressed, the shape of a country can be adapted or simply used as part of a motif, in order to communicate to the reader the part of the world to which the subject is related.

Conclusion

In writing this article, I hope I have not given the impression that somewhere, hidden away, there is a group of cartographers who spend their days drawing maps that will deceive the public. It is only from time to time that one appears, and perhaps it is not so much the question 'Is this a propaganda map?' but 'to what extent has a viewpoint been favoured?'. Often a propaganda map or a map with a bias may be drawn because the cartographer has used the wrong method of presenting his information and, therefore, failed to communicate correctly with the user, who then misunderstood his message. Likewise, the fault may be in the user who may, through no fault of the cartographer, misinterpret the map.

Most of my examples have been German and first appeared in the first half of this century, but some are recent examples from this country, so one must not be led into thinking that they appear only in wartime. In a study of maps a serious effort must be made to find out what the source materials are and precisely how they are used. Maps, like most other means of communication, can be deceptive and in the hands of skilled manipulators they become subversive propaganda weapons.

Bibliography

Maps and Propaganda

Haushofer, K. (1928) *"Die suggestive Karte"* In Haushofer, K., Obst, E., Lautensach, H. and Maull, O. *Bausteine zur Geopolitik* (pp.343–348) Leipzig: Kurt Vowinckel Verlag.

Muir, R. (1974) "The Slickness of the Map Deceives the Eye" *Geographical Magazine* 46 (8) p.436.

Prestwich, R. (1973) *"Maps and the Perception of Space"* In Lanegran, D. A. and Palm, D. *Invitation to Geography* (pp.35–37) New York: McGraw-Hill.

Quam, L.O. (1943) "Use of Maps in Propaganda" *Journal of Geography* 42 (1) pp.21–32.

Skinner, A. (1975) "Maps and advertising" (unpublished) Oxford Polytechnic.

Soffner, H. (1942) "War on the Visual Front" *American Scholar* 11 pp.465–476.

Speier, H. (1941) "Magic Geography" *Social Researcher* 8 pp.310–330.

Thomas, L.B. (1949) "Maps as Instruments of Propaganda" *Surveying and Mapping* 9 (2) pp.75–81.

Weigert, H.W. (1941) "Maps are Weapons" *Survey Graphic* 10 pp.528–530.

Wright, J.K. (1942) "Mapmakers are Human: Comments on the Subjective in Maps" *The Geographical Review* 32 (4) pp.527–544.

Other Sources

Balchin, W.G.V. and Coleman, A.M. (1966) "Graphicacy should be the Fourth Ace in the Pack" *Cartography* 3 (1) pp. 23–28.

Boulding, K.E. (1961) *The Image: Knowledge and Life in Society* Ann Arbor: University of Michigan Press.

Cole, J.P. (1965) *Geography of World Affairs* (3rd ed.) Harmondsworth: Penguin.

Dickinson, G.C. (1973) *Statistical Mapping and the Presentation of Statistics* (2nd ed.) London: Edward Arnold.

Gilbert, M. (1974) *The Arab-Israeli Conflict in Maps* London: Weidenfield and Nicolson.

Harrison, R.E. and Weigert, H.W. (1944) "World View and Strategy" In Weigert, H.W. (Ed.) *Compass of the World* (pp.74–88) New York: Macmillan.

Huff, D. (1967) *How to Lie with Statistics* (9th ed.) London: Victor Gollianz.

Muir, R. (1975) *Modern Political Geography* London: Macmillan.

Pounds, N.J.G. (1972) *Political Geography* New York: McGraw-Hill.

Ristow, W.W. (1957) "Journalistic Cartography" *Surveying and Mapping* 17 (4) p.369–390.

Wirsing, G. (1941) *The War in Maps* New York: German Library of Information.

Yanker, G. (1970) *Prop Art* New York: Macmillan.

'COLOR KEY' PROOFING SYSTEM

... LETS YOU SEE ACCURATE COLOUR PROOFS, EVEN BEFORE THE PLATES ARE MADE-UP!

■ **One of the most helpful** trouble-savers invented for lithographers in recent times is the widely-used, pre-plate proofing system, "Color-Key." It is fairly common today to hear colour printers tell how they've been literally saved from disaster on rush jobs because their "Color-Key" proofs caught serious errors before the plates were made.

■ **We all know** that the conventional methods of colour proofing, such as bluelines, silverprints and brownlines, are good as far as they go. They just don't go far enough for critical colour work. At best, these methods are guess-proofing. A proof should show colour breaks, registration, fit, colour values, copy details, proper stripping and condition of tints and tones. Ordinary methods don't do this with sufficient accuracy.

■ **So it has always been** necessary for the colour printer to process the plates and pull press proofs to really see what he's got. That's a high priced way to discover—and maybe too late—how many mistakes there are to correct.

■ **But "Color-Key" has changed this.** With "Color-Key," you don't have to process plates or wait for press proofs to see how your job will look. As soon as the cameraman has his colour negatives or positives (there's both negative and positive "Color-Key") you can go right to a "Color-Key" proof.

■ **In 20 minutes or less,** you can make a complete, full-colour proof on "Color-Key." It will show you the complete picture of the job with hairline accuracy. With dot-for-dot, line-for-line fidelity to the film negatives or positives. Everything is instantly visible—colour break, register, fit and colour values. Mistakes will pop right out at you—you can't miss them. And the overlay system of "Color-Key" makes it easy to see exactly on which colour flat the error is.

■ **The presensitized polyester base** of "Color-Key" is perfectly stable. It never shrinks, curls or stretches. Positive "Color-Key" is produced in the four process colours. Negative "Color-Key" is available in four process colours and five flat colours, plus opaque white.

■ **"Color-Key" is so accurate** and reliable that many printing buyers now accept "Color-Key" proofs for job O.K. This saves precious time on rush jobs, and both printer and customer arrive at an understanding on the job before time and money are invested in plates and press proofs.

Positive "Color-Key" Streamlines Colour Separation Jobs

■ **Positive "Color-Key"** helps streamline the colour separation processes for cameramen, artists and dot etchers. It gives these craftsmen a quick, accurate check on colour balance with hard dot proofs before contact positives are made, while the copy is still positioned on the board. This is possible because Positive "Color-Key" renders a hard dot proof from the soft dot camera positive. And it holds the dot size exactly the same as high contrast film.

■ **From the Positive** "Color-Key" proof, the artist or dot etcher can make detailed corrections on the *camera* positives before making his *contact* positives. Thus he can cut down on unnecessary contact work, save re-shooting jobs and save quite a bit of time, as well as film.

■ **No new equipment** is necessary for making Positive "Color-Key." It is dimensionally stable, has a clean, clear background, never discolours with age, is easy to mount and has built-in static which makes perfect contact in vacuum frame or in mounted proof. There's no harmful or offensive odour while processing. The four process colours are standardized ink pigment colours, uniform and consistent regardless of exposure.

■ **Positive "Color-Key"** is also used extensively for pre-press proofing, and as a prog proof for pressmen.

How to make negatives without camera or darkroom

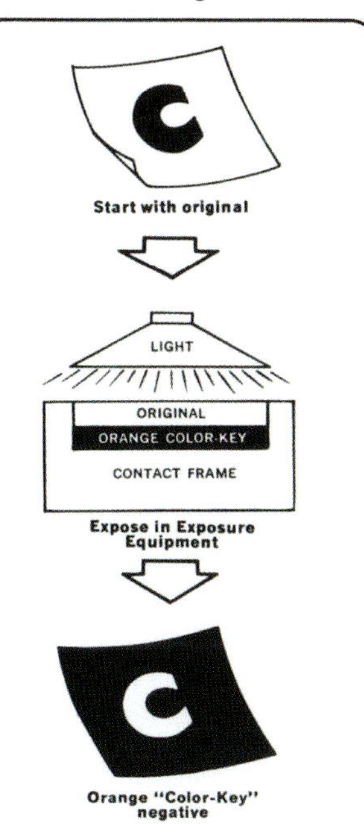

■ **If your shop** could use quick, low-cost negatives that don't require a camera or darkroom, try the new Orange "Color-Key."

■ **No special equipment** is necessary. Your copy can be anything printed, typed or drawn on one side of lightweight paper. Any original (including coarse screen halftones) suitable for diazo reproduction is satisfactory. Simply expose the copy to Orange "Color-Key" in a contact frame or vacuum frame to the same light source used for burning plates. You can work in ordinary room light.

■ **After exposure,** just wipe away the image with "Color-Key" Developer. The transparent background filters out the ultraviolet light like photographic film. There's no drying time, and you get a good quality stable base negative of your original, ready for exposure to a presensitized photo offset plate. You can also use Orange "Color-Key" to make reverses of negative or positive films.

TYPOGRAPHIC LEGIBILITY ON MAPS: A COMPARATIVE STUDY

Tomlinson K.P. Amachree, Hubertus L. Bloemer and Bob J. Walter

Originally published in Volume 11, No 1, pp.27–39

Introduction

Communication of information is the prime function of a map. This communication task is carried out on the map in a variety of ways; symbols, colour, and pattern are examples. Lettering is one of the most important elements in the communication function of a map because it is an ubiquitous map symbol. If lettering is to be efficient and effective in its function it must be legible and readable. Surprisingly, little systematic research has been done to determine legibility in a cartographic context.[1]

The main purpose of this paper is to compare two typefaces: Gill as an excellent example of sans-serif type and Times Roman as an outstanding specimen of serif type. The comparison seeks to discover which of them better aids legibility on maps (i.e. in a cartographic context) as opposed to whatever values these typefaces have in text readability. Legibility on maps has an additional dimension because of the figure-ground; text has no such phenomenon to affect legibility.

One of the most complex problems confronting the cartographer in designing a map is that of making the lettering legible and readable. Understanding cartographic typography is, therefore, fundamental to the map as a communication medium. It is also essential to know in what ways the type used on a map is similar to or different from type as it is used in other contexts, especially in text.

Lettering on a map, which is a subset of all the graphic elements, can be used in a far greater variety of ways than can the letters which make up conventional text (Bartz, 1969b: 128). The major distinction between cartographic use of words (labels) and the use of words in speech is that while one (speech) is mostly concerned with classes of objects or ideas, the other (map) is usually not so concerned. The reality a map depicts has to do with specific, unique place labelling, e.g. it is concerned with 'United States' and not with 'country'. Furthermore, the place labels themselves have no essential connections with one another. The encoding of considerable information, done by both the physical characteristics of the letter shape (type style) and the arrangement of the shapes on the map, is a rare occurrence in non-cartographic typography. If one merely pronounced a word with no comprehension of its meaning, the process would not be called reading. Yet in a sense, this is what happens with a newly encountered map for the unfamiliar names are nonsense words – pronounceable, but having no association or meaning (Bartz, 1969a: 135).

It is apparent that the bases for evaluation of typographic legibility for text and cartographic types must of necessity be different. It is important to point out that there are two broad and very different (yet often intermingled) senses in which the word legibility is used cartographically. One use deals with the map as a total display, while the other is confined to the type which occurs on the map. Neither is identical to the more common, everyday use of the word.

Cartographers have given the evaluation of cartographic type much attention, both philosophically and experimentally. There are as yet no experimental data to answer questions relating to legibility. Many prolific and authoritative writers in cartography – Imhof (1962), Robinson (1960), Eckert (1921), Raisz (1962), Keates (1973), for example – make statements on type legibility, but none has defined what it is.

Legibility is proclaimed as the ultimate aim of lettering (type). Bartz (1969a: 338) says that in itself, the physical variation in type characteristics can be evaluated in the aesthetic, subjective sense, for typography is no more than an arrangement of marks on paper.

It is not type alone that is important, however; rather it is the effect of typographic variation on some communication task that is of central concern to cartographers and map-users. Generally, an increase in legibility is equated with an increase in speed, accuracy, or ease with which the activity involving type is performed. Robinson (1960: 243) states that in cartography the study of lettering as a symbol form is especially important, both because of its universal use and because it is a rather complex and, at times, bothersome element of the map. He (1960: 246) continues that regardless of the kind of map, the lettering is there to be seen and read. Consequently, the elements of visibility and legibility are among the major yardsticks against which the choices and possibilities are to be measured.

Type legibility on maps is a desideratum since lettering is one of the most common map symbols. Lettering can lead to wrong interpretation if not legible and/or properly positioned. It is, therefore, important to search for, identify, and apply the typeface, serif or sans-serif, which aids legibility better whenever a choice arises. In doing this, Robinson and Sale's (1969: 280–281) comments on the use of type are pertinent: that differences in style are best utilized for nominal differentiations; size and boldness are more appropriate for ordinal and interval distinctions.

Cartographic typographic legibility should be judged from a different viewpoint from that of text readability. In cartography, the figure ground of the map should have an effect on the legibility of the lettering. Hence, cartographers in designing maps always take the figure-ground into consideration in selecting the size and design of the typeface to be used.

Evolution of Type

Nesbitt traces succinctly the history of typographic evolution. He (1957: 1) states: 'It is clear that writing was not invented at any set time or place, but grew out of independent origins at different places in various epochs of history [....] The further development of writing grew out of pictures. Again, there was no isolated discovery – picture-writing was born out of the natural urge to imitate'.

Typography developed out of pictures and no isolated discovery was noticeable. As it was not possible to make a clear distinction between the need for expression from the desire to communicate news or knowledge, the dual nature of writing continued through the history of letters. Word writing systems were worked out, notably by the Egyptians in their hierograph, among others. But the phonetic alphabet was said to have followed the Egyptian writing. The picture-writing of cultured ancient peoples survived mostly as relief carving in wood, stone, and some other durable material. These escaped the ravages of time and so man has been able to follow the evolution of writing (typography). Some of the Egyptian cutting dates back to about 200 B.C., although the design was flat (Nesbitt, 1957: 4). The present-day alphabets began with the Greeks about 800 B.C. With their list of twenty-four letters, they provided an alphabet suitable to all Indo-European languages. The Ionian version of the Greek alphabets was standardized in 403 B.C. From alpha and beta, the first two Greek letters, ALPHABET was formed.

The Romans also contributed to the development of typography. The most publicized inscription, the Trajan column in Rome with inscription cut into the base of stone (about 114 A.D.), had its fame from the design of the letters. These letters developed at least through seven centuries of the Roman Empire.

The history of letters is always a part of the general history of styles; with Romans, it followed their architecture. The Romans, possibly the Etruscans before them, gave the names to the present A, B, C letters which were the final result of the 'acrophonic principle'. With this principle at the advanced stage in the development of letters, the word symbol was used to illustrate the first letter of a word; however, the symbol retained the word as its name. 'An example in English would be to illustrate the sound of M by the symbol for Man, because Man begins with M. The Greeks, when they adopted the Phoenician alphabet, followed the Semitic names for the letters quite closely: Alph became Alpha; Bet, Beta; etc.' (Nesbitt, 1957: 9–10).

Christianity had some influence on typography; a good example was the Irish-Anglo Saxon writing. Irish writing was copied from a facsimile page of the Book of Kells found in the church at Kells. The Irish variety of half-uncial (becoming the national hand of Ireland) was based on that carried to Ireland by Roman church missionaries. England is indebted to Ireland for her national hand – the Anglo-Saxon writing.

Although Roman writing styles played a conspicuous role in the evolution of all letters, present day numerals came from Islamic culture – Arabic numerals. Western Europe learned its numerals from Islam, first about the tenth century and later in the twelfth and thirteenth centuries. The new and better system of numbers was widely used by European merchants and mathematicians in the thirteenth century. Since their introduction it has become impossible to suppress them because of their great value to trade and mathematics. The designs changed greatly through time.

The early developments of the alphabet and numerals primarily answered a need to communicate in a written form. In recent decades, particularly since 1950, researchers have focused studies on understanding and evaluating the pscyhophysical aspects of our written languages. Much of this work is conducted to enhance the written communication dealing largely with visual perception of the consumer.

Psychophysical Aspects – Visual Perception

Since maps are utilized as a process of visual communication, they involve a perceptual process depending on the combined activity of the sense of sight and of the reaction in the brain thus stimulated. All perceptual systems have a stimulus, a receptor, and a response. Light from the map becomes stimulus which enters the eye, the receptor, and is converted into meaning by the brain, constituting a response (Keates, 1973: 1). Yarbus (1967: 211) opines that the human eye voluntarily fixates on those elements of an object which carry or may carry essential and useful information, and that the more information that is contained in an element, the longer the eye stays on it. The thought process accompanying the analysis of the information obtained determines the order and duration of the fixations on the element of the object.

This summarizes briefly how people who think differently to some extent also see things differently. Solving problems like visual evaluation of proportions, estimation of lengths, comparisons of angles, likeness and dissimilarity, etc. optimally require macromovements of the eyes. Many visual evaluations are either impossible or done with great difficulty without these movements.

Bizzi (1974: 100) states that the sequence of events in the nervous system that co-ordinates the movements of the eyes and the head in fixating a visual target have been clarified by recent experiments with monkeys in the laboratory at the Massachusetts Institute of Technology. The results showed how reflex sensory feedback, generated by the turning of the head, interacts with centrally initiated programmes and so gives rise to 'co-ordinated' eye-head movements.

Map A

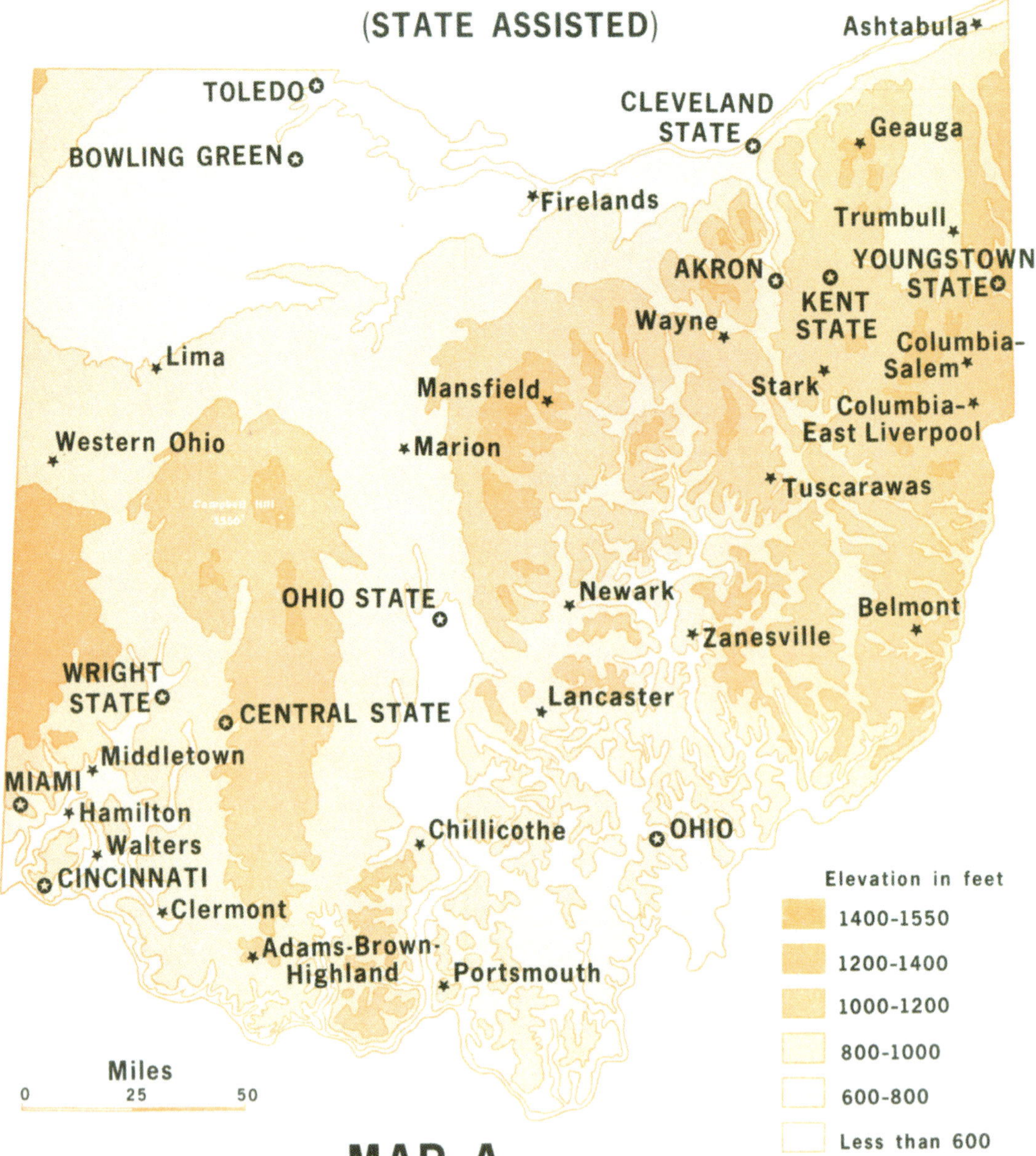

Eye movements, voluntary or involuntary, are essential to vision. Voluntary movements are scanning movements occurring for visualization of targets of any size. Involuntary movements result in the fixing on a point. There are three types of involuntary eye movements: slow drift of the eye; rapid, sudden movements (saccades); and small amplitude, very rapid tremors present at all times (Zusne, 1970: 29).

Estimation of the basic magnitude of shape, size, etc. involves all the characteristics of a discrimination task; e.g. rotating a figure affects its discriminability, or the number and kind of stimulus variables that are simultaneously varied affect discriminability. Length is underestimated when viewed monocularly while, on the other hand, the degree of underestimation of complexity of objects increases monocularly with increasing stimulus complexity and decreasing exposure time (Zusne, 1970: 277).

Visual noise affects discriminability adversely. Constant complexity level with increased noise level increases errors and decreases the rate of information processing. It should be noted, however, that with a constant noise level, increased complexity improves discrimination. Lettering, being a complex map phenomenon, will derive much from this theory where legibility and readability are concerned. In cartographic type legibility, this visual noise level can be introduced by the figure ground. An example would be a map cluttered by too much detail or by the application of too many contrasting bright deep colours. Such situations affect legibility adversely. In most of the extensive series of experiments on shapes preferences as a function of their complexity, scale values of preference and meaningfulness were derived from a pair-comparison data, with symmetry and previous experience as independent variables (Zusne, 1970: 281–282).

OHIO STATE UNIVERSITIES SYSTEM 1972
(STATE ASSISTED)

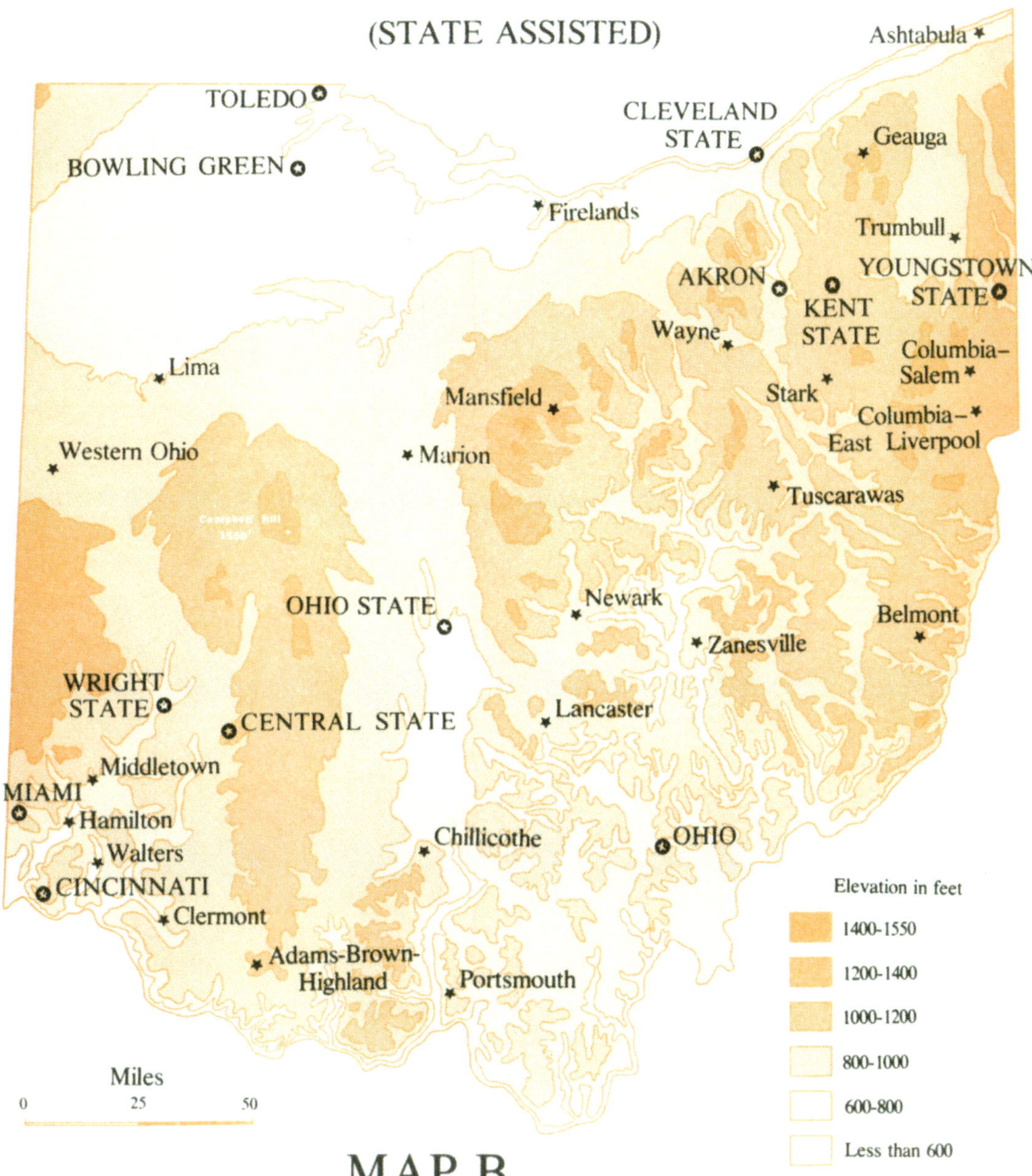

MAP B

It should none the less be borne in mind that all the said experiments regarding visual noise were not conducted in the context of cartographic legibility. They were derived from text readability research. This notwithstanding, they can be applied judiciously to produce effective results on cartographic type usage.

Evolution of Typeface

The use of a particular typeface on maps may enhance or hinder the readability and legibility of the lettering on maps; thus, typeface selection is of critical importance to the cartographer. There are hundreds of different typeface designs, each with its own variation of thickness, strokes, emphasis, serif information, etc. Type design problems arise from the construction of individual letters and figures. Historically, all type design processes have developed from manually placing together individual letters to form required words.

Keates (1973: 202) gives one of the best descriptions of type design. A single piece of type consists of a piece of metal of a certain height, length and width. On the shoulder of this occurs the typeface; this is the part which actually prints. The raised letter itself may occupy all, or less than all, of the shoulder. The shoulder is one plane of the shank, body or stem which controls the minimum space the piece of type can occupy and, therefore, the minimum distance any character may be from adjacent printed matter. The front plane of the body is interrupted by one or more notches or nicks. Being three-dimensional, the body involves three measurements. The type height is fixed and standard; the distance between the front and back planes of the body, i.e. the vertical dimension of the shoulder on the printed image, is measured in points. This specifies the fundamental size of the type. The width of the front plane is the measurement of the type width or set. Whereas type height is standard, point size is varied to give

Map C

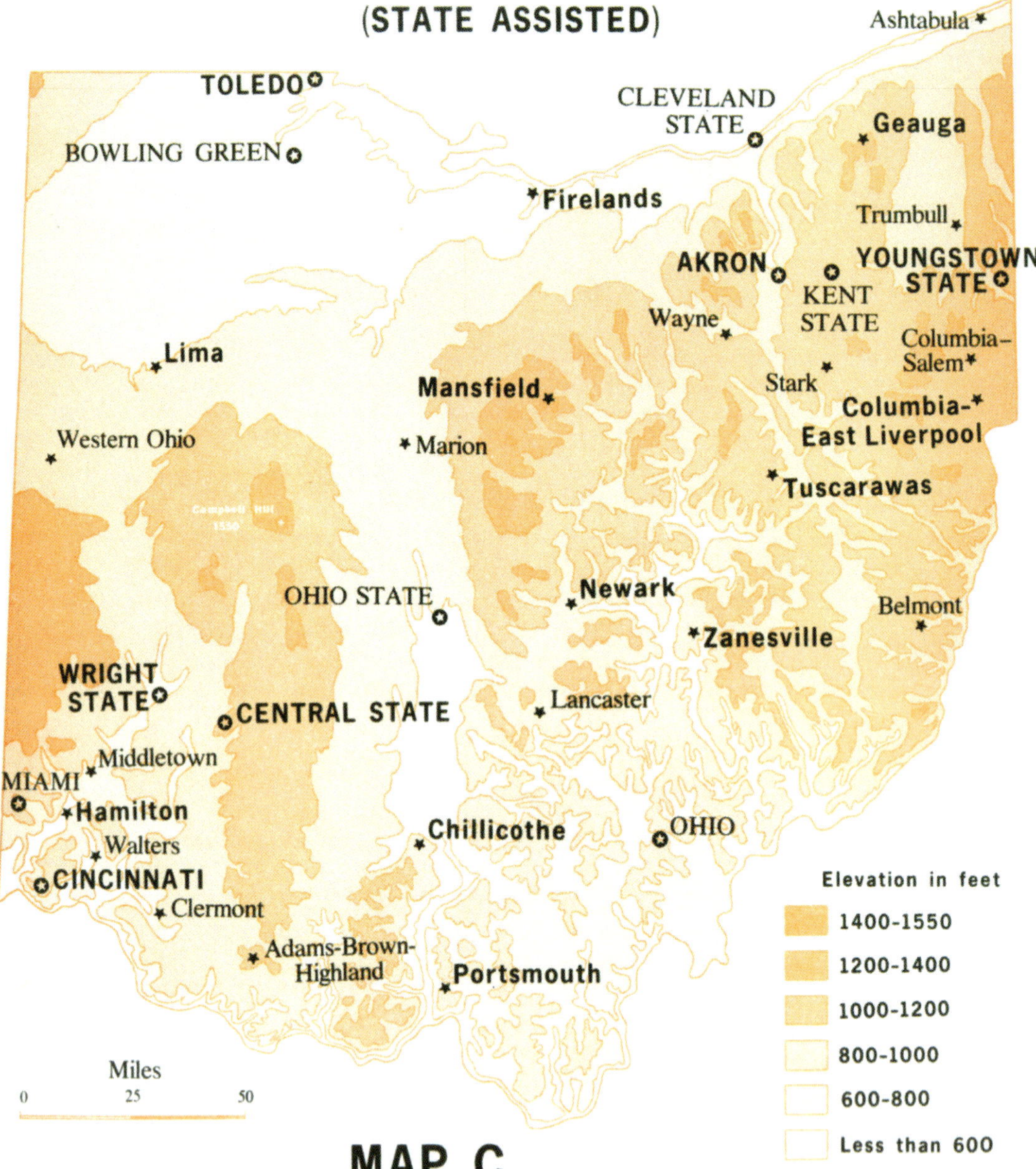

different sizes of type, and set varies with different letters and different designs of type.

Early letterpress printing used the pica, which is not standard, for letter size. The pica slowly evolved as a commonly accepted approximate size. After 1886, a standard unit was introduced when the United States Type Founders Association took a mean value for the pica and divided in into twelve points. In this system, the point equals 0.351 mm (0.13837 inch), and seventy-two points equal 25.27 mm (0.9962 inch). All typefaces use the point as a measure.

Continental Europe used a slightly different point size devised by the French typographer, Dedot. He divided the old French inch into seventy-two points, giving a value of 0.0135 inch in English measure.

The twelve-point em is regarded as standard for general measurement of width, and unless otherwise stated, an em equals 4.15 mm (0.166 inch). For setting width, a number of spaces is supplied with every font (fount) of type. In this case, an em space for the type size is the square of the type height; and an en space is one half of the em. The names are taken from the m, the widest letter of the alphabet, and n, which is about half the width of the m, is regarded as the mean letter width. There are also thick, middle and thin spaces as well as hair space, which is one-twelfth of an em (Keates, 1973: 203).

Only a limited number of different sizes of type are commonly produced; these are six, eight, ten, twelve, fourteen, eighteen, twenty-four, thirty, thirty-six and sometimes twenty-two, forty-eight, fifty-four, sixty and seventy-two points. Some typefaces are 'large on body' and some 'small' by comparison even though they are of the same point size. The width or set is a function of design. Some type designs occupy more space horizontally than others of the same point size. A particular

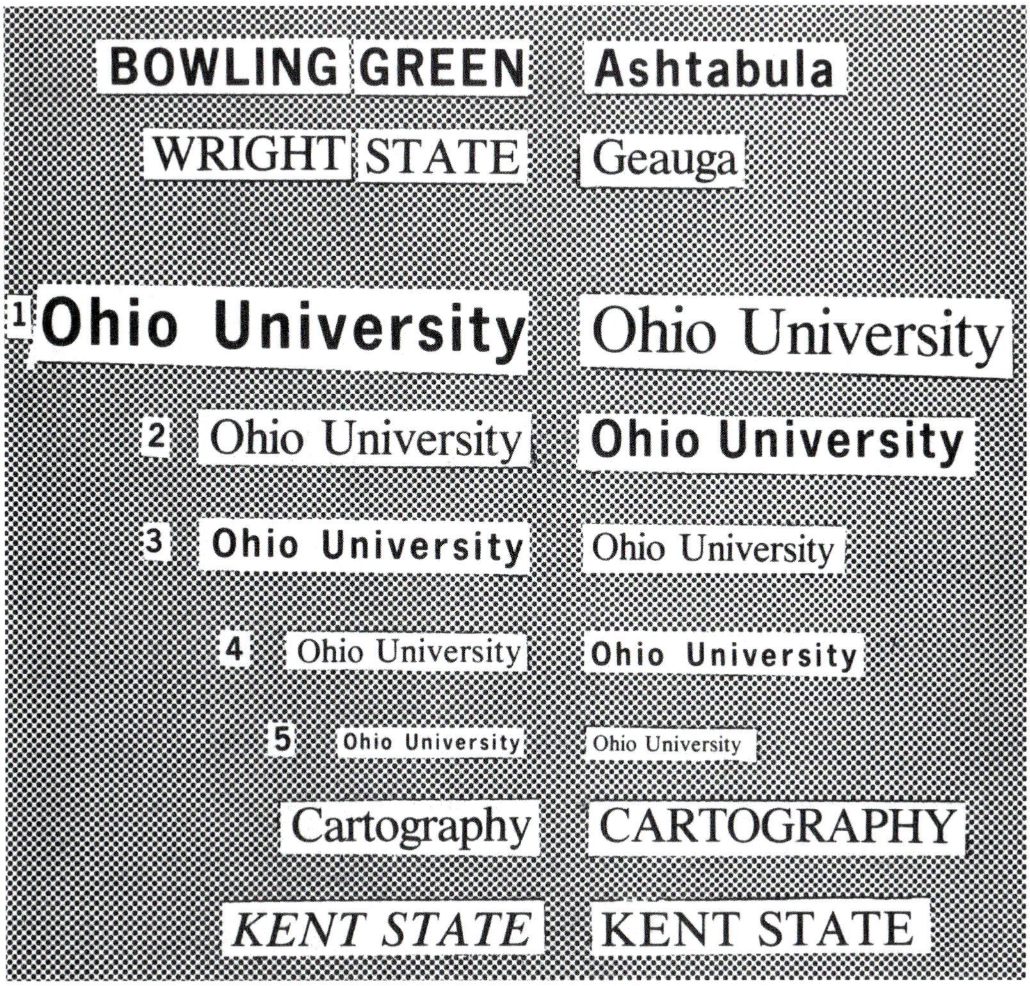

Figure 1
Labels placed on a patterned background to simulate the legibility conditions on a map

type design can be a normal set, in a condensed form which is narrower, or even in an extended form which is wider.

To appreciate the varying proportions of type, the characteristics of letter form needs consideration. Small or lower case letters (like a and e) lie between the base line and the mean line. Other lower case letters with ascenders (like b and k) or with descenders (like g and y) either ascend above or descend below these lines; i.e. to the ascender or descender lines. Capital letters usually rise to a line slightly lower than the ascender line. The x-height varies with designs. If it is restricted, then the design may have long ascenders and descenders. Generally, it is desirable to choose typefaces which are 'large on body' for visibility and easy legibility, on maps where the bulk of type work is small in size, as maps are used as a process of visual communication.

It is the shape and angle of the serif and the way it is bracketed into the stem that makes much of the difference between the different styles of letters. Only a few styles, such as the sans-serif, lack the finishing strokes.

A variety of methods have been employed by previous investigators to determine the legibility of type. The methods used affect the results obtained. Types that appear more legible than others when tested by the distance method often prove less legible when tested by speed of reading.

Burt emphasizes the influences of typeface itself on legibility. He (1959: 5) asserts that legibility (in text) is the result of many factors including the size, the form, the thickness or boldness of letters, the width of the line, the texture of the paper, the ink, the presswork, the lighting, and, what is often overlooked, the intrinsic interest of the subject matter itself.

The valuable contribution of serifs to typographic legibility notwithstanding, the practice of modern microfilming emphasizes that sans-serif type is more suitable for cartographic work planned for storage in microfilm form. Sans-serif capitals seem better in this respect.

It should be noted, however, that in computer typesetting, the Photon Compositor, Keyboard Entry Phototypesetter with Optional Magic Tape Recorder can use both serif and sans-serif capital and lower case letters of different typefaces for text (Photon Type Specimen Book, 1974).

The major concern of the study is to test which typeface better aids legibility on maps: sans-serif (Gill) or serif (Times Roman). Our hypothesis is that (1) serif typeface is more legible than sans-serif on maps and (2) direct reduction of type size affects selection of type for maps. The above cited information with regard to evaluating readability and legibility of specific typefaces indicates that the research to test this phenomenon has been applied to a text situation rather than maps. The following section

Table 1
Abbreviations used in test analyses

A: Architecture Major, Ohio University

AM: Art Major, Ohio University

E: Engineering, all branches, who take courses in graphics, Ohio University

G: Graduates in geography and other fields in Ohio University

GF: Geography faculty, Ohio University

HSS: High School Social Studies Students, Athens High School

OF: Other faculty, Ohio University

UG: Undergraduates including geography majors and others taking geography courses, except those specifically designated, Ohio University

Ag. S: Agree strongly

Ag.: Agree

I: Indifferent

D: Disagree

DS: Disagree strongly

O or f_o: Frequency observed

E or f_c: Frequency expected

df: Degree of freedom

P: Probability of Greater Value than the Specified X^2

Σ: Summation

shows the test results related to lettering typeface applied on maps.

Survey and Questionnaire

Lettering on maps, like other symbols, is statistically a non-quantitative phenomenon (Dickinson, 1973: 17); hence the sampling procedure, a quantitative technique, will concern the views and/or preferences of the test population (Amachree, 1975). It included High School Social Studies students; Geography and other undergraduates and graduates who have been exposed to maps; Art and Architecture major students and faculty knowledgeable in lettering and letter design; Engineering students with graphic art experience; and Geography and some other faculties. The data were collected from a questionnaire structured to compare the legibility of the two typefaces under study.

Table 2
Sample

Test Population	Ag. S	Ag.	I	D	DS
HSS, UG					
G, GF & OF					
AM & A. E					

First, three maps with figure-ground and the typefaces were compared (see Maps A, B and C). Following that, some names from the maps were put on a patterned background, which is akin to map figure-ground, to compare the result with legibility conditions on the maps (see Figure 1). Then, questions testing accepted legibility concepts in text are structured to determine how such text legibility conditions could be applicable to cartographic type situations.

A total of two hundred fifty-four completed questionnaires were obtained from the test population and these constitute the basis for the analysis. The tests used are Chi-square (χ^2) and graphic which includes histograms, frequency polygons and pie graphs. Only the most appropriate test is shown with each result in the following section, although all were used.[2]

The raw data were reduced to categories (Table 1) and the results were collapsed into a matrix as shown in the sample table (Table 2) for the Chi-square test. In the merger, the affinity of the merged groups was a prime consideration. For the histograms and frequency polygons, total number of responses were arranged on a five point value scale. In the pie graphs the percentages for total responses in each case are used to compute the angles of division of the circles.

Data Analysis

The first question dealt with personal background such as age, sex, home town and educational categories. The inclusion of this question was to create a balanced judgment in answers to be returned. The age aspect served to measure the maturity of the test population and the sex question to remove the criticism that the result could have been different if both sexes were tested. Inclusion of the home town was to guard against biased answers should one decide to check one's home town as being more legible. Finally, the educational background was an indication of how exposed to maps and other map-like situations individuals in the test population have been.

The second question stated that the lettering on Map A is easy to read. The histogram (Figure 2) indicates an overwhelming agreement among the test population that Gill sans-serif is legible on maps. This choice is also indicated on the frequency polygon (Figure 3).

Question three stated that the lettering on Map B is easy to read. In this question the Chi-square test was useful and is shown along with the other testing procedures (Table 3). The categories were merged as in Table 2. In addition, the five-point value scale was collapsed into two categories by combining Ag.S and Ag., D. and D.S., and eliminating I. The result shows that the null hypothesis is rejected. A histogram (Figure 4) and frequency polygon (Figure 2) for Map B indicates general agreement that Times Roman lettering is legible on maps, although the proportion stating this is less than with Map A. and Gill sans-serif.

Question 4 compared Times Roman and Gill sans-serif lettering on the same map and stated the lettering was easy to read. The histogram (Figure 5) indicates that a majority of those tested who agreed. Yet the results here are more mixed and less strong than with either Map A or Map B, where there was only one lettering style on a map. The decline in preference for a map with mixed lettering is shown in Figure 3 where all three maps are compared in a frequency polygon. There is a rightward shift in the comparison towards greater disagreement on legibility.

The next two questions compared word pairs of each type style in two situations. In question five, the word pairs were on Map C and it stated that Bowling Green and Ashtabula (Times Roman) are easier to read than Wright State and Geauga (Gill sans-serif). The histogram (Figure 6) indicates general disagreement with this statement. In question six, the word pairs were only a figure-ground (Figure 1). It stated that Bowling Green and Ashtabula (Gill sans-serif) are easier to read than Wright State and Geauga (Times Roman). As can be seen this is the reverse of question five and the histogram (Figure 6) indicates overwhelming agreement with this statement. The

Table 3
Sample table for the Chi-square test

Question 3 Data Processing and Result

f_o	Ag.	D	Total
HSS, UG	104	3	107
AM & A, E	95	6	101
G, GF & OF	13	5	18
Total	212 (0.938)	14 (0.062)	226

f_c	Ag.	D	Total
HSS, UG	100.37	6.63	107
AM & A, E	97.74	6.26	101
G, GF & OF	16.89	1.11	18
Total	212	14	226

df = 2

df = 2 from X^2 Table is 3.023 at P = 0.050

X^2 = 4.364, therefore the null hypothesis is rejected

Figure 4, 5 & 6

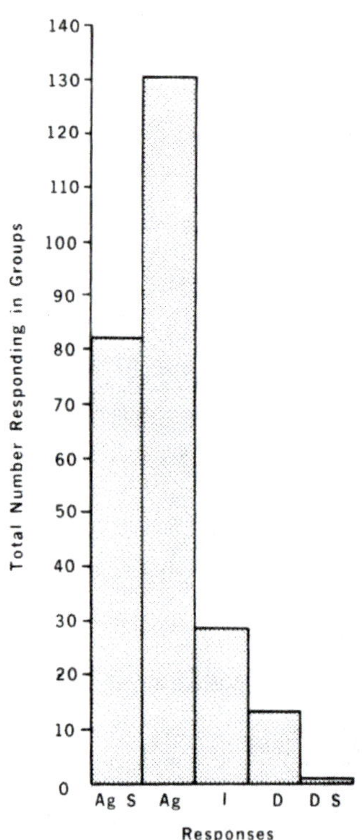

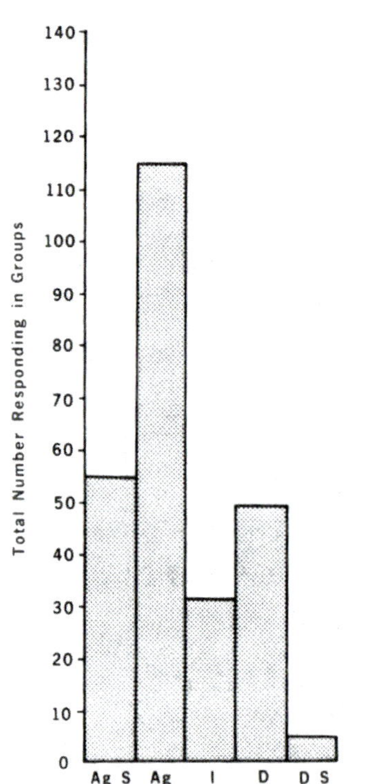

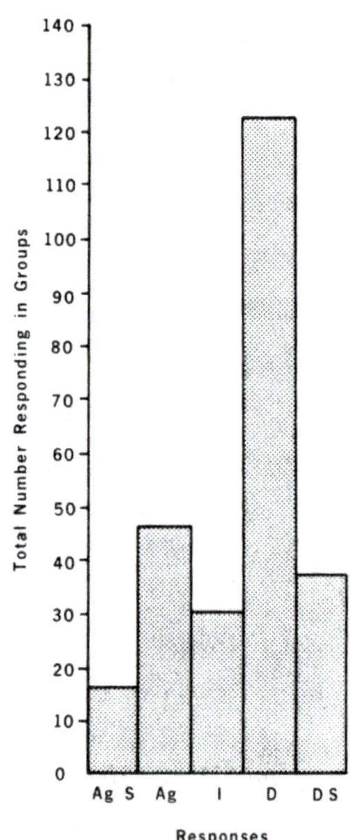

Figure 8

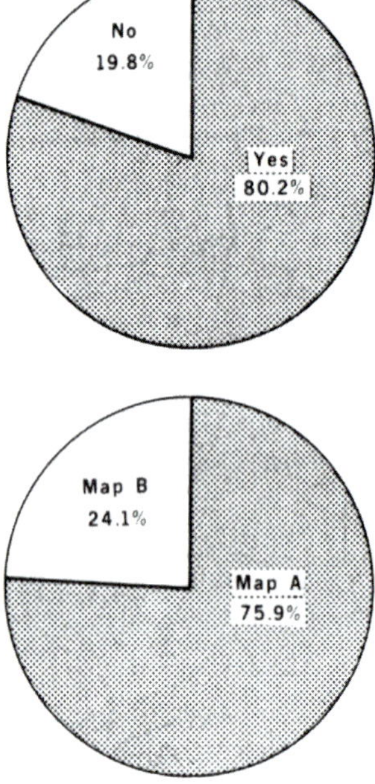

inference to be drawn from these questions is that the test population preferred Gill sans-serif to Times Roman for legibility.

Question seven compared Maps A and B. The pie graph indicates that Map A, with Gill sans-serif lettering, is much preferred in terms of legibility than Map B, with Times Roman lettering (Figure 7). The results here correspond with those of questions five and six in terms of preference for legibility of type style.

Five pairs of words at varying point sizes were compared in question eight (Figure 1). The pie-graph (Figure 8, 1–5) produced such a clear result that it was the most appropriate in this respect and was, therefore, used to determine the results in all the five cases. In examining the pie-graphs, an interesting trend develops. The percentage favouring Gill sans-serif increases from nearly two-thirds preference (62.4 per cent) at twenty-four point, to a high of over seventy-three per cent (73.4) preference at twelve point, and declining slightly to about seventy-two per cent (71.7) at eight point. Times Roman, on the other hand, shows an inverse trend.

The final analysis of question eight indicates that as the point size of both typefaces decreases, preference for Gill sans-serif increases. This indicates, cartographically, one significant result: that for maps requiring reduction, Gill sans-serif should be used as it is perceived as more legible on the reduced map than Times Roman. This is supported by the dominance of Gill sans-serif especially at the smaller point sizes.

The last question, number nine, is a variant of number eight. It tests the legibility of Gill sans-serif and Times Roman on a contoured background at two different point sizes (Figure 9). As the figure shows, words of each type style were arranged in pairs and the test population was asked their preference from each pair. The results are shown in Figure 10.

As can be seen, there was strong preference for Gill sans-serif at each point size. The result supports that of question eight and contrasts with the findings of earlier research on text legibility which concluded that serifs aided legibility.

Conclusion

In recent years maps have become an increasingly important means of communication. Maps are no longer simply regarded as the tool which shows the 'lay of the land' but with the increasing number of thematic maps the utility of the map has taken on a multitude of applications. Within this rather recent development of extensive map utilization, the phenomena of map readability and legibility have come under serious scrutiny by cartographers as well as map users. Type selection for lettering is a natural aspect for investigation concerning legibility and readability.

Figure 9

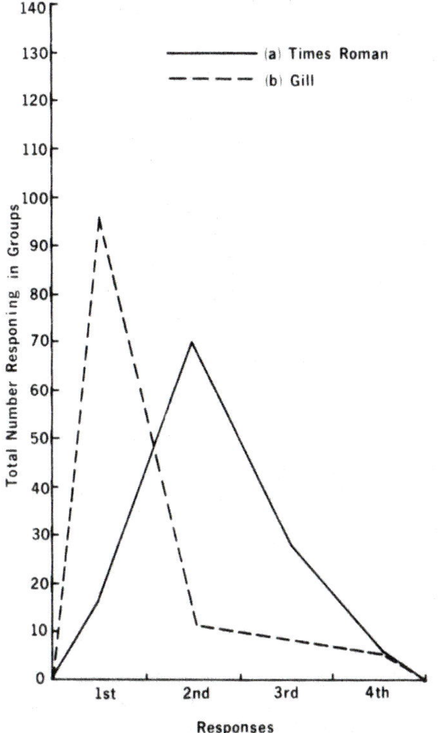

a 18 Point Type

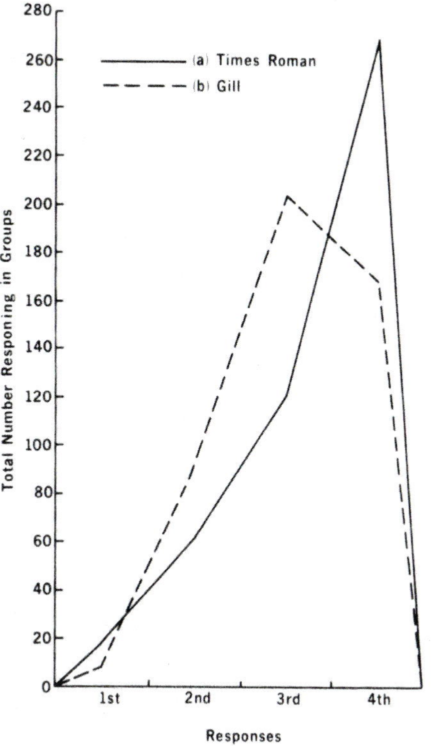

b 12 Point Type

Figure 10

The test results of this study show that the hypothesis which stated serif lettering is more legible than sans-serif was rejected. This finding is of particular significance since it is contrary to earlier studies conducted to determine text legibility and readability. The results emphasize that the cartographer may not assume that text legibility and readability are synonymous with lettering legibility and readability on maps with figure-ground.

The test results further show that sans-serif lettering is increasingly preferred as the lettering size is reduced. This appears contrary to the literature with regard to text legibility and readability where it was stated that the serifs served as leads from one letter to the next – an aspect which should serve a very functional purpose in reduced lettering size.

The fact that sans-serif is preferred by the test subjects throughout this study is indicative of a new trend in lettering perception. One might try to relate the answer to the increasing exposure to computer print-out or lettering which is for the most part sans-serif, thus hinting at a newly acquired familiarity aspect. Whatever hidden dimensions or reasons one may discover, it should not be overlooked that eighteen-point Times Roman lettering has notably less inked area than eighteen-point Gill sans-serif. Regardless of reasons for the Gill sans-serif preference, the fact remains that lettering should not be arbitrarily selected but rather chosen scientifically to best aid legibility and readability of a map. Judicious choice should aid in map communication.

Notes

1. Few studies have been undertaken on type legibility; those already conducted have been done mostly in a non-cartographic context. See Burt (1959) as an example. Bartz (1969a, 1969b, 1970a, 1970b) did some studies with cartography in mind; however, her studies were in part literature reviews. They did not compare the legibility of different type styles. Finally, she (1969b: 136) did emphasize the paucity of research in the field of cartographic type legibility.
2. See Amachree (1975) for the full study and test results.

References

Amachree, T.K.P. (1975) "Typographic Legibility on Maps: A Comparison Between Sans-Serif (Gill) and Serif (Times Roman) Type" *Master's thesis* Ohio University.

Bartz, B.S. (1969a.) "Search: An Approach to Cartographic Type Legibility Measurement" *The Journal of Typographic Research* 3 (4) pp.387–389.

Bartz, B.S. (1969b) "Type Variation and the Problem of Cartographic Type Legibility Measurement" *The Journal of Typographic Research* 3 (2) pp.127–144.

Bartz, B.S. (1970a.) "An Analysis of the Typographic Legibility Literature: Assessment of its Applicability to Cartography" *The Cartographic Journal* 7 (1) pp.10–16.

Bartz, B.S. (1970b) "Experimental Use of the Search Task in Analysis of Type Legibility in Cartography" *The Journal of Typographic Research* 4 (2) pp.147–167.

Bizzi, E. (1974) "The Coordination of Eye-Head Movements" *Scientific American* 231 (4) pp.100–106.

Burt, C. (1959) *A Psychological Study of Typography* Cambridge: Cambridge University Press.

Dickinson, G.C. (1973) *Statistical Mapping and the Presentation of Statistics* London: Edward Arnold.

Eckert, M. (1921) *Die Kartenwissenschaft* (Volume 1) Berlin: Walter de Gruyter.

Imhof, E. (1962) "Die Anordnung der Namen in der Karte" *International Yearbook of Cartography* 2 pp.93–129.

Keates, J.S. (1973) *Cartographic Design and Production* London: Longman.

Nesbitt, A. (1957) *The History and Techniques of Lettering* New York: Dover Publications.

Photon (1974) *Photon Type Specimen Book* Wilmington, Massachusetts: Photon.

Raisz, E. (1962) *Principles of Cartography* New York: McGraw-Hill.

Robinson, A.H. (1960) *Elements of Cartography* (2nd ed.) New York: John Wiley and Sons.

Robinson, A.H. and Sale, R.D. (1969) *Elements of Cartography* (3rd ed.) New York: John Wiley and Sons.

Yarbus, A.L. (1967) *Eye Movements and Vision* (trans. B. Haigh) New York: Plenum Press.

Zusne, L. (1970) *Visual Perception of Form: Analysis of Identification, Discrimination and Recognition* New York: Academic Press.

127 COUNTRIES ON THE MAP–

With Astrascribe cartographic products.

25 years ago the House of Astrafoil revolutionised mapping techniques with Astrascribe tools and remains leader in the supply of high quality photo-mechanical aids for mapping, with:–

Astrascribe – Astrafoil base for positive or negative scribing
Astrascribe 'P' – polyester-base red or orange for negative scribing
Astrascribe Tool Holders and Cutters – also needles to fit any type of machine.
Pre-sensitised Strip-Mask, Drafting Foils and specially-formulated film dyes for work on Polyester Drafting Foils.
For the original Astrascribe mapping aids, at a cost to beat the present financial climate, fill in the coupon.

Please send me full details of Astrascribe aids to map makers.

Name ....

Company..

Address.................................... Tel:............

The House of ASTRAFOIL

DEP, Frith Park, Walton-on-the-Hill, Tadworth, Surrey, Tel: Tadworth 3517.

TITHE SURVEYS & THE RURAL LANDSCAPE OF ENGLAND & WALES

R.J.P. Kain

Originally published in Volume 11, No 2, pp.1–3

When Parliament passed the Tithe Commutation Act in 1836, it put an end to a long and bitter period of dispute over payment of tithes for the support of the Church. The settlement was effected by a minute survey of every piece of tithable land in England and Wales. These tithe surveys cover about three-quarters of the parishes of the country and are one of the most important nineteenth century manuscript sources for historical geographers. This paper describes the surveys and discusses their use for reconstructing mid-nineteenth century landscapes.

The Nature of Tithes

It is not clear when the Church began collecting tithes, the traditional tenth of a farmer's produce given to support the Church, but by the nineteenth century there was certainly great confusion about the manner of payment. This was made locally by each farmer giving to support his own priest. Over the centuries, the peculiarities, ambiguities and irregularities of local customs multiplied. Precedent was piled upon precedent and fresh complications were caused by the dissolution of the monasteries and the enclosure of open fields. By the beginning of the nineteenth century it was no longer possible to discern even the vaguest outlines of a system among a host of local practices and customary arrangements. The courts sanctioned both gross oppressions by the Church and flagrant evasions by tithe payers. The worst abuses were inflicted by obligations in some areas to pay tithes in kind – so many sheafs of corn, so many pigs from a litter. Discontent erupted into violence before a remedy was found in the Tithe Commutation Act of 1836. Parsons were set upon by mobs and tithe barns were burnt down. This was all part of the wider context of agrarian discontent in the first part of the nineteenth century among impoverished and illiterate peasant farmers and labourers. They could not go to court to obtain redress, but they could wield clubs, throw stones and light fires.

The essence of the settlement was the substitution of a money payment, fluctuating from year to year in accordance with the price of grain for all other types of customary payment. This rentcharge was fixed initially by the actual value of tithes collected in a parish and was apportioned amongst its farmers according to the value and type of land which they occupied. All of this necessitated a detailed field survey of parishes from which tithes were payable.

Tithe Surveys

The rural landscape of England and Wales is depicted exactly in these field-by-field tithe surveys. The enquiries of the tithe commissioners covered about three-quarters of the country. The maps drawn for each parish show the boundaries of fields, woods, roads, streams and the position of buildings, while the accompanying schedules of apportionment give the names of the owners and occupiers of each field, its state of cultivation and area. A third set of tithe documents, a series of parish tithe files, provides additional information on the nature of farming practices. In total, the amount of information which the tithe surveys can provide about land tenure, field systems and land use is unequalled by any series of documents. Their accuracy is sufficient to warrant their continued use as evidence in courts of law and their uniformity and comprehensiveness is surpassed only by the Land Utilization Survey of the 1930s. Indeed, they rank as the most complete record of the agrarian landscape at any period.

In the wider context of the history of western cartography, the middle of the nineteenth century marks a high point in the production and perfection of cadastral surveys. Over immense areas of the world, detailed, large-scale plans of land and property were being drawn with meticulous precision. In France, the plans of the Ancien Cadastre recorded fields down to 100 square metres. In Ireland, the Ordnance Survey completed a survey of the entire country at a scale of 1:10,000, and in the United States, the Federal Land Survey was engaged on the colossal task of measuring the whole public domain extending from the Ohio river to the Pacific coast. These and the tithe surveys were in some senses a final reckoning, a concluding statement on a vanishing way of life. In France, the ancién regime was swept away by the Revolution; in the United States the wilderness frontier was being closed and its preliterate society vanquished. In England and Wales, the tithe surveys removed some of the last vestiges of feudal obligations. They converted the traditional rights and privileges of tithe owners into financial assets. The partly communal village world was made over in the image of the land surveyor and government commissioner as they worked over the parishes compiling their tithe surveys.

Tithe Maps and their Use

One of the most important of the tithe survey documents is the map. This was usually drawn at a scale of three chains to an inch which is about 26.7 inches to a mile. Some large parishes were surveyed at smaller four or six chain scales but even so many of the maps cover more than 100 square feet. The specifications for tithe maps were drawn up by Lieutenant Robert Kearsley Dawson of the Royal Engineers who was seconded to the tithe survey in 1836. He fervently hoped that the tithe maps might be assembled together and published as a General Survey of the whole country but for financial reasons the government decided against this. It would have been very expensive to survey areas where maps of sufficient accuracy for tithe commutation purposes already existed. Although the tithe maps did not form an official cadaster, it was important that they were sufficiently accurate to prevent future tithe disputes.

Dawson produced elaborate specifications for surveyors to follow together with a number of diagrams suggesting how systems of internal triangulation of parishes might be laid out, how a parish boundary might be accurately plotted, and how existing surveys might be tested for accuracy. Dawson's military training is very evident in all these matters and there are some close similarities between his instructions and Colonel Colby's Instructions for the Interior Survey of Ireland.

Most tithe maps of open field parishes, like that of Winterbourne Kingston in Dorset, show strip fields by dotted lines and closes by solid lines. It is thus possible to reconstruct the pattern of open field at the time of tithe commutation. In 1848 the parish of Winterbourne Kingston was enclosed and its pattern of fields and landownership totally transformed. Such was the degree of alteration that a complete reapportionment of the rentcharge was needed and comparison of the two maps provides fascinating glimpses of the rural landscape before and after enclosure.

There is a considerable variation in the amount of other detail shown on tithe maps. A few of them use a system of conventional symbols recommended by Dawson to identify different types of land and so they can be read almost as land use maps. Characteristically, inhabited buildings are tinted red on the maps while barns and other structures are shaded grey. In rapidly industrializing areas, like the outskirts of Wolverhampton, tithe maps often distinguished a greater variety of buildings, for instance, chapels and the principal places of manufacture. Most tithe maps also contain some biographical information to help unravel their ancestry. Quite often they bear the name of the surveyor who was usually a local man whose tithe survey work probably extended over a limited area. All their work was tested in Lieutenant Dawson's office in London and maps which passed all his rigorous checks were known as 'first class maps'. Others which failed on some count but which were nevertheless sufficiently accurate for the purposes of commutation were known as 'second class maps'. First class maps can be identified by the presence of the tithe commissioners' seal.

Tithe Apportionments and their Use

A tithe map on its own can provide only a limited amount of information about the rural landscape. Much more can be obtained if the map is interpreted in conjunction with the apportionment roll. Each tithe area on the map (usually conterminous with a field or strip) is assigned a reference number and against this number in the apportionment some more information is given. Firstly, the names of its owner and occupier are stated. This is very valuable information and in this country is really unique to the tithe surveys. It is possible to identify all the fields owned or occupied by particular individuals and, by piecing these together, maps of the pattern of ownership and occupation of land can be produced. It is also possible to work out the size of estates and farms as the sixth column in the apportionment states the surveyed acreage of each field. This can be a long job if it is tackled over a large area but computerized data processing methods help to make it a manageable task. Each tithe area is usually named in the apportionment and these field names can be used to throw some light on the past uses of an area. For example, J.E.G. Mosby in his Land Utilization Survey Report for Norfolk, plotted out all the fields with 'breck' elements in their names. This provided some measure of the incidence of the infield-outfield system of agriculture in the county.

The land use of each field is also recorded in the tithe apportionment. Traditionally this has been the most heavily exploited of all tithe survey information. The tithe surveys distinguish arable and pasture and woodland and note orchards, market gardens and hop grounds. This information used in conjunction with the tithe maps enables the production of land use maps for the mid-nineteenth century. The history of the use of this material goes back to the 1930s and Dudley Stamp's Land Utilization Survey. E.C. Willatts, later secretary to the Survey, wrote a paper in 1933 in which he described the tithe surveys as 'our best data' for the study of land utilization in the past. He advocated the plotting of information on land use for sample parishes to provide an historical dimension to the Land Utilization Survey Reports of each county. Concurrently and quite independently of Willatts, another geographer, H.C.K. Henderson, began a study of the land use of the Adur basin as revealed by the tithe surveys. His study of the changing arable in the Adur basin covers a compact block of country rather than sample parishes and by and large, more valuable results have been obtained by this latter approach.

Tithe data gain in value by aggregation. When the information for one parish is ranged alongside its neighbours, errors and ambiguities tend to cancel out. Since the 1930s successive students have covered larger and larger tracts of country by refining and speeding up the plotting procedures used.

Tithe Files and their Use

One of the problems of tithe land use information is the very crude classification into arable, pasture and woodland employed by the tithe commissioners and their surveyors. Only very rarely are the actual crops grown in the fields recorded in the apportionments. This was done in some parishes and for these it is possible to produce crop maps.

These rare instances apart, the tithe maps and apportionments provide almost no information on agricultural practices. They provide a very faithful record of the end result of the operation of these processes but say little about the reasons for the particular land use or ownership patterns which they reveal. It is essentially a static picture about which explanations have to be inferred from other sources. However, the third class of tithe documents, the parish tithe files do help in this regard as they often contain a printed questionnaire on local agriculture compiled by an agent of the tithe commissioners. The information recorded in these questionnaires was used in conjunction with the field survey recorded in the apportionments and on the maps to calculate a fair rentcharge for the commutation of tithes. They contain descriptions of agricultural practices which can give valuable insights into the working of farms in the mid-nineteenth century. In addition they record the acreage and yield of crops grown on the parish arable and often enumerate the numbers and type of stock pastured on the grasslands. These can be plotted out as choropleth maps and the various distributions related to soil type or used to identify agricultural regions.

A Prospect for Tithe Studies

In 1973 in the preface to A New Historical Geography of England, H.C. Darby said, 'As far as sources are concerned, although much recent work has been done on the Tithe Returns of the 1840s, a comprehensive treatment has yet to appear'. In fact, fully eighty per cent of the material remains to be exploited but I would argue that with modern computerized cartography this is now within the bounds of possibility. Experience over the years has shown that the most effective way of exploiting the tithe surveys is to produce a series of maps of the various landscape elements which they describe. Today a tithe atlas of England and Wales need no longer be considered a pipedream.

Note

This article is based on a lecture given to the Exeter Summer School of the *Society of University Cartographers* in September 1976 and on the manuscript of a forthcoming handbook to the *Tithe Surveys of England and Wales* by H.C. Prince and R.J.P. Kain to be published by William Dawson.

Editorial Note

The handbook mentioned above, and some of the author's other publications directly associated with the subject of this paper are given below:

Kain, R.J.P and Prince, H.C. (1985) *The Tithe Surveys of England and Wales* Cambridge: Cambridge University Press.

Kain, R.J.P. (1986) *An Atlas and Index of the Tithe Files of Mid-Nineteenth Century England and Wales* Cambridge: Cambridge University Press.

Kain, R.J.P. and Oliver, R.R. (1995) *The Tithe Maps of England and Wales: A Cartographic Analysis and County-by-County Catalogue* Cambridge: Cambridge University Press.

Kain, R.J.P. and Prince, H.C. (2000) *Tithe Surveys for Historians* Chichester: Phillimore.

Further Reading

On the nature and history of the tithe system, see Evans, E.J. (1976) *The Contentious Tithe: The Tithe Problem and English Agriculture 1750–1850* London: Routledge and Keegan Paul.

For a discussion of tithe maps and apportionments, see Prince, H.C. (1959) "The tithe surveys of the mid-nineteenth century" *Agricultural History Review* 7 pp.14–26.

On the tithe files, see Cox, E.A. and Dittmer, B.R. (1965) "The tithe files of the mid-nineteenth century" *Agricultural History Review* 13 pp.1–16.

On the use of the tithe surveys to reconstruct landownership patterns, see Kain, R.J.P. (1975) "Tithe surveys and landownership" *Journal of Historical Geography* 1 pp.39–48.

Some studies on reconstruction of the land use of the mid-nineteenth century from tithe surveys are reviewed in Harley, J.B. (1973) "England circa 1850" In Darby, H.C. *A New Historical Geography of England* (pp.527–594) Cambridge: Cambridge University Press.

A recent study reconstructing cropping patterns from tithe file data is Phillips, A.D.M. (1973) "A study of farming practices and soil types in Staffordshire around 1840" *The North Staffordshire Journal of Field Studies* 13 pp.27–52.

An article which examines patterns of land occupation from tithe apportionment is Kain, R.J.P. (1976) "Tithe surveys and the study of land occupation" *Local Historian* 12 pp.88–92.

SOCIETY OF UNIVERSITY CARTOGRAPHERS

Map Preparation: Some Guidance on Fundamentals
by J. Render, Geography Department, Portsmouth Polytechnic

Published 1973 57pp Price 25p

This is an instruction booklet compiled from a series of articles which appeared in the Society of University Cartographers Bulletin, Summer 1970, Winter 1970/1971 and Summer 1971. It introduces the reader to some basic techniques in cartography and includes descriptive chapters on linework, symbols and compilation and plotting. There are fifteen figures in the text and three detailed appendices; with a coloured card cover and a half tone double page spread showing a selection of drawing instruments.

At the time these articles appeared in the Bulletin they created considerable interest amongst members and it was decided to produce them in booklet form as a reference work designed in particular for use by college or university students, trainee cartographers etc.

The booklet is available for purchase (postage included) at the following prices: single copies 25p each, 25-49 copies 24p each, 50-99 copies 23p each and 100 copies and over 22p each. Due to variations in postal charges, prices for bulk overseas purchases will be given on application. 1-4 copies, however, can be purchased at the inclusive charge of 30p each.

If you wish to place an order would you please send the order form immediately to Mr. G.K. Kingdon, Department of Geography, Portsmouth Polytechnic, High Street, Portsmouth. Please note that all orders must be pre-paid, payment to reach Mr. Kingdon as soon as possible, if copies are required for the beginning of the academic year 1973/74.

Map Preparation : Some Guidance on Fundamentals

Please supply .... copy/copies for which I enclose the sum of ....

- for which I have arranged payment of the sum of .... to the Society of University Cartographers account.

Name ....

Address to which booklets are to be sent. ....

....

....

Please note that all orders are to be addressed to Mr. G.K. Kingdon, Department of Geography, Portsmouth Polytechnic, High Street, Portsmouth. Cheques to be made payable to the Society of University Cartographers.

OPTIMAL MAPS

P.E. Sorrell

Originally published in Volume 12, No 2, pp.31–37

During the past ten years or so cartography has undergone a transition process. What has emerged is a discipline re-defined as a communication science as distinct from a rather unwieldy art-science interrelationship, regarded by many as an uncomfortable operational zone. It is not suggested that a communication process did not take place in the past but rather that it was taken almost for granted without consideration given. While it must be accepted that the application of a basic communication model provides enormous impetus towards examining the nature and structure of the particular elements of the subject, in terms of a systems approach rather than as a number of techno-productive operations (many of which are unashamedly the property of other disciplines), it is these techno-productive elements of cartography (which have provided the structuring elements of map making in the past) that still remain as key elements in the systems approach, as can be seen within any of the more expansive cartographic communication models, although in many of the model variations the techno-productive elements have been relegated to a servicing role.

The intent or objective or map making has, it seems, taken on a far different set of dimensions than those which might have been applied a decade ago. Evaluation in terms of time, cost, aesthetics, precision and accuracy has been superseded by the single objective of communicating information as efficiently as possible. This does not infer as a practical proposition that the older criteria must be simply abandoned; they do in fact directly affect any effort to communicate information cartographically. To put it crudely, it might be said that the cartographic product attempts to achieve 'a maximum return from a minimal input'. That might sound a little simplistic or even suspect within the model of cartographic communication, but given enough thought the idea tends to bear fruit. Before continuing in our quest for the elusive 'Optimal Map' it is perhaps worthwhile to familiarize ourselves with the basic structure of the cartographic communication systems and, more especially, the information processing which takes place. Figures 1, 2, 3 or 4 graphically demonstrate the basic concepts of the communication system though they differ somewhat in physical appearance.

From such models it is easily detected that the primary functions of the system are to (i) select a particular set of information; (ii) transform it from a three-dimensional spatial array to a communicable language; and (iii) ensure that a receiver can cope with the message. A number of basic questions must arise from such a premise: (i) who selects what? (ii) what is the structure of the language and how efficient is it? and (iii) what is the decoding and assimilation (graphicacy) ability of the receiver or percipient?

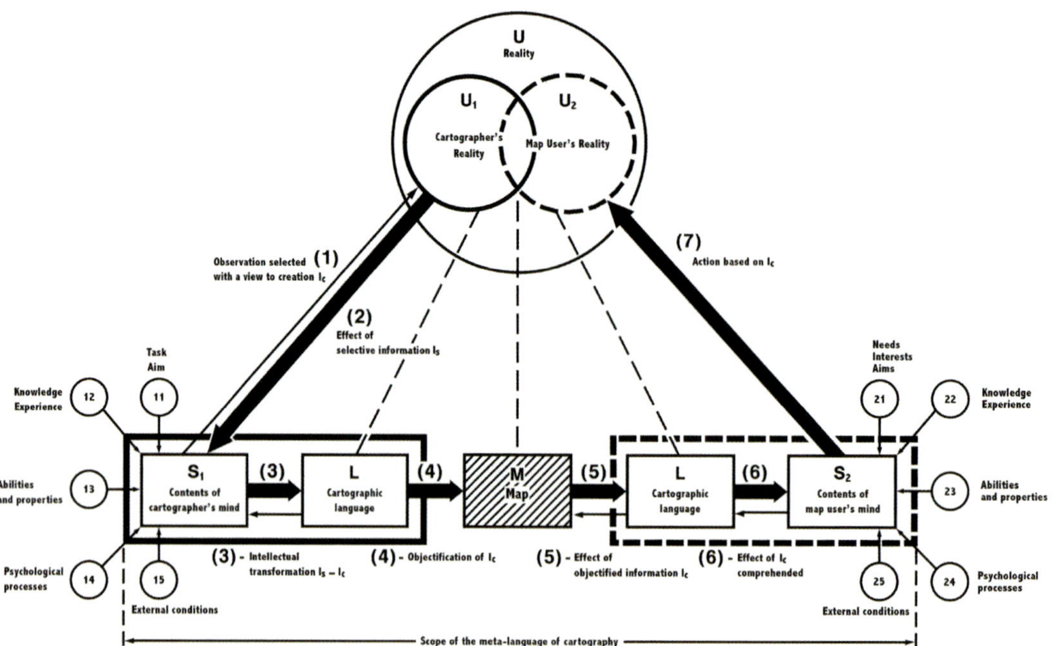

Figure 1
Koláčný (1969)

In addition to these three points any consideration of the percipients' mental response to particular information is of an even more complex nature; Koláčný (1969) summarizes the communication system as follows: 'Cartographic information originates, is communicated, and produces an effect'.

Muehrcke (1972) provides a somewhat more 'up-tempo' arrangement in his definition of the system: 'An active feedback system: data collection, data processing, information display, image processing'.

What is obvious from these various communication models is that the basic concept has changed little as can be seen when comparing the 1969 model of Koláčný's with the 1977 Board model.

What must now be sought is a more realistic operations model within the system. The model by Monmonier (1975) derived from that of Jolliffe (1974) (Figure 5) does in some ways make a more definitive approach to the cartographic processes in the communication system and, to some extent, the structural units of that system, but does it differ so much from the 1967 Board model? (Figure 6). As can be seen from Monmonier's model, the area of direct 'cartographic' involvement is limited to one part of the eight-unit model, the techno-productive or reprographic activity being contained within one other unit. This implies that the area of cartography as we once knew it occupies only one-quarter of the total cartographic communication system. This would correspond well to the view held by Keates (1973), The collection, preparation and classification of data is NOT a cartographic function. The responsibility of the cartographer is to design and produce the map in liaison with the author.

In contrast to this, however, is the view held by Muehrcke (1972) who regards the elements of data collection, cartographic design, construction and reproduction as part of a map production procedure and cartography is defined as being 'the philosophical and theoretical basis, principles and rules for maps and mapping procedures'. Perhaps it is all a matter of titles or the tendency by some to draw away from the manual elements of a system in which human activity is a producer of random noise. Ratajski (1973) is perhaps kinder to all concerned by subdividing the cartographic activity into Cartology and Applied Cartography. Unfortunately, it is precisely because of these labelling processes that the development of a cartographic design model tends to be slow in developing, in that an individual's position and role within the system appears difficult to identify and, indirectly perhaps, is responsible for random and sporadic forays into particular elements of the cartographic process. The characteristics of particular symbols in terms of perception has been a favourite research topic, yet does a design guide exist for such symbols – are the dots on a particular map correlated to design research or the convenience of an appropriate rub down symbol sheet, is the thickness of line work optimally differentiated, or does it depend on the nib sizes that you have available? Psychology and various branches of Physics have been applied, the results have often been of interest, often contradictory and very often based on a very unrealistic sample. Some results have only reinforced intuitive decisions already operating in the map making

Figure 2
Ratajski (1973)

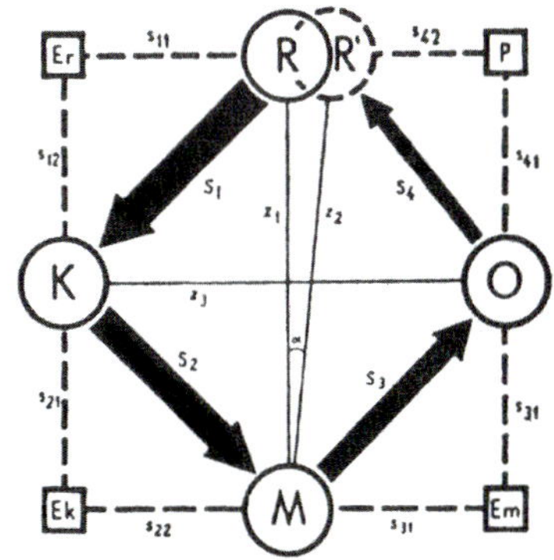

Model of cartographical transmission

R – reality as source of direct information
Er – informative emission of direct source
S_1 – *relation of direct recognition*
s_{11} – *relation of direct emitting*
s_{12} – *relation of direct perceiving*
K – cartographer, sender of message
E_k – *message emission (informative emission of sender)*
S_2 – *relation of message creating*
s_{21} – *relation of message emitting*
s_{22} – *relation of message perceiving*
M – map or message informant
S_3 – *relation of indirect recognition (by means of a map)*
s_{31} – *relation of indirect emitting*
s_{32} – *relation of indirect perceiving*
O – receiver (map user)
P – mental transformation (mental processes)
S_4 – *relation of reality image reconstruction (relation of recognition)*
s_{41} – *relation of recalling*
s_{42} – *relation of imagination*
R^1 – *imagination of reality*
z_1 – *message relation*
z_2 – *relation of identification (relation of map efficiency)*
z_3 – *relation of making available (relation of informing receiver)*
α – *degree of transmission correctness*

Figure 3
Robinson and Petchenik (1975)

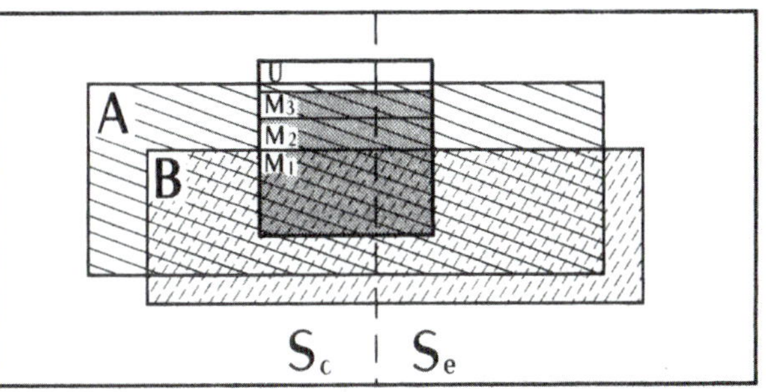

A Venn diagram summarising the cognitive elements in cartographic communication. *(University of Wisconsin Cartographic Laboratory.)*

Diagram elements: C – cartographer; P – percipient; S_c – Correct conception of the milieu; S_e – Erroneous conception of the milieu; A – Conception of the milieu held by C; B – Conception of the milieu held by P; M – Map prepared by C and viewed by P; M_1 – Fraction of M previously conceived by P, M_2 – Fraction of M concerning S newly comprehended by P, a direct increment; M_3 – Fraction of M not comprehended by P; U – Increase in the conception of S by P not directly portrayed by M but which occurs as a consequence of M, an unplanned increment.

Figure 4
Board (1977)

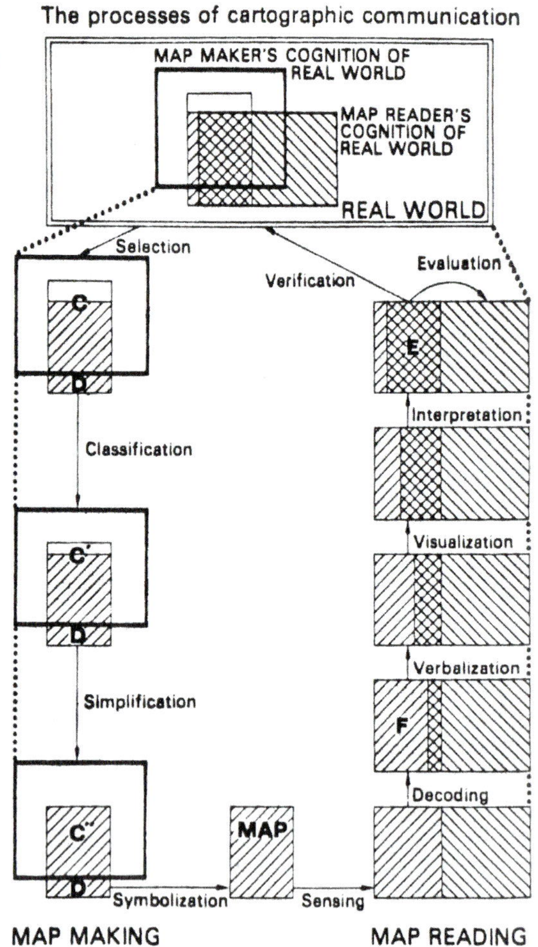

process. A general awareness of the information processing system was felt to exist, but apart from the philosophical interrogation by Robinson and Petchenik (1975) I would suggest that only a loosely assembled framework of the communication model exists at present. In particular, there remain a number of interfaces which fail to interlock, and components based on rather peculiar specifications within some of the more stylized communication models.

What is apparent, however, is the realization that the necessary fabric for an optimal cartographic design/communication model exists already within the cartographic field. The fault of the past has been the apparent lack of an acceptable blueprint, the result of which could be compared with giving a number of children an identical LEGO® set and being presented with an equal number of entirely different models. This does not imply that detailed specifications are required for a particular model, but rather a need for models to satisfy a number of identifiable criteria.

The units of map construction are well-known:

1. Point, Line and Area symbol
2. Shape, size, orientation, texture, colour, contrast (design criteria)
3. Nominal, Ordinal, Interval and Ratio (data ranking)
4. Arbitrary, Semi-natural, Natural (language structure)

All provide a physical expression for a particular phenomenon and all can be regarded as elements of the cartographic language. What appears to be avoided, however, is perhaps best expressed by the proposition 'The whole is equal to more than the sum of its parts'. The implication being, of course, that while each graphic element can be evaluated within a number of dimensions it is in itself meaningless out of the context of the particular message.

If then we can detach ourselves from this fascination towards particular signals, and if we accept that the premise of cartography as a communication science, which has been with us for ten years yet still manages to get an annual spring clean is well-founded, then we must now turn our attention to those units of the system which

Figure 5
Monmonier (1975), modified from Jolliffe (1974)

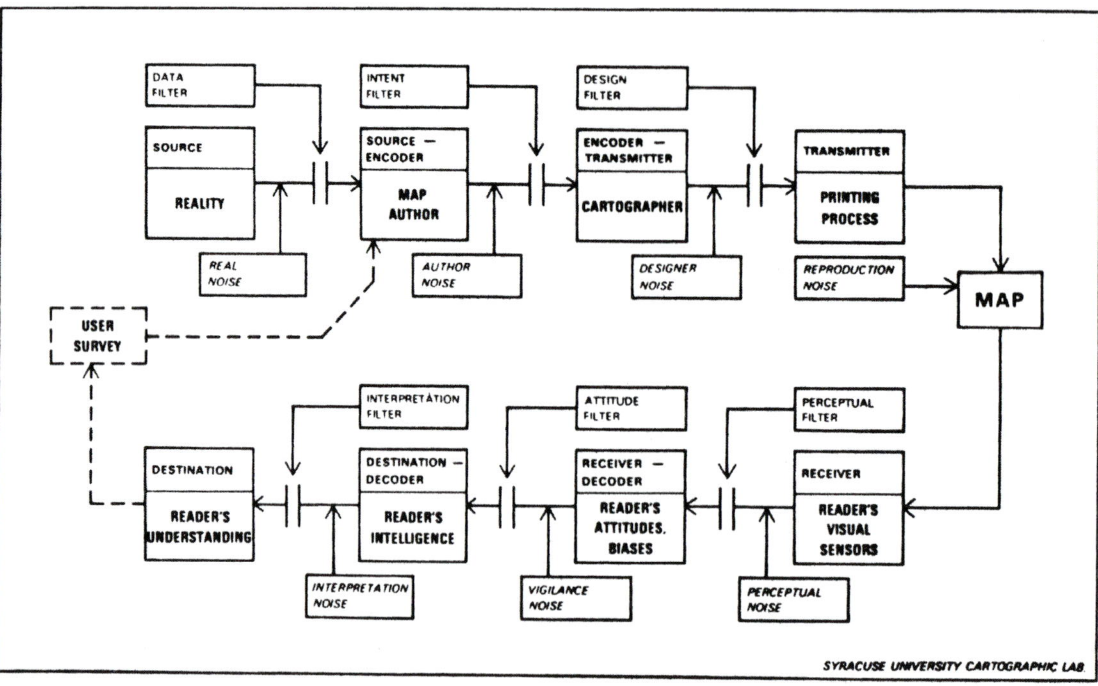

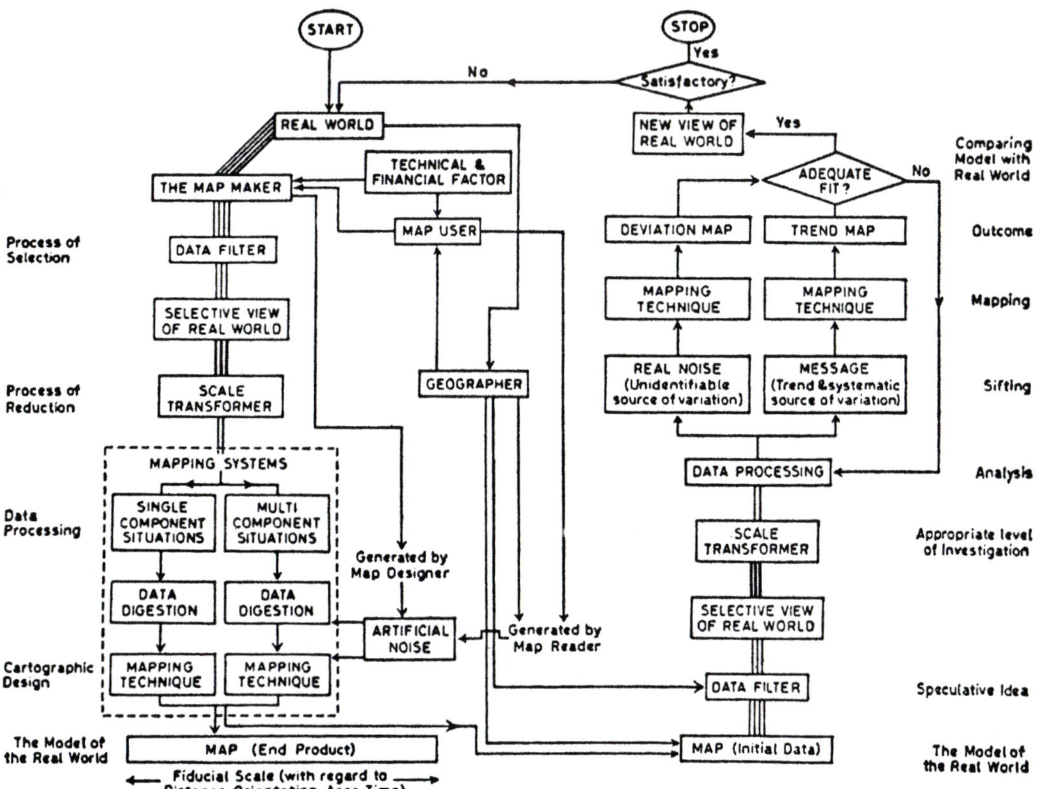

The Map-Model Cycle.

Figure 6
Board (1967)

depend upon the cartographic language.

A basic objective of any communication system is the achievement of a perfect transmission – output being identical to input. While a similar transmission of information is highly desirable within the cartographic system, it is unfortunately hardly feasible, due in part to the 'random' spatial structure of the message compared with the sequential structure of normal 'verbal' communication. Although the perceptual and cognitive evaluation of the latter still leaves much to be desired, in comparison, the decoding and assimilation of the 'map message' is at present almost entirely beyond any quantitative or even qualitative evaluation. If we accept that the cartographic message does reach generally non-graphicate or unsympathetic eyes, surely we must seek out the fault. It would appear that at present both message and receiver have a number of noise-producing defects. Perhaps fault also lies at an earlier process in the system, what of input; it would appear from a number of models that blame can be squarely placed on the shoulders of the 'cartographer's cognition of the Real World' and his 'selection of a particular data set'.

I hope that nobody really believes this to be the case; Monmonier (1975) at least clearly identifies the intervention of other minds within the 'Data Filter' and 'Map Author' activities within the system. The importance of these minds must be stressed if we are to identify the cartographic processes within the communication of spatial information system. There exists a large body of people whose role it is to identify, label and extract particular sets of data on a one-to-one basis at a particular place and time. It might also be recognized that all maps without exception owe their birthright to these minds. It is important to remember that the reaction of each recipient following the decoding process is wholly dependent upon his cognitive map of that particular map area; this in turn has developed from confrontation with other models of the map area at a previous date.

If no cognitive map exists then this first confrontation will provide such a map. In either case a spatial model that provides a referent must exist, or be created. For all the 'cognitive views of reality', it must be realized that the basic underlying structure of the cartographic communication of spatial information is the creation of a referent base map and this is only achieved by reducing subjective intervention to the lowest possible level.

To consider the surveyor's interpretation of the landscape as his cognitive interpretation of the 'Real

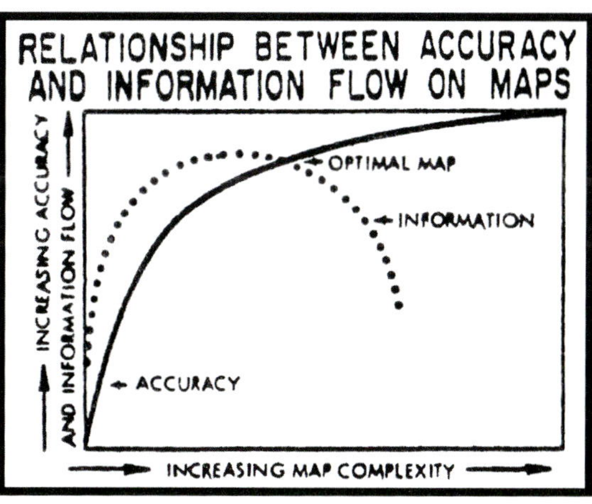

Figure 7
Jenks and Caspall (1971)

World' is perhaps adding a touch of glamour to the very precise activity of identifying particular point locations in terms of distance, bearings and elevation. The principal concern is the creation of a model representing the form of the Earth's surface and particular physical and cultural elements within defined parameters.

The abstractness of such a program is accentuated by modern survey equipment recording angles and distances directly on to magnetic tape, thereby avoiding the plotting of detail onto a two-dimensional surface at a particular scale. Obviously a surveyor develops a cognitive map of the particular area being measured but many aspects of the landscape do not become apparent until mapped, and it is to be hoped that all data collectors, all those that structure data and all data encoders should be entirely objective in their analytical approach to a particular information set and devoid of their own subjective or biased reaction to such an information set.

If a closed communication system is to be developed it is of little use seeking standards for language structure and optimal performance when in addition to an unstable destination, the originator adds additional noise to the system.

If by any act of faith we can rationalize our interpretation of reality, the prime objective of the cartographer must then be the creation of a perfect map – not in the old terms of registration, quality and aesthetics, but in its efficiency as a communication device which implies measurement of both input and output before the map can be regarded as an optimal unit. An Optimal Map has been defined by Jenks and Caspall (1971) as 'the relationship between accuracy of detail, information gained and map complexity', suggesting that the optimum design is where the two curves representing accuracy and information intersect rather than where information content is maximal in terms of units of information (Figure 7).

Expressed in another way, it might be said that the more realistic a map is, the greater is the probability of it increasing in complexity, yet it is generally accepted that the clarity of transmission improves as the more clues/aids are provided. An impressionist graphic for example tends to provide for varied multi-dimensional interpretations. What is required is a point at which the transference of 'information' is maximized, with 'noise' held at an acceptable level at the input stage. This, in fact, is excellent in theory but falters when subjected to quantitative measurement, i.e.

1. What is the information-carrying capacity of the map?
2. How do you measure the percipient's overview of the map?
3. To what degree is the reader's understanding of the spatial relationships influenced by the code signal used?

Replies to each of the above can be summarized as follows:

1. Requires an ability to both evaluate and provide a measurement of information.
2. This problem is even more complex due to each recipient being unique.
3. Requires the development of a cartographic/spatial language and the parallel development of a receiver's graphicacy or decoding ability. Very little can, in fact, be done to affect the recipient's reaction to the decoded message.

Jenks and Caspall (1971) in their paper conclude with the observations that:

1. It can be assumed theoretically that less generalization will be associated with an increased information flow.
2. There is some level at which information added to a map is not perceived.

It is from these points that the cartographic elements of the communication system must be identified in terms of a design model. This design model is not some restrictive device which will constrict free expression of message, for each map is only one interpretation of reality based upon a particular set of data, but rather a logical system of graphic inter-relationships which will provide a framework for the whole field of spatial or chorographic communication.

Muehrcke (1972) defines the major dimensions of cartographic processing as:

1. The study of the relation of map symbols to their referents.
2. The study of the relation of map symbols to one another.
3. The study of the relation of map symbols to their interpretation.

It is only by the summation of these particular relationships expressed in some quantitative way that can provide the evaluation of a map as 'optimal' or not in terms of its effectiveness as a communications device. It should perhaps be pointed out that this does not imply an 'optimal map' in the broadest sense of the term.

The key then is organization, inferring that the inter-relationship between parts may be strongly or weakly coupled; their effect upon each other may be quantitative or qualitative. What we are actually faced with is a list of questions (below) that could have been raised a decade or so ago; it has only been the re-orientation of cartography that has provided the push to examine our product unit by unit and then to reconstruct it to perform optimally.

- Can spatial information be expressed by quantitative units?
- Is all information of equal value in the cartographic model?
- What is the relationship of the parts to the whole?
- How do you measure transmission effectiveness?
- Are results of research concerning perceptual response to a particular symbol applicable within a complex model?
- How do you arrive at an optimal evaluation?

The problem is perhaps highlighted as follows:

'The TOTAL information content of the whole is exactly equal to the sum of the information contents of the parts – when the description of each part includes all possibilities of connection with other parts'…

Bibliography

Board, C. (1967) "Maps as Models" In Chorley, R.J. and Haggett, P. (Eds) *Models in Geography* (pp.671–725) London: Methuen.

Board, C. (1976) "The Geographer's Contribution to Evaluating Maps as Vehicles for Communicating Information" *Paper presented at the 8th International Cartography Conference, Moscow, 3rd–10th August, 1976.*

Jenks G.F. and Caspall, F.C. (1971) "Error on Choropleth Maps: Definition, Measurement, Reduction" *Annals of the Association of American Geographers* 61 pp.217–244.

Jolliffe, R. (1974) "An Information Theory Approach to Cartography" *Cartography* 8 (4) pp.175–181.

Keates, J.S. (1973) *Cartographic Design and Production* London: Longman.

Koláčný, A. (1969) "Cartographic Information: A Fundamental Concept and Term in Modern Cartography" *The Cartographic Journal* 6 (1) pp.47–49.

Monmonier, M.S. (1975) *Maps, Distortion and Meaning* Resource Paper No.75–4 Washington, D.C.: Association of American Geographers.

Muehrcke P. (1972) *Thematic Cartography* Resource Paper No.19 Washington, D.C.: Association of American Geographers.

Ratajski, L. (1973) "The Research Structure of Theoretical Cartography" *International Yearbook of Cartography* 13 pp.217–228.

Robinson, A.H. and Petchenik, B.B. (1975) "The Map as a Communication System" *The Cartographic Journal* 12 (1) pp.7–15.

The 1980s

At a computer trade show in 1981, Microsoft's Bill Gates, in defence of the just-introduced IBM PC's 640KB usable RAM limit, supposedly uttered the statement: '640K ought to be enough for anybody'. Despite the enduring popularity of the legend about the comment, it's hard to find proof that Gates ever said it, and certainly he has consistently denied it. In fact, this was a decade of huge advances in the computer industry and these began to filter through and radically change cartography.

Nevertheless, the world of many cartographers at this time was still very 'old school'. Reading back through Volumes of the *Bulletin* from the early 1980s, you will see adverts for products such as scribecoat and retouching pens. One of the Society's own publications was the compilation of Jack Render's articles entitled 'Map Preparation: Some Guidance on Fundamentals', where he was certainly not writing too much about computers. In 1980, Andrew Tatham took over as Editor of the *Bulletin*, and he started a column called 'Cartographic Design'. Its first contribution was from ever-prolific Jack Render on 'a typographic experiment', which was purely reprographic in scope; manual typesetting was just starting to come into place. In 1981, a first 'Computers in Cartography' column appeared, but it was intermittent for several years before becoming a regular feature. At this time there was a whole series of articles on the theme of 'Cartography at ....' , which were great for seeing what others were doing, and what equipment was in use. It seemed that remote sensing might have been a more important development than computers for practising cartographers to be aware of, although, in retrospect, that turned out to be something of a false dawn.

The decade seemed to produce a significant output of lasting cartographic texts. *Understanding Maps* by Alan Hodgkiss (1981) achieved a lot of exposure early in the decade, followed in the next year by John Keates' book with the same title. There were also English translations published of two classics in the field: Eduard Imhof's *Cartographic Relief Presentation* in 1982 and Jacques Bertin's *Semiology of Graphics* in 1983. These were accompanied by Edward Tufte's *Visual Display of Quantitative Information* (1983). Although narrower in range, *A History of the Ordnance Survey* edited by Seymour (1980), became a standard text on the subject, which many found to be a hugely valuable resource.

One of the first software packages to make an impact in the cartographic field in which I worked (the university sector at the time) was GIMMS. In 1981, it was used by a team at the University of Lancaster to produce a census atlas, and seemed poised to be a big player, as it was taken up by a considerable number of individuals and organizations. At a slightly less productive level, in 1984 the *Bulletin* carried a piece on a circle-drawing routine for the BBC Micro. At a macro scale there was a note in the same Issue entitled 'Goodbye to road maps' highlighting a computerized route-finder from Volkswagen and Siemens. Furthermore, the 1985 Annual Summer School at Middlesex University closed with a presentation on the BBC Domesday Project, a videodisc-based resource that had a huge 'wow' factor at the time.

More pertinently, the Apple Macintosh computer – or Mac – was creeping into cartographers' workspaces, and stories of what MacDraw and CricketGraph, for instance, could achieve started coming through the grapevine. Subsequently, the whole production workflow was changed by the introduction of three hugely influential software packages: Adobe Illustrator, which appeared in 1987, FreeHand in 1988 (from Aldus/Altsys/Macromedia), and CorelDRAW in 1989.

The articles published in the *Bulletin* from the 1980s reflect these monumental changes, but also delve into other subjects and occasionally appear to lag behind, as mentioned above. For this decade, we have chosen a broad range of topics. Map design for children was Herbert Sandford's area of expertise, and he wrote extensively on the subject. The paper included here is part of his work on small scale general purpose maps, and includes a good selection of references. A fascinating account of the cartographic work of Leonardo da Vinci is included by Martin Kemp, Emeritus Professor of the History of Art, University of Oxford, and who asks us to consider how Leonardo's illustrative work 'expressed his sense of the underlying unity of all natural processes'.

The adoption of the Peters projection was a hot topic at this time, and Peter Vujakovic's article analyses both the man and the myth, with the advantage of an interview and correspondence with his subject. Michael Blakemore then provides a review of mapping as it changed from a lineprinters to GIS, arguing cohesively that we should be expecting not merely the emulation of existing mapping, but a whole new approach.

We also offer a series of reflective pieces. As the end of the decade coincided with the 25th anniversary of the founding of the Society we have included two such articles here. There are also two guest editorials that appeared in this decade. Firstly, by John Robertson, who at one point blames the introduction of 'communication diagrams' for introducing too much noise, and pleads for simplicity. Secondly, Roger Anson, who as an educator reviews the situation regarding cartographic training. He wonders about the advance of technology and a possible loss of skills resulting from it.

I took over as Editor of the *Bulletin* in 1988 and immediately moved it into the digital era. I used a Desktop Publishing programme called Corel Ventura to layout the journal and supply camera-ready artwork to the printers, which pleasingly resulted in an award for 'typographical excellence' a year later.

Steve Chilton

References

Bertin, J. (1983) *Semiology of Graphics* (trans. W. Berg) Madison: University of Wisconsin Press.

Hodgkiss, A.G. (1981) *Understanding Maps: A Systematic History of their Use and Development* Folkestone: Dawson.

Imhof, E. (1982) *Cartographic Relief Presentation* (trans. Steward, H.J.) Berlin: Walter de Gruyter.

Keates, J.S. (1982) *Understanding Maps* London: Longman.

Render, J. (1973) *Map Preparation: Some Guidance on Fundamentals* Portsmouth: The Society of University Cartographers.

Seymour, W. (Ed.) (1980) *A History of the Ordnance Survey* Folkestone: Dawson.

Tufte, E.R. (1983) *The Visual Display of Quantitative Information* Cheshire, Connecticut: Graphics Press.

MAP DESIGN FOR CHILDREN

H.A. Sandford

Originally published in Volume 14, No 1, pp.39–48

There have been great advances since the Scott Keltie report of 1886 in the design of maps for children but it is only in recent years that a significant volume of research has been able to confirm or deny the worth of these changes and to suggest lines for future development. This article presents some of the conclusions reached with special reference to the small scale general purpose map.

An Overview of the Problem

Maps for children have their problems and solutions: they should not be merely simplified versions of adult maps (Sandford, 1972; Sorrell, 1974). Those used for jigsaw puzzles and board games, or to illustrate children's books, are generally poorly designed but many atlases attain a high level of validity and effectiveness. Britain led the field until the 1960s but has been overtaken by progress made in Sweden and Germany. Many adults are cartographically unsophisticated and some of the conclusions may have relevance to other than children's maps. Large scale maps will not be discussed. For larger scales children are almost always given adult maps, such as those of the various national topographic surveys, and these present fewer problems. Thematic maps merit separate treatment.

The Information Field

The crux of the problem is to produce a map that the child is able to relate to and integrate with what he already knows, his 'information field'. Unless a child can do this with a map it is of no more use to him than that obstinate piece of jigsaw puzzle that refuses to find its place. In the present context the information field is the child's knowledge of the general geography of the world. He acquires this in three different ways. The foetus develops a perception of the environment of the womb and also a kinaesthetic sense which tells it the location and orientation of arms and legs in relation to trunk and head. By largely visual perception, the baby learns to locate and orientate himself in respect of his home, the toddler his neighbourhood and the child his district: knowledge of this expanding territory is gained at the same time. The child living in town does not usually learn to orientate himself by sun and stars in relation to the wider space (Figure 1).

Running parallel to this direct acquisition of knowledge and location comes a stream of secondhand information from school and the media, mainly from television and mostly imperfect, about parts of the world never to be visited. Coincident with this comes the use of small scale maps of the same and other unknown tracts. Imperfect maps falsify the information field: realistic information fields help the child to read and interpret maps correctly. The child who actually knows about London on the Thames and how cities and rivers are symbolized on maps, sees Manaus on the Amazon and mentally images a very large city on small river – unless the map design anticipates and corrects for this or unless the child has prior knowledge of the Amazon Basin and its geography.

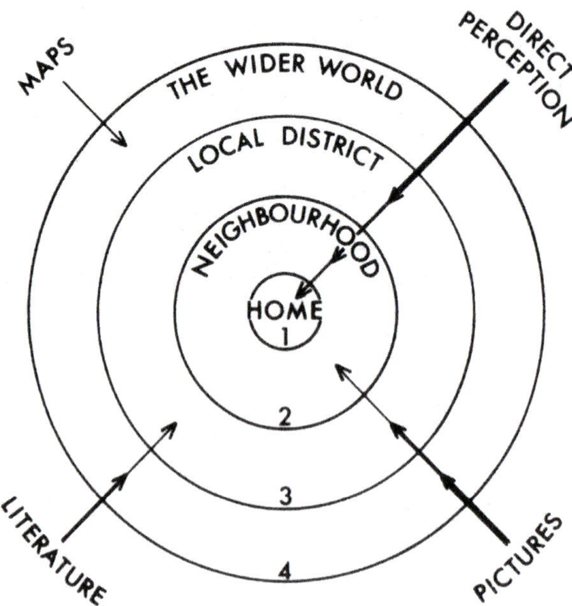

Figure 1 *Inputs into the mental map of the child. Mapped territory of (1) Baby, (2) Infant, (3) Younger child, (4) Older child.*

This general geographical information field is essentially one of landscapes and seascapes, and together these have been called the 'episcape' (Sandford, 1979). The child's information field appears to contain some kind of mental image of the world with its varying scapes and their locations in respect of himself and each other. It is, therefore, a mental map. For a printed map to be useful to a child it must be sufficiently well-matched with his mental map to be located on it. The printed map should add to and if necessary correct, but not falsify, the mental map.

There are actually three kinds of map-like mental constructs that vie with one another and confound the child's comprehension of the real world. Current studies of perception maps are very revealing (Gould and White, 1974) but difficult to evaluate for our purpose as they tend to confuse mental maps, drawn maps and printed maps.

The infant has an essentially topological mental map of his home territory. He comprehends only the simplest relationships such as within and without, big and small, near and far. His mental map is roughly circular, with himself in the middle, and objects and their spacing contract at a distance on a geometric, diminishing scale (Hagerstrand, 1957; Tuan, 1977). Although such maps can be drawn, they are too personal to be readily communicable. No-one ever quite grows out of looking at the world in this way. Does not a hole in a tooth feel enormous? Do not Londoners draw a sketch map of Great Britain with a diminutive Scotland? Children in particular underestimate the sheer size of unfamiliar lands on a map and thoughtful design is called for to combat this.

Piaget has shown that as the child grows older he gradually learns to perceive and think within an imaginary grid or framework that locates objects in their Euclidean relationships in a world of right angles, straight lines and equal-step distances (Piaget and Inhelder, 1956). All things are measurable, though the child can do so only in a proportional manner, e.g. that Paul lives twice as far away as Freya. The child now adopts an imaginary 'viewpoint' simultaneously over every part of the territory mapped – not a bird's-eye view as is commonly stated, but a god's-eye view – and hence, when drawn, his map is communicable to others. It is independent of any standard co-ordinates, orientation or scale, but can be matched with printed maps that have geographical co-ordinates, north at the top and a metric scale. Success in matching, however, depends upon careful attention to map design.

In countries with an educated populace, the acquisition of these mental maps coincides in time with the use of printed maps, largely at school. If, as generally happens, the match between printed map and reality is not achieved and maintained, the printed map is substituted for reality. The child sees many maps and develops, not a mental map of the real world, but a mental map of a paper map. A child can do this more readily than an adult because he still retains his infantile visual memory. By whatever amount and in whatever manner the map he sees departs from a realistic impression of the actual landscape, so will his mental map of the world be defective. Children think that there are few towns in New Zealand (their mental map tells them this and so apparently does their atlas map); that the Kamchatka Peninsula rises like a staircase; that the Ob flows from the Arctic Ocean down into Siberia; that large ships cannot pass through the Straits of Gibraltar and that much of the Sahara is green and fertile. We can readily imagine what symbols on the maps have given rise to these false impressions. That they are a legacy of earlier map reading is shown by the fact that the frequency of error increases in the presence of a map and decreases in its absence (Sandford, 1967, 1970). All too often maps feed incorrect data into the child's information field and falsify his mental map the world. The power of the visual image should not be underestimated (Carpenter, 1976) and it seems that when different facets of the data are in contradiction, the evidence of the printed map is preferred to information gained by other and more direct means.

If the information field is based upon a paper map the child's map reading degenerates into the juggling of marks on bits of paper. Ask such a child where Paris is and he will tell you that it is in the middle of France and lies to the south of Britain. Near enough right, but only because he has been taught that that is what is meant by a red dot on a green patch below Britain on the map page. In other words, he has told us how to find Paris on a map, can do so himself, and be able to draw a map to give us this information. But he has given no evidence that he understands anything about the location of Paris in the real world. He is only beginning to do this when he can point in the direction of Paris from where he is sitting, can give some indication as to how long the journey would take and of his expectations of the environs of the city. Few children can do this. It is interesting that many teachers prefer the simplicity of the paper answer appropriate to a paper world. Similarly, a child does not really know where China is until he can point in the right direction and do so with arm pointing downwards through the Earth (Figure 2). It is found that almost all children behave as though they lived on a flat Earth. One postgraduate geography student insisted that it was correct to point upwards into the sky to the North Pole and even a confrontation with a globe did not convince him otherwise.

The General Map

The printed map that most commonly interfaces with the child's mental map, and which is most often in conflict with it, is the general, name-locating map. These dominate atlases and preponderate as book illustrations, and they serve the purposes of showing the position of places both the local site and the wider situation), route finding and providing background information about areas. They also serve for the study of certain distributions, such as river patterns, though this is the particular function of thematic maps.

In the first decades of this century the general map in a British atlas or book for children combined some 'political' information (countries, cities, railways) with some 'physical' information (mountains, rivers, bays). From the 1930s it became not uncommon to replace this by a pair of page-opposed maps, one political and the other physical. The former was usually politically coloured and the latter bore hypsometric colours in lieu of the earlier 'hairy caterpillars'. When retained as a single general map the political colours were omitted and gradient was added to altitude by means of relief shading (hill shadow or plastic relief). Such maps are now standard in the United Kingdom and can conjure up in the child's mind a reasonable image of land shape. It is much to be regretted, however, that many maps produced for the retail trade retain the pictorial hills of the early eighteenth century, so destructive of the child's mental map of the real world.

The landscape, however, combines land shape or form with land cover, and it is this cover of farm and town, forest

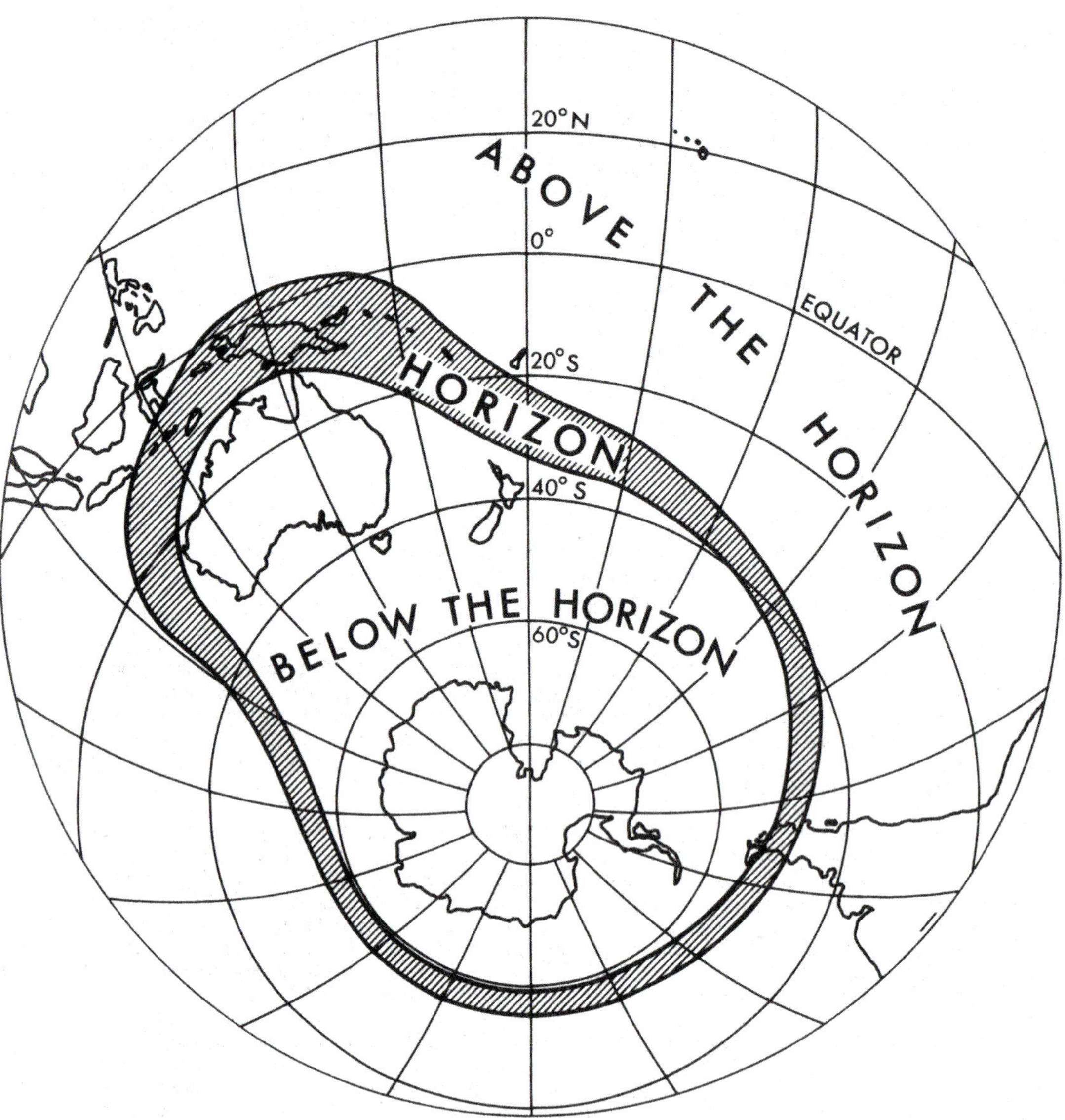

Figure 2
The horizon of the English child's mental image of the world. The child points downwards to only those places that are below the horizon.

and desert, that dominates in the landscape. The standard or conventional child's general map may provide cover information at points (e.g. town stamps) or along lines (e.g. railways), but is largely silent about cover over areas.

The 1960s saw the introduction of a kind of general map that emphasizes cover without neglecting form. At present they require the use of colour and are beyond the scope of line drawing. It is possible that the use of half-tone might be modestly successful in a form reminiscent of a monochrome photomap derived from Landsat imagery. From earlier beginnings associated with Imhof, Lundqvist and Debenham, they have developed in rather different ways and have been variously called environmental, habitat, geographical and so forth. The generic term 'landscape map' seems appropriate.

Ground colours are used over the map to symbolize the cover of forest and tundra, desert and steppe, and relief shading is generally incorporated. The colours are conventionalized but reasonably representative of the natural colours of the Earth as it would be seen from an orbiting satellite, but without the atmospheric 'blueing'. The desirability of thus designing a map that not only portrays the actual landscape but can correctly interface with and, if necessary, correct the child's mental map has implications for all aspects of cartography for children.

Format, Scale and Projection

Decisions about scale, projection, ground area, format and margin are interlocked, and established conventions are often appropriate for children. There are, however, some grey areas about which we now understand more. They mostly concern the printed map/mental map interface, but it will be convenient to mention some related matters more dependent upon the child's visual perception, reading ability and approach to pictures.

A brisk, 'telegraphic' style of title, legend, text and footnote, so often adopted, is not readily understood by children. 'Great Britain – Communications' is better expressed as 'Great Britain – Road, Rail and Air Travel', or, for younger children, 'Our Roads, Railways and Airways'. An increasing number of maps for children are printed upon light-reflective paper. A map is so complex and contains so much that is unfamiliar to the child that it cannot all be appraised or even consciously viewed in twenty minutes (Sandford, 1967) and anything that obfuscates the child's efforts is undesirable.

The maps in a child's atlas are sometimes bled to the edge of the page so as to exclude any white margin. This emphasizes the continuity of the real world beyond the limits of the map, but there is little reason to believe that children are unaware of this, and it is desirable to retain the margin for the clearer numbering of the co-ordinates. In addition, children seem to feel more 'comfortable' with map margins, perhaps because the margins help confine their scanning, perhaps because they resolve some of Arnheim's perception tensions (Arnheim, 1951). Also unknown is the connection, if any, between a margin and a child's very marked tendency to neglect the sides and corners of both maps and pictures (Long, 1953, 1961; Sandford, 1967).

The portrait format, standard for printed books, is not particularly suitable for children's maps. It has evolved in response to the needs of the reader of continuous text: the lines of type are kept short so that the eyes do not slip to the wrong line when making their swift saccadic progress down the page. As a visual image, however, the map is more comparable to a picture. The initial movement of the eyes tends to be towards the centre while peripheral vision extends towards the map or picture frame. As we have two eyes our field of view is elliptical and this gives a preference for the landscape format. Australia lends itself well to this but South America and many other territories do not.

As the child becomes more familiar with the distribution of land and sea, the coastline becomes a more important frame of reference for reading the map than the map frame itself, whether or not there is a margin. The map should be sufficiently small for the frame, whether of map or coastline, to be within the peripheral vision. This frame serves as a control over the child's scanning, during which his eyes fixate at first one feature, then another, in a rarely achieved endeavour to image the map as a whole (Phillips *et al.*, 1978). During the course of scanning enough is maintained in the immediate memory to enable a holistic image to be approached if not attained (Sandford, 1967). Although his vision is more acute than that of the adult, the child is more near-sighted and may approach the map more closely. The field of view is measured as the angle subtended at the eye and so the requirement is for even smaller maps than an adult of the same mental age would require. There is no absolute criterion, and the 10 in x 9 in (254 mm x 229 mm) format suggested by the Geographical Association (1953) gives some guidance. Wall maps should be proportionately larger to subtend the same angle at the eye. By this criterion many current table maps are too large and, by virtue of their contained detail, many wall maps too small.

The area of territory within the map frame is critical to the child's ability to locate it within his mental map of the world. The dominant frame of reference for this mental map is the coastline. The optic nerves and their pathways to the brain are genetically programmed to pick out the junctions between contrasting colours and textures. Land and sea provide the strongest such contrast on most maps. The drawn coastline reinforces this contrast but there is no implication in favour of further reinforcement by unrealistic wave-like or cliff-like symbolism. It is, therefore, of the greatest help to the child if the area mapped can be extended to include some coastline. A map of Bolivia, for instance, might deliberately be extended from seventy degrees west to seventy-two degrees west so as to take in a length of the Pacific coastline: the map scale would be decreased only insignificantly. Children like the blue colour of the sea (Sorrell, 1974). It seems to reassure them, though the reasons for this are obscure.

For an adult map the amount of detail required largely decides the scale. It may not be too great an exaggeration to say that on a map for a child the extent of territory within an appropriate format should decide the scale. It can be far smaller than generally it is, and the detail thereby excluded is usually only a matter of surplus and unnecessary names. The scale should be presented both visually and quantifiably. A same-scale inset of the home region, as of the British Isles on a map of Africa for the United Kingdom market, provides children with a good visual indication of scale. When the mapped area is large a superior method is to use an inset globe with the territory mapped clearly shown, and this also serves to locate the area. If distances have to be quantified the child benefits from being given the required information in three forms: scalar line, representative fraction and verbal statement. The norm of having north at the top of the map is best retained as it helps locate the map within the child's mind.

There is no ideal projection for maps of the whole world for use by children. We need to present a coastline that can be matched with the child's mental map and also a map frame that confirms his concept of a round world. The usual factors governing choice of projection (Maling, 1973) lose some of their relevance. The shape of map frame, the extent of territory covered and the outline of the continents become the major considerations. We are further limited to projections commonly used in order to avoid confusion in the formation of the mental image. It is, therefore, helpful to the child to show the whole of the Earth and to do so 'in the round' as by the use of page-opposed pairs of azimuthal projection hemisphere maps as in the *Modern School Atlas* (Philip, 1966). Inset globes help to assert the roundness of the Earth, and most helpful in the *Schoolatlas* (Meulenhoff Educatief, 1978) is a series of hemisphere maps that take the child round the world in a suite of overlapping 'space views'.

By these criteria the interrupted Mollweide projection has merit but the uninterrupted Mollweide sacrifices shape to suitability for showing trade routes. It is common for the poles to be omitted and for a Hammer or modified Gall to be used to reduce distortion. With these and others an inset globe is particularly helpful.

The commonly adopted projections for small regions present no problems, but maps of North America and Eurasia are liable to show strong torsional distortion in the corners. This distortion itself is not the main cause of the difficulty experienced by children in recognizing the British Isles, for example, in such cases. The difficulty is largely the conflict between left-right or 'horizontal' names (suggesting one orientation) and a coastline that follows another. Widely spaced and rather unclear parallels and meridians are not able to resolve this conflict. It is resolved, however, if all floating names (names of points and patches as distinct from names of linear features such as rivers and ranges) are aligned and curved with the geographical co-ordinates. Sorrell (1974) suggests that names should be horizontal to facilitate their reading but much comprehension is thereby lost.

Oblique 'space views' with varying orientations are increasingly used for adult maps of landscape type and their high degree of realism commends them for the use of children when quantified mapwork is not required.

Both children and adults need geographical co-ordinates for use with a gazetteer but it is the child who has greater dependence on them to inform him of the true shape of land distorted by projection and perhaps also to help control his scanning of the map. He finds their numbering less helpful in locating the area, however. Older children need geographical co-ordinates also for the information they provide as to local time, the seasons, length of day, scale and so forth, and they are commonly an external examination syllabus requirement. A failure to appreciate that the geographical co-ordinates serve purposes other than for use with an attached gazetteer has led them to be dropped from many maps in books and to their replacement in many children's atlases by alpha-numerical or similar co-ordinate systems. These are almost devoid of information and quite idiosyncratic, their use being limited to a single map and to a single function. In some cases the alpha-numerical system makes use of parallels and meridians already marked on the map, rather than imposing a new grid. There is then merely the absence of useful information, but if a new grid is imposed it is positively harmful in giving a false impression of shape.

Children undoubtedly find geographical co-ordinates difficult in use but they are helped by closer and clearer drawn lines. The majority of candidates in a Certificate of Secondary Education examination thought that the United States of America lay northwest of Mexico on a Mercator map of the world! To help prevent such errors some atlases draw directional arrows within the map frame or indicate north, east, south and west in the margin.

On children's maps the legend requires especial care: an adult can usually guess his way through a map. Few legends on maps for children explain the typographic code and this may be one reason why children are so insensitive to typeface differences. It is difficult to place a name so that it is clearly attached to a feature and, therefore, helpful if the child is able to gain from the typeface some anticipation as to the kind of feature being named. On a map of Asia most children who noticed the name, referred to the 'Karaganda Steppes', though the places were actually the town of Karaganda and the Kirghiz Steppes, the latter being in italics.

Mapping Land Form

There has been developed no entirely satisfactory way of depicting both altitude and gradient – and hence relief – on a general map for children. Conventional hypsometric colours mislead the child (Patton *et al.*, 1977). The Prairies may be visualized as more barren than the Sahara and the contours separating them take on the character of the risers of a flight of stairs. If the contours are reinforced and the colours strongly contrasted the problem worsens, while the irrational class intervals commonly adopted further distort the appearance of land shape (Sandford, 1976). For contour maps *per se* the indications are for a small vertical interval and a strictly geometric scale. When hypsometric colours are employed, the indications are for them to be weakly contrasted and almost to diffuse. Gradient shading (hill shadow) is very 'visual' but is frequently too generalized for realism, or has either incorrect or spurious details. Success doubtless depends more upon the practitioner than the method, but airbrush does not seem to succeed so well as pencil and both are surpassed by the method of photographing obliquely illuminated relief models. Gradient shading does not always succeed at larger scales, regional differences in particular becoming lost, and the more extensive use of physiographic symbols for children's maps might be considered. The method was systematized by Raisz (1931) and they are much used in America as in the *Picture Atlas of our Fifty States* (National Geographic Society, 1978). This has been called the landscape method of relief portrayal and, in subdued tones, could be a useful way of adding form to cover on a landscape map.

Mapping Land Cover

Land cover is most nearly complete on landscape maps and these make the best match with the child's mental map. The style associated with Debenham and now best known from American publications is highly generalized and airbrush colours merge into one another. This appropriately represents the zones of transition that are the small scale reality, but the maps tend to be too generalized and even stylistic to present a true image. In this mode of landscape map it is common to show the 'natural' or the 'potential' plant communities, omitting farm and town, as in the *Children's Atlas of the World* (Usborne, 1979). Gradient shading is normally added.

The style of landscape map developed in Sweden and which has now spread to Britain as the *Third Atlas of the Environment* (Bartholomew, 1978) uses Landsat colour imagery to provide greater precision to the mapping of both farmed and unfarmed areas. Transitional zones are successfully shown at smaller scales, and the artwork involved could (but does not always) show the firmer boundaries between different uses of the land at larger scales, the 'fences' of Coleman (1969). This style of landscape map generally incorporates gradient shading and is the most successful so far produced for young children.

The *Alexander Weltatlas* (Ernst Klett, 1978) is the best example of an alternative approach in which a rigorous, if not entirely satisfactory, classification of scapes is derived from more varied sources of information. Colour patches are bounded and imply fences even where none exist. A child must allow for this but it is perhaps no greater a problem than allowing for false transitional zones. Current exemplars of this approach are, because of their complexity, more suited to the older child, but the method is not age dependent, the use of texture to complement colour is helpful and the cover is eminently quantifiable.

All landscape maps so far published to a greater or lesser extent neglect the most important scape of all – the townscape. They do not even recognize townscapes as a mappable category of landscape but commonly employ the town stamps appropriate to a political map. Thus they come to show field and meadow extending to the very core of great cities. A town of no more than 100,000 people may be shown at its true-to-scale size and shape on a map at 1:5,000,000, quite a typical scale for a school atlas map of, say, France. A quarter of the world's population lives in towns of this size and larger, and the Third World is fast approaching a degree of urbanization hitherto associated with the more industrialized nations. It would be helpful to the child for landscape maps to reflect this reality.

Generalization

The degree of generalization is of paramount importance. During recent decades increasing editorial generalization has so reduced the amount of detail of hill, river and town, that an unrealistically simplistic view of the world is provided. Lamentably, this is sometimes accompanied by an inattention to accuracy, whether or not associated with the increased smoothing of river and coastline which is also sometimes evident. There is evidence that this is an invalid line of development and that only a tolerably detailed map can reflect complex reality (Sandford, 1978). A subdued but detailed backcloth of physical information appears neither to obscure nor to be obscured by overlying 'political' information. There is no need to omit towns merely because there is no room for their names: a cartographer does not normally name all rivers and hills, lakes and islands. Children benefit from a reduction in names and not in landscape detail.

On political and general maps most towns are shown by town stamps and these indicate the sites of their centres. Selection is based mainly upon the census returns for administrative districts, but employing a lower population limit where towns are few (Dixon, 1967). Such a statistical base has neither meaning nor significance for the child, and the varying rigour in selection distorts the true distribution of urbanization. Rank-size plots of towns, usefully summarized by Garner (1967), reveal that the character of settlement is one of ever larger numbers of ever smaller communities. This characteristic could be faithfully reflected if only the more significant of the towns were named. If necessary a minor variation in the town stamp could indicate which are and which are not named.

Jay (1954) suggested that towns and other features should be selected according to their significance to school geography. There could be merit in selecting the towns to be named along, such lines, though defining 'significant' in a rather wider sense. On a map for children of the relevant age, this would enable us to show the significant cross-channel terminals of Cherbourg and Dieppe while discarding the names of the larger towns of Bourges and Aix.

Significant towns that are too small in extent to be mapped in their true-to-scale size and shape could still be shown by a town stamp, and for such towns this remains a valid device. When its extent can be shown, the built-up area is more appropriate than the administrative area, which would show a Sheffield inflated by that city's extensive tracts of moorland and a shrunken London with integral parts of its urban tract outlying as detached entities. On general maps that show relief by hypsometric. colours these may be preserved, if it is required to show land form beneath the urban sprawl, by the use of a bounding line rather than a colour patch, and for large towns and cities it may be helpful to indicate the Central Business District and to have some line of communication penetrating the colour patch.

This approach to mapping settlement is valid for conventional political and general maps, but, as it delineates the townscape, is precisely the approach required if the general map is of the landscape type. In fact, the landscape map solves all the problems associated with mapping towns in a manner suited to children.

Place-names

If town names have not been reduced sufficiently it is possible that the reduction in the names of physical features has been carried too far. Children are developing as verbalizers. The instant of perception involves a classification of what is perceived in terms of a named class of phenomenon. Children remember and think in terms of proper names and class names, and it is difficult for them to comprehend a map in the absence of names, let alone communicate their observations. With the present generation of children's maps this seems to apply more particularly to the physical content o£ geography. It is difficult to describe the position of Moscow on many children's maps if the River Moskva, the Valdai Hills and even the Urals are not named. In addition, children's scanning is very largely a progress from name to name (Sandford, 1967).

Customary solutions to the problems of aligning names with the geographical co-ordinates and of placing them in relation to the features nominated (Keates, 1974) are largely effective for children. Current changes in name spellings, however, are contra-indicated. With so much of the map that is unfamiliar or even mystifying, the child is greatly frustrated if he must cope with strange spellings, spellings not used at home or even at school. Until the new farms become part of his actual or potential vocabulary they should not be used. Overseas, London is variously spelled as Londres, Llundain, Lontoc, Londyn and so forth. People overseas do not use a foreign language (English) when referring to our capital city, and there is no merit in anticipating, as do some children's maps, our possible adoption into the vernacular of El Qahira and Beijing for Cairo and Peking (Sandford, in prep.).

The best choice of typeface remains a largely unresolved question. Most typographic research relates to continuous text but names on maps are isolated, lie on strong backgrounds and are generally unfamiliar, foreign names with exotic pronunciations. Adults read largely by the pattern of ascenders with few fixations of the eyes, while a child often needs to study each syllable or even each letter. Research into cartographic typography per se has generally involved stylized, fictional maps (Bartz, 1970) and are of doubtful relevance. When children are tested with published maps it is found that sans serif typeface is more effective (Sandford, 1978) and there are indications that uncommon typefaces, italics and faces with thin strokes should be avoided. Despite the child's high visual acuity, small image sizes are also contra-indicated. To be balanced against these constraints is the desirability of employing a variety of faces and image sizes as a typographic code but present knowledge does not give any very clear guidance.

Conclusion

Small scale coloured maps for children are approaching a very high standard of realism to meet their special needs and their ultimate form may be that of a Landsat colour mosaic photomap, appropriately adjusted and annotated and with an indication of the relief. The fine visual impression of landscape maps could advantageously be complemented by a similar treatment of seascapes (Sandford, 1978) when the required techniques have been developed. The problem of producing a reasonable representation of scapes in monochrome may never be solved however.

The accompanying map (Figure 3), questions and answers are intended to be more than an amusement. They are a true reflection of some types of error found in the course of research involving some 3,000 children in six counties over 15 years.

Figure 3
The map of Mythica as seen by a child. Item from index: 60 Po Yang (L) Mythica D4.

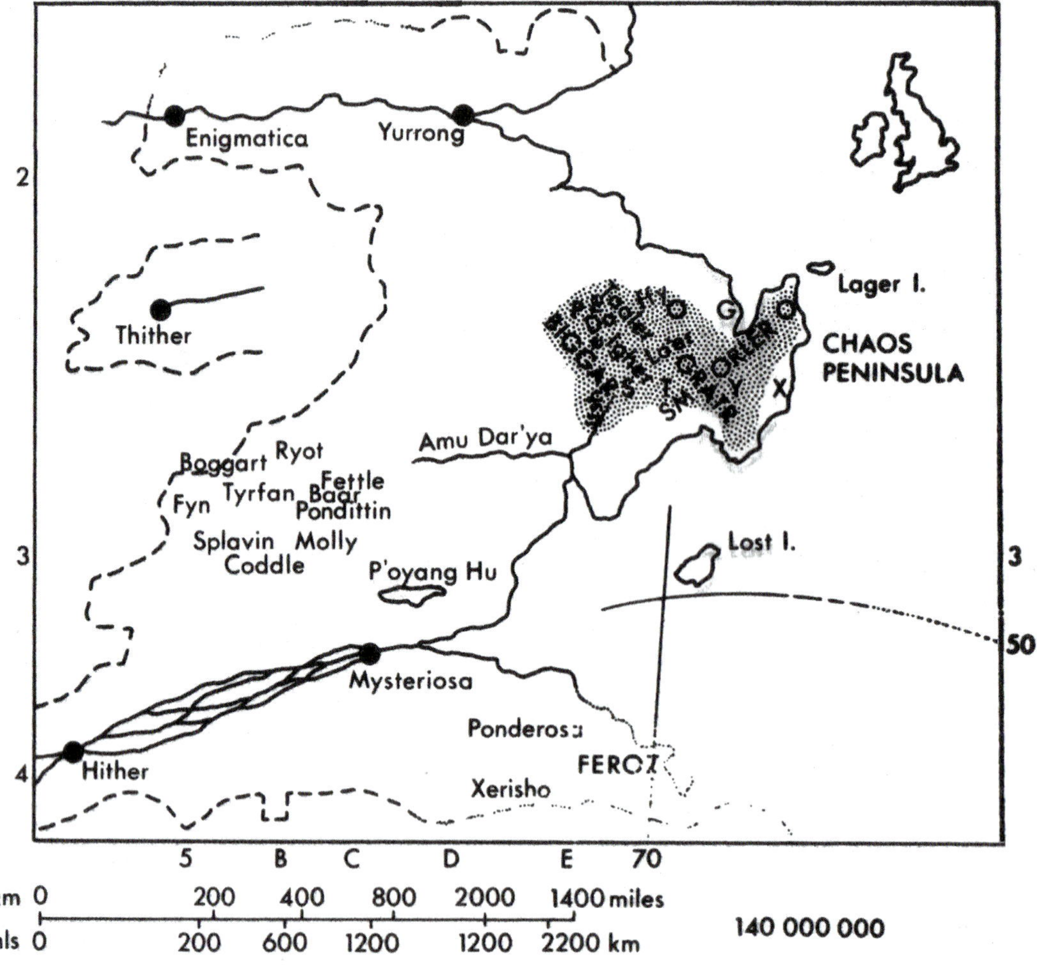

Answer the following questions:

1. What is the direction to Enigmatica from Mysteriosa?
2. Name the highest mountain in Chaos Peninsula.
3. Give the distance in kilometres between Hither and Thither.
4. What is the direction of Yurrong from Britain?
5. Give the latitude and longtitude of Lost Is. to the nearest degree.
6. Which is the bigger, Lost Is. or Lager Is.?
7. Use the Index to locate Lake Poyang.

8. Use the gazetteer to find the river Oxus.

9. Which is the shortest route from Hither to Mysteriosa, road, rail, river or canal?

10. Name the largest town in the central part of Mythica.

Answers to Questions about the Atlas Map of Mythica:

1. NNW or NW? Quite wrong! You misread the question which is "What is the direction from Enigmatica to Mysteriosa?" Try again. SSE or SE? Wrong again! This is an area south of the equator and mapped by a conical projection. The correct answer is SOUTH. You cannot clearly see the co-ordinates and do not understand them anyway.

2. You are suffering from an acute case of perceptual overload.

Congratulations if you found Itshere.

3. 1200 km? No – your eye has slipped from kilometres to miles on the scale line.

2000 km? No – you thought the scale line started at 0 but it actually extends further to the left with smaller divisions.

1800 km? Wrong again! You did not notice that Thither is in an inset. The approximate answer is 600 km.

4. West? Sorry, but the British Isles is enclosed by a rectangle and on the map is printed

"Same scale inset of the British Isles".

5. Your answer is incorrect as you cannot clearly see all the co-ordinates, you cannot interpolate between them, you do not know whether you are north or south, you invariably confuse west and east and you give the longitude first because of your training with grid references on O.S. maps. And in the question you read longtitude for longitude.

Lager Island? GOOD! Some of the boys and girls in your class cannot compare sizes and they think Lost Island looks bigger!

7. You have found it. Excellent. You have found it though it is Lake Poyang in the question,

Po Yang (L) in the index and P'oyang Hu on the map. You also realized that the letter/number co-ordinates were related to the geographical co-ordinates of this conical projection. (Or did you abandon the index and scan the map? I don't blame you!)

8. Not in the index? Sorry, but it's now called the Amu Dar'ya.

9. The railway? Very good. Some boys and girls in your class are still guessing what the symbols mean.

10. Sorry. Another incorrect answer. You have chosen a river. All the typefaces in the atlas look the same to you.

Score Evaluation:

2 out of 10 or higher score. Excellent: move to the top of the class.

1 out of 10. Pass mark. This is a good average mark for boys and girls.

0 out of 10. A little below average but you are showing great promise.

References

Arnheim, R. (1951) *Art and Visual Perception* London: Faber & Faber.
Bartholomew (1978) *Third Atlas of the Environment* Edinburgh: John Bartholomew & Son.
Bartz, B.S. (1970) "Experimental use of the search task in an analysis of type legibility in cartography" *The Cartographic Journal* 7 (2) pp.103–112.
Carpenter, E. (1976) *Oh, What a Blow that Phantom Gave Me!* St Albans: Paladin.
Coleman, A. (1969) "A geographical model for land use analysis" *Geography* 54 (1) pp.43–55.
Dixon, O.M. (1967) "The selection of towns and other features on atlas maps" *The Cartographic Journal* 4 (1) pp.16–23.
Ernst Klett (1978) *Alexander Weltatlas* Stuttgart: Ernst Klett Verlag.
Garner, B.J. (1967) "Models of Urban Geography and Settlement Location" In Chorley, R.J. and Haggett, P. (Eds) *Models in Geography* (pp.303–350) London: Methuen.
Geographical Association (1953) "Essentials of a good secondary school atlas" *Geography* 38 (1) pp.33–35.
George Philip (1966) *Modern School Atlas* London: George Philip and Son.
Gould, P. and White, R. (1974) *Mental Maps* Harmondsworth: Penguin.
Hagerstrand, T. (1957) "Migration in Sweden: A Symposium" In Hannerberg, D., Hägerstrand, T. and Odeving, B. *Lund Studies in Geography* Series B (13) Lund: The Royal University of Lund.
Jay, L.J. (1954) "Significant placenames in school geography" *Geography* 39 (1) pp.28–32.
Keates, J.S. (1973) *Cartographic Design and Production* London: Longman.
Long, M. (1953) "Children's reactions to geographical pictures" *Geography* 38 (2) pp.100–107.
Long, M. (1961) "Research in picture study: The reaction of grammar school pupils to geographical pictures" *Geography* 46 (4) pp.322–337.
Maling, D.H. (1973) *Coordinate Systems and Map Projections* London: Philip.
Meulenhoff Education (1978) *Schoolatlas* Amsterdam: Meulenhoff Educatief.

National Geographic Society (1978) *Picture Atlas of Our Fifty States* Washington, D.C.: National Geographic Society.
Patton, J.C. and Crawford, P.V. (1977) "The perception of hypsometric colours" *The Cartographic Journal* 14 (2) pp.115–127.
Phillips, R.J., Noyes, E. and Audley, R.J. (1978) "Searching for names on maps" *The Cartographic Journal* 15 (2) pp.72–77.
Piaget, J. and Inhelder, B. (1956) "Child's Conception of Space" London: Routledge.
Sandford, H.A. (1967) "An Experimental Investigation into Children's Perception of a School Atlas Map" *M.Phil. thesis* University of London.
Sandford, H.A. (1970) "A Study of the Concepts Involved in the Reading and Interpretation of Atlases by Secondary School Children" *Ph.D. thesis* University of London.
Sandford, H.A. (1972) "Perceptual problems" In Graves, N.J. (Ed.) *New Movements in the Study and Teaching of Geography* (pp.17–28) London: Maurice Temple Smith.
Sandford, H.A. (1976) "Bringing the atlas to life" Spectrum (*British Science News*) 141 pp.2–5 London: Central Office of Information.
Sandford, H.A. (1978) "Taking a fresh look at atlases" *Teaching Geography* 4 pp.62–65.
Sandford, H.A. (1979) "Things maps don't tell us" *Geography* 64 (4) pp.297–302.
Sandford, H. A, (in prep.) "Analysis of atlas maps" In Mills, D. (Ed.) *Geography in the Middle School* Sheffield: Geographical Association.
Sorrell, P. (1978) "Map design – with the young in mind" *The Cartographic Journal* 11 (2) pp.82–90.
Usborne (1979) *Children's Atlas of the World* London: Usborne.
Tuan, Y. (1977) *Space and Place* London: Edward Arnold.

One of the many cartoons by Alan Bartlett to appear in the ***Bulletin*** *which captured the flavour of many of the Society's Annual Summer Schools (e.g. Rotterdam in 1980).*

LEONARDO'S MAPS AND THE 'BODY OF THE EARTH'

Martin Kemp

Originally published in Volume 17, No 1, pp.9–19

Leonardo's maps, made in the context of his activities as a town planner, military and hydraulic engineer, exploit techniques known in medieval and Renaissance cartography. His map of Imola must be regarded as one of the supreme products of these techniques. But more than this, his maps breathe the spirit of the microcosm-macrocosm analogy, through which he expressed his sense of the underlying unity of all natural processes.

Introduction

Any historian of art concerned with the history of techniques of representation inevitably comes into contact with the history of cartography. This contact generally goes no further than an uneasy awareness that the story of map-making has important and largely unexplored links with the history of pictorial representation. I do not, in this paper, claim that I will progress beyond this state of imperfect awareness, but rather hope to illustrate one aspect of the art-historical material which bears upon these questions. I also hope to pose a few problems for further investigation.

During the course of my studies into the interactions between art and science, I have become aware of a number of potentially fertile points of contact between the histories of art and cartography. These points, broadly speaking, concern techniques of representation. The first clear point of contact, which like all the others has yet to be elucidated properly, is the relationship between the Ptolematic revival in 15th century Italy and the rise of linear perspective in painting.[1] Linear perspective was, after all, founded upon a grid – a grid projected into space. In some methods it also used a grid on the picture plane through which the perspective projection was seen. The second point of contact is really a development of the first, and concerns techniques for projecting spheres and the exploration of conic sections during the 16th and early 17th centuries. This was the period when the developing science of projective geometry, cartographic techniques, and the more mathematical concepts of pictorial perspective seem to have intermingled in a complex way.[2] The basic texts have been relatively little explored, certainly from the point of view of the history of perspective. The third point which I have encountered comes in the later 18th century with Lambert and the birth of 'free perspective' as it was called – the perspective of curved surfaces, which eventually led to the creation of non-Euclidean geometry.[3] The fourth point is somewhat more surprising, and concerns colour theory. In the mid-18th century, Tobias Mayer was the first designer of a colour pyramid; that is to say a coherent colour solid for exploring relationships between hue, tone and saturation of colour.[4] He did this in connection with his desire to reform the colour printing of maps. Perhaps these issues have been clarified in the history of cartography, but within the history of art they remain predominantly hazy and, as yet, only projects for exploration.

All these are large subjects within the technical history of representation and I hope they may receive proper investigation in the future. My aim in this paper is both narrower and broader than this kind of programme. It is narrower because I am focussing upon one man, Leonardo da Vinci. It is broader in that I will emphasize the intellectual context in which his maps occurred, and this will take us far outside the technical history of mapmaking or even the technical history of perspectival representation.

It is not surprising that Leonardo made maps. From the earliest to the last phase of his career, he exhibited an absorbing interest in the structure and functions of the earth. His earliest dated drawing depicts a precipitously hilly landscape. It is inscribed 'day of Holy Mary of the Snows, 5th August, 1473', and is the earliest known landscape in the history of Western art to carry a date. The topography is generally recognized as that of the Valdarno but its precise location has yet to be identified with certainty. It was drawn when Leonardo was 21. Amongst the last major productions of his career as a draughtsman are a series of extraordinary drawings in which comparable landscapes are overwhelmed by cataclysmic vortices of raging air and water. These so-called 'Deluge Drawings' illustrate in an extreme form that the surface of the Earth is in continual flux, responding to the inexorable forces which brought both life and death in their train.[5] It is within this context of a consuming interest in the living Earth – its functions and all its associated phenomena – that we have to set Leonardo's own map-making. I would argue that an understanding of his cartographic activities depends upon understanding the intellectual, even philosophical context from which these maps originated.

In more general terms there may be a moral here for those concerned with the history of any technical subject, namely that its history is grossly distorted if we do not take into account the conscious aspirations and the implicit assumptions which motivated its practitioners. This applies

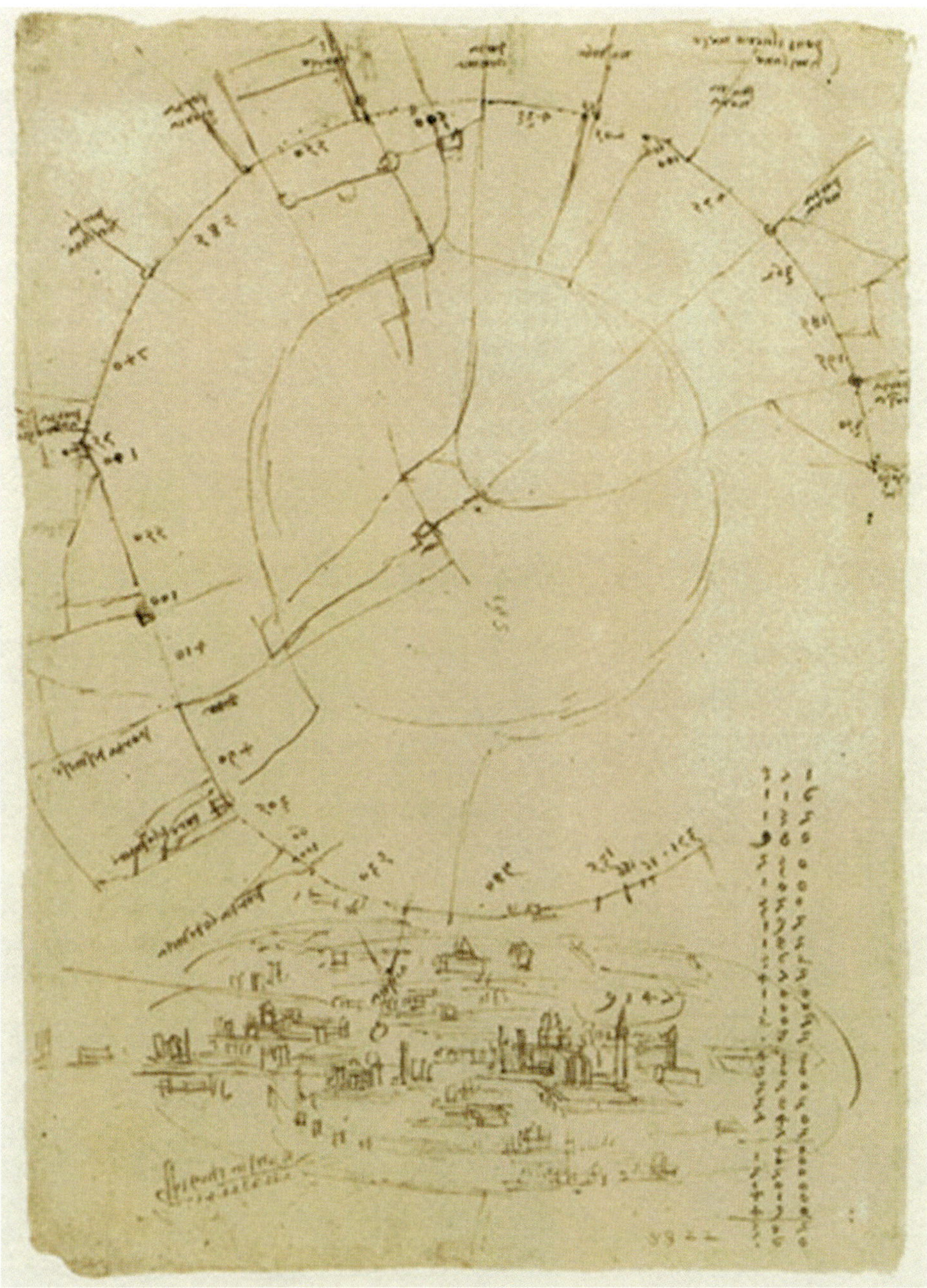

Figure 1
Leonardo da Vinci, Sketch-Plan and View of Milan, Milan, Biblioteca Ambrosiana, Codice Atlantico, 73r-b (image courtesy of the University of Virginia)

no matter how metaphysical or remote from our own vision of science these motivations and assumptions may be. There has been a tendency in the history of technical subjects to ignore the metaphysical base in favour of the descriptive outlining of technical advances. However, all this is jumping the gun. Let us begin by establishing what Leonardo actually accomplished in the field of cartography.

I will look first at what Leonardo produced, or at least a typical selection of his maps and map-like representations, asking why and how they were made, in as far as I am competent to provide answers. I think three main practical concerns can be discerned behind his functional maps. (I will distinguish here, although it is not a sharp distinction, between functional maps made for a specific purpose and maps made in the more general context of his exploration of nature.) The functional maps concern firstly town planning, secondly military engineering and military planning generally, and thirdly hydraulic engineering.

Town Planning

His activities as a map-maker in town planning are the least significant cartographically and can be dealt with reasonably briefly. The illustrated example (Figure 1) is probably the earliest of his surviving plans of a town. It is not a very substantial drawing, but it does at least introduce the two main varieties of technique he was to use in portraying the surface of the earth: the flat map-making technique in the upper drawing, and the 'mountaintop' view in the rough sketch below. The subject of the drawing is a scheme for radial extensions to the City of Milan. The city, as so often in early town maps, approximates to a circle, which it did not in reality. The idea was for a ring of extra settlements which would grow out from the main city gates of Milan. Medieval and Renaissance cities generally

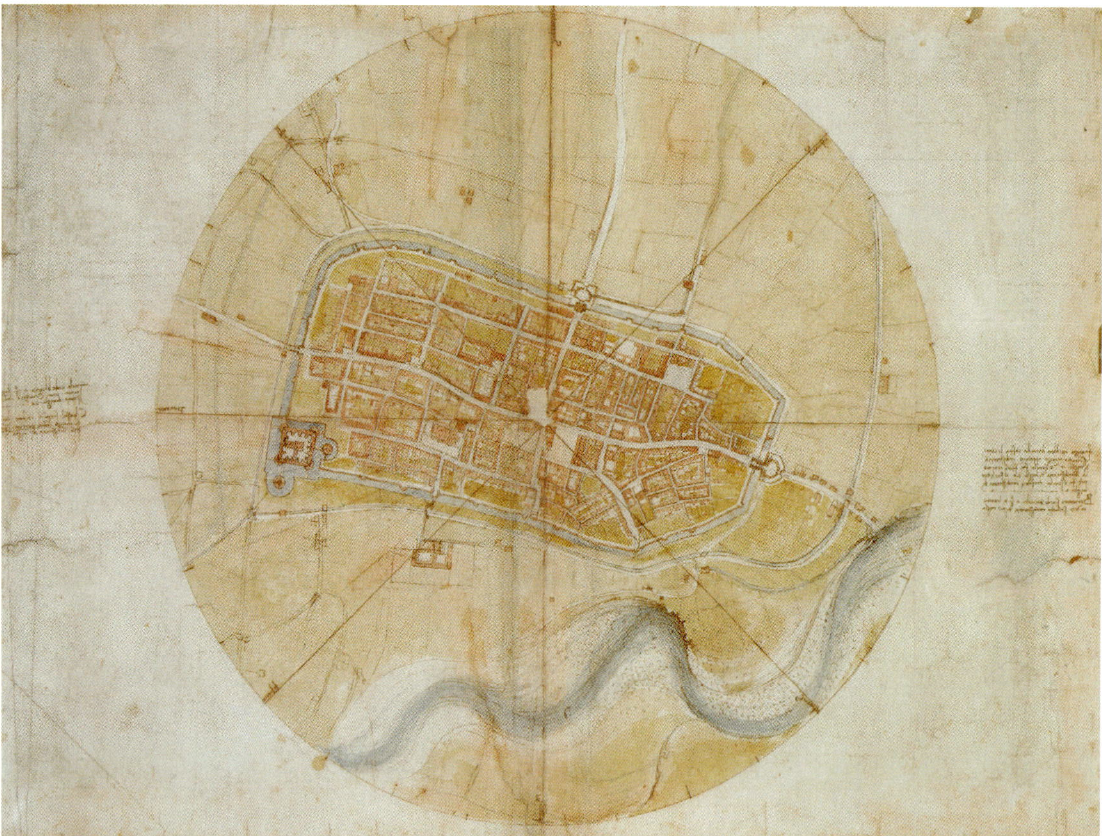

Figure 2
Leonardo da Vinci, Map of Imola, Windsor, Royal Library, No.13284

developed in this centrifugal manner, the town walls gradually increasing in circumference to enclose the growing population. The flat sketch or map tells us little apart from the fact that he had made or taken over some measurements of distance. There are indications around the circumference of figures which represent paced-out distances. We will see much more precise evidence of such pacing-out techniques in his most substantial later map.

The 'mountain-top' perspective – I call it this rather than a 'bird's-eye' view to distinguish it in terms of its angle of view – is not related at this stage to a great knowledge of earlier map-making techniques, but to relatively standard ways of portraying towns in artistic representations. There is an early image of Rome from 1414 by Taddeo di Bartolo, which uses a comparable approximation to a circular plan, in that the walls literally encircle the town.[6] Leonardo's 'mountain-top' view also depends upon the kind of printed images of cities which were beginning to be produced in Italy during his own lifetime. One example is a woodcut view of Florence produced in 1486, known as the 'Catena Map' from the chain decoration around its border.[7]

Leonardo's town plans never reached the point of realization and we do not know if he actually presented to his patrons any of his schemes as measured maps, designed according to systematic surveys. This sketch is the kind of slight evidence we have to rely on for his use of map-making in the context of town planning.

Military Engineering and the Imola Map

When he came to work as a military architect, Leonardo was certainly required to produce results, above all in 1502 and 1503 when he was working for Cesare Borgia. He was Cesare Borgia's 'general architect and engineer', serving Cesare in the huge sweep of territory in central Italy which Cesare was subjugating on behalf of the Pope -and himself. We know that on the 18th August 1502, Leonardo was granted free access to all Borgia strongholds, and the relevant document goes on to say that he could requisition the services of local men for what was called 'measurement and estimation'.[8] He could, in other words, command a gang of people to undertake such measurements as he wished to have made.

Of his work with Cesare and his map-making activities on the Duke's behalf we have one major surviving product, the map of Imola (Figure 2). In cartographic terms this is his main claim to fame. We know that Cesare Borgia was bottled up in Imola by some of his enemies in October 1502. I suspect that the map dates from this time. It is not difficult to reconstruct the technique by which it was produced. The most obvious element is the superimposed wind-rose, which consists of 64 finely-drawn radial lines, of which eight are emphasized and named after the winds. This immediately gives a clue as to the techniques and the instruments used. Leonardo's source for this technique probably lies not so much within the specialist history of cartography as with a theorist whom he studied in other areas of his work, namely Leon Battista Alberti. Leonardo almost certainly knew Alberti's *Descriptio urbis Romae*, the description of Rome which was completed in the 1440s. He also knew a vernacular book by Alberti, *Ludi Matematici,* which was written in the 1450s and dealt with mathematical 'recreations' of various kinds. It included substantial sections on measuring, estimating and

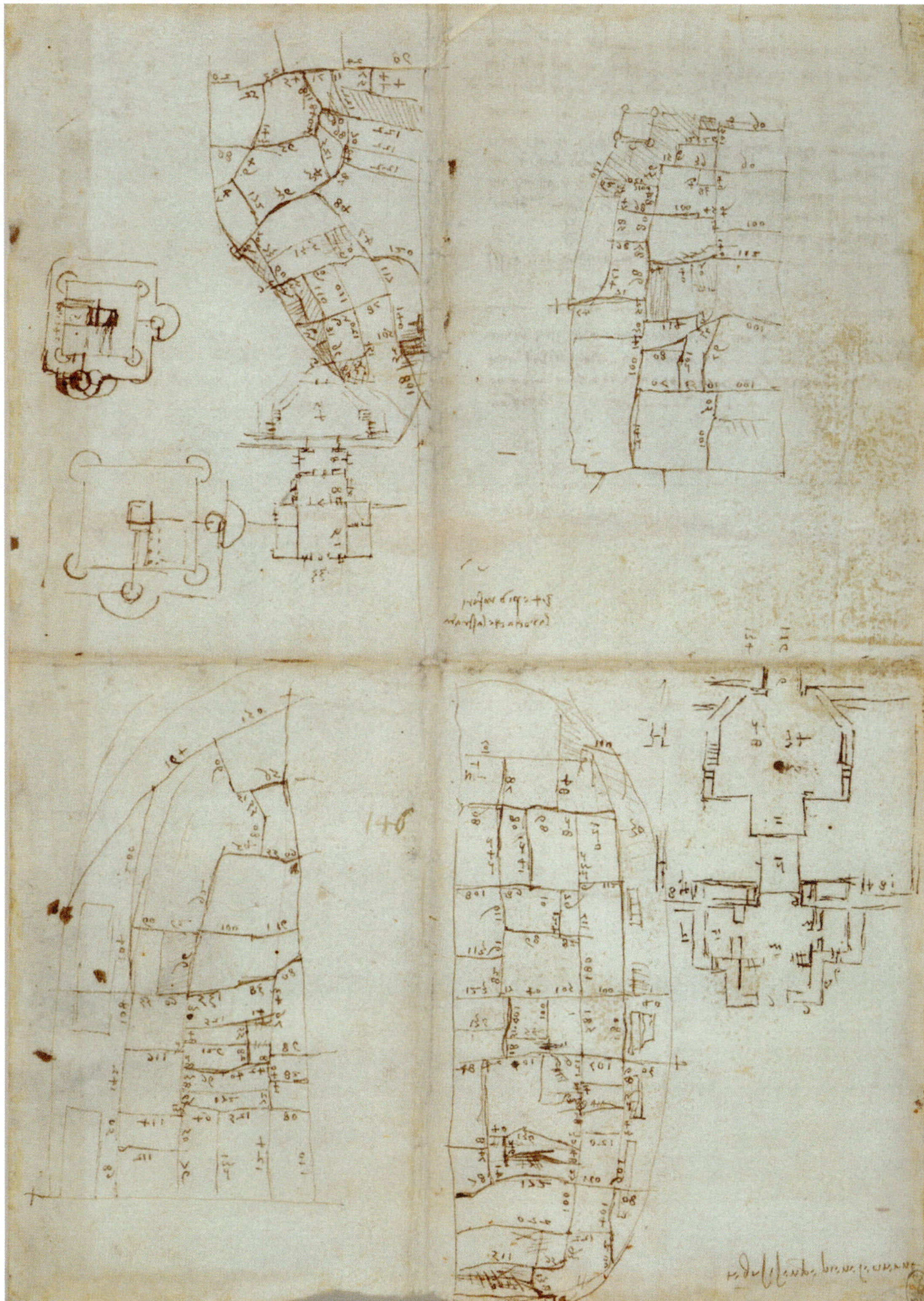

Figure 3
Leonardo da Vinci, Preliminary Drawings for the Map of Imola, Windsor, Royal Library, No.12686r.

surveying.[9] Alberti insisted upon the use of mathematical instruments to achieve properly proportioned maps. To do this he used an astrolabe (perhaps a mariner's astrolabe, rather than a complex astrolabe of the heavens) which he placed in a horizontal position. A comparable set-up is shown in later illustrations of map-making techniques.[10]

Alberti does not claim to have originated this technique. He noted that gunners had used an astrolabe or similar instrument to take vertical and horizontal bearings in the quest for accuracy in their particular science. He also made it clear that he was using techniques which were well-known in maritime map-making. The mariners' Portolan maps of the period were characteristically designed with a series of interlocking wind-roses. From a central vantage point, Alberti took bearings of the most prominent landmarks. In the *Descriptio* he also explained that he supplemented this measurement of angular bearings with the long-winded method of 'pacing-out distances'. The later treatise, the *Ludi Matematici* records the taking of sightings from more than one vantage point:

'Go to a place which has been seen from the first one, and direct your instrument flat and in such a position that it lies in the line of that same number through which you saw it first on your instrument, that is, place it so that a ship which had to navigate from the first to the second place could go along the same windline.'[11]

What he was describing, rather awkwardly, was the basis for a process of triangulation. The internal evidence of Leonardo's Imola Map does not indicate unequivocally whether he used the first or second method. He certainly did utilize paced-out distances, as a series of preliminary sketches show (Figure 3), but these could well have been combined with sightings from more than one location. A comparable series of preliminary drawings for lost or unexecuted maps of Urbino and Cesena contain sets of measurements and compass bearings which suggest that he used a method of multiple sightings comparable to that described before 1520 in the famous letter to Leo X, which is associated with Raphael's scheme for surveying ancient Rome.[12] Raphael's technique (and by implication Leonardo's) used an instrument similar to that later illustrated in Vincenzo Scamozzi's *Idea dell'Architettura* (Figure 4).

Leonardo's ultimate aim was to produce maps which

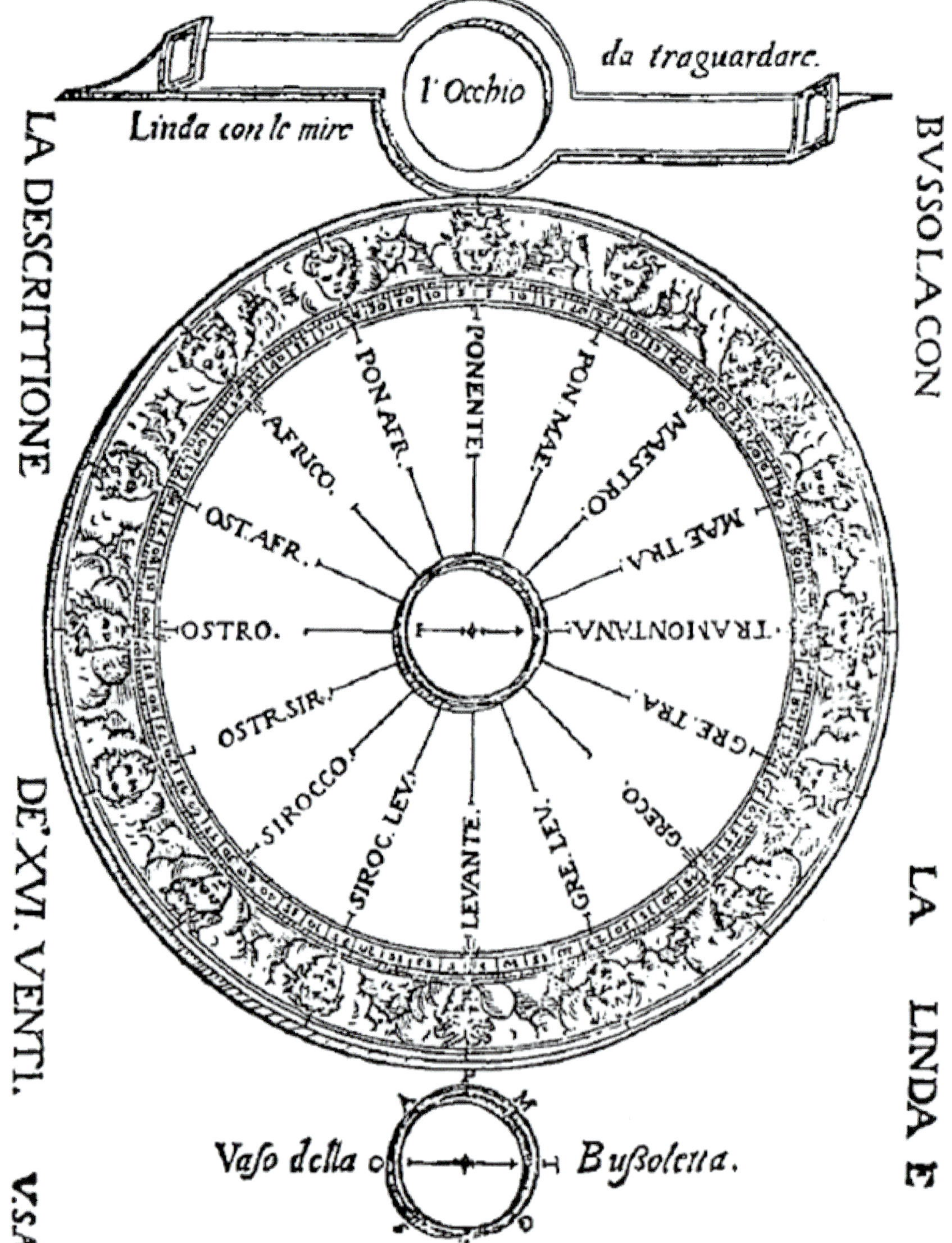

Figure 4
Vincenzo Scamozzi, Components of a Surveying Instrument (Sight-Vane, Graduated Ring and Magnetic Compass, from Idea dell'Architettura (Venice, 1615, I, ii, 14, p.143).

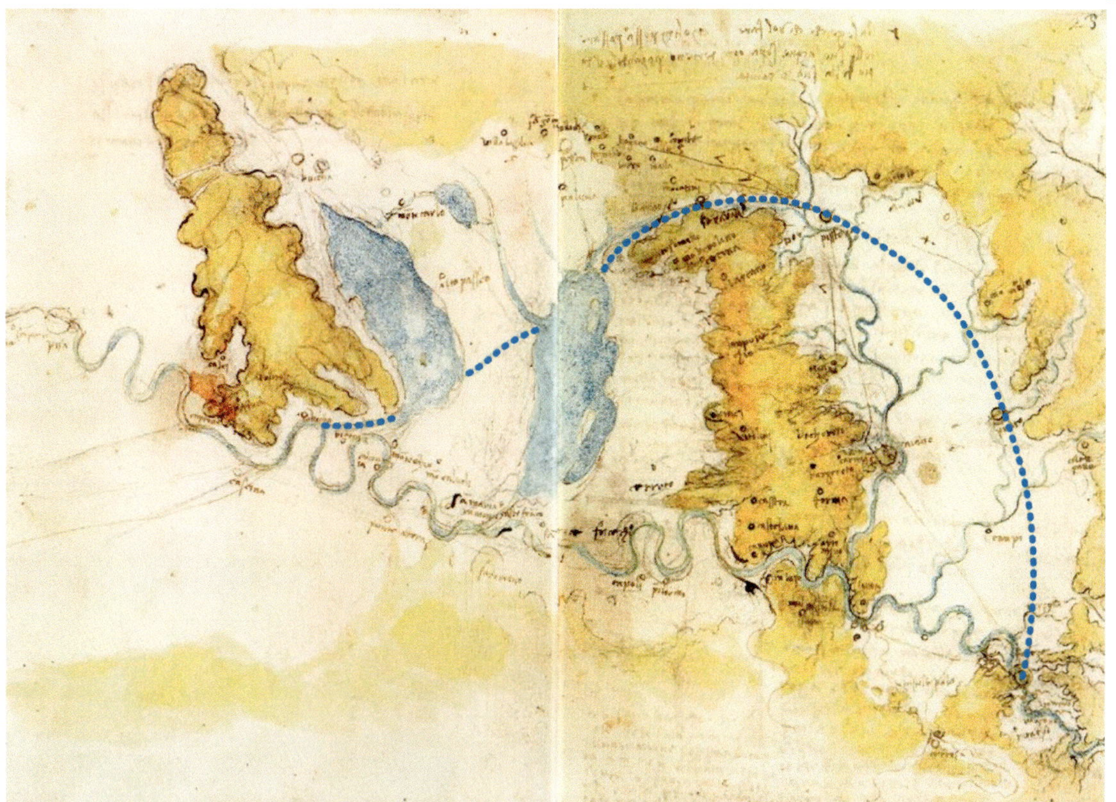

Figure 5
Leonardo da Vinci, Map of the Arno Valley between Florence and Pisa (Leonardo's indication of the proposed route of the canal has been emphasized by the dotted line). Madrid, Biblioteca Nacional, Codex Madrid 11, 22v-22r.

were properly proportioned throughout. Although he apparently did not follow some of his predecessors and contemporaries in using a Ptolemaic grid, he did place a linear scale in the margins of some of his maps, so that any required distance could be accurately calculated.[13]

The Imola Map and, we may suspect, lost maps of Urbino and Cesena are devoted to the kind of precision which Cesare needed in planning the defences of the cities in his possession, in working out escape routes and in designing fire lines. It also records the bearings and distances of other cities, noting that Imola 'sees' Bologna and other cities on certain bearings and at measured distances.

However, it is not simply a drily objective record; it stirs with a perceptible sense of life. There is a sense of that vitality and movement which Leonardo always perceived as an integral facet of nature. This is visible above all in the River Santerno as it moves in a series of percussive parabolas through the landscape, biting into the higher slopes on the Imola side and sweeping out on to the lower ground on the side away from the town. This relates very much to a theme I am going to take up later, the theme of the body of the Earth as a living organism. It also relates to his feeling for the motion of water in the Earth, and to the hydraulic engineering projects which provide the third of our categories of functional maps.

Hydraulic Engineering

The best documented of Leonardo's projects of hydraulic engineering concerns a great sweeping canal which was to by-pass the unnavigable section of the Arno just below Florence.

The great Arno Canal was a project with which he seems to have been involved between 1503 and 1505. This was not, I should emphasize, Machiavelli's military project for diverting the Arno round Pisa. Leonardo's involvement in Machiavelli's scheme seems to have been very marginal, and he was not responsible for the considerable fiasco which resulted.[14] The scheme I am discussing was designed for economic benefits in peacetime, in that it would link Florence to the sea by a navigable waterway. The idea was for a great sweeping canal in a virtually complete semi-circle (Figure 5) striking north from Florence and passing through Prato and Pistoia. These cities would benefit substantially from the collection of tolls. If this should seem hopelessly Utopian we should remember that he had spent many years in Lombardy where there were already major networks of canals serving just this kind of function. Leonardo's projected canal joins the Arno some way inland of Pisa and is in this respect quite distinct from Machiavelli's scheme. In his sketches the meanderings of the river arc somewhat schematized – to indicate simply that there are meanderings rather than to give an accurate and clear representation of the actual course of the river. He was primarily concerned with the accurate description of the course of the new canal. He devised a convention of colours to indicate the different ground levels in the area where his projected canal would travel.

There is clear evidence in his mapping of rivers that he did attempt to make accurate measurements from place to place. The most substantial records appear on his plans of rivers in Brescia, on which he has noted direct measurements between the various settlements on the banks. A less elaborate example is provided by a map of the Arno between Pisa and Empoli (Figure 6).

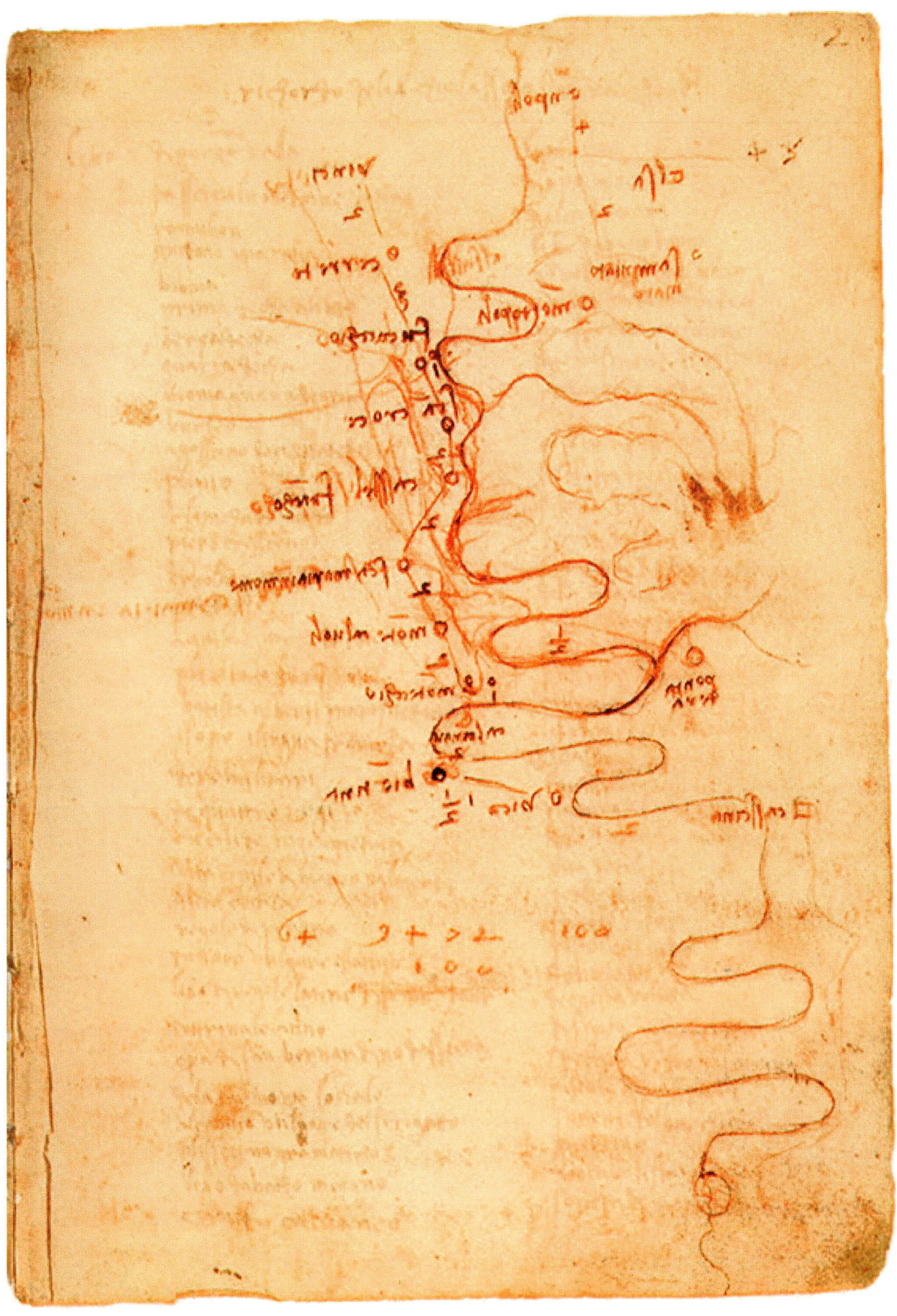

Figure 6
Leonardo da Vinci, Sketch-Map of the Arno between Pisa (bottom) and Empoli (top), Madrid, Biblioteca Nacional, Codex Madrid II, 2r.

Leonardo's project did not, particularly in the troubled circumstances of the time, stand a great deal of chance of being realized. However, his ideas about this project were not limited to the immediate implications of this particular canal. In a characteristic manner, his thoughts spiralled onwards, away from the immediate task in hand. Since the 1480s, he had expressed a keen awareness of the remote past of the Earth, the great transitions it had undergone, the huge shaping processes which had been at work and had resulted in the present configuration. He was also aware of the continuing nature of these processes.

His study of the fossil record had led him to dismiss the Biblical Deluge as the cause of fossils on high mountains. Looking intensively at the Arno Valley, as he was bound to do in making this scheme, studying its levels and its water courses, he came to realize that its present appearance was but a transitional stage in the Earth's development. In a series of notes and drawings concerned

Figure 7
Leonardo da Vinci, Map of Umbria and Tuscany, showing Arezzo, (left), Perugia (top right) and Siena (bottom centre), Windsor, Royal Library, No.12278.

with the Arno Valley and adjacent features, he speculated as to how the landscape might have arrived at its present state.[15] The drawings are not in this instance flat maps but bird's-eye views of relief features (Figure 7). This particular technique may perhaps be related to some of the printed German maps of the late 15th century which combine three-dimensional description and flat mapping in a hybrid manner.[16]

While perusing his studies of the landscape around the Arno, he became convinced that it had originally been the site of two huge lakes at different levels. He wrote that the Arno had once been dammed by huge rock barrier which 'formed two huge lakes, the first of which is where we now see the City of Florence flourish, together with Prato and Pistoria'.[17] That is precisely the line of his canal – the relatively low ground he intended to use as his route in a northerly arc from Florence. He also explained that 'in the upper part of the Valdarno, as far as Arezzo, a second lake was formed, and thus emptied its waters into the aforementioned lake'.[18] The illustrated map (Figure 7) shows a considerable area of water south of Arezzo, which may be intended to indicate the flood plain of the river Chiani and certainly suggests the ancient presence of a further lake in this extensive valley. The immense processes of change which arise from the inexorable cycles of fluid motion in the body of the Earth, have gradually resulted in the present state of affairs in the valley. He made it clear that such processes will continue. He anticipated, for example, that the Mediterranean Sea will eventually be transformed into no more than a major river.[19]

The Body of the Earth

Leonardo's paintings contain many images which evoke processes of geological change. The Mona Lisa, for example, clearly depicts two lakes at different levels. These are not literal descriptions of the ancient lakes around Florence, but the landscape as a whole is very much concerned with a sense of geological time. The landscape in the Mona Lisa is an image of the Earth in continual flux and transition, an image which dominates Leonardo's later conception of the world. Until the very last phase of his thought, and certainly when he painted the Mona Lisa, Leonardo was an avid subscriber to the ancient view of the microcosm and macrocosm. The human body was seen as reflecting in miniature the processes of the world as a whole. He wrote: 'This Earth has a spirit of growth and its flesh is the soil. Its bones are the successive strata of the rocks which form the mountains. Its cartilage is the tufa stone. Its blood is the springs of the waters. The lake of the blood which lies within the heart is its ocean. Its breathing is by the increase of blood in its pulses, and even so in the Earth the ebb and flow of the sea'.[20] This image of the Earth's vascular system does not only concern the superficial motion of water on the Earth, but also the penetration of water throughout the porous structures which he believed to exist within the Earth.

A transcription I have made from a relatively faint diagram in the Codex Leicester (or Hammer) illustrates this porosity (Figure 8). In common with a number of ancient and medieval authors, Leonardo was especially concerned to explain how springs of water could emanate from high places in mountains. Initially he drew an analogy with the sap oozing from the cut top of a vine, or blood pouring from a severed vessel in the head. By the time he wrote the Codex Leicester he no longer considered that this was an

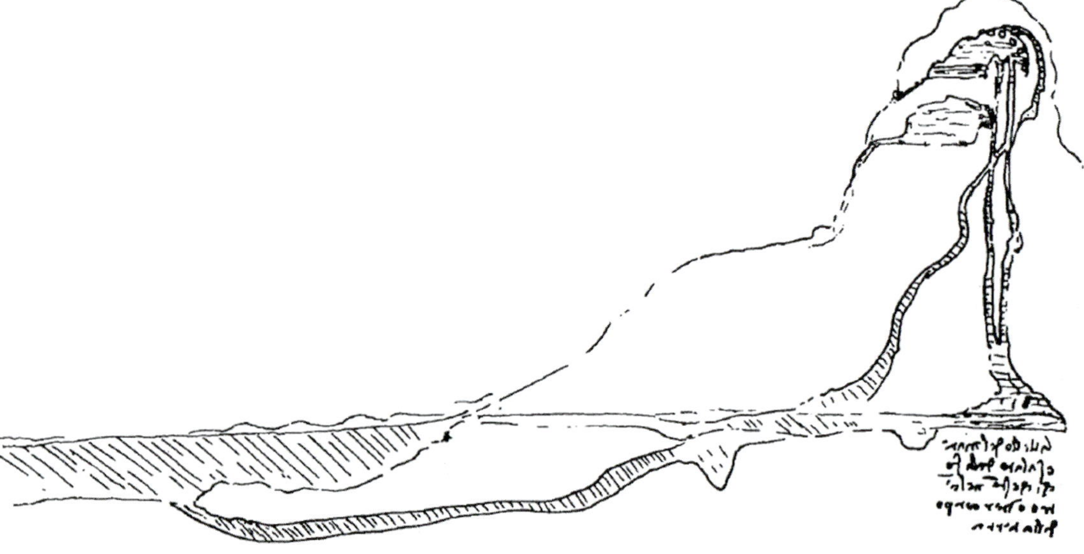

Figure 8 *Leonardo da Vinci, Veins of Water in the Body of the Earth, transcribed from Codex Leicester 3v, Armand Hammer Collection.*

entirely satisfactory way to explain natural phenomena.[21] He therefore turned his attention to a more rigorously physical explanation. He speculated that it may be the result of distillation – the high chambers in the mountains acting as retorts in which the water condenses. However, he also eventually abandoned this explanation, and at the end of his life was in a state of what we may call educated but doubted as to why this phenomenon occurred. His consideration of these questions has certainly given fresh impetus by his activities in designing the Arno Canal. At the same time he was also taking up again the theme of the vascular system in the human body – what we may legitimately describe as his mapping of the internal rivers of the human body. In one incredibly complex, composite diagram, he heroically attempted to show the irrigation systems for air, blood and other fluids within the outlines of a female figure (Figure 9).

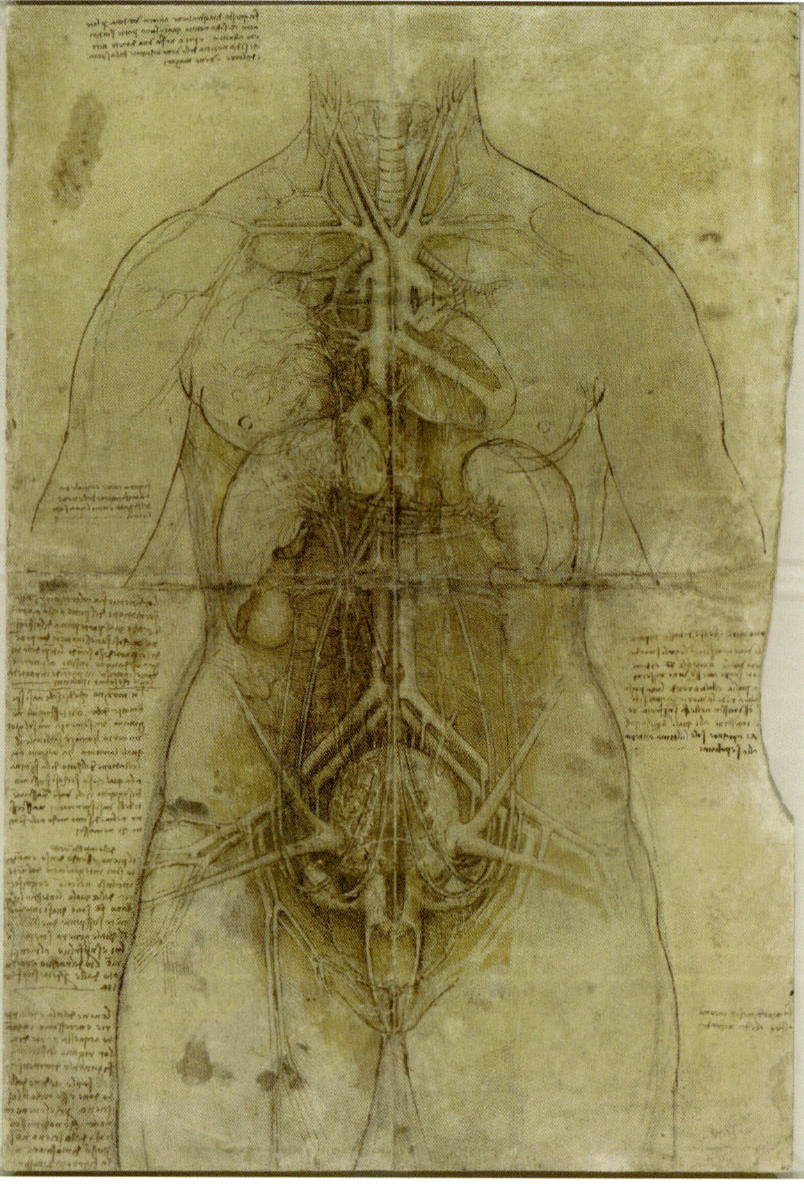

Figure 9 *Leonardo da Vinci, Vascular, Respiratory and Urino-Genital Systems in the Female Body, Windsor, Royal Library, No.12281.*

The popular concept of the microcosm, had been directly expressed in at least one medieval map to my knowledge: the Opicinus de Canistris Map, drawn in the 1330s.[22] The outlines of the countries around the Mediterranean form the shapes of interwoven figures, somewhat like a child's puzzle drawing which contains hidden forms. From the point of view of technical map-making this may seem little more that a diversion or novelty on the designer's part. However, this apparently flippant invention expressed in an idiosyncratic way a deeply-held conviction about the unity between the human body and the nature of the earth. For medieval philosophers this concept expressed the essential unity of God's creation.

We should not dismiss such ideas simply as curious diversions in the history of technical subjects, but acknowledge that they were capable of providing profound motivations for many of the early investigators. Much of the early research into the physical appearance of the world was motivated by metaphysical notions of the analogy between the Earth and the human body, and the essential harmony of all created forms.

In Leonardo's case the analogy was founded on the underlying causes of the processes of the Earth and the processes of the body. He sought, for example, the universal laws which governed the movement of fluids through channels or pipes, formulating for the first time the fundamental law concerning the volume of liquid passed at a constant speed as proportional to cross-sectional area. An extraordinarily beautiful drawing of the lungs shows a rigorous bifurcation in which the sum of the cross-sectional areas at each stage is equal to the cross-sectional area of the main tube.[23] This is a spurious regularity in anatomical terms but one which he saw as necessary given the universality of natural law. The microcosm analogy led him to draw fundamental parallels between the motion of water, movement of air, the growth of plants, and generally with every form of circular motion, including the vortices which ceaselessly erode the banks of rivers.

Once the human engineer had gained an understanding of such matters, he could work in harmony with the natural world, rather than forcing it to act against its inherent nature. In excavating a canal, or in diverting a river, the engineer must understand how the water would behave and he should learn to use the currents as aids in his excavations. He could not, like the Florentines when they attempted to divert the Arno around Pisa, push water around against its natural tendencies without disastrous results.

It is this sense of the surface of the world reflecting the laws of its inner life which emerged in every stroke of Leonardo's pen when he was drawing a map. His maps are never inert. They are never static. They always express the living forces of nature. This vision is not altogether incompatible with some of the recently published satellite images of the Earth's surface, which to my eye possess something of the same kind of life and vitality. I am not saying that Leonardo is 'anticipating' modern discoveries. Indeed much of my own writing on Leonardo has been devoted to showing that he can best be understood if we do not try to force him into the mould of a modern scientist.

What I am saying is that his maps reflect a timeless response to nature as a living whole, as perceived both intuitively and analytically – or, in other words, poetically and scientifically. In this sense, art and science can be seen as cognate activities, not opposed tendencies in the human mind.

Notes

1. Most recently, K.H. Veltman, “Ptolemy and the Origins of Linear Perspective”, *La Prospettiva Rinasci mentale* ed. M. Dalai Emiliani, Florence, 1980, pp.403–7.
2. See R. Sinisgalli, “Gli Studi di Federico Commandino sui Planisfero Tolemaico come elemento di rottura nella tradizione della teorica prospcttica della Rinascenza” *La Prospettiva Rinascimentale*, pp.475–485.
3. J.H. Lambert, *Die Freie Perspektive*, Zurich, 1759.
4. T. Mayer, “Die Mayerschen Farbendreiecke”, *Gottingische Anzeigen*, 11, 1758.
5. A conspectus of Leonardo's landscape drawings is contained in A.E. Popham, *The Drawings of Leonardo da Vinci*, London, 1946. See No.253 for the Val d'Arno drawing and Nos.290–6 for the 'Deluge Drawings'.
6. Illustrated in J. Gadol, Leon Battista Alberti. *Universal Man of the Early Renaissance*, Chicago, 1969, p.166. A comparable map of Milan is shown in The Renaissance, ed. J.H. Plumb, New York, 1961, p.170.
7. Illustrated in The Renaissance, ed. J.H. Plumb, p.131, and P.D.A. Harvey, *The History of Topographical Maps*, London, 1980, pp.68–9.
8. L. Beltrami, *Documenti e Memoire riguardanti la Vita e le Opere di Leonardo da Vinci*, Milan, 1919.
9. Gadol, *op.cit.*, p.167 ff.
10. For example, Cosimo Bartoli, *Del modo di misurare* (1564) in Gadol, op.cit., p.173.
11. Gadol, *op.cit.*, p.175.
12. See N. De Toni, “Leonardo da Vinci e i relievi topografici di Cesena”, *Studi Romagnoli*, 8, 1957, and C. Pedretti, *A Chronology of Leonardo da Vinci's Architectural Studies after 1500*, Geneva 1962, pp.32–3, and Appendix I.
13. For the scales on Leonardo's maps, see Windsor, Royal Library Nos.12675, 12676, 12679 and 12685r. Ptolemaic systems in the 15th century are discussed by L. Bagrow, *Die Geschichte der Kartographie*, Berlin, 1951, p.64 ff, and revised ed., by R.A. Skelton, London, 1964, p.77 ff.
14. See C. Pedretti, *Commentary to The Literary Works of Leonardo da Vinci* (ed. J.P. Richter), 2 vols., 1977, 11, p.174 ff.
15. *The Notebooks of Leonardo da Vinci*, ed. E. MacCurdy, 2 vols., 1951, I, pp.316–7. The major note is in the Codex Leicester, Armand Hammer Collection, folio. 9 r.
16. For example, the Nuremberg Chronicle and von Breydenbach maps illustrated in Bagrow, *op.cit.*, pp.77 and pp.80–1, and Skelton *op.cit.*, p.90. 17. Notebooks *loc.cit.*

17. Notebooks *loc.cit.*
18. Notebooks *loc.cit.*
19. Notebooks, *op.cit.*, p.325.
20. *The Literary Works of Leonardo da Vinci*, J.P. Richter, 2 vols., London and New York, 1970, 11, p.178. See also M. Kemp, *Leonardo da Vinci: The Marvellous Works of Nature and Man*, p.107 ff. and p.258 ff.
21. M. Kemp, "The Crisis of Received Wisdom in Leonardo's Late Thought", *Leonardo e l'Eta della Ragione*, ed. E. Bellone and P. Rossi, Milan, 1982, pp.27–42.
22. Bagrow-Skelton, *op.cit.*, p.220.
23. M. Kemp, "Dissection and Divinity in Leonardo's Late Anatomies", *Journal of the Warburg and Courtauld Institutes*, 35, 1972, pp.205–6.

Editorial Note

The original images in this paper have been replaced with full-colour versions for this publication.

Bibliography

The references and bibliography relating to the vast literature on Leonardo have been deliberately limited. The books listed below provide an introduction to the primary material and sources of further bibliographical guidance.

MacCurdy, E. (Ed.) (1938) *The Notebooks of Leonardo da Vinci* Jonathan Cape: London.

Richter, J.P. (Ed.) (1970) *The Literary Works of Leonardo da Vinci* (3rd ed.) London: Phaidon (with Commentary by C. Pedretti, 1977, Oxford: Phaidon).

Clark, K. and Pedretti, C. (1968) *The Drawings of Leonardo da Vinci in the Collection of Her Majesty the Queen at Windsor Castle* (2nd ed.) London: Phaidon.

Kemp, M. (1981) *Leonardo da Vinci: The Marvellous Works of Nature and Man* London: Dent.

Kemp, M. (1982) "The Crisis of Received Wisdom in Leonardo's Late Thought" In Bellone, E. and Rossi, P. (Eds) *Leonardo e I'Eta della Ragione* (pp.27–42) Milan: Scientia.

Pedretti, C. (1962) *A Chronology of Leonardo da Vinci's Architectural Studies after 1500* Geneva: Droz. (See especially Appendix I, on surveying methods.)

Popham, A. (1946) *The Drawings of Leonardo da Vinci* London: Jonathan Cape.

Acknowledgement. The drawings from Windsor are reproduced by gracious permission of Her Majesty the Queen.

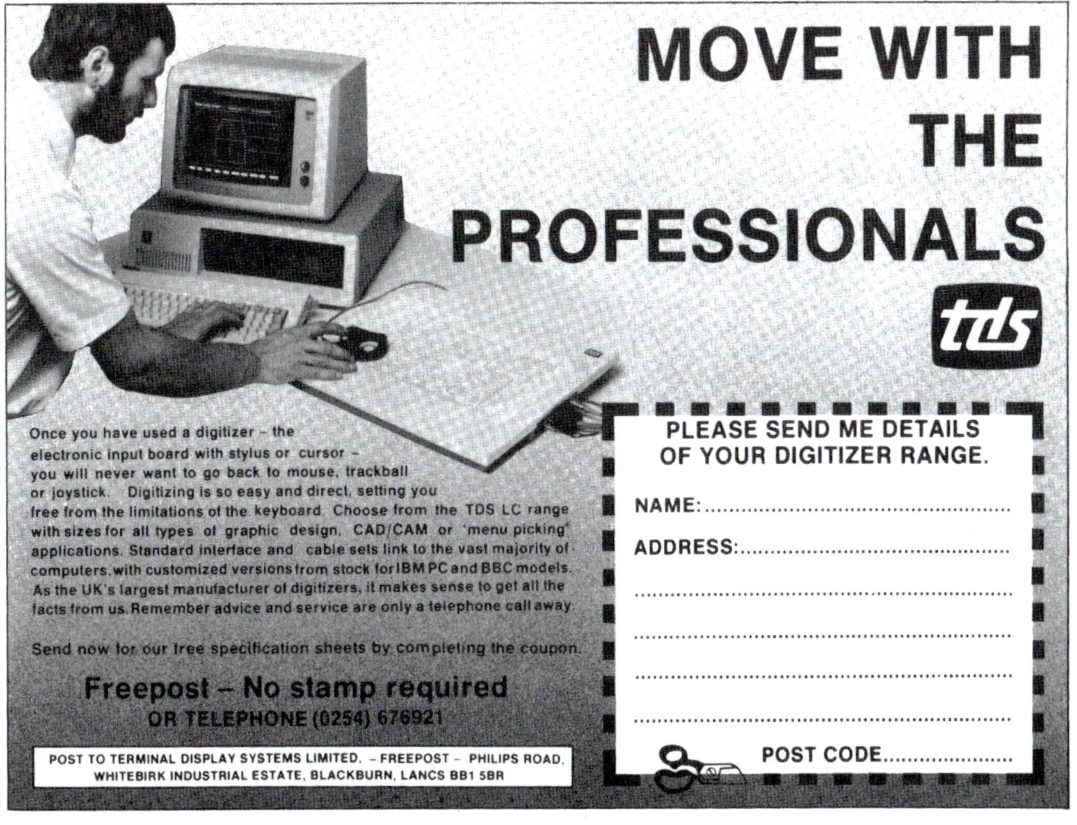
MOVE WITH THE PROFESSIONALS
tds
Once you have used a digitizer - the electronic input board with stylus or cursor - you will never want to go back to mouse, trackball or joystick. Digitizing is so easy and direct, setting you free from the limitations of the keyboard. Choose from the TDS LC range with sizes for all types of graphic design, CAD/CAM or 'menu picking' applications. Standard interface and cable sets link to the vast majority of computers, with customized versions from stock for IBM PC and BBC models. As the UK's largest manufacturer of digitizers, it makes sense to get all the facts from us. Remember advice and service are only a telephone call away.
Send now for our free specification sheets by completing the coupon.
Freepost - No stamp required
OR TELEPHONE (0254) 676921
POST TO TERMINAL DISPLAY SYSTEMS LIMITED. - FREEPOST - PHILIPS ROAD, WHITEBIRK INDUSTRIAL ESTATE, BLACKBURN, LANCS BB1 5BR
PLEASE SEND ME DETAILS OF YOUR DIGITIZER RANGE.
NAME:
ADDRESS:
POST CODE

SIMPLICITY IS THE KEYNOTE — A GUEST EDITORIAL

John C. Robertson

Originally published in Volume 17, No 2, pp.61–63

Back in the halycon days of his youth, it was the proud boast of the author that he had danced in Hollywood. (But then so had Mickey Mouse and Donald Duck!) A particular incident occurred during a ballroom competition. A judge, officiating at the event, heard calls from the side of the floor of 'KISS John. KISS'. Eventually his curiosity impelled him to inquire as to the exact nature of this exhortation and was informed that it meant 'Keep It Simple Stupid!'

This exhortation could equally well be aimed at many cartographers today. Much of the philosophy appears to be 'This is a complicated situation – and here is a complicated picture to prove it!' This is the antithesis of what any form of communication aims to do. Yet many learned articles try to stress that cartography is a form of communication. Surely the skill of the communicator is to take the complicated situation and show simply the essential spatial elements? This is certainly not done and indeed the very articles and diagrams on cartographic communication, which try to formulate the process, have themselves become tremendously complex and largely unintelligible. If that is the state of play of the theory, what chance does the actual product have?

In judging a situation the layman [...] *seldom bothers to get at the facts. Such judgement always begins by looking at a map. People are bored and confused by maps.*[1]

This latter statement is bound to be challenged immediately by cartographers. They presume that because they spend much of their professional lives producing and using maps, and are therefore familiar with them (i.e. 'cartographically literate') others are the same. This cartographic myopia is the very reason why many maps are not successful. This attitude was also well illustrated at a recent meeting of the British Cartographic Society when the new London Bus Map was presented.[2]

This map contains some innovative approaches which are the result of fairly painstaking research, investigation and trial. Yet the general reaction from the audience was of the nature 'I can use the old map perfectly well, there is no need for the new one'. Even the presentation of user findings by London Transport and experiments by a TV company did not move the cartographic establishment. Therefore, not only are their opinions right in the first place, they cannot be altered by mere facts pertaining to potential map users! Cartography shows itself to be insular, inward-looking and resistant to change. There is even evidence of cartographic despotism. They know best, not the customer – an attitude that would send people like Marks and Spencer to the wall very quickly.

Cartography has been defined as 'the art, science and technology of making maps [...]'.[3] What seems to have happened is that the science and technology have undoubtedly developed, though not always in the right direction, whilst art seems to have been forgotten. The graphic art approach is so different from cartography. The latter still displays the overflow from the 'topographic mapping mentality' i.e. to squeeze as much as possible into the space available, subject to scale – a particular 'god', adherence to whom often mitigates against the effort towards clear communication. This all-encompassing approach is justified in the actual production of topographic maps because of their many potential applications.

On the other hand, the graphic artist would take the approach of 'as little as possible' commensurate with still permitting the user to make the appropriate identifications. This means no unnecessary lines, symbols, names, patterns, or even complexity of shape. Simplicity is the keynote! Anything extraneous that interferes with communication of the message is 'noise'. Getting rid of noise is often the method by which a map becomes successful i.e. the simplification of a complicated situation to its essential elements. Many of the communication diagrams shown in the cartographic journals are themselves full of noise, contradicting their own existence.

Another offshoot of this topographic mapping mentality can be reflected in cartographic competitions or at any august gathering round a printing press at the output end. Thread counters immediately appear and maps are subjected to intense, eagle-eyed scrutiny of lines, tones, registration, etc. How many map users carry thread counters or normally look at a map from that distance? Many of the older generation, a large proportion of map users, have less than perfect eyesight. Consequently many of the subtleties of 6 pt lettering and fine line variations are indiscernible to them. In any case, small changes in line thickness, or colours have to be visually transferred from the adjacent situation in the legend, where they may be just distinguishable to the graphic complexity of the actual map situation, where they often are not.

If cartographers were honest with themselves what they are actually looking at, in many cases, is excellence in reprographics not cartographics – and indeed they often are excellent. A look at the prize winners at the 1983 BCS

Conference at Exeter will make that assertion self-evident. But where is the innovation, the graphic 'pizazz', the sheer excitement of looking at a map? Some will argue that a map is purely a functional tool and does not need dressing-up. But graphic impact is the name of the game, a vital part of the function of a map, often during restricted exposure to decision makers such as planners, town councils, etc.

Look around at other areas of graphicacy in the 80s – advertising, TV, record and book covers, even such mundane things as bus liveries. Look at the exciting innovations, the dynamic graphics invoking a compulsion to observe. Where are the similar 'visually magnetic' maps? Some do exist, often produced by graphic artists, but fall down by ignoring a few basic tenets of cartography. Given some incentive and thought, cartographers can produce visually exciting maps.[4]

'Innovation', the scream will go up, 'what about computers? – magic!'. Whilst this is definitely a blossoming area in the technical sense there is still much to be desired graphically. In some cases this is because cartographers have fought shy of the situation and left it to the computer people. They in turn produce some real visual botch-ups which are accepted, and even lauded, despite basic tenets of cartography, such as logically arranged visual levels, or name hierarchies, being blissfully ignored.

Take a close look at the cornucopia of articles on computer cartography, assisted or otherwise, that appear in journals and papers. Most of them are involved with the nuts and bolts of computer draughting or data manipulation. What is produced by geographers in particular is often graphically hideous and even misleading. Where is the application of the exciting computer graphics such as are seen in films like Tron or War Games? There is a quiet development in the U.S. of the video disc, allowing fourth dimension maps but almost unnoticed so far. The video has replaced the book in many homes as the media for information and entertainment. The atlas is basically a book. Therefore where is the video atlas?

Where is the evidence of computer graphics being used as an editorial design tool? Despite the awe-inspiring presence of this booming technological innovation someone still has to make decisions on graphic content and presentation of a map. The computer may well largely replace the draughtsman. It will not replace the cartographer. Rather it will give a freedom of design and interaction even taking into account the desires and tastes of the actual or potential map user.

The computer, in many ways, is as important a development as the printing press was in the 15th century in terms of overall impact on the field of cartography. But just as the product (map) and the producer (cartographer) had to change considerably to accommodate that long ago development, so they have to change again to accommodate this new era. Yet much of the current effort seems devoted to using computers to reproduce existing types of maps, often poorly. Not the best adaptation of the technique. For example, type styles, a hangover from the Letraset period, are often glaringly unsuitable.[5]

Much is made of the 'advances' in education, but is it education or training that is required, particularly with the introduction of the computer-based subjects? This dichotomy was highlighted way back in 1968, and not really resolved since.[6] Cartography, as an industry, has not come to terms with the proportions of education (thinking) required of a cartographer vis-a-vis training (doing). This applies equally to the traditional methods as to the computer-assisted. Much of this has to do with the long-standing confusion as to where a draughtsman ends and a cartographer begins (rather akin to the definition of hills and mountains). The production of 'chiefs' has increased whilst 'Indians' disappear, or at least prove elusive to find. Another problem that isn't new.[7]

In common with many other scientific fields of endeavour, education levels in cartography have expanded upwards to degree and postgraduate status. Only time will tell whether or not the appropriate end-product is emerging in terms of employability. Textbooks are largely disappointing cohesions of rhetoric and jargon,[8] though recent publications by Cuff and Mattson[9] and the ICA[10] go some way towards resolving this situation. Meanwhile, basic training through day release is suffering from geographical locational problems.

Again, there is evidence of lack of cohesion in both fields of cartography and education. On the initiative of R.W. Anson a meeting was held, attended by many of those concerned, to offset this attitude and seek national policies in cartographic education.[11] But it is an uphill struggle as each educational institute battles for survival and is therefore reluctant to give up, or modify, any courses whilst students continue to enrol. And enrol they will when their immediate alternative is unemployment!

It is unfortunate that, at a time of economic decline, maps and mapping, both at a personal and a national level, are regarded as items on which money may be saved. This is very short sighted policy as: a) maps will be needed again and they cannot be produced, or updated, overnight; b) it could be argued, in resource terms, that in a depression the need for knowledge and awareness of spatial distributions, e.g. the very unemployment it causes, is even more vital than in the days of plenty.

What about the commercial producers, also struggling to survive in a competitive market, both at home and abroad? Should it be so competitive? Should there be half a dozen road maps for every region, several school atlases for Nigeria? Just as the shipbuilding industry had to amalgamate to survive is there not a case for a BRITCART combine? This could pool resources and approach the market in a more logical manner. That is not to say choice for the customer should be eliminated altogether. Is the fact that many cartographers spend a great deal of their time with other cartographers in their work situation conducive to the general introspective outlook? Should more consideration not be given to Alvin Toffler's views on 'task forces' where there is closer liaison between fellow scientists and professionals?[12]

In conclusion, there are many aspects of cartographic products and practitioners that need a long hard look, especially as to their viability and usefulness. International conferences never seem to produce the necessary revolutions, or even reasonable advances, except in minutiae (e.g. what did emerge from Moscow '76?[13] Even people who went there don't seem to know). Perhaps a basic approach to the various aspects of cartography can initially be regarded in the light of that exhortation, made to the author in the midst of his Terpsichorean endeavours, to 'Keep It Simple Stupid'. This is necessary for maps to fulfil the art aspect of their definition, to produce an 'elegant graphic shorthand':

Words following words in long succession, however ably selected [...] *can never convey so distinct an idea of the visible forms of the earth as the first glance of a good Map* [...] *In the extent and variety of its resources, in rapidity of utterance, in the copiousness and completeness of information it communicates, in precision, conciseness, perspicuity in the hold it has upon the memory, in vividness of imagery, in convenience of reference, in portability, in the happy combination of so many and such useful qualities, a Map has no rival.*[14]

The map must also be functional. It must reveal not conceal. Otherwise, as Snoopy from the Peanuts cartoon strip once claimed, 'Of course I can read maps – I just don't understand what all the lines and squiggles mean'.

Notes and References

1. Wouk, H. (1971) *The Winds of War* London: Fontana.
2. BCS meeting, London, 10th January 1984: A. Holmes 'The London Bus Map'.
3. International Cartographic Association.
4. For example, *Melbourne after dark* Melbourne, Australia: Cartographies International, 1980.
5. Robertson, J.C. and Bertacco, J. (1981) "Quality and user acceptance of computer generated graphics" *Paper presented to the 9th Annual URPIS Conference, Geelong, Victoria, Australia.*
6. Robertson, J.C. (1968) "Education and training for cartographers-dovetail or dichotomy?" *The Bulletin of the Society of University Cartographers* 3 (1) pp.6–12.
7. Robertson, J.C. (1977) "Where the Hell are the Indians?" *Proceedings, Seminar on the Education of the Technician Cartographer, Canberra, Australia, June 1977.*
8. For example, see the review of Arthur H. Robinson's "Early thematic mapping in the history of cartography" by Denis Wood (1983) in *Cartographica* 20 (3) pp.109–112.
9. Cuff, D.J. and Mattson, M.T. (1982) *Thematic maps: Their Design and Production* London: Methuen.
10. ICA Commission on Continuing Education in Cartography (1984) *Basic Manual on Cartography (Part 1)* (in press).
11. British National Council for Geography under the auspices of the Royal Society, organized by BCS, 'Cartographic education for the future' London, 18th October 1983.
12. Robertson, J.C. (1978) "Future shock and its applications to cartography" *Proceedings of the 3rd Australian National Cartographic Conference, Brisbane, Australia, October 1978.*
13. ICA, 8th International Conference, Moscow, 1976.
14. Greenough, G.B. (1840) Presidential Address to the Royal Geographical Society, 25th May 1840. *The Geographical Journal* 10 pp.xliii–lxxxiv.

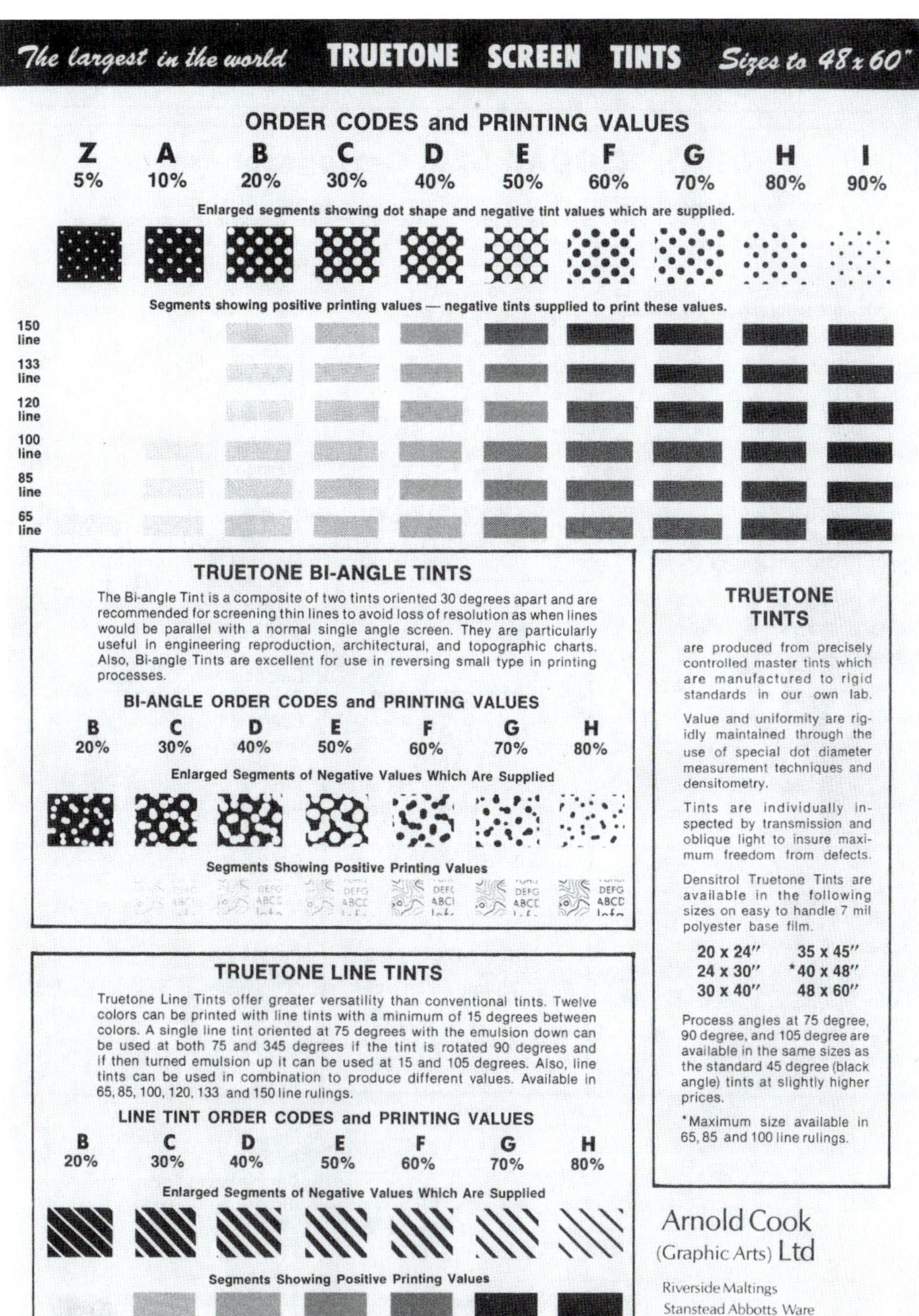

The largest in the world TRUETONE SCREEN TINTS Sizes to 48 x 60"
ORDER CODES and PRINTING VALUES
Z 5%
A 10%
B 20%
C 30%
D 40%
E 50%
F 60%
G 70%
H 80%
I 90%
Enlarged segments showing dot shape and negative tint values which are supplied.
Segments showing positive printing values — negative tints supplied to print these values.
150 line
133 line
120 line
100 line
85 line
65 line
TRUETONE BI-ANGLE TINTS
The Bi-angle Tint is a composite of two tints oriented 30 degrees apart and are recommended for screening thin lines to avoid loss of resolution as when lines would be parallel with a normal single angle screen. They are particularly useful in engineering reproduction, architectural, and topographic charts. Also, Bi-angle Tints are excellent for use in reversing small type in printing processes.
BI-ANGLE ORDER CODES and PRINTING VALUES
B 20%
C 30%
D 40%
E 50%
F 60%
G 70%
H 80%
Enlarged Segments of Negative Values Which Are Supplied
Segments Showing Positive Printing Values
TRUETONE LINE TINTS
Truetone Line Tints offer greater versatility than conventional tints. Twelve colors can be printed with line tints with a minimum of 15 degrees between colors. A single line tint oriented at 75 degrees with the emulsion down can be used at both 75 and 345 degrees if the tint is rotated 90 degrees and if then turned emulsion up it can be used at 15 and 105 degrees. Also, line tints can be used in combination to produce different values. Available in 65, 85, 100, 120, 133 and 150 line rulings.
LINE TINT ORDER CODES and PRINTING VALUES
B 20%
C 30%
D 40%
E 50%
F 60%
G 70%
H 80%
Enlarged Segments of Negative Values Which Are Supplied
Segments Showing Positive Printing Values
Available in 65, 85, 100, 120, 133 and 150 line rulings.
TRUETONE TINTS
are produced from precisely controlled master tints which are manufactured to rigid standards in our own lab.
Value and uniformity are rigidly maintained through the use of special dot diameter measurement techniques and densitometry.
Tints are individually inspected by transmission and oblique light to insure maximum freedom from defects.
Densitrol Truetone Tints are available in the following sizes on easy to handle 7 mil polyester base film.
20 x 24"
35 x 45"
24 x 30"
*40 x 48"
30 x 40"
48 x 60"
Process angles at 75 degree, 90 degree, and 105 degree are available in the same sizes as the standard 45 degree (black angle) tints at slightly higher prices.
*Maximum size available in 65, 85 and 100 line rulings.
Arnold Cook
(Graphic Arts) Ltd
Riverside Maltings
Stanstead Abbotts Ware
Hertfordshire SG12 4HG England
Ware (0920) 870991

THE CARTOGRAPHER'S ROLE: REFLECTIONS, REALITY AND REVERIE — A GUEST EDITORIAL

Roger Anson

Originally published in Volume 19, No 1, pp.1–5

More than two decades ago, as a recently qualified Geography graduate, I made my first hesitant (and not particularly successful) attempts at editing a map which was intended for eventual commercial publication. My appreciation of the complexities of presentation and production were, at that stage, virtually nil but I was lucky enough to have been employed by an 'old school' cartographer who soon started to supplement my formal geographical education with a rather more relevant 'on the job' training. This included all aspects of compilation, draughting, photomechanical working, printing production, marketing and distribution, etc. I shall always be grateful for the extra time so generously given to correcting my naïve opinions and improving my practical abilities and commercial awareness.

My route to an acceptable level of professional competence was not unique, and probably reciprocates that experienced by many similarly qualified members of the mapping community. In the absence of any formal courses of cartographic education in the UK we learned the principles and practices of mapping from skilled practitioners many of whom had, formerly, made an important contribution to the work of military survey during the period 1939–1945. The majority of these worthies, whose importance should not be underestimated, have now retired from the professional arena and it is perhaps appropriate to consider how developments, in many instances initiated as a direct result of their experience and foresight, have lead to changes and modifications which have affected all areas of our discipline. The efforts of a comparatively small number of 'innovators', during the 1950s and early 1960s, have thus influenced the shape and style of Cartography as we know it today.

A commentary of this sort will scarcely be relevant unless made with reference to, and in comparison with, events of the recent past. Further, it must incorporate a certain amount of crystal-ball gazing relating to the possible importance and development of today's operations in the future. Of course prediction of any sort is notoriously unreliable, and I have yet to meet a cartographer who is in receipt of any form of occult- or divine-guidance! The problems facing our profession are perhaps more difficult to solve than those confronting many others, because mapping is a subject which is likely to be radically modified, virtually overnight, by some unexpected event or unanticipated development. In 1942 the American John K. Wright asserted in his article "Map Makers are Human" that 'Maps are drawn by men and not turned out by machines'! Very obviously the situation has changed somewhat drastically during the last four decades, as is demonstrated in the contribution made to The Times on 23rd April, 1985, by his British namesake (John Wright) entitled 'Why maps are taking the high-tech route to the future'!

However, it is true to say that Cartography, which was formerly generally considered to be a discipline founded upon convention, is now enjoying a remarkably rapid rate of development that shows no signs of decreasing despite the economic difficulties which are impinging upon so many areas of commercial life. Techniques and even points of view that were taken for granted when I entered the profession are being, or in many instances have been, questioned and superseded by new or alternative methods and ideas. Not long ago the cartographer was performing operations and applying skills essentially similar to those used by his counterpart a generation earlier – today the discipline is in the midst of a revolution and little of its complex field, and those of associated practices which contribute to map making (survey, photogrammetry, photography, printing, computer science, etc.), remains unaffected by change.

During the lifetime of the SUC *Bulletin* man's general preoccupation with 'information', 'quantification' and 'communication' has greatly influenced our subject. It is now generally accepted by those who are spatially aware that a well researched, designed and produced map is certainly one of the best methods for the graphic communication of information; and that the cartographic process is concerned with data collection and manipulation, information display and image processing, rather than merely drawing maps! Maps and graphics have become ever more important in scientific debate, and the cartographer is increasingly involved with aspects of perception psychology, new techniques for data collection, machine processing and systems analysis.

Developments to date have created a need for a different breed (in terms of twenty years ago) of technologists and technicians who can service and operate complicated equipment and techniques. Experiments relating to new and quicker methods of map production have resulted in investigations into photo-and electrostatic-mapping; the use of colour, infra-red, false-colour, and satellote-photography; the employment of results obtained by seismic sensing; and the application of 3D graphic techniques. In particular the use of digital mapping and computer-assisted cartography have caused the most exciting and significant advances to date, and obviously still more developments will take place in the future. All of these elements are now vital ingredients of both UK and international cartography, whether relating to military, civil, public, private or academic map production. Even the underdeveloped areas of the so-called Third World are instituting mapping programmes based on these advances and their application to specific projects.

In the writer's opinion there are three major areas of technological innovation which have caused profound changes in map making capabilities and responsibilities during his working lifetime. These are the development of dimensionally stable, polyester plastic base materials and their addition to the available reprographic media; the extension of the use of photogrammetry – and ultimately its adaption to facilitate the exploitation of imagery obtained by environmental remote sensing; and finally the perfection of photo-lithography enabling the speeding up of the printing process and the simplification of multicolour reproduction.

It is an unfortunate but inescapable truth that, whether we like it or not, wars act as a major stimulus to the advancement of cartographic development - witness the contemporary documentary programmes and literature associated with the recapture and defence of the Falkland Islands. In the case of World War II, one of the results was a previously unprecedented demand for maps. As a direct consequence mapping technology was developed rapidly, and new organizations were established to produce maps and service military needs. Some hundreds of non-specialists with a geographic or graphic background became 'instant' cartographers, and brought to the discipline techniques and skills which had previously been unthought of or dismissed by the traditionalists within the profession. These changes of personnel and approach did much to accelerate production. In addition, innumerable individuals from both military and civilian background came, as a direct result of their wartime experiences, to realize the immediate value and potential of maps as media of communication. Post-war there occurred an increase in the number of maps produced, many illustrating topics previously never considered viable or appropriate. The overall experience demonstrated that maps were (i) surprisingly complicated things both to use and also to produce efficiently, and (ii) essential tools in many aspects of everyday life. The perfection of television throughout the 1950s further amplified the value of maps in the illustration, for the general public, of news stories relating to geographical areas previously considered merely as 'over there'!

Thus by the mid 1950s, in the UK and throughout the Western World, there was an evident consumer requirement for maps and a burgeoning technology enabling production agencies to meet the demand. The situation has not changed some thirty years later and in fact 'affluence' (demonstrated by car ownership, foreign holidays, etc.), and most recently increased leisure time (perhaps in part resulting from unemployment), have also led to extended print-runs.

It is, I think, viable to review developments up to the present – and perhaps also into the future – under distinct headings. These are 'Cartography as an Academic Subject', 'Employment in Mapping', 'The Role of the User', and 'Technological Factors'.

Cartography as an Academic Subject

As A.H. Robinson *et al.*, (1977) stated in the article "Cartography 1950–2000":

'A recognized need plus a useful technology is the basis for the development of a scientific field. When this occurs the required number of practitioners tends to grow to the extent that trainers become necessary to keep a steady workforce. Experimenters also become necessary to keep the technology current by observing or forcing developments. The trainers and experimenters require education which, in turn, demands scholars, teachers and a structured curriculum. [...] *Meanwhile conceptual and technological advances continue to take place. When all of these conditions are met, an identifiable scientific field can be said to exist.'*

Within the UK all of these provisos have been satisfied and further progress is being made with regard to the refinement of approved schemes. Close liaison with members of the cartographic profession (resulting in the offering of job related courses of study) has led to the establishment of a full range of options at various levels of academic attainment, and to suit different employment requirements. These consist of schemes sponsored and approved by the Business and Technician Education Council (BTEC) at Ordinary and Higher Certificate and Diploma levels (previously developed by the Joint Committee and the Technician Education Council (TEC) between 1969 and 1983); together with BA/BSc Degree and Honours Degree programmes, and postgraduate studies. Although the UK has lagged somewhat behind the USA and a number of countries within Western Europe in the arranging of courses, it is now safe to assume that the necessary foundations for the future development of Cartography as an identifiable academic discipline are firmly in place. Today the trend is deliberately flexible in order to take account of continuing developments, but it is generally directed towards the identification of the subject as a distinct discipline – an intellectual art, science and technology – rather than its appearance merely as an adjunct to studies centred around Geography and/or Surveying, and consisting primarily of training in the use of draughting equipment and techniques. This view is strengthened by the apparent attempts being made by organizations involved in national mapping to establish a professional cartographic cadre with a distinct career structure. It is vital that this impetus be sustained in order to ensure that genuinely spatially and cartographically aware personnel maintain control of map production, rather than allowing computer scientists and geographers (often with a low level of aesthetic appreciation and production awareness) to make 'take-over' bids!

The last 25 years have witnessed previously unparalleled advances in academic cartography. A range of international scientific, technical and technological literature has been developed; there has been a, rapid and significant awakening of interest in all facets of the discipline within the UK; learned societies such as the SUC and BCS have been established and have their counterparts worldwide; and stimulating debate has been, and is still, taking place amongst practitioners – particularly within the International Cartographic Association (established in 1959) with regard to the essential nature of the subject. The ICA has also been responsible for the publication of a variety of internationally generated texts relating to the whole field (Basic Cartography for Students and Technicians, Vol.1, 1984; Vol.2 to be published in 1986), and also specific aspects of the subject and their applications within a variety of kindred areas.

Employment in Mapping

As has been previously suggested, the current educational provision was established primarily in response to professional demands and likely staffing requirements. Students fall into two categories – those in employment but allowed day-release facilities to study at local Colleges of Further Education or Technology; and others who attend full-time course at Colleges of Higher Education, Polytechnics or Universities. Entry requirements for the various courses are clearly documented but, due in part at least to the employment situation, potential demand at all levels far outstrips the number of places available. As a direct consequence the major employers' advertised minimum entry qualifications for trainee, basic grade draughtsmen (who are normally required to attend a BTEC Certificate course at a local College) are far exceeded by persons enquiring. Five applications for each available opening is not unusual, nor is an approach from a PhD holder with respect to a post which, fifteen years ago, would have been allocated to a school leaver with four relevant 'O'-levels.

A direct result of this situation has been a significant increase in the numbers applying for full-time BTEC Diploma courses, and also to degree courses in Cartography or Topographic Science/Surveying and Mapping Sciences. However, in both of these situations demand is again significantly greater than the number of places available. Within the last decade the overall size of the profession has decreased, largely as a result of the government's policy with regard to the rationalization of the civil service, from more than 4,500 to circa 4,000 persons nationwide.

From the foregoing it would appear that employment opportunities are somewhat restricted! However, this is not necessarily the case provided that job applicants are appropriately qualified and motivated. The accompanying table illustrates statistics relating to the employment of holders of the Oxford Polytechnic Diploma in Cartography for the period 1973–1982 (inclusive). Comparable statistics for our first graduate students (1984) demonstrate a successful employment rate of 85 per cent within six months of the completion of studies. In the present year the number of advertised opportunities would seem to be up by some 10 per cent on the same period during 1984, and the volume of requests for copies of Careers in Cartography (4th edition, BCS, 1984) shows no signs of slackening.

OXFORD POLYTECHNIC **DIPLOMA COURSE IN CARTOGRAPHY**

Employment of Qualified Students – 1973–1982 (inclusive)

Year	Initial Enrolment	Exam Failure (after resit)	Females	1st job in Cartography	Employment in Cartography (December '82)
1973	12	1	7	11	8
1974	17	–	13	15	11
1975	17	–	10	15	13
1976	17	–	11	17	14
1977	20	1	12	17	15
1978	15	1	10	13	11
1979	17	–	9	17	16
1980	18	–	13	15	15
1981	20	–	16	20	19
1982	19	1	14	16	16
TOTALS	172	4	115	156	138

Conclusions (as at 31st December, 1982)

1. At least 82% of the total number of successful Diplomates (during the census period) were still in gainful cartographic employment at a level commensurate with their qualification.
2. Of the 12 successful Diplomates who opted *not* to enter the cartographic profession, 9 were actively employed and 4 of these were making careers in areas which could be said to relate back to their cartographic education i.e. journalism, publishing, advertising, general graphics.
3. Women found it no more difficult than men to gain employment.

Obviously there are still numerous people who are map oriented and anxious to make a career in this area. However, it is now very much a 'buyer's market' and employers will continue to recruit the best qualified applicants available. In consequence, school-leavers (or others considering making an approach based on their interest and possession of the minimal number of passes at GCE 'O'-level or equivalent) should be prepared, at least initially, for possible disappointment.

The Role of the User

There is increasing evidence, in all areas of mapping, of a growing concern for the map user or reader. Whereas in the 1950s the goal was merely to produce a graphic which was technically well-produced and aesthetically pleasing to its originator, today – at least in theory the map maker (technician/technologist) creates the display envisaged by the author (cartographer/ cartographic scientist) who may well have been in contact with intended consumers, or who should at least be sensitive to the intellectual capabilities of the expected user. There is a lessening concern for the map acting as a storage device for various sorts of spatial data, and an enhanced awareness of the need to produce an efficient medium of communication.

It has become increasingly possible to store detailed, accurate, specific data within a computer and to regurgitate this, in a graphic form, as and when required. The precise location of general data can also be effectively mapped at small scales if necessary. Many products are now 'temporal' or 'ephemeral' in that they are of the 'throw away' rather than the 'storage' variety. Regrettably this has resulted in the publication of a number of low accuracy and less aesthetically pleasing products than was formerly the case. However, the increase in the number of thematic representations has meant that illustrations may be deliberately compiled to give a general impression of a specific idea to a particular user group, and in consequence accuracy with regard to location may be of lesser concern.

All of these points relate very much to the user's stated requirements and the method of data input. The cartographer must be responsive to the user in terms of his interpretative abilities and academic status, colour and type preferences, and even with regard to specific reference points enabling his personal orientation. There is obviously a rich field of potential research here, and the varying emphases displayed in national atlases are perhaps a worthwhile starting point with the propaganda element playing an especially significant part!

Technological Factors

Mention has already been made of the significant advances which have taken place in mapping technology over the last quarter of a century. Although this has rendered obsolete a number of the problems which previously confronted cartographers it has, simultaneously, engendered a new and larger array of considerations requiring attention and research. The major current and likely future problems are:

(i) How to deal with the vast amount of available data in terms of its storage and eventual employment;

(ii) The discovery of the most appropriate method for the effective application and exploitation of remotely sensed imagery;

(iii) The development of techniques by which the evolution of faster and more accurate methods of map reproduction and data display can be assured; and

(iv) Offsetting the considerable expenses relating to both research and hardware operating costs, whilst also paying the specialist staff necessary to undertake and maintain its employment.

This last point is possibly overstated as production costs have been kept down, and even reduced in some cases through a combination of technologically based factors which allow the elimination of certain traditional production tasks. It might perhaps be more realistically suggested that the difficulty is one of recruiting suitably experienced personnel with an appropriate cartographic background.

It would appear that, at present, we are tending to use new technology simply to duplicate the traditional map maker's former tasks and skills – obviously attitudes must change and a more open-minded approach be adopted. I foresee that, in the future, more and more computer-driven devices will be employed and that, certainly within the topographic mapping sphere, automatic draughting will become the norm and possibly be used in association with photographic base material. Paper-based maps will probably be gradually phased out and temporary map display – involving the use of a CRT (cathode ray tube) interfaced to an appropriate digital data base – will become increasingly important. Materials intended to be kept longer will probably be stored on disks or microfilm and reproduced to order. The moving map display (à la James Bond) is already in an advanced state of experimental development, and is another technique which should be in regular use by the general public before the end of the millennium.

By the year 2000 it is anticipated that the user may not want, or be willing to pay the cost of, a flat paper map. Not only may it be considered ecologically wasteful to generate a graphic on paper or plastic, but also the user may simply not wish to have it lying around and requiring storage.

In my working life to date I have already witnessed profound changes in terms of cartographic theory, practice, philosophy and approach. A completely new technology has become available and different conceptual attitudes to the subject's methodology and the application of techniques have necessarily been evolved. At last the user is starting to play a part in influencing map content and appearance. In my opinion the cartographer who tries to ignore this important individual, and his ever more sophisticated and increased demands, will be unemployed (or even considered unemployable) within the next ten years! Even if no further breakthroughs occur before the end of the century, cartographers will be very 'pushed' in their efforts to make optimum use of the technology that we have available to us today. The cartographer of the future must, of necessity, be a highly educated, well trained, flexible and adaptable individual, capable of utilizing the facilities offered by the application of new technology and of contributing to the exciting changes which are taking place in areas potentially applicable to mapping. However, in spite of the foregoing it is to be hoped that the envisaged 'super cartographer' will retain some aspects of humanity, and that a knowledge of traditional mapping skills will not be completely lost to future generations.

References

Anson, R.W. and Ormeling, F.J. (Eds) (1984) *Basic Cartography for Students and Technicians* (Volume 1) London: International Cartographic Association and Elsevier.

Robinson, A.H., Morrison, J .L. and Muehrcke, P.C. (1977) "Cartography 1950–2000" *Transactions of the Institute of British Geographers (New Series)* 2 pp.3–18.

Wright, J.K. (1942) "Map makers are human: Comments on the subjective in maps" *Geographical Review* 32 pp.527–544.

FROM LINEPRINTER MAPS TO GEOGRAPHIC INFORMATION SYSTEMS: A RETROSPECTIVE ON DIGITAL MAPPING

Michael Blakemore

Originally published in Volume 19, No 1, pp.1–5

Through its comparatively short history digital mapping has gone through many stages of technical innovations and software provision. In the light of current developments in geographic information systems this study reviews the key literature of the early years to assess why certain suggested developments were pursued with great energy by others, yet some far-seeing proposals largely were ignored.

Introduction

An examination of the huge literature on digital mapping would indicate a rapidly expanding field of study. In Geo Abstracts Section G, concerned with Remote Sensing, Cartography, Photogrammetry and Surveying, there now are some 3,500 abstracted studies per year, a great many of which use computer assistance as a fact of everyday life. Such a maturity has emerged over a comparatively short time; indeed the time period is less than 30 years. In the late 1950s and early 1960s a relatively small number of papers were produced. Those were pioneering days, times of heady enthusiasm with a new technology, times to break the old moulds in geographical and cartographical publications, and times when the digital computer was to give impetus to that quaintly named movement the 'Quantitative Revolution'. The emergence of digital mapping was to become identified with a number of pioneering specialists whose influence still exists today, and it was to set the scene for a long period of aggravation between those who grasped the implications of the new technology, and those who shunned the quantitative and digital areas, or who viewed the technology as presenting a threat to their trade guilds of traditional manual map production. The latter, of course, has been one of the saddest outcomes, since it fragmented the cartographic world into academic and manual sectors, and in any case I have yet to hear of a professional cartographer who has lost a job to a computer.

This review assesses those early papers both for the ways in which they presented digital mapping to a largely non-technical audience, and in the ways their suggestions were assimilated or ignored by cartographers and geographers.

Early Days

In terms of absolute date, Waldo Tobler's paper of 1959 must be one of the earliest seminal studies. Here he examined the possibilities of automation and cartography, and utilized analogies of maps as computer input processes to communicate the operations needed to convert a map into digital format and redraw it. Technology available to him was, by today's standards, prehistoric. Punched cards were the only easy data input mechanism, and his description of a digitizing process laid down the fundamentals for all later vector digitizing operations:

'First, an appropriate map of the United States was chosen, and a transparent overlay with an orthogonal grid was placed over the map. The coast line was then broken into straight-line segments, on the principle that any curve can be approximated by a series of straight lines' (Tobler, 1959: 532).

The plotting of this map, stored in outline on 343 punched cards, took some 15 minutes. On a modern plotter that would be no more than a few seconds. What Tobler put in the mind of the new generation of digital cartographers, was the ability to digitize a map in outline form, to scale it, add symbols, perhaps some text, and replay it on a plotting device. This can be termed the 'reproductive' form of digital mapping, where the digitized outlines merely form a template on to which other material is to be placed. It has, until relatively recently, remained the most popular form of digital mapping. Towards the end of his paper Tobler notes some of the limitations already apparent in this type of operation, foreseeing that 'the heuristics of generalization, balance, and contrast seem complex and difficult to define' (Tobler, 1959: 533). In this instance he acknowledged that digital cartography comprised a strongly deterministic set of processes, and little opportunity seemed to be available to allow professional cartography to flourish using computer mapping; in essence one could not stray outside the parameters supplied in a mapping package.

Nowhere was this to be more apparent than in the rapid growth of lineprinter mapping that was to follow. Tobler used a graph plotting device that was available on precious few computer systems. What was ubiquitous was a lineprinter, and the choice was a stark one for so many – if you wanted to become a computer cartographer then lineprinter mapping was the only opportunity. It must be the case that lineprinter styles caught on not because they were seen as being good cartographic practice, but because of a fear of being left behind. Look at the perceived credibility which lineprinter maps brought to Departments. How many map libraries are stocked full of 'A Computer Atlas of...'? The material may have been trite, the visual aesthetics below the capability of a child, but the magic word 'computer' added sales credibility. In another paper Tobler (1965) put the lineprinter in the context of other output devices, stressing 'these maps are very crude and have several disadvantages, including an inability to print between the lines and a restricted choice of symbols' (Tobler, 1965: 34). In his concluding sections he stressed again the problems of computerizing behavioural processes like generalization, the rather crude techniques of automatic contouring, and noted that 'we can expect more maps, cheaper maps, perhaps better maps, more useful maps, and more imaginative maps; possibly less artistic maps (Tobler, 1965: 37).

In so many of his predictions Tobler was looking beyond the level of technological infatuation that grew up around digital mapping. Kao (1963: 533) proposed that geographers should become concerned with analytic or numerical information, and presumably then to downgrade their involvement with descriptive forms. Little surprise then that anti-quantitative groups quite rightly criticized an obsession with strictly numerical data. The computer could even be a panacea for many geographical problems, and looked forward to the 'intelligent' computer which 'must be a special purpose machine designed to reproduce partly or wholly the nervous system of the human brain' (Kao, 1963: 542–543). The ability of computers to store and replay outlines had earlier prompted Sherman, to enthuse about a 'Universal Map Data File' which he thought would be developed in the near future. It would:

'be compactly stored on tape, for which a series of programs are available for processing and extracting any particular type of data [...]. *Equipment requirements would be large, however, one center possible at the UN Headquarters could possibly provide the master service for the world'* (Sherman 1961: 14).

Brave New World indeed! A year later Harbaugh (1962) outlined the algorithms for irregular interpolation of contour maps, using paper tape as data input, and lineprinters for map output. His system was constrained to a specifically rectangular study area, and the rapid developments which occurred in the field of terrain mapping and modelling are epitomized both in the extensive range of monographs produced by the Kansas Geological Survey (see for example the contouring techniques of McIntyre, Pollard and Smith, 1968), and in the guidance systems of modem weapons whose ability to navigate vast distances and kill significantly more people is due in no small part to the willingness of many digital cartographers to sell their souls to agents of death.

The mid 1960s saw a gradual maturity of outlook among the· experts, while the majority were glad to be able to produce lineprinter simulated maps of the reproductive form. Colwell and Robinson (1966) examined the potential of computer cartography for producing animated sequences. Beforehand the need to make slight adjustments to a great many maps was prohibitively expensive. Some cartographers pursued this line, for example Moellering (1973), but found that this form of cartographic expression, like the use of colour, has been fair game for conference presentations, but impossible to publish in classical formats such as this journal. Even now, with colour displays becoming the norm, the inclusion of a colour map in a journal paper usually causes a sharp intake of breath from the publisher. How ironic that the more exciting forms of digital cartography are not amenable to publication in the existing channels of communication, and how frustrating that academic promotion boards see a journal paper as being more worthy than a film, or video (clearly the sign of a dilettante artist), or a set of computer code (obviously a dabbler in machinery).

Three papers from the mid 1960s illustrate the extent to which digital mapping had progressed in less than 10 years. Petrie (1966) gave an extensive account of the available equipment, ranging from photogrammetric use of digital hardware, through line digitizing systems, graph plotters, flatbed plotters, and data storage in the form of punched cards, paper tape and magnetic tape. Of the latter he noted that it was expensive, had a storage density of 200 characters per inch, and 6 million characters could be stored on a reel. Nowadays 1,600 is the density used mainly for transfer between machines, and significantly higher densities are available. This paper has been word processed on an Apricot microcomputer, with 3-inch disks each storing over 0.75 of a megabyte. Such is the increase in storage capacity. In terms of proper developmental systems for cartography, Bickmore's paper (1966) stands out as an excellent appraisal of the potential. His study was the formulation of the Experimental Cartography Unit at the Royal College of Art in London. It was a reasoned review, for example drawing upon US Air Force studies which estimated that automated methods could lead to cost savings of 70 per cent when compared with manual techniques. There was a clear advantage to be gained in storing together data of different environmental, graphic and statistical types; surely a foretaste of integration in geographic information systems which the modern-day manifestation of Bickmore's Unit now pursues. He even addressed the educational needs, though his warnings seem sadly to have been ignored in the British educational sector:

'There is no dispute that bad programmers waste valuable computer time; the same will apply to automatic cartography. There is an urgency about the need to build up an informed body of expertise in this new kind of information handling' (Bickmore, 1966: 4).

Lastly, among the early studies, there is clear evidence that the specialists were well aware of the behavioural problems that still have to be solved. In a discussion of papers by Tobler, and by Bickmore and Boyle, one commentator notes the difficulty that computer programs would have in coping with subjective generalization;

'Imagine a gorge with a river and a road and a railway. First we plot the river, then we display the road. The railway is displaced further and finally the contours are moved. This presents a very difficult problem for the machine to solve' (Anon., 1965: 33).

The way it was overcome was to digitize from existing maps sheets where the generalization process had been carried out by manual means. Derived mapping, as it became known, has formed the mainstay of digital cartography throughout the years, and it is only relatively recently that primary mapping, mostly by photogrammetric means, though increasingly by remote sensing, has been on the increase. The problem then is how to integrate vast data banks for example of photogrammetrically derived data with often more geodetically accurate primary data, which also may come from the new range of total surveying stations. The meeting point for these data is in the overlay techniques of the modem powerful geographic information systems, considered later.

Early Systems and Software

Without a doubt one of the most influential early digital mapping packages was SYMAP, designed by Howard Fisher and launched in 1964. Orientated towards mainframe computers with lineprinter output it proved to be a success story; indeed it still is one of the more widely used packages. Wolf (1969) notes that the relative speed of automatic lineprinter mapping made it an attractive prospect for marketing management, while Gould (1970) in an extensive paper on computing in geography, stresses SYMAP as an effective teaching aid. The package, however has a number of drawbacks. Its size demanded large computer systems on which it could run, and the output using combinations of lineprinter characters was aesthetically not pleasing. Neither did it have any facilities for spatial searching, and although its interpolation algorithm for contouring was to be an influential one, its facilities still were for reproductive styles of mapping.

To overcome the size limitations some attempts were made to produce small lineprinter mapping systems. Scripter (1969: 133) notes that SYMAP was 5,000 cards in length and was 'beyond the immediate means of most colleges' who could not afford machines with large memory capacity. To widen the community who could experience digital mapping Scripter wrote his own small lineprinter program for choropleth mapping. The situation whereby packages exist but resources do not permit many to mount them was as big a problem then as it is now. The new generation of commercially developed geographic information systems are beyond the purchasing power of all but a few. For this reason, at the time of writing, the ARC/INFO system is mounted in the UK on only one university machine, in Birkbeck College, and is intended by the Economic and Social Research Council as providing a national resource.

These years were a time when many became addicted to the new technology. Also there was an element of institutional and national chauvinism in a desire to develop one's own software rather than purchase it. Many North American institutions started to develop their own in-house software, AUTOMAP at the CIA being one (Schmidt, 1969). Gaits (1969) describes two systems LINMAP and COLMAP, developed in the Urban Planning Directorate of the Ministry of Housing and Local Government. LINMAP was a lineprinter package, but had additional facilities of statistical manipulation and spatial retrieval which SYMAP lacked. To achieve colour mapping in COLMAP tapes were written of colour separations in red, blue and yellow, read by a typesetter, which output layers on to film or paper, and these were used to make printing plates. These long processes now look so tedious given modern colour equipment. The multiplicity of different systems, different data structures and encoding methods was to lead to chronic problems in transporting digital cartographic data (see Baxter, 1980).

While the lineprinter was a ubiquitous resource in the late 1960s and early 1970s, other plotting devices were not. Gould (1970) noted that graph plotters were increasing in number, but they were relatively slow, expensive, and in great demand. Even in the mid 1970s I can remember waiting for three days for a simple plot to clear the queue. Monmonier (1968) was one of the pioneers in transportable packages for what he termed the 'digital incremental plotter'. His package, called GIPSY, allowed base maps to be defined as digital coordinates, but with no other map information. A simple structure of code, x, y, where the code was 1 if the point was the end of a line and 0 if not, was used. This emulated the basic move and draw routines of early graphics packages. A border could be added to the picture, text placed on the map, and point symbols of the square, circle and tetrahedron types scaled and drawn. In a later paper Monmonier (1970a) discusses algorithms for shading polygons on a plotter, and GIPSY reached a wider audience in Monmonier (1971).

All these developments still followed the line of simply reproducing digital map data with limited scale change, and the addition of thematic data in forms such as choropleths or point symbols. Even early three-dimensional packages replayed the digital information, and made relatively little extra use of it other than interpolation. Thus, a logical next development was to evaluate the technology, and identify applications areas.

Applications Area and Evaluation

Alongside the developments in what can be termed 'mass market' systems such as SYMAP, there were specialized investigations into wider aspects of digital mapping. Voisin (1968) cast a critical commercial eye, since in Rand McNally they needed to justify the cost-effectiveness of the technology before implementing it. He was quite cynical of one aspect, noting that by 1977 no 'single type of traditional map had been created' using digital mapping techniques. Those that had been produced were different types of maps, such as transient lineprinter types, and specialized thematic maps. Additionally,

'A private producer must be able to justify the larger portion of its development and expenditures in a specific return through money saved or in the development of a new product or services heretofore unavailable, which in themselves will afford additional income possibilities' (Voisin, 1968: 78).

In the commercial world of today the cost-effectiveness is well proven, with bureaux such as CACI, PINPOINT,

ESRI, INTERGRAPH, SYSCAN and many others providing mapping and intelligence facilities at a quality and speed not possible before.

There were early investigations into the utility of using UK Ordnance Survey maps in digital formats. Howard (1969) assessed the 190 sheets of the old 1:63,360 series. First estimates were that the sheets contained about 42,300 metres of contour detail, and that these comprised one third of the total detail on the maps. When digitized there would be in the region of 2.37 million coordinate pairs, requiring at that time 1,300 reels of magnetic tape. The sheer magnitude of data still is a problem for the Ordnance Survey.

An extensive assessment of digital mapping for hydrographic charting was produced by Boyle (1970), and the facilities were intended for the Canadian Hydrographic Service. Here the system was one of the most sophisticated of the time. Interactive digitizing (and how many people still digitize off-line even now?), high-quality flatbed plotter, software to snap sheet edges together and compensate for changes in orientation in digitizing, a concern for the possibilities of data interchange, and a developmental scanner system for digitizing. Bell (1972) outlines similarly high-quality developments in the Experimental Cartography Unit, and hints at future areas of research, since while it was now a trivial operation to replay digital information, 'some criteria for assessing what constitutes a good map are aesthetic and hence unprogrammable' (Bell, 1972: 714).

At this stage the range of expertise broadens further still. The early 1970s saw a rapid expansion in the use of reproductive computer mapping, but for a few a trend away from this was occurring, towards proper analytical use of the data so laboriously captured by digitizing. The latter process was to lead to a consideration of the integration of disparate data sets, while an off-shoot was a semi-philosophical concern with the impact the new technology was having on cartographic and geographic practice.

Reflection and Reaction

Monmonier (1970b) identified a worrying reaction to digital mapping as being a reluctance by many to accept it into the body of traditional cartographic techniques. One reason, stressed earlier, was an understandable, though not necessarily justified; fear that computers would replace human beings. The fear is the same as the reaction of scribal guilds to the advent of woodblock printing processes in the 1400s. But was the negative reaction to digital mapping solely the result of protectionism? Monmonier posed several other reasons. First was the view that most people had (and the digital cartographers did not dispel) that computer assistance implied purely quantitative processes, and so it quickly became subsumed among the statistical and mathematical trends. Balchin and Coleman (1967) had noted this when they discussed the success of numerical maps where 'weather maps originate as instrument readings, contour maps as observations' (Balchin and Coleman, 1967: 122). They hinted at a lack of 'intelligence' in digital mapping, and this was taken up by O'Callaghan who discussed the real limitations:

'Since the foundation of the design rules are based upon the human perceptual functions, about which little is known, it does seem however, that a completely automated situation is remote' (O'Callaghan *et al.*, 1971: 100–101).

He saw the possible answer as being in the development of 'Artificial Intelligence' which would aim to emulate the behavioural decision-making of human cartographers. Perhaps this is why so much digital mapping has been in the derived and reproductive fields. By digitizing existing maps (as has been done by the Ordnance Survey) the behavioural bottlenecks are being by-passed. Awkward decision-making processes have been carried out already by manual cartographers. The reliance on derived mapping to a great extent contradicted Bickmore's (1968) view that digital mapping should overcome the obsolescence of many topographic map sheets which were out-of-date before even being printed. A long up-dating process would be required to correct the digital versions of paper maps, although when complete, digital maps supposedly would not suffer from the problems of areas of interest always lying on sheet edges. Bickmore was quick to note that sheet edges would not really match perfectly and that geometric processing would be required. He was also conscious of the different accuracy standards used in digitizing, and a need for what he called quality statements was vital if databanks were to be fully integrated:

'Even with the most respectable of maps the positional reliability of one feature, the outline of a building or a river, for example, will be greater than that of another, such as a vegetation or climatic boundary' (Bickmore, 1968: 227).

Fuzziness of boundaries is not something that can be overcome in many modem systems, even though Bickmore predicted the problem in 1968. The presence of doubt or imprecision indeed became unpalatable for many promoting automation as being a process that was transforming cartography from an art into a science. Morrison (1972) examined philosophical change in thematic cartography. He agreed, quite rightly, that the computer was forcing cartographers into thinking of different products, and of what constitutes a map. It was 'supposed to be cost effective with decreased cost per map, and increased speed in the production of map', although he was considering only the cost of using applications programs, not the cost of developing them in the first place. Monmonier (1970b) added to this the not inconsiderable purchase and running costs of hardware. Morrison seemed to think that automation enhanced the credibility of cartography:

'It is easy to envisage how cartography can thus become independent of geography, history or any other discipline that might need a map of distribution, and take its place among other disciplines of communication' (Morrison, 1972: 7).

So cartography was to become a 'science', and artistic elements were downgraded in importance. As it became a science so it embraced mathematical, statistical and geometric techniques, and the emergence of early geographic information systems built upon this somewhat positivist philosophical foundation. Applications used the numerical power of computers, typical of items being Fryer, Smith and Macleod, who at least used geometric techniques to match sheet edges, but wrote

'This system recognizes that the computer is better at repetitive operations, data storage, retrieval and calculations, while the craftsman is superior in aesthetics and judgement' (Fryer, Smith and Macleod, 1974: 128).

Discarding the non-scientific added to the myth that computers would do only mechanical repetitive tasks. It was reinforced by the fossilization of programming styles into FORTRAN, or even BASIC. Most existing mapping systems are FORTRAN based, not because it is the best language, but because the ad hoc educational system, noted earlier, forces new researchers to think along old lines. In the early 1970s few would have learnt to program in a language other than FORTRAN, and even today it holds its influence.

Conclusion

It is a salutary lesson to find that the early papers in digital mapping contained so many predictions of the problem areas. The lack of any ability to assimilate the intelligence and skills of professional cartographers was to be a major reason in the dominance of reproductive mapping. The alliance with quantification led many to seek numerical answers to mapping problems, when clearly the demand was for some method of coping with logic and intuition. Much is being written at present about the advent of 'fifth-generation' computers, and 'expert systems' which aim to emulate behavioural factors. 'Artificial Intelligence' as it is also termed, may provide digital cartographers with tools to develop mapping systems that learn from the cartographers that use them. This may sound insidious, but it is an attempt to cope with the following type of problem that a professional would discard as being obvious and trivial. If a map is to be reproduced at 1:25,000, the background paper is off-white, rivers are drawn in blue using a pen nib of 0.2 millimetres, and the rivers are quite dense in certain areas, and the text font is to be Helvetica medium, colour black, what colour combinations are suitable for the inclusion of histograms at key sites representing river flow over 12 months? No readily available mapping systems give unsophisticated users advice on possible options. What is more, even if a user achieves a successful map, the program does not remember the combinations of lines, colours, pens, etc. It is up to the user to remember it. Another user may come along the next day with the same problem, and unless the two converse, the second will not be able to use the 'rules' derived by the first. 'Expert systems' use languages, such as PROLOG, which allow the program to store these rules, and so offer 'advice' to other users.

As such, digital mapping may at last be trying to take advice from the professionals rather than expecting the human experts to mould to the restricted facilities of many existing systems. At the same time, there still remain many serious problem areas: the problems of matching sheet edges, of overlaying distributions derived from different coverages such as land use, geology, soils, and remote sensing data types. Many existing geographic information systems use numerical techniques to cope with these, and in many cases use only the geometric properties of the lines in their calculations, and not the actual cartographic attributes. As these systems embrace artificial intelligence we may at last move towards the situation where digital mapping no longer is trying just to emulate the hands of manual cartographers but is actually trying to use the cumulative professional expertise stored in their brains.

References

Anon. (1965) "Record of the discussions on the papers of Waldo R. Tobler, and D.P. Bickmore and A.R. Boyle" *International Yearbook of Cartography* 5 pp.30–33.

Balchin, W.V. and Coleman, A.M. (1967) "Cartography and computers" *The Cartographer* 4 pp.120–127.

Baxter, R.S. (1980) "The transfer of software systems for map data processing" In Freeman, H. and Pieroni, G.G. (Eds) *Map Data Processing* (pp.223–246) New York: Academic Press.

Bell, S.B.M (1972) "Development of an interactive graphics system for automated cartography" *Proceedings, ONLINE 72 Conference* 2 pp.711–730 (London: Online Conferences).

Bickmore, D.P. (1966) *Experimental Cartography Project* London: Royal College of Art.

Bickmore, D.P. (1968) "Maps for the computer age" *Geographical Magazine* 41 (3) pp.221–227.

Boyle, A.R. (1969) "Data banks and the computer analyst" *Canadian Surveyor* 23 (2) pp.111–115.

Boyle, A.R. (1970) "Automation in hydrographic charting" *Canadian Surveyor* 24 (4) pp.519–537.

Coiner, B.J. (1967) "Line-simulated map" *Surveying and Mapping* 27 (3) pp.459–464.

Cornwell, B. and Robinson, A.H. (1966) "Possibilities for computer-animated films in cartography" *The Cartographic Journal* 3 (2) pp.79–82.

Cude, W.C. (1962) "Automation in mapping" *Surveying and Mapping* 22 (3) pp.413–436.

Fryer, J.G., Smith, D.R. and Macleod, D.G. (1974) "An automated system for thematic mapping" *Cartography* 8 (3) pp.122–128.

Gaits, G.M. (1969) "Thematic mapping by computer" *The Cartographic Journal* 6 (1) pp.50–68.

Gould, P. (1970) "Computers and spatial analysis: Extensions of geographic research" *Geoforum* 1 pp.53–69.

Harbaugh, J.W. (1962) “Direct printing of contour maps of facies data by computer” *The American Association of Petroleum Geologists Bulletin* 46 (2) p.268.
Howard, S.M. (1969) “A cartographic data bank for Ordnance Survey Maps” *The Cartographic Journal* 5 (1) pp.48–53.
Kao, R.C. (1963) “The use of computers in the processing and analysis of geographic information” *Geographical Review* 53 (4) pp.530–547.
McIntyre, D.B., Pollard, D.D. and Smith, R. (1968) “Computer programs for automatic contouring” *Computer Contributions* 23 (Kansas Geological Survey).
Moellering, H. (1973) “The computer animated film: A dynamic cartography *Proceedings of the Association of Computing Machinery Annual Conference* pp.64–69.
Monmonier, M.S. (1968) “Computer mapping with the digital incremental plotter” *The Professional Geographer* 20 (6) pp.408–409.
Monmonier, M.S. (1970a) “Shaded area symbols for the digital incremental plotter” *Tijdschrift voor Economische en Social Geografie* 61 pp.374–378.
Monmonier, M.S. (1970b) “The scope of computer mapping” *Special Libraries Association Geography and Map Division Bulletin* 81 pp.2–14.
Monmonier, M.S. (1971) “Plotter mapping – GIPSY 2 and SURGE 2” (Computer Contribution No.11) Dept. of Geography, University of Nottingham.
Morrison, J.L. (1972) “Automation's effect on the philosophy of thematic cartographers” *Proceedings of the Fall Convention of the A.C.S.M.* S-26.
O'Callaghan, J.F., Stanton, R.B. and Barter, C.J. (1971) “Problems in automated cartography” *Cartography* 7 (3) pp.93–101.
Petrie, G. (1966) “Numerically controlled methods of automatic plotting and draughting” *The Cartographic Journal* 3 (2) pp.60–73.
Rentmeester, L.F. (1968) “The universal cartographic data base” *The Canadian Cartographer* 5 (1) pp.42–49.
Schmidt, W.E. (1969) “The AUTOMAP system” *Surveying and Mapping* 29 (1) pp.101–106.
Scripter, M.W. (1969) “Choropleth maps on a small digital computer” *Proceedings of the Association of American Geographers* 1 pp.133–136.
Sherman, J.C. (1961) “New horizons in cartography: Functions, automation and presentation” *International Yearbook of Cartography* 1 pp.13–19.
Tobler, W.R. (1959) “Automation and cartography” *Geographical Review* 49 (4) pp.526–534.
Tobler, W.R. (1965) “Automation in the preparation of thematic maps” *The Cartographic Journal* 2 (1) pp.32–38.
Voisin, R.L. (1968) “Automation in private cartography” *Surveying and Mapping* 28 (1) pp.77–87.
Wolf, J.S. (1969) “SYMAP: Computer graphics for marketing management” *Journal of Marketing Research* 6 pp.357–358.

ARNO PETERS' CULT OF THE 'NEW CARTOGRAPHY': FROM CONCEPT TO WORLD ATLAS

P. Vujakovic

Originally published in Volume 22, No 2, pp.1–6

This paper reviews the work of Arno Peters to date, including *The Peters Atlas of the World*, which its publishers claim 'represents the greatest single advance in map-making in over 400 years'. The paper includes material from an interview given to the author by Peters during a visit to the UK to publicize the Atlas.

Introduction

The publication by Longman of *The Peters Atlas of the World* earlier this year, represents a significant milestone in Arno Peters' promotion of his 'new cartography'. Peters has been mounting an attack on the bastion of 'traditional' cartography since the 1970s.

Cartography's Iconoclast?

In 1952 Professor Peters published his *Synchronoptische Weltgeschichte* (*Synchronoptic World History*). This was an attempt to provide an objective, 'universal' world history, in which the emphasis was not on Europe, but gave equal weight to other world cultures. He was attempting to bring equality and balance to the treatment of history (Peters, pers. comm.). It was during this period that he became increasingly interested in global maps and cartography. In his history he portrays cartography as one of the important factors in the formation of human awareness:

'Maps have been made for almost five thousand years and for almost the last three thousand they have been instrumental in forming our global concept' (Peters, 1983: 149).

In seeking for the causes of national arrogance and xenophobia he claims he was continually led back to world maps as a major influence on people's view of the world around them. His belief that a Eurocentric view of the world is still a potent image is given credence by recent studies (e.g. Saarinen, 1988; Saarinen *et al.*, 1988).

During the preparation of an atlas volume to accompany his history he became disillusioned with existing global maps, which were '[...] worthless for an objective representation of historical situations and events' (Peters, 1983: 146). This convinced him that a revision of cartographic practices was long overdue. The cartographic profession, by its retention of old precepts derived from a Eurocentric world view, is seen as incapable of developing an egalitarian global map.

The results of Peters' review were the development of his own global projection (first shown to the Hungarian Academy of Sciences in Budapest in 1967 (Loxton, 1985)) and a dissertation, *Die Neue Kartographie* (*The New Cartography*) (1983).

The New Cartography describes the emergence and development of global maps up to the Mercator projection of 1569. According to Peters the need for a new cartography must be viewed in relation to the Mercator map '[...] which has dictated our geographical world concept for the last four centuries' (Peters, 1983: 56). Peters' view of Mercator is far from totally negative. He sees it as a vast improvement on previous maps, especially in its use of a rectangular grid. However, it is the lack of 'fidelity of area' which disqualifies it as a universally acceptable global map. Peters (pers. comm.) claims that his own solution (the so-called 'Peters projection') is in fact derived from principles used by Mercator, but he has been able to add fidelity of area to Mercator's good points.

He is dismissive of other projections which are an attempt to overcome problems inherent in the Mercator world map. Other equal-area projections are seen as being achieved by abandoning important features of the Mercator; such as parallel lines of latitude and longitude ('fidelity of position').

Under the chapter title 'Taking Stock' Peters (1983) attempts to strip away some of the 'myths' of traditional cartographic teaching. He concludes that the teaching consists '[...] of half truths, irrelevancies and distortions' (p.102). His critique focuses on ten 'myths':

1. Fidelity of Angle.
2. Incompatibility.
3. The Arbitrary or Compromise Map.
4. The Teaching of Projection.
5. Tissot's Indicatrix.
6. The Scale.
7. Equatorial Orientation.
8. Rounded Grid Systems.
9. Greenwich.
10. Thematic Cartography.

A number of these so-called 'myths' are not new to (or disputed by) many cartographers and geographers. For example, Peters' observation that the use of a representative fraction or a linear scale on a small-scale map of a large area without qualification is extremely misleading (myth 6). However, while it is true to say that this is accepted by cartographers, Peters is correct in pointing out that such scales continue to be used (see for example the *Family Atlas of the World*, published in 6 parts and free to over one million households and organizations taking *The Sunday Times* in 1988).

Other of his observations are open to dispute on cartographic grounds. Robinson (1985) provides an important critique of Peters' arguments. He describes the 'myths' as straw men used by Peters to condemn the cartographic profession, and his arguments as absurd and spurious. For instance, Robinson notes:

'To Peters, the term winkeltreu does not denote (as it does to all cartographers) the precise property of conformality, namely, that at each point the Scale Factor is the same in all directions. Instead, he asserts that it simply means angles on the globe are retained on the map' (p.105).

Peters follows his critique of traditional cartography with his own catalogue of 'attainable map qualities' (Table 1) which forms the basis of the 'new cartography'. This catalogue includes 'fidelity of area' (area distortion equal to zero) and 'fidelity of axis' (a rectangular graticule); without these a global map cannot have the quality of 'universality' (a single projection that can be used for all general maps of the world or parts of it).

Table 1. Attainable map qualities. (Adapted from Peters, 1983).

	Mercator 1569	Hammer 1892	Peters 1974
Fidelity of Area	✗	✓	✓
Fidelity of Axis	✓	✗	✓
Fidelity of Position	✓	✗	✓
Fidelity of Scale	✗	✓	✓
Proportionality	✗	✗	✓
Universality	✗	✗	✓
Totality	✗	✓	✓
Supplementability	✓	✗	✓
Clarity	✓	✓	✓
Adaptability	✗	✗	✓

Peters (1983) compares his projection with eight others (including Mercator and a number of other equal-area maps, e.g. Hammer's (1892) projection. It comes as little surprise to find that only the Peters projection fulfils all ten categories (Table 1). No other projection quoted scores higher than four. (It is worth noting that Peters does not include Lambert's (1772) cylindrical equal-area projection in this list, although it is mentioned elsewhere in his book).

Peters (1983) then explains the construction of his own projection and discusses the wider attributes of a 'new cartography'. These range from a repositioning of the zero meridian and a decimal grid system to a 'New World Concept' and 'New Attitude' (Peters, 1983; New Internationalist, 1983; Stalker, 1989). Peters sees his 'new cartography' as the basis of a new, objective, egalitarian, global concept.

Peters (1983) concludes with a final broadside at the cartographic profession. He claims that it is totally incapable of developing an egalitarian world map due to '[...] its retention of old precepts based on the Eurocentric global concept [...]' (p.149), while the revolutionary character of his '[...] new cartography lies in its defeat of the ideologies which have hitherto stamped all worlds maps' (p.150).

Peters' message and map were rapidly accepted outside of the cartographic profession; particularly by organizations involved in world development issues. With the publication of his world map on the covers of the two Brandt reports (1980; 1983) it became a symbol of concern for development and of the North-South divide (Vujakovic, 1987).

The Shock of the New

Against a background of increasingly visible support for the Peters projection, cartographers and geographers began to take the Peters phenomenon seriously. In an article entitled 'Map Wars', Stalker (1989) claims that '[...] some academic cartographers are provoked into fits of rage by the very mention of the Peters projection, accusing it of all sons of sins' (the author has experienced this very reaction from a member of the editorial panel of a respected cartographic journal). Stalker is probably right in suggesting that some of the criticism is due to the closing of ranks against an outsider.

The reaction against Peters' map focuses on a number of key issues. Objections have been raised as to its cartographic validity (Maling, 1974; German Cartographical Society, 1985; Loxton, 1985; Robinson, 1985). Even, its claim to be an equal-area projection has been disputed. Maling (1974) states that measurement of the graticules of the Peters projection unveiled in 1973 shows that the spacing of the meridians is consistent with a cylindrical equal-area project of standard parallels at 46°20' North and South, '[...] whereas the spacing of the parallels corresponds to some other variant [...]. In other words, Peters' projection is not equal-area' (p.510).

Its originality is also disputed by Maling (1974), Loxton (1985), Robinson (1985) and Baker (1986). Maling and Loxton see the projection simply as a variant of the Lambert (1772) cylindrical equal-area projection (with standard latitude at the equator). Other variants have been produced using different standard latitudes; for instance Behrmann (1910) chose 30° North and South (believing that this displayed the least overall angular distortion). Robinson and Baker claim that the so-called Peters projection was first presented to the British Association for the Advancement of Science by James Gall in 1855 and published later that year in the *Scottish Geographical Magazine* (Gall, 1855). Gall's Orthographic projection is a cylindrical equal-area projection with standard latitudes at 45° North and South, which closely corresponds to the Peters projection. Peters (1983) does acknowledge the existence of Lambert's projection, but objects to its distortion of Europe. He claims not to have been aware of the Gall variation until recently and has not as yet seen documentary evidence (Peters, pers. comm.).

Other critiques have been concerned with whether the Peters map really does provide a better alternative to existing maps. Bain (1984) assesses its importance as a general educational aid. He is not convinced that it is any better than other existing equal-area projections, many of which are less distorting of continental shapes than Peters'. This problem of the severe distortion of is a major source of dissatisfaction with the projection amongst professional cartographers. Robinson (1985) suggests that Peters' '[...] landmasses are somewhat reminiscent of wet, ragged, long winter underwear hung out to dry on the Arctic Circle' (p.104). However, both Bain (1984) and Vujakovic (1987) note that this distortion is perceived as a benefit by some of Peters' supporters, as it challenges contemporary 'world views'. Vujakovic (1989) has focused on the role of the Peters map in development education. He concludes that Peters' insistence that certain map qualities must be retained (e.g. his 'cult' of 'fidelity of area') may actually be hindering the use of appropriate maps in development education. The authors of two recent thematic atlases have argued against its use as a global base map (Kidron and Segal, 1981; Crow and Thomas, 1983), preferring other equal-area projections.

While many of the points made by Peters' critics appear valid, there does seem to be an unwillingness to acknowledge any contribution by him to the cartographic debate. There can be no doubt that Peters has raised public awareness of the importance of maps. Stalker (1989) believes that it is this which has most galled the cartographic profession. He quotes Arthur Robinson:

'The real danger is not the projection [...] *but instead the long-term harm that can be done to the profession as a consequence of the techniques Peters is using to promote his map. He is clever'* (p.l09).

It is interesting to note that the National Geographic Society has recently replaced its old global map (Van der Grinten) with one produced by Arthur Robinson (*The Sunday Times,* 1st January 1989: p.3). This new projection is supposed to be a more realistic world view (even though it still suffers from area distortion). In an article introducing the new projection (Garver, 1988), the National Geographic's chief cartographer studiously manages to avoid much of the debate about the desirable qualities of world maps that Peters has generated.

The Peters Atlas of the World

The Peters Atlas of the World (1989) is the culmination of ten years of work by Arno Peters. It is the latest manifestation of his crusade to supplant traditional cartography with his 'new cartography'. The publication of the atlas offers an ideal opportunity to study the practical application of Peters' cartographic principles. To date this has been limited to an examination of the occasional and partial use of the Peters projection by his supporters (Vujakovic, 1989).

The atlas is divided into two major parts; a topographic map section and a thematic section. Each part will be reviewed separately before turning to the wider implications of the atlas, which its publishers, claim '[...] is set to become the standard recommended quality atlas'.

i) Topographic Section

The topographic section consists of 'The World in 43 maps at the same scale'. The use of the same scale for all of the topographic maps is one of a number of key innovations that are claimed for this section, based on arguments put forward by Peters in 'The New Cartography' (1983). Terry Hardaker (the Cartographic Editor) argues that all previous world atlases have been totally inconsistent in their use of scales; he declares:

'We have come to accept as "natural" a representation of the world that devotes disproportionate space to large scale maps of areas perceived as important, while consigning other areas to small-scale general maps' (Peters, 1989: 6).

Consistent with Peters' arguments against the use of linear distance scales on world and small-scale regional maps, the scale he uses is defined by the quality of 'fidelity of area'. All of the topographic maps in the atlas have an equal-area scale (one square centimetre on the map equals 6,000 square kilometres in reality). One-sixtieth of the Earth's surface is displayed on each map. This is supposed to represent a fairer and more equitable view of the world.

This 'challenge' to people's misconceptions concerning the relative sizes of counties and regions is a valid one and to some extent it may work. The compilers of the atlas ask us to compare the British Isles (p.32) with Madagascar (p.47). It is readily apparent that the atlas does provide a very different emphasis to that seen in 'traditional' world atlases published in Britain. However, problems are also immediately obvious; for example, many large countries and regions are never seen in their entirety on a single map. This makes certain types of geographical comparisons more, rather than less, difficult. The Soviet Union for instance, is spread over nine separate maps and the United States over five maps.

Another problem caused by the single scale is over emphasis on human settlement with small populations in sparsely settled regions. Contrast some of the sparsely populated semi-arid states of Africa with parts of densely populated Europe. For example, Botswana and France have similar land areas, but very different population sizes (approximately 1 million and 55 million respectively in the mid 1980s), yet France appears to have only twenty-five more settlements within the categories used in the atlas. A major problem here is the lack of a lower limit for the smallest class of settlement shown (less than 100,000). All forty- nine settlements shown for Botswana fall within this category (the total urban population only just exceeded 160,000 in 1981), while France would appear to have only twenty-nine! This is obviously meaningless. These problems clearly run counter to Peters' intention of producing an objective world atlas. His simplistic classification of infra-structure ('communications') also causes similar problems to informative and objective comparisons between countries and regions.

Many authors have been highly critical of the extreme distortion of the continents on Peters' global map (e.g. Bain, 1984; Robinson, 1985). This problem is also to be found in the topographic section. Although the grid for each topographic map has been recalculated to remove the worst distortions of the world map, problems still remain. For example, while the map of north-west Europe (p.32) now shows the British Isles with minimal distortion compared to their shape on the global projection, Iceland is still very badly distorted (elongated along its east-west axis). An interesting and unfortunate product of recalculating the individual grids is that some land masses have become more distorted at the larger scale! New Zealand, which was relatively undistorted at the global scale, is now very much worse (p.79). Such distortions also condition our perception of other geographic features, for instance, island chains, river systems and mountain ranges. A good example of this problem can be seen on pages 30–31, where the southern tip of South America and the Falkland Islands are severely distorted. It is also well illustrated by the Soviet and Canadian island groups of the Arctic Ocean (pp.12–13 and 50–51). An attempt is made to resolve this problem of distortion by the use of a new (unnamed) projection to display the polar regions (the final eight maps of the topographic section). Peters also discusses this problem in 'The New Cartography' (1983: 115) and describes the method by which the polar regions can be redrawn. Unfortunately this new projection lacks a number of the attributes Peters claims are key to the 'new cartography' (e.g. 'fidelity of axis'). The use of a different projection also compromises his claim of 'Universality' and 'Adaptability' for the Peters projection (Peters, 1983).

Another 'innovation' is the use of colour to represent variation in 'ground cover', rather than elevation. Browns and greens are used to indicate bare ground and dense vegetation respectively, while thin or scattered vegetation is shown in intermediate colours. NOAA-AVHRR (USA) satellite imagery was used as the basis of a land cover survey for the atlas by the Remote Sensing Unit of Bristol University (Lloyd and D'Souza, 1987).

The NOAA-AVHRR images do not appear in the atlas, but provided the information on ground conditions which was then transferred to the topographic base maps by traditional hand colouring. Hardaker claims that this makes it '[...] the most up-to-date statement available of world vegetation distribution' (Peters, 1989: 6). Peters (pers. comm.) claims that the satellite imagery was not included because it would have required a detailed explanation and detracted from his aim of keeping the atlas as simple as possible. He noted his disquiet at the way in which photographs and satellite images were being incorporated into modem atlases at the expense of maps. Peters says he wants to retain the feeling of old style atlases, while using developments in computer cartography to improve accuracy of content (Peters, 1989: 7). However, it can be argued that Peters has missed an opportunity to show the dynamic nature of global environmental systems. The AVHRR imagery provides a dramatic picture of temporal changes in ground cover conditions at world and regional scales, e.g. the Sahel (Tucker and Justice, 1986). A small section devoted to this in the topographic or thematic section would have been valuable, especially with increasing public concern for global environmental issues such as 'desertification' and 'deforestation'. As it stands, the atlas simply provides a 'snap-shot' of conditions at one moment in time, hence, only a partial view.

To give the maps the impression of relief, photographs were taken of three-dimensional models of the Earth's surface. The photographs were used as base maps and enhanced by hand shading. Elevation is simply given by the use of occasional spot heights.

ii) Thematic Section

The thematic section of the atlas consists of 146 maps, each with a single theme, representing over 40,000 individual pieces of factual data (Peters, 1989).

The publishers make impressive claims for this section. In their 'press information' they announced that the maps '[...] provide an unrivalled reference resource of factual information: a complete and in-depth picture of today's world'. However, a number of problems with regard to both the cartography and the geographical content of this section are readily apparent.

Firstly, the appropriateness of the Peters projection as the basis for all of the thematic maps is open to question. For instance, the maps showing the movement of continental masses over the past 560 million years (pp.98–99) are good examples of inappropriate use. The continental shapes are very severely distorted in many instances (especially that of Antarctica), making relationships between the land masses difficult to understand. Peters is effectively surrendering the flexibility of cartography to sustain his own 'myth' that his projection is universally applicable. This contrasts with Peters' use of a compromise projection in the topographic section.

Another factor which is immediately obvious is that a Euro-Afrocentric projection is used for every one of the thematic maps. Yet Peters sees 'Supplementability' as one of the ten desirable qualities of a world map (the ability to 'cut' the map so that any continent can be placed centrally). It is surprising that an atlas which seeks to '[...] make possible a fundamental change in our conception of the world' (Peters, 1989: 3) should not use this facility to challenge our supposed Eurocentric global concept, which is Peters' major reason for developing his 'new cartography' (Peters, 1983). Peters (pers. comm.) unconvincingly claims that he has not used other 'cuts' because this is an added 'problem' for users of an atlas which already contains a wide range of innovative features. However, non-Eurocentric projections are now commonplace in modem atlases and are unlikely to represent a problem to most users. It should also be noted that this facility is very rarely used by adopters of the projection (Vujakovic, 1989).

The absence of a range of symbolization is equally surprising in a thematic atlas. The data is almost invariably displayed as choropleth or isarithmic maps. Peters sees the principle of one theme per map '[...] represented by simple grades of colour [...]' as enabling the user to more easily understand the data and compare it with the other maps. Unfortunately, the final product is rather monotonous and suffers from the problem of assuming homogeneity within the basic sub-divisions of the map in the case of the choropleths.

The choice of colours used appears to be arbitrary and tends to confound some of Peters' basic aims. For instance, the topics 'Mineral Resources' and 'Industrial Products' each show 16 maps on a double page, however, different hues and tonal ranges are used for each one. This results in bias towards strongly coloured maps. The logic of Peters' 'objective' position would seem to demand the use of a single colour and tonal range.

It is clear that a single base map has been used throughout production of the thematic section, resulting in poor definition of lettering on the smaller scale maps following reduction.

The geographic content of the thematic section is interesting and varied. Topics chosen range from 'Natural Dangers' to 'The Status of Women'. Peters (1989: 97) claims that 'No interpretation or evaluation of information has been undertaken [...]' in order not to detract from the user forming '[...] an objective and unprejudiced personal picture'.

A number of specific problems are readily apparent. The source(s) of the data used for the individual maps is never given; without this the readability of some of the maps must be in doubt. The atlas also states that where official figures are unavailable, the 'leading experts' in the various fields concerned where consulted. However, there is no indication of which maps this refers to or how this provides satisfactory data. The lack of dates for information shown on the maps is another major shortcoming, especially where rates of change are mapped. For instance, the maps of 'Population Growth' (p.131), 'Economic Growth' (pp.166–167) or' Inflation' (pp.174–175) are meaningless unless the period concerned is stated.

Definitions for individual themes are not always clear. For example what is meant by 'Social Professions' (pp.168–169)? Other misleading maps include that entitled 'Urbanization' (pp.160–161). Urbanization is 'The process of becoming urban' (Johnston, 1986). However, the map simply shows percentage of the population who live in cities (no date). This does not indicate whether the process of urbanization is occurring, or rather the reverse trend of counter-urbanization! Many of the themes seem to be arbitrarily chosen and in some cases pointless. Are two large maps really needed to show us that only the USSR and USA have achieved interplanetary and manned space flight? What is the logic of the animals included (or omitted) from the page of maps on 'Hunting'? Why are kangaroos included, but not whales or any of the 'big cats'?

iii) The Atlas

The publisher's claim that the atlas represents an 'epoch-making' advance in our overview of the world, and is the greatest single advance in cartography in over four hundred years, are far from confirmed.

The atlas has achieved certain of its authors' objectives. Peters is convinced that much of the effectiveness of atlases still comes from traditional forms of workmanship. His aim was to produce an effective merger of traditional and modern techniques in order to retain the '[...] good feeling of handling an old atlas' (Peters, pers. comm.). The craftsmanship, of the topographic section in particular, is not in dispute and does capture something of the beauty of the Earth. However, both of the major sections are flawed. The problem can all be traced to the application (or not!) of the principles of the 'new cartography'. Peters dogmatically resists the use of any projection but his own in the thematic section. Yet in other circumstances he seems prepared to compromise his claim of 'universality' (that it is '[...] possible to unite in one grid system all the cartographic qualities which should be retained when converting the features from the rounded surface of a globe onto a flat map [...]' (Peters, 1983: 82). For example, by using a different projection to show the polar regions. This latter example also undermines his claim that projections need no longer be taught (myth 4).

The main lesson which this atlas provides is one that cartographers and geographers have already learnt, that is that map production is about flexibility, adaptability and compromise. There is no one correct answer!

Conclusion

The Peters' 'new cartography' has generated a great deal of controversy. The publication of the Peters Atlas will add further fuel to the debate. His ideas have been accepted and adopted by a wide range of organizations, including various sections of the United Nations (e.g. UNICEF). In the UK the main visible support is from the voluntary sector concerned with overseas development issues. The Peters phenomenon poses cartographers and geographers with a dilemma. Peters and his supporters have raised general awareness of the importance of maps in forming people's 'global image' to a degree which neither of these professions have managed in recent years. Yet the means by which he has done this are open to question. The atlas is evidence that his 'new cartography' is far from infallible in practice.

The reaction of the professions has been to vilify Peters, or to ignore him, in the hope that he may go away. This has not happened. It is time that his contribution (flawed as it may be) is recognized and used as a basis for constructive development, rather than continued defensive criticism.

Peters' sin is not a failure to defer to another cartographer in the naming of his projection, but his insistence that it represents the only, equitable, universally acceptable world map. His sin is not that he has questioned the bases of traditional cartography (correctly or not), but that he is seeking to replace it with his own dogmatic cult of the 'new cartography'.

Acknowledgements
I should like to thank Professor Arno Peters for allowing me to interview him during his short stay in England in March 1989. Thanks are also due to Longman (the publishers of *The Peters Atlas of the World*) for arranging the interview with Professor Peters.

References

Bain, I. (1984) "Will Arno Peters take over the World?" *Geographical Magazine* 56 (7) pp.342–343.
Baker, A. (1986) "Computers bring Map Projections back to life" *Geography* 71 (4) pp.333–338.
Brandt, W. (1980) *North-South: A Programme for Survival* London: Pan Books.
Brandt, W. (1983) *Common Crisis* London: Pan Books.
Crow, B. and Thomas, A. (1983) *Third World Atlas* Milton Keynes: Open University Press.
Gall, J. (1885) "Use of cylindrical projections for geographical, astronomical and scientific purposes" *Scottish Geographical Magazine* 1 (1) pp.119–123.
Garver, J.B. (1988) "New Perspective on the World" *National Geographic* 174 (6) pp.910–913.
German Cartographic Society (1985) "The so-called Peters Projection" *The Cartographic Journal* 22 (2) pp.108–110.
Johnston, R.J. (Ed.) (1986) *The Dictionary of Human Geography* Oxford: Blackwell.
Kidron, M. and Segal, R. (1981) *The State of the World Atlas* London: Pan Books.
Lloyd, D. and D'Souza, G. (1987) "Mapping NOAA-AVHRR imagery using equal-area radical projections" *International Journal of Remote Sensing* 8 (12) pp.1869–1878.
Loxton, J. (1985) "The Peters Phenomenon" *The Cartographic Journal* 22 (2) pp.106–108.
Maling, D.H. (1974) "A minor modification to the cylindrical equal-area projection" *The Geographical Journal* 140 (3) pp.599–600.
New Internationalist (1983) "The New Flat Earth" *New Internationalist* 123 (May) pp.22–24.
Peters, A. (1983) *The New Cartography* New York: Friendship Press.
Peters, A. (1989) *The Peters Atlas of the World* London: Longman.
Robinson, A.H. (1985) "Arno Peters and his New Cartography" *The American Cartographer* 12 (2) pp.103–111.
Saarinen, T.F. (1988) "Centering of Mental Maps of the World" *National Geographic Research* 4 (1) pp.112–127.
Saarinen, T.F., MacCabe, C.L. and Morehouse, B. (1988) "Sketch Maps of the World as Surrogates for World Graphic Knowledge" *University of Arizona, Department of Geography and Regional Development Discussion Paper* 88-3 pp.1–27.
Stalker, P. (1989) "Map Wars" *New Internationalist* 193 (March) (Wallchart edition: “The Globe at a Glance").
Tucker, C.J. and Justice, C.O. (1986) "Satellite remote sensing of desert spatial extent" *Desertification Control Bulletin* (UNEP) 13 pp.2–5.
Vujakovic, P. (1987) "The Extent of Adoption of the Peters Projection by 'Third World' Organizations in the UK" The *Bulletin of the Society of University Cartographers* 21 (1) pp.11–15.
Vujakovic, P. (1989) "Mapping for World Development" *Geography* 74 (2) pp.97–105.

THE SUC: 1964 TO 1989

A. Carson Clark, Terry Garfield, Jack Render and Eila M.J. Campbell

Originally published in Volume 23, No 1, pp.1–4

The following mini-papers were presented as keynote speeches on the occasion of the SUC's 25th Anniversary Summer School, at Milton Keynes (OU), on 4th September 1989. They review the founding of the society and some of its history.

Life Before the SUC

What was the standing of the University Cartographer before the formation of the Society of University Cartographers? He or she worked in comparative isolation, described as one whose ability was 'to assist the Academic Geographer in the preparation of maps', and they had to possess a high degree of 'manual dexterity' – a derogatory term used frequently by academics to emphasize their view of the inferiority of all non-academic staff. It should be borne in mind that most, if not all, were at that time employed in technical staff roles in the Universities, some years had to elapse before the great institution the 'Polytechnic' was to appear.

These and other attitudes led to the feeling of isolation experienced by all of these lone cartographers, usually only one per university Geography Department (where at that time most were employed), with the exception of some of the London universities, where the situation was much better due to the numbers of cartographic staff employed. To give a personal illustration, my entry into university life came after two years as a Surveyor with the Forestry Commission, three years on Ordnance Survey large-scale drafting, and I was currently attending a College of Art course in order to improve my design and artistic skills. I was a professional cartographer accustomed to discussing and sharing with colleagues in the general way in which people of the same profession enjoy sharing and discussing.

Nothing in my previous work experience could have prepared me for the new situation I was to experience on joining a university. Far from being treated as a professional cartographer, at 24 years of age the new boy was also tea boy and odd-job man.

Not surprisingly I can now see with the benefit of hindsight that my new colleagues were equally unsure of themselves. The only positive lead came from the excellent Head of Department, Professor F.J. Monkhouse, under whose tutelage I was greatly encouraged in my profession and role in the department. In the course of 10–12 years the totally unsatisfactory situation had been transformed into a modern, efficient, fully co-ordinated carto-photo-repro unit, meeting all the varied needs of an expanding department. In these early days the lone cartographer was actively discouraged from making contact with any other similarly placed cartographer, and only when a publication was shared between authors working in separate universities was it at all possible to speak with one's opposite number. Then one could discuss questions of style, design and lettering to be used on the maps, which were mostly illustrations prepared for publication in textbooks at this time.

So, as can be seen, there was a tremendous need to get together with other professional colleagues and to establish some form of uniformity of style and design. Early on in the attempts to form some sort of society of cartographers' training and qualifications formed an integral part of our aspirations.

The head of my department, Professor Monkhouse (ex-Liverpool University) had earlier introduced me to Alan Hodgkiss, who was also later to play a major role in the establishing of SUC. With approval obtained from my Head of Department, I wrote to Alan and sent some suggestions for possible training and qualifications, and invited him to send his ideas in return. Alan's Head of Department at Liverpool was Robert Steele, and he agreed that a co-operative arrangement between Southampton and Liverpool's University cartographic staff should exist.

When some sort of standards and training ideas were formulated, an approach was made to the Institute of Science Technology, an Institute already recognized by the Universities. The resulting course became known as the Southampton Cartographic Training Course. The course covered a brief history of cartography, basic principles and elements of cartography. I leaned heavily for source material on Professor Robinson's famous work *Elements of Cartography*. I wrote to him telling him how much I appreciated his excellent work and he returned his good wishes for the venture. The course was used for training each new trainee cartographer in the Southampton Department of Geography, and just a few other departments in the UK were able to make use of this facility. There were many difficulties in the organization and administration of the course. Although the primary element was by correspondence, on a practical level all attempts to get even half-day get-togethers with interested parties failed. It had to be in our own time and no funds were available to assist participants.

This was just one example of various attempts at

achieving some sort of training which would enable university cartographers to achieve a consistent standard of cartographic illustration. To quote from Arthur Robinson:

'The map must be legible, the symbolism or notation must be suited to the objective of the map, and the whole must be fitted together to make an efficient communication'.

Many were the disappointments and frustrations in the years prior to the advent of the first meeting of the Society of University Cartographers, and none of us could see how exactly we should move forward. Gradually more and more contact began to be made, particularly with those working in isolated situations. At one stage we estimated that around 30 cartographers must be employed in the various universities.

Reactions and feedback began to come back to me, and Alan Hodgkiss heard that the subject had been raised within such an august body as the Institute of British Geographers at their annual meeting. Some felt we might be a militant or at least subversive group of people who should be carefully watched. However, this was not the view held by Frank Monkhouse and Robert Steele, and in 1963 when a questionnaire was prepared by myself and Alan Hodgkiss it was countersigned by our respective Heads of Department. Responses to the questionnaire were generally encouraging, and this was probably the most important move towards the events of 1964 which led to the establishment of the SUC.

It should be noted that many of us felt that as founder members of the British Cartographic Society (in 1963) that maybe we should find our particular niche in the larger organization, but this was not to be. That Society was established with a far wider catholic taste in mind.

Early in 1964, when all the information regarding cartographic staff within the UK had been assembled, we had a somewhat surprise invitation from John Keates to a week's course in cartography at the Department of Geography at the University of Glasgow. This was attended by 21 cartographers, from 14 different university departments. During the course of this week those present decided to form a New Society to be named the Society of University Cartographers. Much discussion took place on the name for the society, but SUC was born. Those of us who founded the Society look back with gratitude to the three cartographers, John Keates, Gordon Petrie and Michael Wood, who provided the stimulus and gave us a push in the right direction.

The first Issue of the Society's *Bulletin* was published in November 1964. The covers were made of blue dyeline paper and the pages were simply stencil duplicator pages, but it represented a real landmark to those of us who had aspired to have our own society and had at last achieved at least the beginnings of it. The *Bulletin* was published without financial support under the very capable Editorship of Alan Hodgkiss. The Chairmanship was in the respected hands of Roy Versey, and the Secretary was of course Terry Garfield, who has remained a constant guide and very hard-working member of the Society.

At a later date the Wallis Award was generously provided by David Wallis (of Survey and General Instruments) to be presented for the best piece of cartographic work submitted by a member of the Society. This was again a tangible demonstration of our aim to improve cartographic standards within the universities.

Looking back today those of us who had a small part to play in seeing the establishment of the Society of University Cartographers would still like to pay tribute to all the many people that encouraged us to press on. Only the present day inheritors of SUC can decide if it was worth it.

A. Carson Clark

The Early Years

This will inevitably be more about the early years than the present day. There was an upsurge in cartographic interest in the country in the early sixties, a cartographic awakening if you like. A significant cartographic event took place in September 1963 when a meeting was held at Stamford Hall (Leicester University) by invitation of Professor Pye, Head of the Geography Department, of interested members of the cartographic profession from all over the country.

At this meeting in which John Keates, Lecturer in Topographic Science at the University of Glasgow, was a leading protagonist, it was decided to form the British Cartographic Society. There are several people present today from that meeting, and no doubt one or two more throughout the week. John Keates was later to become Editor of the BCS journal and eventually President of the BCS.

In the following September he organized a 'SUMMER SCHOOL IN PRACTICAL CARTOGRAPHY' at the University of Glasgow, details of which are referred to elsewhere. Following informal discussions in the Senior Common Room one evening John Keates suggested we ought to form our own group, and thus the SOCIETY OF UNIVERSITY CARTOGRAPHERS was formed, its stated aims being 'to promote and maintain a high standard of cartographic illustration'. The meeting was held on 25th September, and 17 members from 10 institutions were present – it lasted two-and-a-half hours.

The first Chairman was Roy Versey from University College London. I was in the combined office of Secretary/Treasurer (as membership grew a separate office of Treasurer was created and Guy Kingdon from Portsmouth Polytechnic was appointed to that post in 1969). He was Treasurer for many years, and it was largely due to his efforts that the Society was financially viable. Sadly, Guy died in 1987 – he would have been extremely pleased to be present at this 25th anniversary. We were a small but enthusiastic band; without the enthusiasm we could not hope to carry on. With 36 members in 1965 (seven of whom are still members today), we were to grow steadily to our present number of 214. From the beginning we tried to encourage overseas members. Oliver Dixon from Ife, Nigeria, was a founder member. Later he became a lecturer at Portsmouth Polytechnic, and is now in commercial cartography. A branch was formed as SUC CANADA – and for a time there was much interest in Africa and Australia. Certainly cartographic interest was strengthened in these continents and North America, and cartographic organizations were formed there.

Leicester was to figure significantly in these events. I was originally Secretary/Treasurer, then Secretary for 11 years, and then Chairman for four years. The SUC owes a lot to Professor Norman Pye and the interest and help that he was prepared to give to the Society.

The *Bulletin* has been with us since that year, although the numbered Volumes didn't start until 1966. Alan Hodgkiss from Liverpool put his heart and soul into the job of Editor. In the early *Bulletins* I included short descriptive notes on publications that had proved useful at Leicester, and which I thought would be of interest to members. From these small beginnings the Reviews section has developed and I have been Reviews Editor for the duration. The *Bulletin* was well received and has been the flagship of the Society over the years. For several years the *Bulletin* itself was typed in Leicester and the Reviews section alone for a longer period. Proceedings of the Summer School were published for several years and a Newsletter was started early on.

Occasional one-day meetings were specially arranged, and day and half-day visits. These were organized by willing members in various parts of the country. Anne Lowcock from Manchester was particularly interested in this idea and was always a willing organizer. At this juncture I ought to express our thanks to Anne for becoming our Archivist, collecting together and keeping an invaluable photographic record of SUC events over the years.

The highlight of the Society year is the Summer School, and the first one was organized by Alan Hodgkiss at Liverpool. It was intended from the beginning to include in this a practical element, a day of field excursions/visits pertaining to our links with Geography, and a special occasion in the Annual semi-formal dinner with speeches. We have had a series of outstanding Summer Schools all over the country, with two particular highlights being Rotterdam and Dublin, plus of course, the exciting venture to the United States by kind invitation of Noel Diaz (of UCLA) in 1987.

Rotterdam was a tremendous step forward into the unknown, made possible by the growing international flavour of the membership and the fact that Chris Moore (ex-Portsmouth Polytechnic) was the Chief Cartographer at the Economic Geography Institute. We were made very welcome by the Royal Dutch Cartographic Society and shared a joint day's programme with them.[1]

The Summer School has been held in 24 different venues to date, with Leicester the only place to have hosted twice, bringing the total to 25. From the first introductory meeting in 1964, the question of training and qualifications has always been at the forefront of our thinking, as the following minute shows:

'Training of Juniors: much discussion arose on this topic. It was one of the most important functions of the Society, as a good training scheme would ensure a high standard of cartography in the future. It was decided to look fully into workable schemes, to obtain ideas from everyone and to see if a nationally recognized scheme could be adopted. All Heads of Departments would be informed of this at a later date. Ideas on the training of Juniors should be pooled and modified accordingly. Mr Versey suggested that ideas be sent to the Secretary. There was much discussion on the question of who would do the training and how it could be done. Some standardization was obviously needed'.[2]

It may seem incomprehensible now, but we were actually in the process of setting up our own training scheme, with all its attendant problems. Standards were in our mind when the Wallis Award was implemented in 1975, and Dr Roy Boud (Leeds) was the first winner. Here again Leicester comes to mind as the only department to have hosted three winners.

Training eventually became recognized as a national issue, and due to the efforts and encouragement of Professor Kirk, formerly at Leicester, then Queen's Belfast and then Secretary of the IBG, we were represented at the original discussions at the Royal Society. Throughout the life of the Joint Committee we were represented through Carson Clark and Biki Wilson. I would like to record here the debt that the SUC owes to Mrs Wilson, for her efforts in the educational field over the years on our behalf.

Towards the end of 1978 Professor William Kirk wrote a final report on the dissolution of the Joint Committee for Education in Cartography. In it he wrote that 'those individuals involved in the founding discussions in the mid-sixties should, I think, be reasonably satisfied with the outcome'. These sentiments can also be applied to the founding members of the Society of University Cartographers. The Society today is an active, lively body which has its own part to play in influencing the development of its own particular sector of cartography and fulfils this role, I think, with success.

We have been extremely fortunate in our choice of Honorary Presidents – they have all been good friends of the Society and extremely helpful. Our Chairmen, Officers and all Committee Members have always been devoted workers for the Society but without the members we would not exist and throughout our first 25 years they have been particularly supportive and our thanks are due to all of them.

Finally, we should place on record the help received actively in the organization of the Summer Schools, and continually behind the scenes, from our Geography Departments and Departmental Heads. We were originally created through this geographical interest, and I hope this primary link continues and flourishes.

Terry Garfield

Notes

1. See paper by Garfield, T. (1980) "Cartography in Universities and Polytechnics in the British Isles with Special Reference to the Society of University Cartographers" *Proceedings of the SUC Summer School, Rotterdam, 1980.*
2. From the minutes of the meeting held on 25th September 1964.

SUC Productions

Throughout the history of the Society the *Bulletin* has had an important role as link to members of the Society. From the first issue described by Carson Clark to the present the Bulletin has gone from strength to strength. It now has an international circulation of 500 copies, being roughly half to members and half to subscribers.

In the 25 years the Editorship has passed from Alan Hodgkiss to myself, and then on via Andrew Tatham and Bob Parry to its present Editor, Steve Chilton. Throughout this period the common link has been the Reviews section, which has always been edited by Terry Garfield. Such is the growth of the Society that there are now separate posts of advertising and subscriptions managers. The other common feature has been the constant appeals from editors for more contributions, although there has always eventually been enough material to produce the issues.

By the time the *Bulletin* was edited by myself at Portsmouth Polytechnic the production had improved to one of photo-reduced typescript. We even managed to include some 'colour supplements' and advertising leaflets. Well remembered are the label-sticking sessions and the parades to the Post Office to distribute the *Bulletins*. All the while economy of production was important in such a small society, with the thrift of Treasurer Guy Kingdon always a spur to greater efforts. In this respect we have always been fortunate in receiving so much goodwill and active help from our host departments.

There have been several spin-off products from the Society over the years, including the booklet (Map Preparation) and the videos produced by John Hunt at the Open University. Whilst it has been easy to look back on the last 25 years, it is surely impossible to speculate on the cartographic world of 25 years hence, but it is an interesting exercise, and we will have to repeat this process of looking back.

Jack Render

President's View

It is always difficult to look back objectively down the avenue of time. Academician I.P. Gerassimov of Moscow once told me that one should look back only to go forward with greater alacrity. I find it difficult to believe that the Society of University Cartographers (SUC) is 25 years of age because it still retains perhaps the most important attribute of youth – enthusiasm.

As you well know, the SUC was founded within a year of the British Cartographic Society (BCS). Several of those who founded the SUC were members of the BCS, but they believed that cartographers working in university departments of geography and geology – 25 years ago no technical college or polytechnic had a cartographer – needed a society of their own, one that would cater for their particular needs. One must remember that in 1964 few, if any, cartographers or cartographic draughtsmen/women in the UK had received formal training in the work that they were expected to carry out. Most of the more senior cartographers of the day had 'trained' themselves. In the larger departments, the more experienced cartographers trained their younger colleagues. At the time, there were no full-time or part-time cartography courses in institutions such as Luton Technical College or Oxford Polytechnic. The young university cartographers had no corporate identity, and there was little done for them. Those working on their own without another cartographer-colleague often felt isolated. I mention all this because I believe that in 1964 the SUC fulfilled both the need for, and the want of, an organization that would bring together young cartographers, most of whom were working in university departments of geography, give them a corporate identity and increase their professional status. The SUC has achieved all this.

Members of the SUC have always emphasized the friendly and informal nature of the Society. There is no reason why the Society should change even if it decides to drop the word 'University' from its name in order to extend full membership to 'all persons engaged in cartography, cartographic reproduction and the care of maps in the fields of education, local government and the private and public sectors'. But change it will. The Society may well have difficulty in convincing young cartographers working in, for example, local government or the private sector that they should join the 'Society of Cartographers' rather than the BCS. The newly named society might be well advised to campaign for members among young cartographic trainees/students before they qualify, and perhaps give them a reduced subscription during the first three years that they are in post!

There is need for the SUC to be different from the BCS and to retain a uniqueness – unpopular as the word may sound to some of you. At present, the SUC is notable for its annual Summer School and its twice yearly *Bulletin*. The Summer Schools have always been excellent. However, as the years have passed, the Summer Schools have changed – the workshop element has diminished and the annual 'summer' gatherings are now more akin to a conference or a symposium than to a Summer School. The 1970s and 1980s have seen fundamental changes, as digital and automated cartography has become dominant.

The SUC has always been closely associated with geography and geology departments in institutions of higher education. It is thus not surprising that the four Honorary Presidents to date – in order of serving: Robert Steele, Eila Campbell, Bill Kirk and Chris Board – have held posts in Geography Departments during the time of their presidency. We are all very conscious of the honour the SUC afforded us in inviting us, and each of us has very much enjoyed the association with you all. Two of your Honorary Past Presidents (Steele and Kirk) were presidents of the Institute of British Geographers, and two (Board and Campbell) are, or were, Chairmen of the British National Committee for Geography's Subcommittee for Cartography – i.e. the UK's link with the International Cartographic Association.

The SUC enjoys a high standing in the international world of cartographers. Mainly it is known because of the *Bulletin*. Unfortunately, over the years few members of the SUC have been able to attend ICA Conferences, partly, if not entirely, due to the high cost. But in 1991, as many of you will know, the ICA will meet in Bournemouth, and hopefully some of you will be able to go. Attending an ICA meeting is an interesting experience. One meets cartographers from all parts of the world. Here I should perhaps emphasize that papers read are not 'elitist' in any way. If you have a paper which falls within the Bournemouth Conference's terms of reference, I urge you to offer it in good time. There will be a poster sessions as well at the conference.

Before I end I should like to wish the Society another 25 years of success, and express the hope that I shall be alive and fit enough to attend your 50th Anniversary meeting in 2014!

Eila M.J. Campbell

INTRODUCING GIMMS-PC

A microcomputer version of the GIMMS mapping and graphics software package is currently under development for the IBM PC/PS range (including good compatibles).

This is a *full implementation of GIMMS* with the full system available as on mainframe and mini computers.

COMPUTER REQUIREMENTS

IBM PC/PS Range (or compatibles) MS-DOS or OS/2

(1) IBM PC/AT or upwards (i.e. 80286 or 80386 processor)
(2) Floating Point co-processor (i.e. 80287 or 80387 co-processor)
(3) At least 3 Mbytes of memory (Extended memory for MS-DOS)
(4) MS-DOS 3.3 (or later) OR OS/2 operating systems
(5) Hard disk with at least 5 Mbytes available space
(6) Parallel port with 25 pin D connector

GIMMS-PC has been in use at existing mainframe sites since March 1989, and a version of GIMMS-PC with menu enhancements will be available for general distribution during the summer of 1989.

For further information and prices, please contact:

Marlene Ferenth, Sales & Marketing, GIMMS LTD,
30 Keir Street, Edinburgh EH3 9EU, Scotland
Tel: (031) 668 3046 Fax: (031) 668 2104

25 YEARS — WHERE NOW?

John C. Robertson

Originally published in Volume 23, No 1, pp.5–8

The following is a personal view of cartographic education, which parallels the first 25 years of the Society of University Cartographers. These are personal views, and other members of staff at Oxford Polytechnic would probably express different views.

Introduction

Twenty-five years of the Society of University Cartographers (SUC) were enthusiastically celebrated at the Milton Keynes Summer School. Twenty-five years is also the approximate period of existence for many of today's courses in Cartography, e.g. Luton, Oxford and Glasgow. Although much has been achieved, cartographic education stands at a vital crossroads with the advent of computers and GIS. It is time to review the situation. The key question is 'Where are we going?' Do we know? I am not sure that we do.

Status

Difficulties arise from the ongoing lack of professional status for Cartography. Even the attitudes of other academic staff reflect a rather 'poor relation' image. They are often not convinced that Cartography is a real academic subject. Indeed some will openly regard it as 'Mickey Mouse'. But Cartography may not be alone in that regard. In a recent novel it was asserted that 'there was more Mickey Mouse in Academia than in the whole of Disneyland' (Rivers, 1987).

In many ways cartography is still trying to establish itself in the public eye. People often have only a hazy notion of what we do. 'What is that you teach, photography?' Even when they are aware of maps their linkage to the map producer is equally vague. This can be expressed in the words of a poem (actually a song but we don't have singing journals yet):

'People ask me what do I do
How do I spend my day
I'm away from nine till five
What do I do anyway
I tell them I'm a cartographer
They always say "What's that?"
I tell them I'm a cartographer
A man who makes a map
"A man who makes a map" they say
What a funny thing to do
"A man who makes a map" they say
"I would just buy one if I were you"

(J.C. Robertson)

Even my current Head of Department, a Civil Engineer by profession and not bigoted by any means, has still asked in all seriousness, 'Is there an academic basis for Cartography?'. It is not always easy to articulate the necessary response. He has an open mind. He is prepared to be convinced. But isn't it a little sad that he NEEDS to be convinced?

What, How, Why

Those involved in teaching cartography know WHAT they teach. It is mainly a matter of syllabus construction. Endless articles on 'Cartography at'... Oxford, Luton, Melbourne, Omopoggo University, etc... testify to that. Subject to individual variations they know HOW to teach the material. After all, just like driving, we are ALL brilliant teachers! We are certainly better than anyone else we listen to, because they have faults. But do we really know WHY? No doubt some will turn to the aims and objectives stated in submissions to CNAA as being evidence that they do. But these are often stated in very pompous terms, and drafted together in a peculiar form of academic jargon. You know the sort of thing:

MODULE XYZ

'The student should achieve an understanding of the basic principles of cartographic representation of spatially orientated data and its subsequent percepto-psycho recognition by actual or potential users, with particular regard to any possible discordance that may occur during the ensuing application process'.

Fine. But does that mean in actual teaching? Usually any interpretation one wants to make of it. Syllabi are seldom the children of academic purity. Limits are imposed by time, cost, equipment, and, especially, the abilities and interests of the staff available. Also, after ten years of a basically anti-education Government, reduced resources have led to much of the co-operation with service subjects disappearing. It is 'man mind thyself' and the defence of Full-Time Equivalents (the numbers game), hence funds and sometimes jobs. But much of the lack of direction in education stems from a wider difficulty of definition of cartography and its professional status.

In Times Past

Looking back over some material I had written at intervals,

some of the quotes or queries could still be made today. The problems are not new, though some may well be exacerbated. By no means do I claim to be any sort of all-seeing 'Guru of Cartography', but there is an ominous consistency of statements made in various places at various times. For example:

'How many geographers consider themselves to be overall authorities on maps, without whose aid they would never be produced?' (Robertson, 1965).

'If you are thinking of launching into a career in Cartography, be warned! There is no well-defined beaten track to follow. You will have to hack your own way through the scrub and bushes of ignorance and prejudice' (Robertson, 1967).

'This lack of appreciation of cartographic principles is both rather widespread and disturbing [...]. There is a great need for a crusade from the cartographic front into the ranks of the visual applicators to instil in them the principles of the discipline' (Robertson, 1968).

'It must also be faced that the market is, by the nature of the high degree of specialization involved, extremely limited. Thus any duplicity between courses at various Institutions and/or over supply of graduates would seriously weaken the standing of cartography [...] to say nothing of being extremely unfair to the graduates themselves' (Robertson, 1976).

'The very concepts of education and/or training are in themselves fundamental. Training implies an ability to perform a set function without the necessity of understanding either the ins and outs of the operation or the place of that function within the total framework. Dogs, horses and monkeys can be trained. It is not possible to educate them (exceptions may be made for Lassie, Mr Ed and Cheetah!). To educate means to instil an understanding of the concepts involved, to put learning processes to work and to permit possible innovations and advances' (Robertson, 1977).

'There then is a very clear message for cartography and cartographers if they are prepared to listen to it and act accordingly. Some hidebound attitudes may well have to be discarded and some brave steps taken into a future largely unknown...' (Robertson, 1978).

'Again there is evidence of Jack of cohesion in both cartography and education [...] it is an uphill struggle as each educational institute battles for survival and is therefore reluctant to give up [...] any courses whilst students continue to enrol' (Robertson, 1984).

Getting There

Why has Cartography arrived at this position? Why, after twenty-five years as, at least in some quarters, a recognized academic subject, do the graduates produced from the system suffer, in the main, from being classed as second-class citizens? The British Cartographic Society (BCS) must shoulder some of the blame. It was the Society that could have, and should have, put Cartography in Britain on a professional basis. That they didn't was in some measure due to concerted opposition by other professions, particularly geographers. But the nature of development over time of the BCS was also a contributing factor.

Firstly there was the 'era of the Brigadiers'. No problems for them. They were 'officers and gentlemen' and therefore people of good standing by right. But they did little to enhance the position of the 'troops'. Also dominant was the O.S. attitude that only they produce REAL mapping. Anything else is a Toytown Assembly Kit which only has credence if it has the label 'Based on the Ordnance Survey by kind permission of... '.

They were followed by the 'giants of industry', many of them men of great ability and long experience. But once again they were in a comfortable personal position with expense accounts and directorships beckoning and didn't have the need to strive for overall standing of the profession.

Latterly BCS has been the stage for members of the academic world. Certainly efforts by such as John Keates and Michael Wood must be regarded as notable and significant. But the main result was to galvanize the people in the industry to get moving and upgrade courses, etc. An important achievement, but they have not managed to convince the world at large.

Upgrading

In any case, is the continual upgrading of qualifications necessarily the answer? An example can be taken from surveying. In the early days people bashed in pegs for a couple of years or so whilst training on the job to eventually become a Licensed Surveyor. Then it was decided that a Certificate Course was required; then a Diploma; then an Ordinary Degree and eventually an Honours Degree. This is fine if supply equals demand. But if competition between levels culminates in over-production at any level, graduates end up taking lesser demanding, and lower paid, jobs despite their qualifications. In the meantime, the person of lower academic ability, who may have been quite happy bashing in the pegs, hasn't got a job at all. Thus the system produces two types of dissatisfied people.

This could equally apply to Cartography and I find it disheartening to consistently meet graduates of three to five years ago who have now opted out of the industry. Complaints usually are a lack of any career structure and a 'flat' salary profile. When an Honours graduate with three years' experience is earning £8,600, even in a genuine part of the cartography industry, something is wrong. It is not the fault of the employer – they have to survive in a fiercely competitive world where people are not really prepared to pay for maps.

Aims and Needs

Therefore it could be questioned whether there is room for an Honours course in 'traditional' cartography producing large numbers. Yet the question of numbers as related to job availability produces glaring examples of Institutional immorality. I have been told at other Institutions as well as my present one, that, quite frankly, it is 'not our worry' whether or not jobs are available at the conclusion of any particular course. I believe that it most certainly is if the course is classified as 'vocational'. This raises the continued unresolved dilemma in Cartography of vocational versus educational, as well as technical versus academic. Cartography can be offered as an education course but it would be very different in aims, objectives, content and structure from those presently run.

There really needs to be national aims and targets particularly in terms of overall numbers of graduates produced. But who is to set these targets? Who decides? If it were agreed that there was over supply, who closes? In theory CNAA should control this but that august body also displays a propensity for a rather vague notion of what cartography is all about. Should it be the industry itself through BCS? Probably, but imagine returning from a conference to your own Institute saying it has been decided, by cartographers, to reduce the number of courses and this one should be deleted. Very popular!

The attitude of individual Institutes in a 'market-orientated' education industry is now ultra-competitive, not cooperative. It is the second round of the Graduate FA Cup. It is Oxford VERSUS Luton. The winners meet Glasgow in the next round! Attempts have been made by SUC to institute national training at technician level. But many members come from the 'wrong side of the academic blanket' and do not have the clout. As a result they are not usually involved in planning and policy. This is very wrong as it undervalues their role.

Progressive Studies

Queensland are experimenting with a novel form of progressive education. If converted to Cartography it would mean that ALL students interested in Cartography would go to College A for two years. A certain number would qualify with a HND whilst the rest would progress to Polytechnic B for a year. Some would graduate with an Ordinary Degree (or Diploma) whilst the remainder would go on to University C to emerge after a further year with an Honours Degree or go on to Post Graduate. There are many fascinating educational advantages to the system but it is almost impossible to even imagine British Institutions dovetailing to that extent, especially now. And what about 1992? By then maybe all cartographic education will be concentrated at Enschede. I do not necessarily propose this, but the accountants who now run education might.

Society Membership

BCS should have become, in part at least, the professional society. Certain jobs should require 'capital M' Membership. Then educational aims and objectives could be much more clearly defined. Australian cartography went through this phase in the late '70s. There were many, often acrimonious, debates within the Institute particularly from the entrenched draughtsman fraternity. It finally revolved round the Government gradings and the Civil Service Commission asked a fairly simple question 'What does a cartographer do that a draughtsman doesn't?' In the end computerization of the mapping process provided the key. Job definition then led to education aims, though it is interesting that today there are now no pure cartography degrees in Australia, but only cartographic streams within degrees in Land Information Systems. But the amount of change over the last six years in the Cartography course at my previous Institute, the Royal Melbourne Institute of Technology, contrasts with relatively slow change at my current one.

Computers Are Magic

Are computers and, in particular, GIS the answer? Maybe in part, but there are also dire warnings for Cartography. Whilst the 'push button' boys churn out maps with seeming ease there are some dangerous precepts emerging. The public are being conned into thinking that any map produced by a computer, especially a complicated one, is 'magic' and, by implication, correct. Yet often the very basic principles of Cartography are blatantly ignored. A badly designed map will still be badly designed even when produced on a computer. Indeed easy choices of such items as lettering make an awareness of design principles more vital than ever. The old computer adage applies GIGO – Garbage In, Garbage Out.

Direction

Where is the direction that Cartography and cartographic education should be heading? Even the latest definition of Cartography by ICA could almost be a definition of a GIS. Maybe this is correct as it could be considered that a map is a GIS. Are we back to age-old problems where maps are merely tools? But tools are made by craftsman not fellow professionals. Old attitudes of a second-class undertaking are emerging compounded by the computer men. Cartographic education is generally starting to move in that direction as evidenced by recent developments at Kingston and Luton. The latter has certainly long held a well-deserved reputation for producing the kind of people the industry wants, suggesting their direction may well be the right one. But I still feel haunted by the spectre of national over-production swelling the ranks of unemployed graduate cartographers.

Conclusion

In conclusion, many of the difficulties for cartographic education are concomitant with the position of professional status. Those involved in cartography have a general feeling that it is a justifiable professional calling; but they have yet to convince others. Maybe I am wrong by even talking at Milton Keynes, or publishing in this journal. That is more or less preaching to the converted. Maybe this article should be aimed at journals for geographers, planners, environmentalists and so on, persuading them as to the importance of cartography and cartographers. On an even wider basis there needs to be a sort of cartographic evangelical task force spreading the word.

Twenty-five years of SUC has justifiably been a celebration of achievement. I regret adding a sour note that all may not be fine and dandy. But things change, recently very rapidly with the advent of computers. Education, if anybody, must be responsive to change. I have posed questions but not given many, if any, answers. But if one believes something to be the case, even though it might eventually be proven wrong, personal professionalism says 'stand up and be counted'. Divorcing professional status and educational aims brings problems. Britain has by and large attempted to upgrade educational qualifications and hope professional recognition will somehow follow. This is the wrong way round and as a consequence, over-education for the existing, and immediately forseeable, market may well be taking place.

Cartography had its glory days, even marked with Royal patronage, in the Age of Discovery. A famous name from that era was Christopher Columbus. But if you think about it, Columbus was a man who, when he set out didn't know where he was going; when he got there, didn't know where he was; and when he got back didn't know where he had been. Is cartographic education in Britain in a similar situation? The reader may agree or disagree with my views as they see fit – but at least think about it!

References

Rivers, C. (1987) *Intimate Enemies* London: Futura Publications.

Robertson, J.C. (1965) "Are we map conscious?" *GE* (Geography Department, Edinburgh University).

Robertson, J.C. (1967) "A cartographer in government service" *GE*.

Robertson, J.C. (1968) "Education and training for cartographers – dovetail or dichotomy?" *The Bulletin of the Society of University Cartographers* 3 (1) pp.6–12. (Also delivered as paper to Symposium of the British Cartographic Society, Brighton, 1968).

Robertson, J.C. (1976) "Trends and developments in the education of cartographers" *Proceedings of the 2nd Australian National Cartographic Conference, Adelaide.*

Robertson, J.C. (1977) "Where the Hell are the Indians? *Proceedings of the Seminar on Education of the Technician Cartographer, Canberra.*

Robertson, J.C. (1978) "Future Shock and its Application to Cartography" *Proceedings of the 3rd Australian National Cartographic Conference, Brisbane.*

Robertson, J.C. (1984) "Simplicity is the Keynote" (Guest Editorial) *The Bulletin of the Society of University Cartographers* 17 (2) pp.61–63.

The 1990s

In 1995, I began my undergraduate degree course in Cartography and Geography at Oxford Brookes University (at what had been Oxford Polytechnic before 1992 and Oxford College of Technology before 1970). By this time, of course, advances in personal computing and the development of vector graphics software had meant that WYSIWYG cartography was more of a reality than a dream and it was rare to find mapmaking outfits that still relied on 'traditional' methods. Nevertheless, the modular Cartography course (in which geodesy and spherical trigonometry were compulsory) encouraged students to appreciate both manual and computerized methods of cartographic production; first we were trained in using drawing ink on Herculene drafting film and scribecoats before moving on to Apple Macs to create the same map of France.

One of the core texts for the course was the sixth edition of *Elements of Cartography*, which had been published that year and had long been established as 'The Bible of Cartography'. The content had seen much expansion since the first edition and unlike some earlier editions that had perhaps shown some reluctance to keep abreast of new technology and its application within cartography, this latest (and last) edition with Arthur Robinson at the helm was not afraid to embrace GIS, remote sensing, and automated cartography. Indeed, this reflected the wider reality of what was expected of the modern cartographer – to be competent in the acquisition, extraction and interpretation of spatial data as well as in its design and visualization.

Nevertheless, by the start of the decade, a new 'nature of maps' was in the making. Brian Harley, a British historical geographer who had been appointed Professor of Geography at the University of Wisconsin-Milwaukee in 1986, published several papers (e.g. Harley, 1989; 1990; 1991) that would champion a new paradigm and challenge the view of maps as objective, value-free documents. Taking up Harley's cause after his death in 1991, Denis Wood's *The Power of Maps* of 1992 argued that 'maps serve interests' to a wider public, and on my brother's recommendation (who had studied Geography at the University of Sheffield some years before), I read it – and was hooked. Along with other texts, such as the popular *How to Lie with Maps* (Monmonier, 1991), maps – and the choices behind their representations – became more accessible for scrutiny and critical examination.

Of course, many have criticized Harley, some arguing that he did not go far enough in his reading of French philosophers Jacques Derrida and Michel Foucault (e.g. Belyea, 1992) and others (e.g. Keates, 1996) that he overestimated how much autonomy cartographers have and offered no practical guidance. Certainly, the approach provided no help towards the discipline's progression along a *positivist* scientific trajectory, even if it served to renew society's fascination with maps. For cartographers actively creating maps in university Geography departments, however, there was little to celebrate. Maps were being seen as subjective and biased representations that removed – rather than lent – an air of authority to the presentation of rigorous academic research. Despite Harley's call for greater ethical responsibility, maps were treated with greater suspicion. The overall demand for academic cartography declined and the availability of satellite imagery – supposedly free from the convoluted politics of representation perceived in maps – became more attractive as the map's more 'objective' cousin.

In the process of exploring and visualizing spatial data, however, maps were crucial and GIS modules – and courses – became more and more popular in UK universities. The development of GIS software to work within a Windows-based environment in the early 1990s (e.g. Esri's ArcView) broadened the appeal of GIS and ensured its users no longer had to dabble in programming languages such as UNIX in order to perform a range of spatial analyses. The publication of Alan MacEachren's *How Maps Work* in 1995 brought a greater understanding of how maps operate as tools and fostered their (re)adoption by those using maps to explore and analyse spatial data.

The papers we have chosen from this decade reflect the changing approach to maps and provide an insight into how this was perceived within the professional context of university cartography. There is a defence of why cartographic design and its associated skills are important, a call for cartographers to see how then can apply their expertise to web design, and a vision of a world without maps – and hence cartographers! These papers sit alongside others which show how maps are being used to explore and present data in new ways (e.g. cartograms from census data), demonstrate new approaches to better understand map design, show how maps are being used to illustrate stories in the press, and to map other planets.

Gary Brannon, who moved to the University of Waterloo, Ontario, Canada as a university cartographer from the UK in 1969, contributed several thought-provoking papers to the *Bulletin* during the 1980s and

1990s. Two particularly poignant articles are included in this section and provide a special glimpse of the atmosphere of uncertainty at the time, particularly of how technological advancement was perceived as a threat to cartography. All papers in this book are products of society and it is revealing to appreciate them from a different perspective today.

Providing a somewhat different appraisal of the status of cartography, a constant theme running through the contributions made by Mike Wood is his emphasis on the enduring relevance of cartographic design – whatever technological and societal changes may bring. But this does not imply that cartographers should sit on their laurels and he encourages greater integration with other disciplines and sharing of expertise for cartography to flourish. Perhaps the following article, by Danny Dorling, exemplifies exactly this. British Academy Fellow in the Department of Geography at the University of Newcastle-upon-Tyne when he presented this paper at the Summer School at that university in 1994, Danny has since published hundreds of papers and books in collaboration with many authors. The paper here touches on themes of social inequality in Britain and the visualization of that data – topics that he continues to explore in his new post as Halford Mackinder Professor of Geography at the University of Oxford.

Henry Castner, Emeritus Professor of Geography at Queen's University, Kingston, Ontario, presented his paper on map design at the Society's Summer School at Nottingham Trent University in 1996. The basics of visual perception are covered in the second edition of *Understanding Maps* by Keates which was published the same year and this paper provides some especially useful examples to illustrate how map design is influenced by what and how we see.

Nick Tasker, who was to become President of the Society, was a Cartographic Analyst at the UK Hydrographic Office when his article encouraging cartographers to appreciate the transferability of their design skills was published in the *Bulletin*. Accelerated by a decline in the demand for the creation of new maps, the remit of university cartographers was sometimes broadened to incorporate the design and maintenance of departmental websites. Nick encourages readers to see this as an opportunity, not a threat.

At a time when maps are more widely used now than at any point in history (Parsons, 2014), Gary Brannon's second contribution – a vision of the world without maps – is so strikingly bleak that we begin to question the circumstances that gave rise to such a hopeless assessment. Yet it is difficult to remember what the world was like before the changes in the next decade that were to so dramatically alter the availability and accessibility of spatial information to such a far-reaching extent.

The next article, by Peter Vujakovic, was one of the first surveys of how maps are used in the media, in this case, five 'broadsheet' newspapers. The innovative use of cartograms to display the findings, together with the 'corrected' maps showing the range of Taepodong missiles from North Korea set the tone for the follow-up paper published in Volume 43 ('New Views of the World: Maps in the United Kingdom 'Quality' Press in 1999 and 2009).

Although Bob Parry had worked with Chris Perkins to produce two editions of the monumental *World Mapping Today* – a nation-by-nation survey of (principally topographic) mapping – by 2000, his thoughts were literally out of this world when contributing the article we reproduce here. Bob's survey is a fascinating and concise – yet comprehensive – insight into planetary mapping, a subject not encountered very often in the *Bulletin* and other general cartography journals. Slightly closer to Earth, the year 1999 saw the launch of IKONOS, the first satellite to offer commercially available imagery at the resolution of 1 m, and with it a taste of things to come.

Alexander J. Kent

References

Belyea, B. (1992) "Images of Power: Derrida/Foucault/Harley" *Cartographica* 29 (2) pp.1–9.

Harley, J.B. (1989) "Deconstructing the Map" *Cartographica* 26 (2) pp.1–20.

Harley, J.B. (1990) "Cartography, Ethics, and Social Theory" *Cartographica* 27 (1) pp.1–23.

Harley, J.B. (1991) "Can There Be a Cartographic Ethics?" *Cartographic Perspectives* 10 pp.9–16.

Keates, J. (1996) *Understanding Maps* (2nd ed.) Harlow: Longman.

MacEachren, A.M. (1995) *How Maps Work: Representation, Visualization, and Design* New York: Guilford Press.

Monmonier, M. (1991) *How to Lie with Maps* Chicago: University of Chicago.

Parsons, E. (2014) "Ambient Location" *Public lecture given as part of the 50th Anniversary celebrations of the Society of Cartographers, University College London, 25th July.*

Robinson, A.H., Morrison, J.L., Muehrcke, P.C., Kimerling, A.J. and Guptill, S.C. (1995) *Elements of Cartography* (6th ed.) New York: John Wiley & Sons.

Vujakovic, P. (2009) "New Views of the World: Maps in the United Kingdom 'Quality' Press in 1999 and 2009" *The Bulletin of the Society of Cartographers* 43 (1&2) pp.31–40.

Wood, D. (1992) *The Power of Maps* New York: Guilford Press.

IS MAP DESIGN A LOST ART?

Gary Brannon

Originally published in Volume 25, No 2, pp.7–8

As the profession of cartography prepares to enter the 21st-century, the emphasis placed on map design appears to be in serious decline. The author contends that today's cartographic entrant often lacks this fundamental skill, and that this trend will inevitably lead to utilitarian maps and safe, uninspiring delineation.

The Importance of Design

The crucial test of any map is its ability to communicate effectively. When a map is said to have failed in this respect, the resulting postmortem will most often lay the blame, not on the accuracy of the data nor the craftsmanship of the cartographer, but squarely at the door of poor design. Design is the most difficult of skills to teach, and the most important to learn.

Design can be loosely defined as being the organization of graphic elements within a space whose dimensions are finite. The objective of cartographic design is to represent data, either qualitative or quantitative, in a manner which can be readily understood by the reader. The cartographer's dilemma, with respect to design, is that he or she is often confronted with a vast array of variable combinations. Lines, type, symbols, tones, textures, hues and, perhaps, statistical data, must all be brought together to form a cohesive image. Add to these elements the fixed parameters of sheet size and scale, and the problem becomes so complex that it can often only be solved by a series of measured compromises. It is the cartographer's task to take this melange, this superfluity of information and skilfully present it in a meaningful graphic form. The skill that is required is design, and good design skills must be cultivated through study and practice.

Illusion and Communication

We often confuse aesthetics with good map design. Certainly an attractive cartographic product is pleasant to behold, but it must also be functional; communicating effortlessly with the reader. It must convey the necessary information in a clear, straightforward manner. It should be simple, uncluttered, legible and unambiguous. Repetition is a valuable ally of the cartographic designer. The repetitive use of symbols, for example, enables us to recognize many map features without recourse to a legend. Such symbols have become loose conventions that can be used as a framework for communication. The ability to communicate in a graphic form, like so many other creative processes, is partly intuitive and partly learned. It demands that strictly objective factors, such as the width of lines, the size of symbols and the placement of type, be carefully balanced with subjective considerations like clarity, balance, harmony and, yes, aesthetics. Occasionally the solution to a cartographic problem will be immediately apparent. Though more often it will require a compromise. While there are a few conventions in cartographic design, there are in fact no hard-and-fast rules. What rules there are, are merely guidelines to alert the unwary from straying too far into the world of abstraction.

Map design has much in common with illusion. A cartographer must take a piece of our real world, scale it down many hundreds or thousands of times, reduce highways and rivers to the thickness of an inked, scribed or computer-generated line, work magic with symbols and patterns, tactfully add labels where they are required, yet somehow still preserve the illusion of terrestrial reality. It is no easy task.

Art or Science?

The question often arises as to whether cartography is a technical or a pictorial art. Much has been written about the subject and the debate within factions of the profession seems endless. The great Swiss cartographer Eduard Imhof (1895–1986) expressed a popular sentiment when he wrote '[...] Artistic talent and aesthetic feeling, the sense of proportion and harmony of colour and form are indispensable in the production of a beautiful – and by this I mean an easy to read and expressive map. Theoretical cartography is thus a technical science with a strong artistic trend'.

The science of cartography is rooted in mathematics, governed by geometry and scale. The art of cartography is similarly governed, but by the geometry and dimensions of the printed page, the proportions of symbols and type, and the value of percentage tints.

The making of a map is a highly selective process. The degree of generalization or the emphasis placed on any particular feature are decisions made by the cartographer at the time of preparation and are governed by the ultimate purpose of the map. It would be virtually impossible to produce a 'perfect' map. Even if it could be done, who would there be to judge it? For each map user might interpret the information it contained in an entirely different way. Clearly the best that a cartographer can ever hope to achieve is to present an abstraction of geographic reality, in a systematic manner, following the general rules of cartographic convention without becoming handcuffed by them, and conveying the necessary information in a way that preserves graphic harmony and integrity.

Design in the Computer Age

The computer age, while facilitating the easy preparation, revision and updating of maps, has done nothing for their design. Rather the reverse has happened. When maps were habitually produced by manual methods – each line etched or drawn with consummate precision; each name placed according to a precise and innate sense of logic; each symbol an essential part of the whole – they lent validity to Eduard Imhof's assertion that 'artistic talent and aesthetic feeling' were indispensable elements in the production of any worthwhile map. Computers have dehumanized cartography to a startling degree, reducing art to science almost overnight. While new entrants to the profession soon seem to master the technical skills needed for computer graphics, the nuances and subtleties of cartographic design seem increasingly to be a lost art. Excessive generalization is accepted as the inevitable price of progress, type fonts are limited by cost and printer compatibility, screens and patterns by their availability. It is cartography by compromise rather than design.

One need only peruse the wonderful English county maps of Christopher Saxton (1542–1606), John Speed (1552–1629), and Richard Blame (1660–1705) or the North American masterpieces of Venetian cartographer Antonio Zatta (1757–1797), replete with muted colours, distinctive yet legible lettering, and bold, stylized depictions of relief, to appreciate fully the aesthetic legacy that early cartographers bestowed upon us. These are beautiful maps, yet they are functional, expressive and easily read.

The Decline into Mediocrity

The maps of the 20th century have become increasingly utilitarian and, to a great extent, entirely devoid of character. This can be largely blamed upon the development of 'modern' techniques: scribing, typesetting, photography, and the computer. As the process of map-making became faster, simpler and more cost-efficient, aesthetics soon began to take a back-seat and design became a matter of safe, uninspiring delineation.

Examples of cartography's wholesale decline into mediocrity are all around us. Pick up any commercial road map and compare it with another, or any modem atlas with its neighbour. The striking similarities, which are surely no accident, simply provide more evidence that the age of homogeneity and mass-production has arrived.

Although design has always been a neglected facet of cartographic education, it was in the past somewhat ameliorated by the fact that many of those entering the profession had both an artistic bent and an instinctive talent for design. Today's cartographic entrant, however, being the product of an electronic world, often lacks this fundamental skill. Cartography in the 21st century seems certain to follow the well-trodden path laid down in the last five decades of technological change. As more and more maps are produced digitally, design skills will inevitably continue to diminish and aesthetics will no longer be in the vocabulary of the map-maker. At that point, I fear, map design will become a lost art.

References

Brannon, G. (1992) *Practical Cartography* Waterloo: Escart Press.

Imhof, E. (1963) "Task and Methods of Theoretical Cartography" *International Yearbook for Cartography* 3 pp.13–25.

WHITHER MAPS AND MAP DESIGN?

Michael Wood

Originally published in Volume 27, No 1, pp.7–12

Although times and techniques have changed, the fundamentals of map design have not. Textbooks may carry explanations of principles, but effective design methods have largely remained the domain of the cartographer. This paper reviews ways in which maps and map-making are changing and increasing in importance in the wake of new technologies. The search for a fuller understanding of basic design principles is thus more urgent to support both new map-making software and the non-cartographic users of such systems. Expert practitioners may be able to contribute to this search.

Introduction and Overview

Over the centuries cartographers have produced maps of great variety. Some of today's maps are well designed; many are adequate and some are just poor. Perceptive thinkers in the 1930s (Eldridge *et al.*, 1933) sensed these inadequacies and called for radical improvement, but no common framework existed to support such a fundamental change of attitude.

Although some cartographers advanced the techniques of design in the early 20th century, not until the 1960s was the study of 'map design and perception' given serious consideration. The concept of cartographic communication emerged as the preferred theoretical approach and is still accepted by many cartographers. Although research into perception and design continued, the trend was toward cognitive issues and the map-reading process.

The period, since the 1950s, saw the growth of special purpose and single-sheet maps which offered more scope for design experimentation than had been available from working with conventional map series. The 'look' of some maps began to reflect both growing ideas about cartographic design and knowledge acquired from the field of general graphics. The latter, more commercially - orientated, discipline also provided short-cuts for the production of high-quality images with construction aids such as adhesive symbols and lettering (e.g. Letraset). These new products provided opportunities for the more adventurous to escape the restrictions of manual drawing. Greater variety of type style resulted but, as product ranges increased and fell in price, some uniformity crept in.

The eventual arrival of desktop computers (especially the Apple Mac), and versatile software such as: Adobe Illustrator and Aldus FreeHand, were to trigger the start of a new period of expansion, especially in small cartographic units. Productivity rates increased dramatically and system versatility encouraged widespread improvements in design output. However this new 'ease-of-use' may have also widened the gap between the extremes of map quality ('very good' to 'very poor'). As in the early days of Letraset, more of the cartographically inexperienced tried to become 'instant' mappers, often with frightening consequences! In some cases, however, talent was also unleashed and a new generation of 'super-cartographers' emerged.

There have always been such people. Imhof was one, not only a trained topographic scientist, but also skilled in the physical production of map images and artwork. The superlative has been added here to describe those, today, who not only have cartographic knowledge and experience but also have mastered the full potential of their computer systems (Brown 1992). They are helping raise the general status of map design as a living and growing activity, although often their contributions are like islands of magic in a sea of mediocrity. Such specialists are still in the minority and concern is growing over the problems which can emerge when non-cartographers take advantage of 'user-friendly' systems for digital mapping and GIS. However, just as the 'intelligent' car may eventually take over the main responsibilities of navigation and manoeuvre, expert systems and on-hand 'amplified intelligence' (Weibel and Buttenfield, 1992) will provide support for these mapping packages and systems. Still incomplete is a fully-structured body of expert knowledge on the problems of perception and design. If the earlier calls for map improvement had been heeded and a structured programme of research had emerged sooner (Robinson, 1952), firmer foundations may have been laid for contemporary developments in the new technology. Researchers into expert systems must not only gather what is known but carry out further research to confirm and expand that knowledge.

An Approach to the Subject of Map Design

What is a map?

'Designing a map is one of the cartographer's core concerns'.

Although apparently a simple (and obvious) statement, this sentence contains three words, 'design', 'map' and 'cartographer', different interpretations of which could alter the treatment of the subject of this paper!

The need to define terms has become more urgent in recent decades with the emergence of separate scientific disciplines and sub-disciplines. One of the earliest responsibilities of the International Cartographic Association

(ICA) was to prepare a multilingual dictionary (Meynen, 1973). To quote from Professor Salichtchev's foreword to that publication 'To know these terms, their significance and their actual meaning is indispensable for mutual understanding between cartographers in different countries'.

Twenty years have passed since that book appeared. The world of cartography has changed and at least some of the early definitions have come under further official scrutiny, e.g. 'map':

(1973) *'A representation, normally to scale and on a flat medium, of a selection of material or abstract features on, or in relation to the surface of the Earth or of a celestial body'.*

(1992) *'A symbolized image of geographic reality, representing selected features or characteristics, resulting from the creative efforts of cartographers, and is designed for use when spatial relationships are of special relevance'.*

Searching for satisfactory definitions can be a thankless task. Exceptions are inevitable and the reader may already have identified a map type which does not match the above criteria. For example what of fantasy maps? Does their lack of geographic reality exclude them from the concern of the cartographer? In recognition of these difficulties the 1992 version is classified as a 'working definition' and still open to discussion.

Study of cognitive (mental) maps was one growth field in the 1970s and 1980s, as was digital mapping. Both types rest uncomfortably within the above definitions. In the context of Geographic Information Systems, for instance, Moellering (1984) suggested two sub-categories of maps, 'real' (directly viewable and having a permanent, tangible reality, e.g. the traditional paper map) and 'virtual'. The latter he subdivided further into three types:

I: Directly viewable; no permanent tangible reality (e.g. map on a computer screen).

II: Not directly viewable; with a permanent tangible reality (e.g. map stored on a laser disc).

III: Not directly viewable; no permanent tangible reality (e.g. database stored on a magnetic disc).

Although some cartographers are happier with strict definitions Vasiliev *et al.* (1990) re-examined the problem in their brief but challenging paper 'What is a map?'. They suggest that the common idea of categorization, in which 'all members of the set are equally good exemplars of the set' may not be appropriate for 'maps'. More satisfactory might be the category which has central characteristic members (prototypes, such as topographical maps) and a radial structure comprising other objects which are related to the prototypes in some way. In a limited but imaginative experiment to find what made something 'map-like' they identified the following sub-groupings of map-like elements:

a) Correspondence with locations in geographic space.

b) Graphic nature.

c) Symbolism, crafting, and controlled generalization.

d) The prototype effect (as described above).

e) Apparently having a Use or Function.

In the growing environment of GIS-type databases the viewer is able to switch rapidly between graphic modes and viewpoints (e.g. from contour image to an oblique view of a landscape-digital elevation model), but effectively remain within the same data bases. Also the complexity of the display map may be under greater control if zooming and panning facilities are available and if separate 'layers' of information can be switched on and off at will. Interactive modification of the screen specifications of separate symbol elements can also increase the scope of map use. The elusive and dynamic nature of such virtual maps and map related images has introduced new problems for the graphic system designer. These and other characteristics may necessitate a greater flexibility of definition for 'maps'.

Does the 'real' map have a future?

Concern for clearer subject definition may suggest some insecurity within the discipline. The rise of GIS and access to new categories of 'virtual' image may imply the demise of at least part of the raison d'etre for traditional maps. However, no matter how effective a database facility might be in satisfying routine sub-tasks (e.g. distance and area measurements, numbers of features in a zoom) derived previously from the map face, the preferred mode for sophisticated exploration, analysis and presentation is often still visual. During an examination of the requirements for an interactive spatial information system for military use Taylor *et al.* (1984) observed that:

'Spatial arrangements [...] *do not have any inherent significance. Their significance derived from the association of their parts into meaningful structures. Skilled tacticians* [...] *military* [...] *or chess* [...] *can remember meaningful arrangements of units much more easily than random arrangements. They remember groupings suited for functions such as* [...] *area defence* [...] *and can reproduce these structures as parts of the overall situation.'*

These conclusions about the power of maps and spatial imagery have been resoundingly endorsed by leading thinkers and writers, e.g. Denis Wood (1992) and Stephen Hall (1992). The latter, in his book 'Mapping the Next Millennium', describes how visual mapping has been of fundamental importance across science, from astronomy to medicine. The recent emergence of data visualization by computer, in which the viewer/system-user is offered high levels of physical interaction with both data and the graphic process, has further confirmed this view. When an investigator uses such a ViSC (Visualization in Scientific Computing) system the process of probing, scanning, 'flying' round and through the screen image, similar ease of access can be offered to the relevant database. Interaction with the image effectively means interaction with the data from which it is being derived (Earnshaw and Wiseman, 1992).

Interactive map use has traditionally been restricted, largely, to what might be called a 'visuo-cognitive dialogue' between percipient and paper image. Although dynamic, graphically-based interaction with numerical datasets may describe the activities of some scientific users in the future, there will still be a continuing demand for designed maps with pre-selected data. Even computer-using exploratory investigators will need to employ what might be called 'freeze-frame' techniques, from time to time, and, perhaps with a paper copy, contemplate stages of the animated image. These images, whether static or dynamic, still require design.

Is there such a thing as 'map design'?

Although their definition may have been stretched in the recent past, 'real' (hard copy) maps are still firmly established in some areas of use. Map image creation also remains a visual-graphic process. Some theorists have emphasized the common principles which underlie most design procedures, but there are clear differences between fields. The technical challenges faced by a car designer, for instance, are obviously quite distinct from those of an architect or a cartographer. Differences between design fields arise from the materials, problems, requirements and magnitude of the separate activities. Architects might be able to appreciate the broad aspirations of map designers but overlap between their fields is too narrow for them ever to consider trading places! Closer similarities exist between maps and other graphic products, and common requirements such as contrast, legibility and balance, apply. The constraints of cartographic design and associated specialist background knowledge, however, can be underestimated by graphic artists required to create the occasional map. It is assumed that the map is just another specialized design problem. Some such designers produce successful maps. Possible reasons are: a) the relative simplicity of the map requirement, b) the designer has acquired, at an earlier stage, sufficient map knowledge to avoid making elementary mistakes, c) he/she is under the direction and guidance of an experienced cartographer. The design method, especially for prototypical map types, is quite distinctive and complex, embodying, as it must, difficult procedures such as cartographic generalization.

The Experience of Designing

The author can remember first discovering, within Microsoft Word for Windows, the simple technique of introducing minute adjustments to the 'leading' between selected lines of text (i.e. 'Format/Paragraph/Spacing after a line/set to 0.3 mm/OK'). Having struggled in the past with real 'cut-and-paste' operations when assembling booklets and small leaflets, this was like a dream come true! Once the overall scheme has been selected or created for a map or text layout, the most difficult design activity is often the introduction of many small, final-stage adjustments to satisfy the intuitive eye of the designer. These are more limited in the case of text design, e.g. between headings, sub-headings and paragraphs, but in maps they can escalate in number and complexity.

After identification of the purpose, areal coverage and the most likely uses/users predicted for a proposed map, the basic content is selected, plotted and generalized. Important considerations in the design of the map face include the following:

- The desired appearance and style of the map.
- The nature of the symbols (abstract or pictorial).
- The dimensions and colours of the symbols.
- Allocation of name groups and text specification.
- Applying the symbols and locating the text.
- Refining an overall design strategy (such as visual hierarchy) which may have been introduced at the planning stage. This could involve small adjustments to symbol specifications and positioning.
- Final adjustment to all aspects of the map.

Potential changes or adjustments tend to decrease in magnitude but increase in number from the beginning to the end of the design process. Experienced designers, however, will anticipate many such changes in the earlier stages. Keates (1989) observed that raising the quality of a map design from the 'merely legible and competent' to an 'aesthetically attractive composition', requires talent. Only an expert may be sufficiently aware of the necessity to introduce these small changes. Unfortunately there are few real experts although education and practice can advance the level of expertise.

Map Design – Theory into Practice?

Earlier cartographic creations had something in common with architecture – form often preceded function. Historical maps were frequently low in reliable information but they seldom suffered from a lack of 'design', with elaborate cartouches, etc! Route maps such as that by Taylor and Skinner (1776) were among the exceptions, where the nature of predicted use must have played a part in the final form of the product. Old maps are often admired for their creative artistry but, for technical and other reasons, were ruled by convention. It was difficult, especially within map series, to escape from this condition. In common with other disciplines, as the restrictions of available techniques and materials were lifted, the scope for more sophisticated design increased. Unfortunately, unlike the fields of publishing, architecture and art, there was no immediate mass exploitation of these new possibilities within cartography. Convention had a firm grip on the cartographic profession, even amongst cartographers employed in ephemeral publishing such as newspapers. It is still present in some areas today.

Although the watershed of British graphic design lies in the 19th century, accompanying the expansion of printed materials such as books, the conscious concept of 'designing things' is more recent. Publications about design methods only began to appear in the 1950s. Prior to that, designing was just what the professional 'designers' (architects, product creators) did (Jones, 1981). Cartography was never considered as part of this 'design' community. Not until the last decade or so, as publicity about graphic design became absorbed by wider audiences such as practising cartographers, have 'design' effects or techniques appeared in map products. Theorists attempted to isolate the essence of designing things and translate it into procedure. Definitions followed, e.g. at a general level, design is:

'Goal-directed problem solving' (Archer, 1965).

'Decision-making in the face of uncertainty with high penalties for error' (Asimow, 1962).

In the cartographic context, 'uncertainty' might be the common problem of trying to identify exact map-use tasks, specific user groups and narrowly-defined circumstances of use. The 'look' of a map to different people will also be uncertain. Commercially the 'penalties' could be criticism, leading to low sales volume. In some contexts the simpler the requirement for a design the simpler the task, although the designer of a new company logo might disagree! A product designer could be requested to create a hammer for driving in small nails of particular length, diameter and hardness, to be used in a restricted space and for timber of specific type. The anticipated user, too, might even be specified as to hand-size and strength! These criteria will help determine the size, weight, shape and composition of the hammer head and the specifications of the handle. Logically the more general the task to be performed and the broader the characteristics of the user the more difficult the designer's challenge. Equally some design requirements are limited (perhaps like the 'hammer') while others are immensely complex, as they embody many subordinate and inter-related design problems. Similar challenges face the map designer. A simple map to explain the route between a hotel and the local airport cannot be compared with one which must set rock type, lithology, communications and names against a background of mountain topography. Incidentally, the suggestion that striving after such image complexity is counter-productive to the user is challenged by Tufte (1990): 'High density designs allow viewers to select, to narrate, to recast and personalize data for their own uses'. 'Data-thin displays' reduce the scope for pattern-seeking. Such dense data displays present real challenges to the cartographic designer.

It could be argued that most functional maps will be criticized primarily for their content (or lack of it), and that expert users will struggle to comprehend images of almost any quality or complexity if motivated to do so. This, however, does not excuse map makers from failing to seek effective approaches to design. Higher education began to include aspects of logical map design in cartography courses in the 1960s following calls from writers such as Robinson, in *The Look of Maps* (1952). With the increasing work of cartographic theorists (many of whom were geographers), textbooks of cartography began to spread the new gospel of 'communication'. This approach, derived originally from transmission theory, has been developed over many years. In summary the model of cartographic communication states that the cartographer selects information from his/her cognitive realm and encodes it (the 'message') symbolically into map form. This is then decoded and interpreted by the user. But it is common knowledge that the vast majority of existing maps (topographic, navigation, thematic series, such as geology and even orienteering maps) are compiled mainly (if indirectly) by the users. Cartographers are primarily facilitators. The dictionary definition of 'communicate' is to 'succeed in conveying meaning to others'. Unlike a telephone message where single, serially-structured ideas are actively transmitted to a listener, a map only represents selected (spatial) information in a passive way. To quote Taylor *et al.* (1984) 'space passively exists' but 'language actively moves towards a goal'. Map communication, where it exists, is activated and heavily influenced by the map reader. Attempts at communication must exist at the most basic level: design of symbols to express intended meaning. These must be coded effectively (and explained in the key) if a map is to have any value as a data source. As the map content increases, however, the possibility of creating individual and specific messages decreases (other than at a very general level, e.g. 'here is the topography of the Alps'). Even a simple specific map showing the distribution of towns in Scotland of over 10,000 people will pose more questions to the reader than offer messages. There is seldom a guarantee that the information eventually interpreted (received) was even contemplated by the cartographer (Keates, 1982). This does not mean that the cartographer has failed. Most maps provide a spatial graphic data source containing myriads of potential 'messages' awaiting interpretation by the reader. This is part of the magic of maps!

The concept of communication remains within the literature. It helps focus attention on the separate stages of data selection, map-making and map use. This in turn has reflected parallel research being made into these individual processes. Unfortunately many experiments in which map samples are presented to user groups (normally students!) for visual analysis, suffer from the variety in the nature and complexity of the individual observers. Average responses will only be average for an experimental group and may match none of the individual members of that group. Great care must therefore be applied when using the results from such experiments to modify the designs of maps for the wider community.

The model of cartographic communication was developed mainly by scholars, not directly involved in design research (Dobson, 1985) and was nurtured in the environment of special-purpose maps. It was believed that, if students learned the basics of the subject they could become satisfactory cartographers. But without constant practice in solving new map design problems this is but an educators dream. Isolated exercises, provided within geography degree programmes, do not offer such experience. When university/college trained students enter the profession of map making their training merely provides the foundations. The proving ground of practice and experience is yet to follow. Skill and competence in map design take time to nourish unless ignited by the catalysts of talent and motivation.

Map Design Today

Although no fully-structured body of knowledge exists within the discipline of map design, everyone involved in mapping is now aware of the importance of such knowledge. The criteria of purpose, geographical area and task must all be defined before the design process begins. Available technology has always controlled the methods involved in map production and today the computer has transformed many of the necessary skills and continues to do so. The softest transition has been exchanging pen or scriber on drawing table for mouse-and-cursor on computer screen. Some practitioners have found this changeover difficult while others (in large offices and small) have seen huge burdens of tedium lifted from their shoulders and time released to follow the much more important aspects of designing. *Aldus Magazine* recently highlighted four Americans most of whom developed their interests within traditional cartography but are now using the graphics software, FreeHand, to make maps (Brown, 1992). Pat Dunlavey is an orienteer who makes orienteering maps. He can use up to 120 Freehand 'layers' to isolate every object type, giving him immense control over design issues within the system. He has made imaginative use of his knowledge of files, formats etc. in order to create his specialized maps; Penn. State University's David DiBiase, (with a background in singing and songwriting!) has become an innovative map creator. While he enthuses about mapmaking as a powerful means of expression he is aware of the many ethical issues of cartography. His concern includes the fact that computers put powerful, inexpensive software 'in the hands of people who may have little or no understanding of the principles of mapping'. He is also an active proponent of scientific visualization techniques by which patterns of numbers can be revealed through animation; John Parsons of Eureka Cartography in Berkeley only transferred to computer methods in 1989. Early experiences of trying to handle complex map images led to a design philosophy of keeping things simple. This was partly a design strategy but also helped to keep files small to facilitate printing. His observation that the real art of cartography is 'knowing more what to leave out than what to put in' must be close to the hearts of many map makers; and Dick Furno whose transition from the offices of the National Geographic (where some maps might take a year to complete) to the Washington Post (where the pace is so fast maps are sometimes required within 15 minutes) was salutary. He has been very critical of what he called 'maps as newspaper art' which compromised the ability to show information. He thus strives, even in the sometimes restricted environment of the newspaper page, to keep geography 'as real as possible'. Although some of these practitioners are operating in a semi graphic-design environment, all have a firm grounding in mapmaking. This underlines the importance of domain knowledge in the sometimes hard-to-define discipline of map design.

And finally... whither maps and map design?

Maps, real or virtual, are entering a new era of significance especially within GIS. As the technology expands, facilities such as hypertext, multimedia and even hypermedia (a grand integration of many elements: graphics, text, pictures and video), are offering new challenges. Some people have suggested that cartographers, many of whom have had to become expert in the 'new' techniques of the past, such as photography and printing, should also embrace these new technologies and add them to their virtual tool box. But should these additional skills become part of cartography's core? The most logical scenario is tor cartographers, who see the advantage of creating maps within interactive hypermedia systems, to extend their knowledge as needed. The creation of animated environments such as these will require many new skills much closer to those of the video journalist or even the film-maker. It is too soon to widen the cartographer's core curriculum into such fields. The need for scaled, abstracted and symbolised graphic images remains central to the world of science, geographical and otherwise.

Although perhaps not fully acknowledged by all scientists, mapping within scientific visualization is growing in importance. Querying a database may provide many answers which were obtained previously (often painfully and unsatisfactorily) from paper maps. But, whether for private investigation or when trying to communicate ideas to others, 'maps' will always be an essential interface. They offer a generalized, yet information-rich representational environment which can be grasped almost simultaneously and acted upon by the human ability of 'spatial intuition'. Such maps should always be clear and legible. In the words of Bertin (1981), they should be maps 'to be seen', (i.e. appreciated quickly) and not require painstaking 'reading' to extract their content. Consequent on these new priorities within data visualization, and on the growing status of graphic display, a much fuller understanding of the map design process is desirable. This knowledge could then be added to the inbuilt intelligence of map-related software in the future. Users will become more polarized between those who simply read maps (on screen or in printed form) and those who use them (and their underlying databases) more interactively. The latter will have access to spatial data visualization systems founded on GIS. These interactive users may have control over the nature and quantity of map content but be relatively unconcerned with design aspects, provided the information is legible. This new software will require integrated expert-system control. Commonly agreed criteria of graphic design procedures must be applied to the images which appear on the screen.

In the immediate future, however, the world of special purpose 'real' maps will, as in the past, be enriched by 'super-cartographers'. The talented specialists of today have acquired the benefit of experience – having tested their design solutions in the 'market place'. For them the working environment of the computer is near-perfect. Interaction and feedback permit previously unconceived-of control over design processes, thus removing almost all barriers to creativity.

Leaving aside for the present the expanded palette of multimedia techniques, two areas of continuing research, although related, can be separately targeted: a) to identify the most appropriate and effective default specifications to be used in ready-to-run mapping systems, and b) to extend this investigation in an attempt to isolate what is accessible about the fundamentals of design within the field of cartography. The latter would improve the traditional cartographic curriculum and also provide input for what has been called 'amplified intelligence' (Weibel and Buttenfield, 1992) which can be programmed into 'ideal' mapping packages (like having an expert cartographer looking over your shoulder). Where is this knowledge? Some of the framework has obviously been distilled into textbook form over recent decades and, at this level, has wide acceptance. The problem is translating principle into practice. But is there more? Greater understanding will emerge from a closer analysis of the perceptual/cognitive processes. Analyses which began almost forty years ago have provided some of the foundational scientific knowledge. However this still leaves the puzzle of the map in which the design is 'right'; e.g. the map which has been created by what I have referred to as a 'supercartographer'. That extra knowledge must exist in the minds (and hands) of experts!

Amongst practising cartographers there is a spectrum of expertise. Many have a high level of technical knowledge, skill and design ability. Perhaps fewer possess what is described as 'a real talent' for design. Too often in the past the scientific search for the elements of such knowledge has been controlled by non-practising cartographers, mainly academic geographers. To discover the secrets beyond basic rules (no matter how well explained by perceptual studies of graphics) enquiries will have to probe further; to involve the map maker-designers themselves. However, Shepard (1978) has warned that 'especially creative people' are often too 'right-brained' (i.e. talented in non-verbal abilities such as spatial perception and intuition) to be able to share their knowledge easily. No matter how great their special ability 'nothing could be more inimical to their whole approach than to try to articulate it in words'. Therein lie both dilemma and challenge. Can there be a meeting of minds? Without it any solutions will be incomplete and remain the products of 'left-brained' thinkers (i.e. logical, sequential, analytic… seldom innovative and creative). In a scientific context this search might have become the responsibility of the 'knowledge engineer' (e.g. analysing the procedures of experts in medical diagnosis). Another approach was used by Betty Edwards (1979), who became intrigued by the problems of teaching representative drawing to school students. She overcame these problems by trying to understand the artist's ways of thinking and then translating this knowledge into a teaching scheme. She, however, is an artist herself. The challenge to distil out the essence of 'cartographic design' is, in some ways, even greater.

References

Archer, L.B. (1965) *Systematic Methods for Designers* London: Council of Industrial Design.

Asimow, M. (1962) *Introduction to Design* New York: Prentice-Hall.

Bertin, J. (1981) *Graphics and Graphic Information Processing* (trans. Berg, W.J. and Scott, P.) Berlin: de Gruyter.

Brown, C. (1992) "All over the map" *Aldus Magazine* (May/June) pp.14–24.

Dobson, M.W. (1985) "The future of perceptual cartography" *Cartographica* 22 (2) pp.27–43.

Earnshaw, R.A. and Wiseman, N. (1992) *An Introductory Guide to Scientific Visualization* Berlin: Springer-Verlag.

Edwards, B. (1979) *Drawing on the Right Side of the Brain* Los Angeles: J.P. Tarcher.

Eldridge, A.G., Abrams, A.W., Jansen, W. and Shyrock, C.M. (1933) "Maps and map standards" In Whipple, G. (Ed.) *The Teaching of Geography* (32nd Yearbook of the National Society for the Study of Education) (pp.396–405) Bloomington, Illinois: Public School Publishing Company.

Hall, S.S. (1992) *Mapping the Next Millennium: The Discovery of New Geographies* New York: Random House.

Jones, J.C. (1981) *Design Methods* New York: Wiley.

Keates, J.S. (1982) *Understanding Maps* London: Longman.

Keates, J.S. (1989) *Cartographic Design and Production* (2nd ed.) London: Longman.

Meynen, E. (1973) *Multilingual Dictionary of Technical Terms in Cartography* Wiesbaden: Franz Steiner Verlag.

Moellering, H. (1984) "Real maps, virtual maps and interactive cartography" In Gaile, G.L. and Willmott, C.J. (Eds) *Spatial Statistics and Models* (pp.109–132) Hingham, Massachusetts: Reidel Publishing.

Robinson, A.H. (1952) *The Look of Maps* Madison: University of Wisconsin Press.

Shepard, R.N. (1978) "Externalization of mental images and the act of creation" In Randhawa, B.S. and Coffman, W.E. (Eds) *Visual Learning, Thinking and Communication* (pp.133–189) New York: Academic Press.

Taylor, G. and Skinner, A. (1776) "Survey and Maps of the Roads of North Britain or Scotland" (A portable atlas) London.

Taylor, M.M., McCann, C.A. and Tuori, M.I. (1984) *The Interactive Spatial Information System* DCIEM Report No.84-R-22 Ontario: Defence and Civil Institute of Environmental Medicine.

Tufte, E.R. (1990) *Envisioning Information* Cheshire, Connecticut: Graphics Press.

Vasiliev, I., Freundschuh, S., Mark, D.M., Theisen, G.D. and McAvoy, J. (1990) "What is a map?" *The Cartographic Journal* 27 (2) pp.119–123.

Weibel, R. and Buttenfield, B.P. (1992) "Improvement in GIS graphics for analysis and decision-making" *International Journal of GIS* 6 (3) pp.223–245.

Wood, D. (1992) *The Power of Maps* London: Routledge.

MAPPING AND GRAPHING THE SOCIAL STRUCTURE OF BRITAIN

Daniel Dorling

Originally published in Volume 28, No 1, pp.7–18

This paper presents some results of the author's work on mapping and graphing the social, economic and political structure of Britain by stretching the limits of cartographic and graphic design convention. Visualization techniques have been used to analyse the results of the last three British censuses and also mortality, election, employment and housing records.

Introduction

A New Social Atlas of Britain is being produced from these datasets which contains over two hundred pages of colour maps. Most of the maps use projections which distort space to highlight urban areas across the nation. The aim of this work has been to show how localities have changed over time, visualizing British society partitioned into over ten thousand similarly populated neighbourhoods (wards). In total, over four million individual variables are presented in this atlas. The atlas and all the maps and diagrams shown here were produced on an Archimedes home micro-computer.

The argument for presenting so much information in a visual form is that conventional quantitative statistical techniques mask the complexity of the spatial patterns in British society which computer assisted cartography and adapted traditional graphics are able to illustrate and clarify. Illustrations of various solutions used in *A New Social Atlas of Britain* are given in this paper and a rationale is presented for the techniques used. Some of the possibilities and some of the problems in visualizing the large amount social data which is available for a country, nationally, across many localities are also discussed through examples, as are the difficulties of presenting unconventional images to a wide audience.

Drawing a New Social Geography of Britain

In many parts of the world information about very small localities stored on computers has become available for research. This information often contains hundreds of statistics about people living in thousands of localities. This paper illustrates how a researcher can visualize these statistics to attempt to understand more of the localities of one country: Britain. An aim of this work is to show people facets of the society in which they are living which they would not otherwise recognize. The methods which are

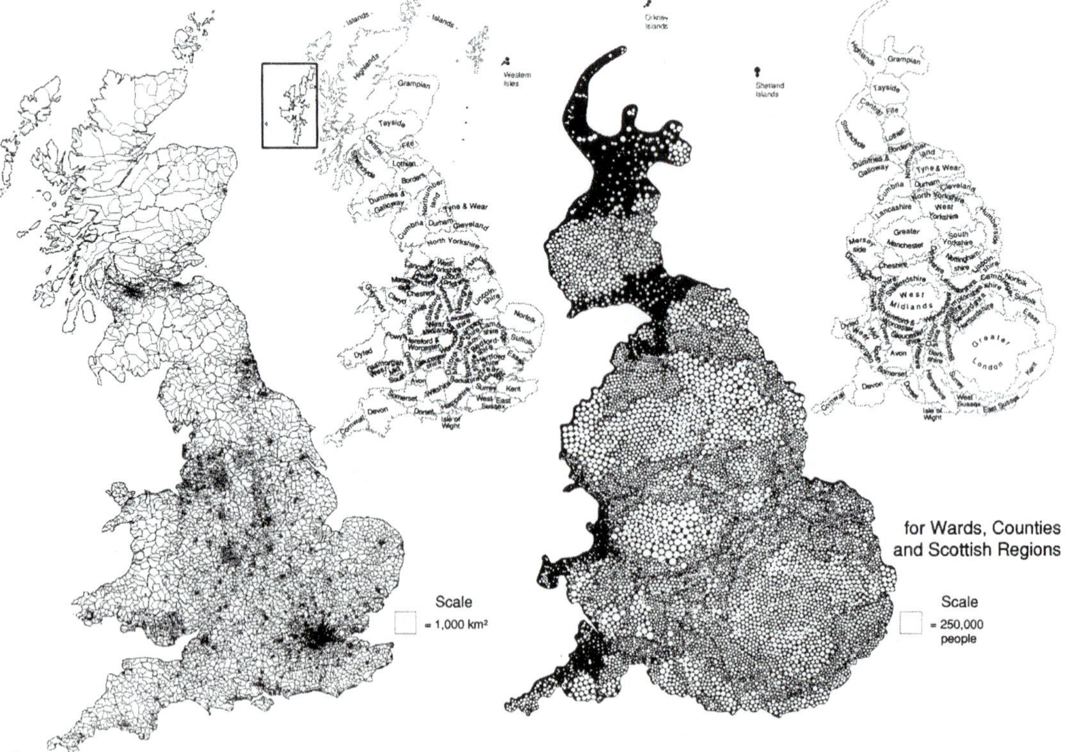

Figure 1
A traditional map and a population cartogram of Britain

presented here could be applied to any part of the world for which the information is available, and the problems which are faced in dealing with this information are also universal.

The traditional means by which information about thousands of localities is presented is in equal-area choropleth map form. Unfortunately traditional maps distort this information by over-emphasizing statistics which relate to rural localities and under-representing the characteristics of the majority of the population who live in towns and cities. This situation applies to Britain, but an even stronger case could be made in many other parts of the world where rural areas are more sparsely populated and urban areas are more densely populated.

Figure 1 shows a traditional map of the ten thousand localities in Britain for which most local statistics are available: local government wards. The figure also shows an equal population cartogram in which each of these ten thousand localities is represented by a circle (the area of which is proportional to the number of people who live there). County and region boundaries are also included in the figure so that these places can be identified. On the cartogram the statistics relating to each person are given equal visual weight. Thus the majority of space on the cartogram represents localities within the largest cities in Britain.

Cartograms can also be used effectively when only a few hundred areas are being mapped. A key advantage of using cartograms in these cases is that the illustrations can be reduced to a small size while the areas in which most people live remain visible. This means that many cartograms can be placed side by side, allowing temporal trends in geographical patterns to be compared. Figure 2 shows two cartograms of Britain based on local authority districts. A detailed key to these cartograms is given as an appendix. The first cartogram shows the changing proportion of the population of each area born in the New Commonwealth in the 1970s, the second shows this trend in the 1980s. The first cartogram demonstrates how the mainly black immigration of that time was confined to the cities of London, the Midlands, Manchester and Yorkshire. The second cartogram reflects the mortality of the largely white older New Commonwealth born population. These inferences cannot be made from the cartograms themselves but are possible if enough related information is examined. To be able to present the quantity of information required to draw parallels such as these, between different temporal trends, the methods of graphical presentation need to be compact, as well as being fair to the population presented.

Making Area Proportional to Population

The principal of making area proportional to population can be extended from maps to diagrams of all kinds. This is often done intuitively. Figure 3 shows how the numbers of people immigrating to Britain from different countries changed each year from 1955. The vertical scale of the figure shows how many people were entering the country each year and the horizontal scale shows which year the information refers to. Thus the area of the band of colour representing each area of origin is proportional to the total number of people who immigrated from that area over the period shown. This diagram contains a great deal of information presented in a compact and just way. Unfortunately, to be reproduced in this *Bulletin* all the illustrations presented here have had to be converted into black and white, but even with this limitation it is possible to show a great deal of detail.

Traditional maps which draw places in proportion to land area can be useful, particularly when the population of interest lives in very remote areas. A good example in

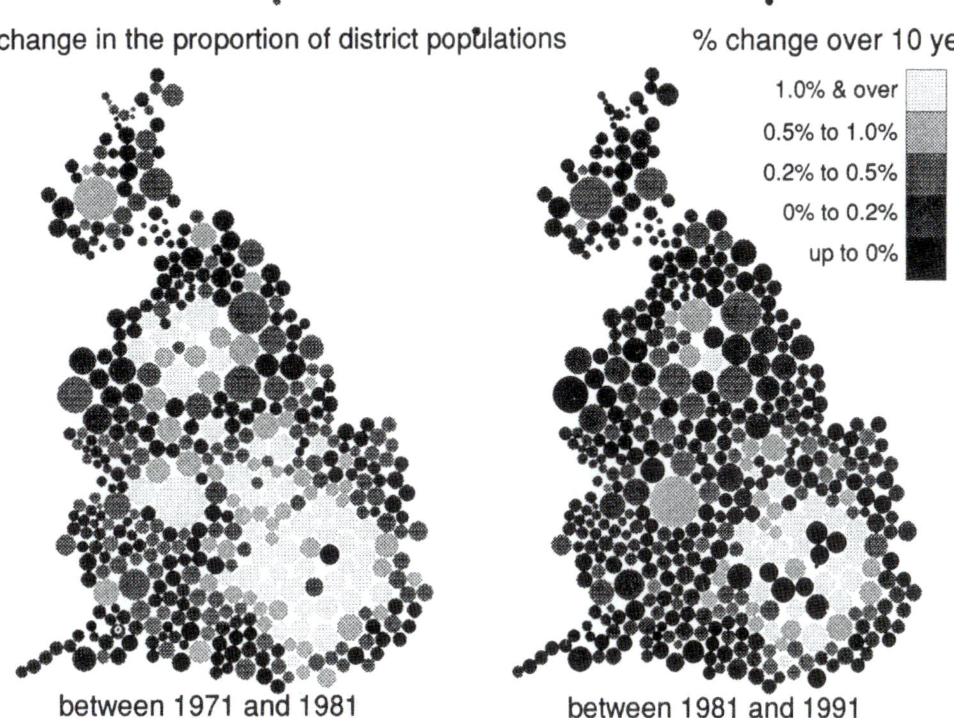

Figure 2
People resident in Britain, born in the New Commonwealth, 1971–1991

Figure 3 *Residents born outside the UK by year of entry and country of birth, 1989–1991*

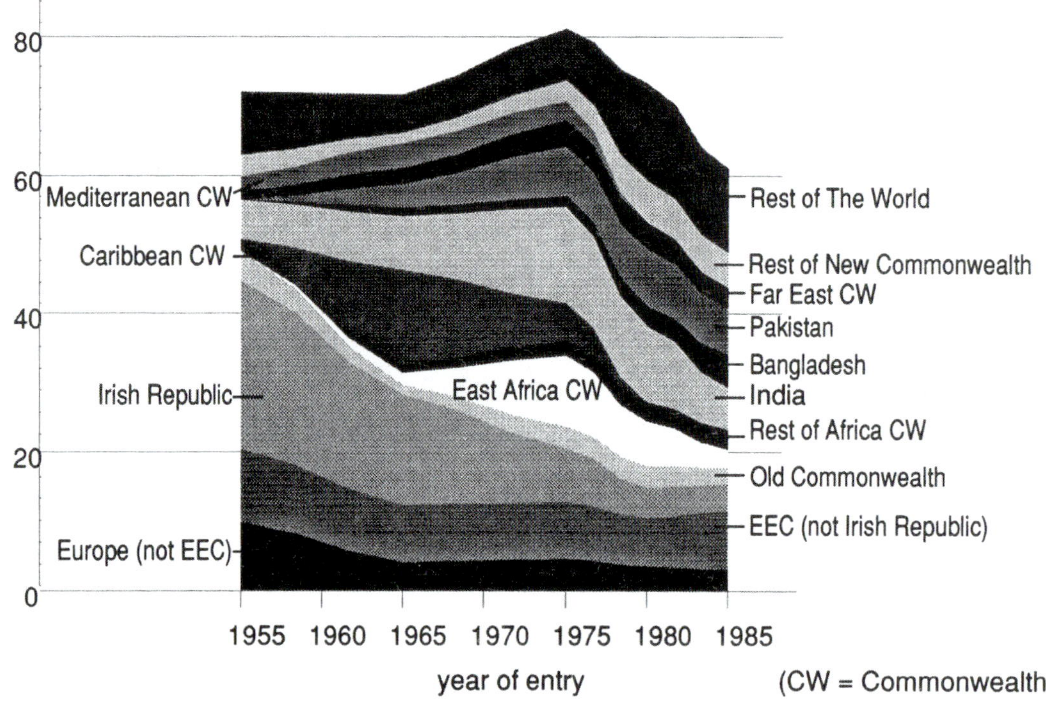

Figure 4a *Residents speaking Gaelic or Welsh, 1991*

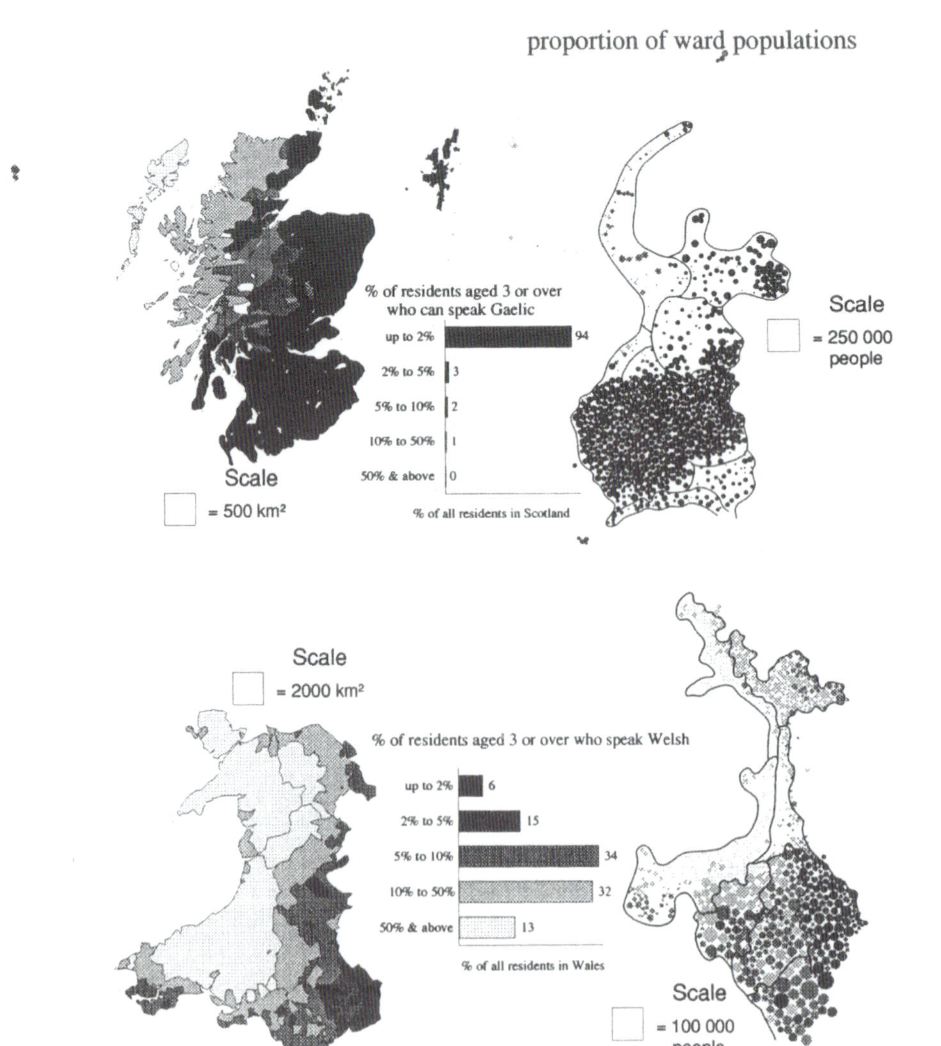

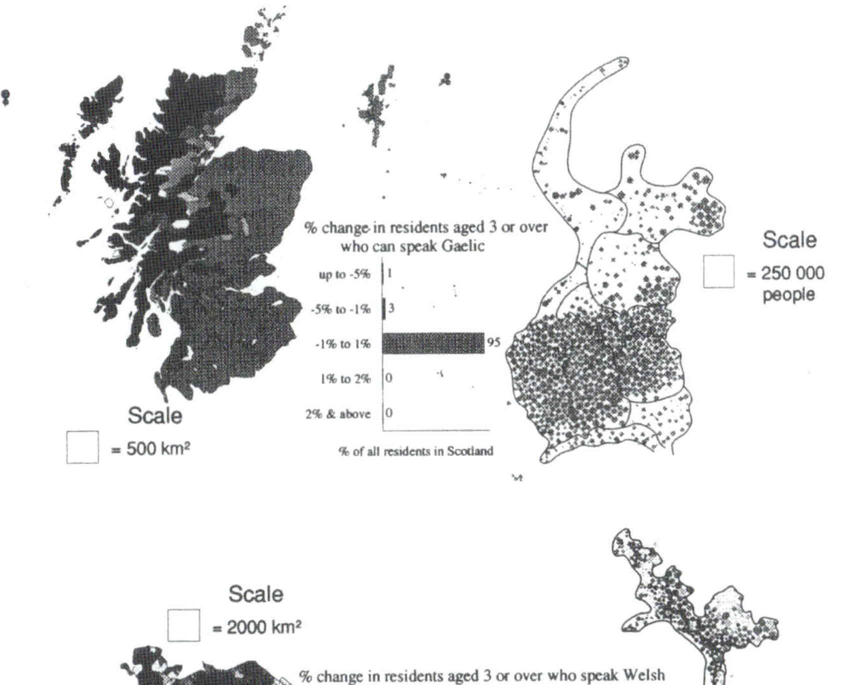

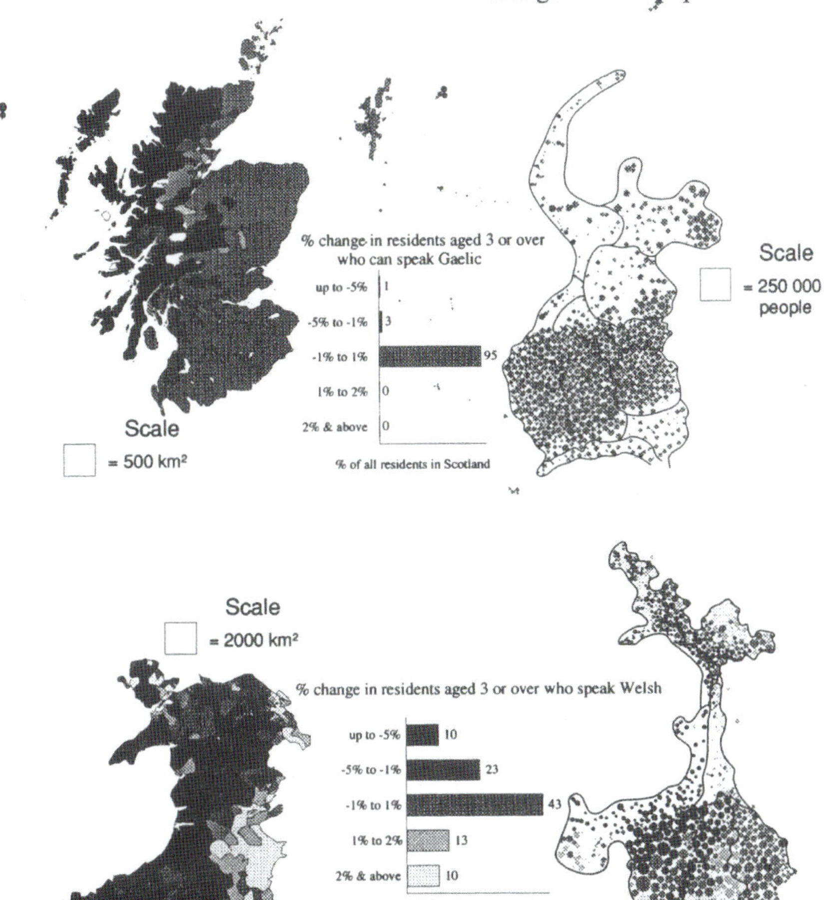

Figure 4b
Residents speaking Gaelic or Welsh, 1981–1991

Britain is of people who can speak Gaelic or Welsh. Figure 4 presents maps and cartograms depicting their distribution in Scotland and Wales, showing also how that pattern has changed over the last decade. The map of Gaelic speakers in Scotland in 1991 reflects how the proportion increases from less than one person in fifty speaking Gaelic in the South East of that country to over half the population being able to speak it in the North West. However, the equivalent cartogram (in its key) shows that 94 per cent of the population of Scotland lived in wards where less than 2 per cent of people could speak Gaelic. A similar, if less extreme pattern is seen in Wales. The cartogram of Wales shows the geography of change to be very different to that suggested by the map. Welsh speaking is not declining in most of Wales (by population) because most of the Welsh live in the Southern Valleys where the effect of teaching Welsh in schools has had much influence.

The next two figures illustrate how the principle of making the components of graphics about people proportional to the numbers of people can be extended. Figure 5 shows how many workers in each industry were in various occupational groups in 1991. This type of graph is often given as an option on computer packages but is rarely used when it would be appropriate. The graphs shown in Figure 6, in contrast, cannot be created easily with standard packages. In the main graph the vertical scale shows the proportion of all workers in each industry while the horizontal scale shows the proportion of those workers who are male or female, full-time, part-time or self-employed. Thus the total area of the bars is in proportion to the size of the workforce. For each industrial group a population pyramid is also drawn, again with its area in proportion to the total number of workers in that sector. These graphs can be difficult to read and they are certainly not simple to label. However, again they contain a great deal of information and can be argued to present it fairly.

Complexity and Simplicity in Visualization

Occasionally very clear patterns are found in quite complex data. Often it is only after producing many graphics that these patterns are evident to the researcher. This is where the use of computers to visualize social data is most advantageous. Figure 7 presents another pair of district level cartograms. The cartogram on the left shows the year in which unemployment was highest in Britain between 1979 and 1993 in each district. The North/South divide can be seen to be abrupt. The cartogram on the right shows what that highest level of unemployment was in each district. Here the urban/rural divide can be seen to be

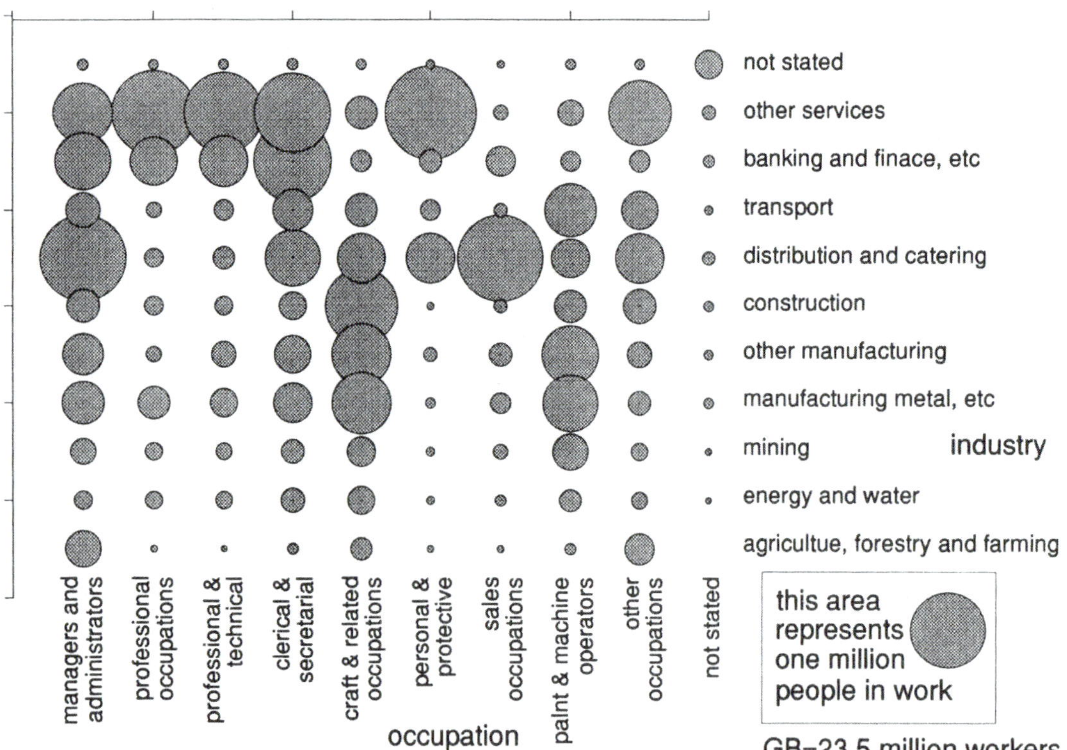

Figure 5 People in employment by industry and occupation in Britain, 1991

most acute. Low levels of maximum unemployment circle the districts which make up the major cities (in which over a fifth of the workforce has been unemployed at one time or another). To understand these cartograms a new geography of Britain has to be learnt (again, see the appendix for a key to places on these maps). That appears to be harder for researchers who are most familiar with the traditional representation of this country. British cartographers, in particular, are often averse to these images, whereas people unfamiliar with Britain often expect to see London as the largest, rather than the smallest, place on the map!

When large quantities of data are being analysed standard types of graph often require modification. An example is the scatter-plot, in which overlapping dots are obscured. When there are many dots to plot, and many

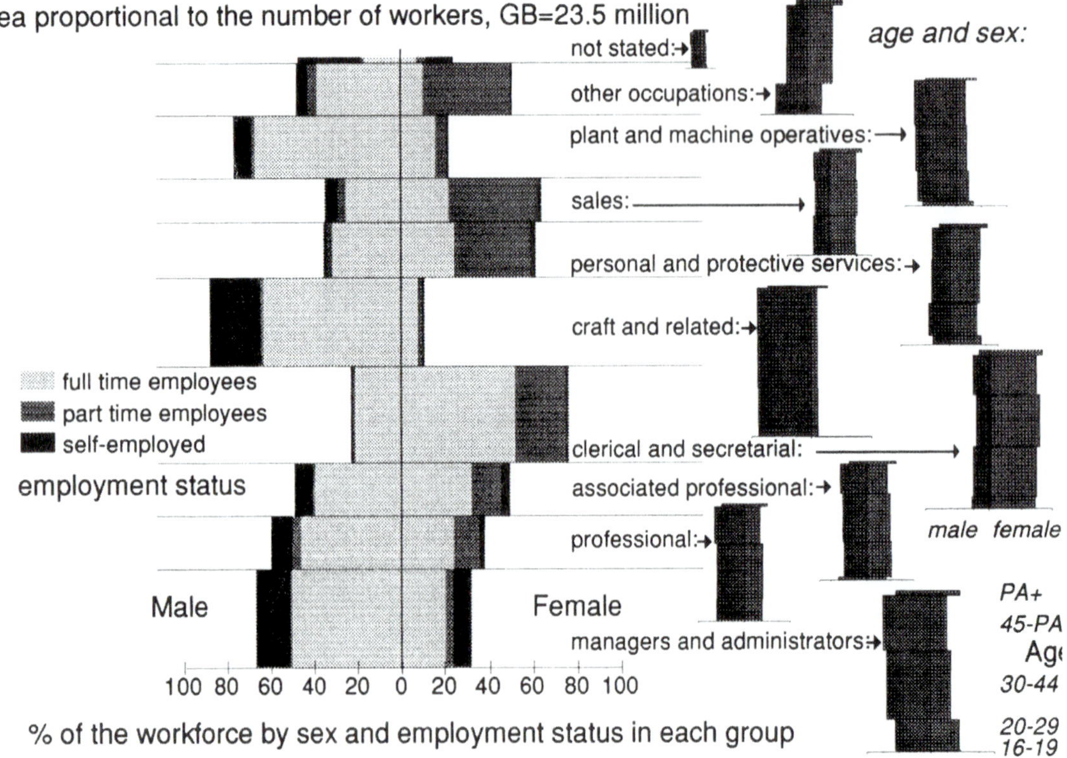

Figure 6 Residents in work in Britain by age and sex for each occupational group, 1991

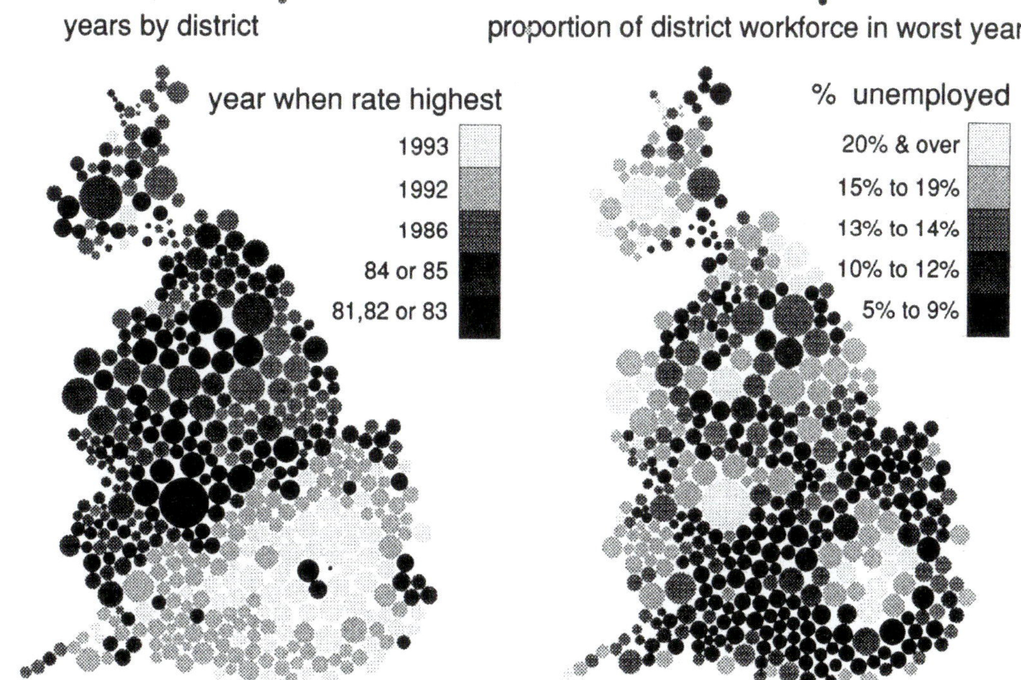

Figure 7 Year and rate of highest unemployment levels in Britain, 1979–1993

overlap, a misleading impression of the distribution can result. Figure 8 shows one method of dealing with this problem. In this figure a dot is plotted to show the relationship between the changes in house prices and unemployment in 459 districts each plotted ten times to show the situation in each of ten years. The dots are drawn with their areas proportional to the population affected. When two dots would overlap their populations are amalgamated and a single larger dot is drawn. The effect is reminiscent of the shading in old newsprint and the aim is similar, to darken certain parts of the paper more than others (in this case the parts which represent the experiences of more people). Thus, although a rough negative relationship can be discerned, it is apparent that most people in most years have experienced very little change in unemployment and have seen modestly rising housing prices. In the original version of this graphic the dots are coloured to show in which year these changes were strongest. That is not possible in black and white here.

It is possible to show the geographical distribution of average housing prices across ten thousand areas simultaneously. This is done in Figure 9 where the average

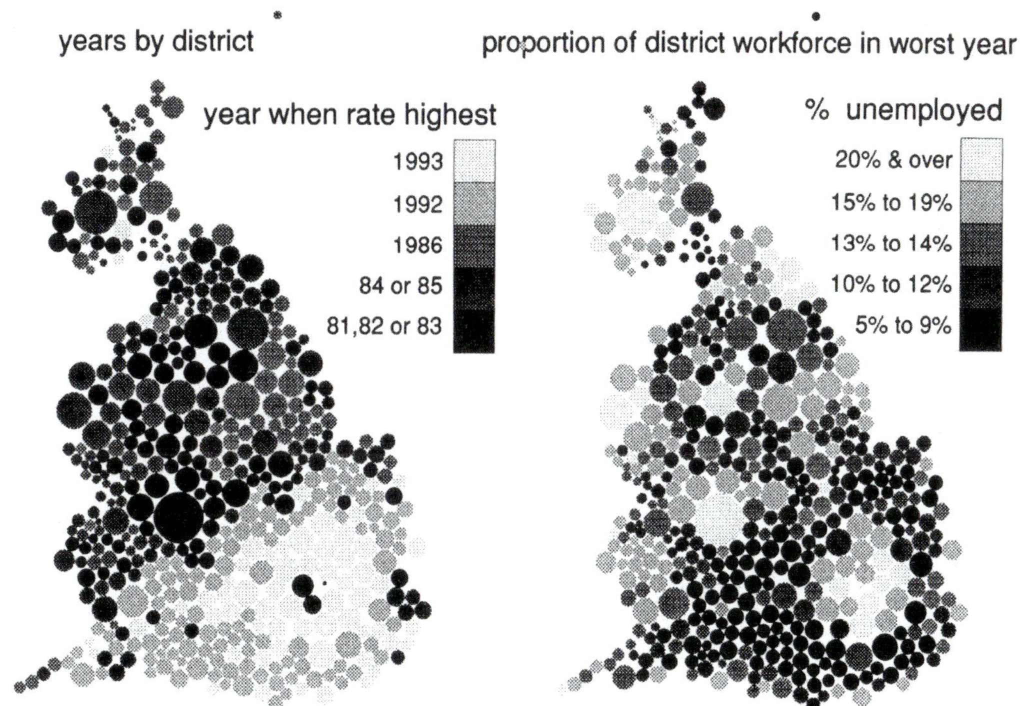

Figure 8 Unemployment and house prices, 1981–1991

Figure 9
House prices by 1991 average ward housing price

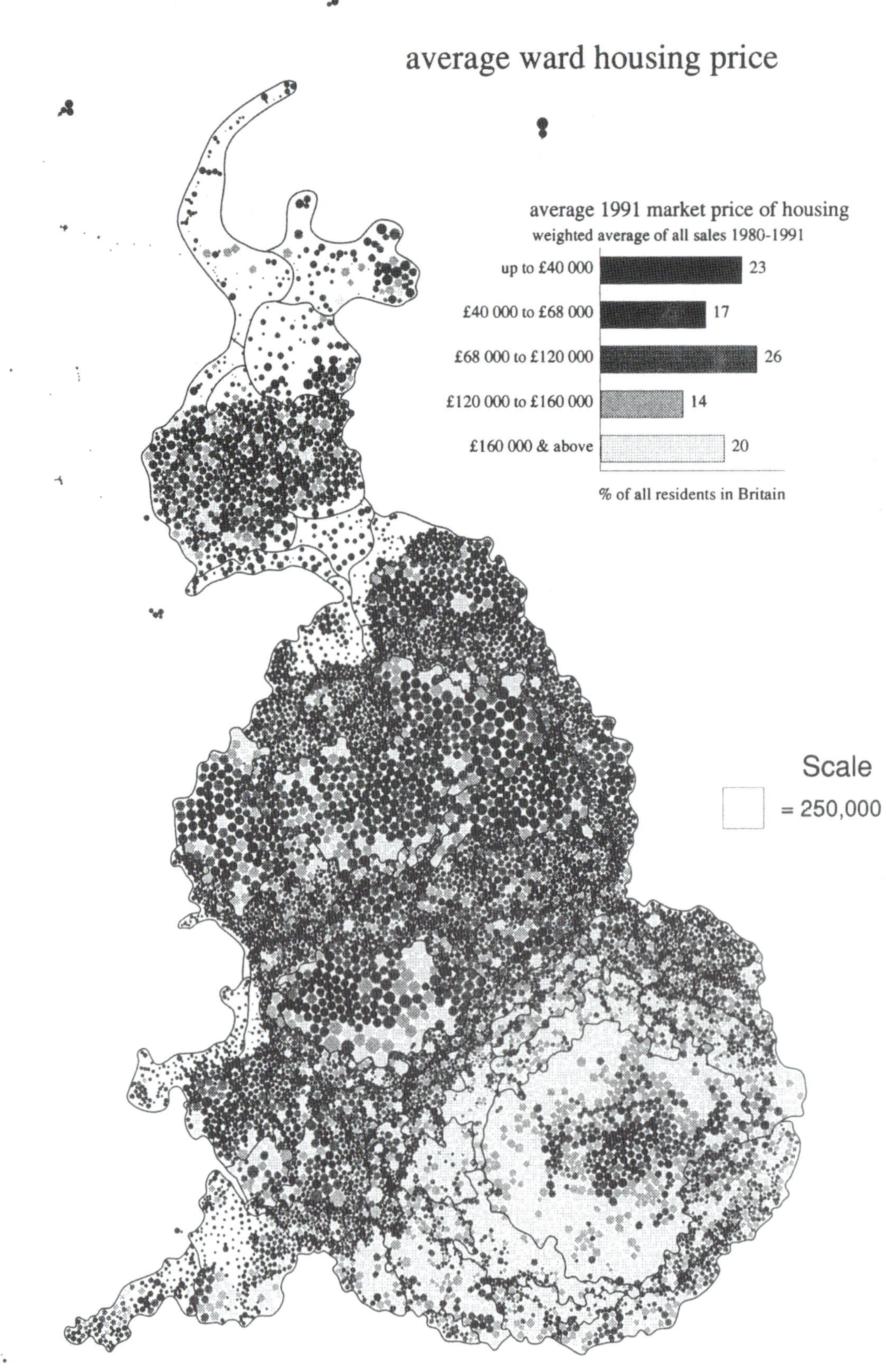

prices which buyers paid for houses in each ward over the 1980s are shown at 1991 prices. The prices are calculated from the mortgage book of a major building society. Every sale is linked to a ward through the postcode of the address of the property and a weighted average of the sale prices is made with the weights allowing for inflation up to 1991. Given that in some wards the size of this sample is quite low, it is remarkable how even the pattern is. This reflects how rigid local housing markets tend to be. The huge differences between the centre and the suburbs of London,

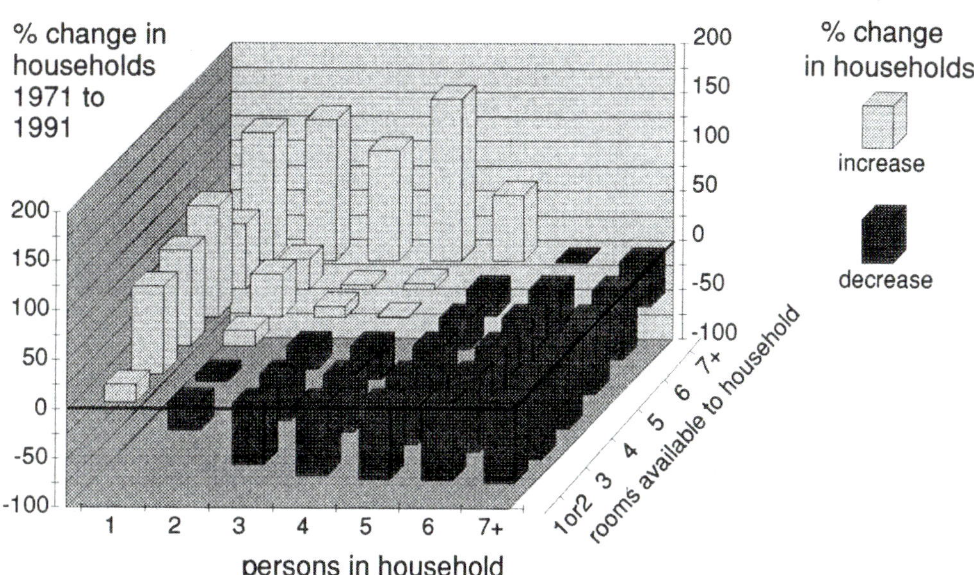

Figure 10
Persons and rooms per household in Britain, 1971–1991

or the West and East sides of Birmingham, are immediately apparent (Figure 1 acts as a key to this figure). The pattern of low rates of unemployment in Figure 7 can be seen reflected in the high levels of average housing prices in much of the South East. Through presenting graph after map after table after figure, given enough patience and space, readers can form their own views as to what facets of society appear to be interrelated most strongly and how. This is preferable to asking them to accept the results of statistical tests which disguise the prejudices and assumptions of their authors more than do a series of picture, each concentrating on a relatively simple subject.

It is important that the form of graphics used in a social atlas varies if the reader is to remain alert. Occasionally there is merit in using 'three-dimensional' charts. Figure 10 gives an example in which the almost exclusive rises in one person or in seven-plus room households in Britain are emphasized. It is when a diagram or topic appears dull that pretty graphics have their most valuable uses. Figure 11 shows an opposing case. Here the chances of people dying by two causes of death by sex and single year of age are displayed, and how those chances are changing. A very simple form of graphic can be used in this example as the contents are of greater interest to many readers and because quite a lot of data is being portrayed which might only be confused by further embellishment.

Finally, Figure 12 shows how a collage of cartograms can be used to depict a whole series of changes, in this case the results of all general election in Britain since 1955. These cartograms do not actually show who won in each

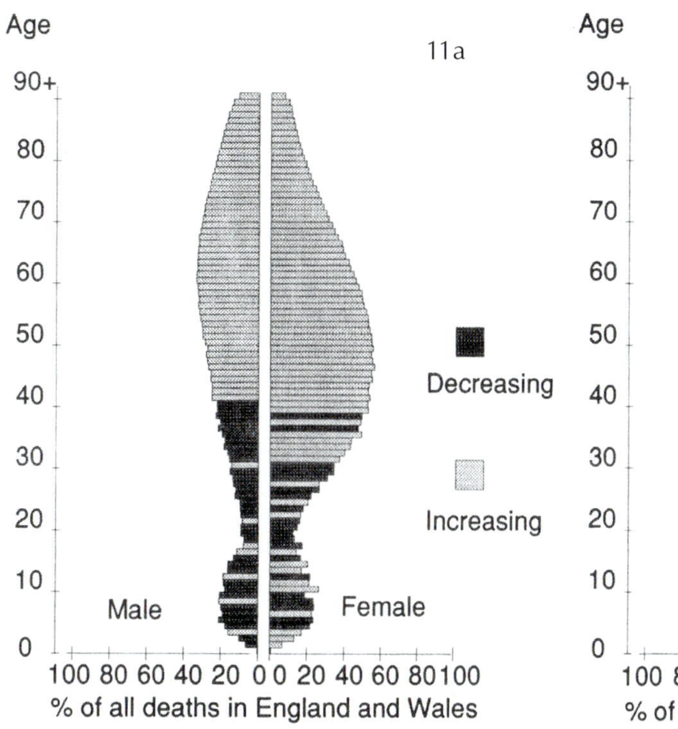

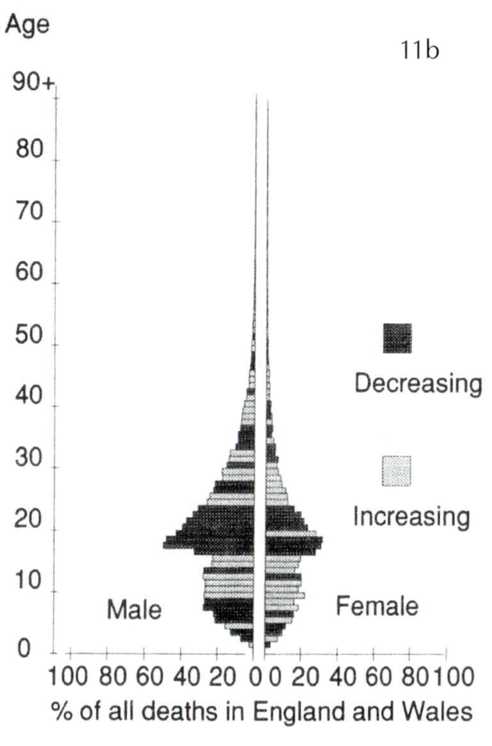

Figure 11a
Mortality from cancers between 1981 and 1989

Figure 11b
Mortality from traffic accidents between 1981 and 1989

Figure 12
Party placed second in each general election in Britain, 1955–1992, by constituency

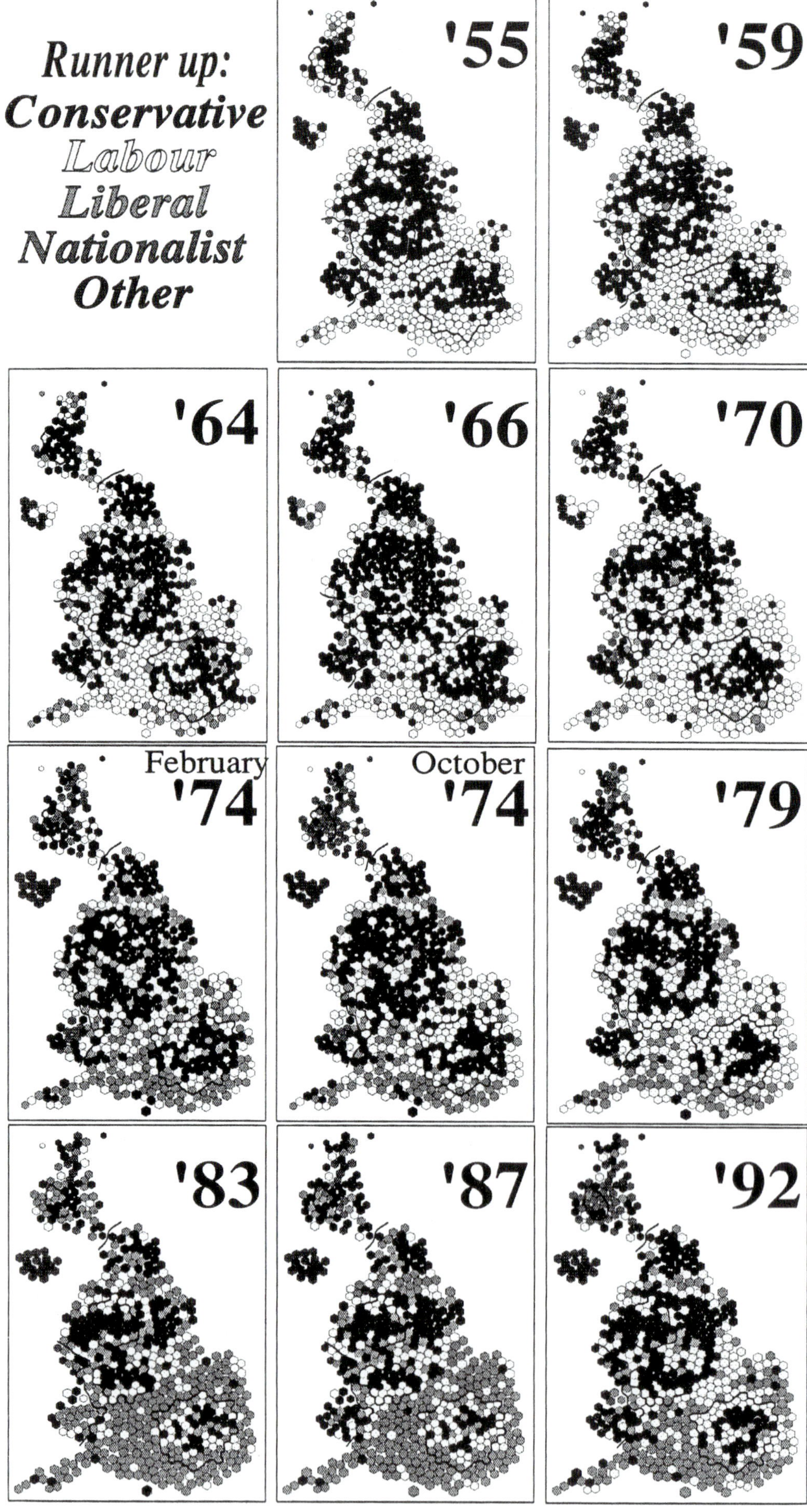

Appendix: Key to Places on the District Population Cartogram

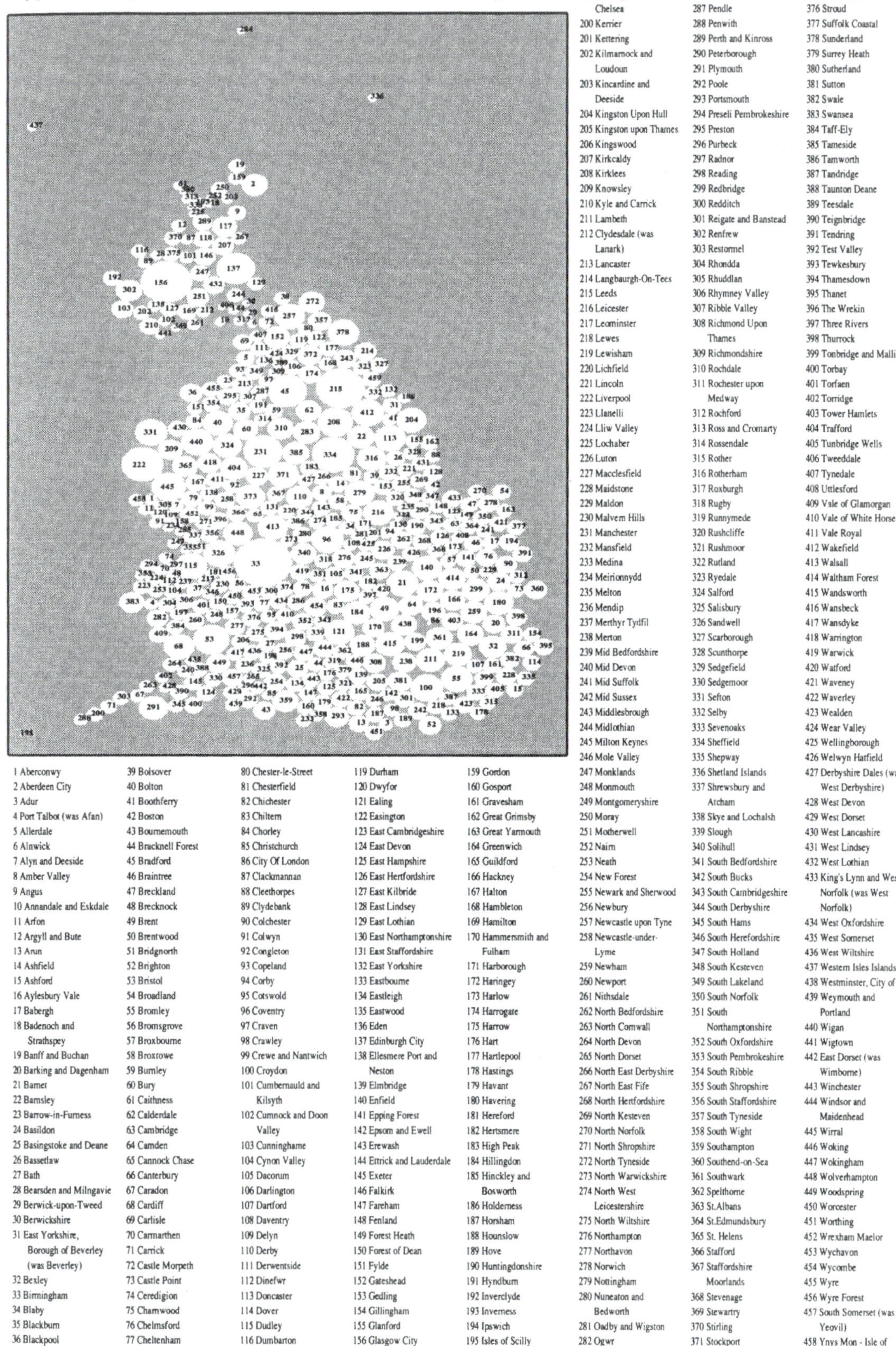

1 Aberconwy
2 Aberdeen City
3 Adur
4 Port Talbot (was Afan)
5 Allerdale
6 Alnwick
7 Alyn and Deeside
8 Amber Valley
9 Angus
10 Annandale and Eskdale
11 Arfon
12 Argyll and Bute
13 Arun
14 Ashfield
15 Ashford
16 Aylesbury Vale
17 Babergh
18 Badenoch and Strathspey
19 Banff and Buchan
20 Barking and Dagenham
21 Barnet
22 Barnsley
23 Barrow-in-Furness
24 Basildon
25 Basingstoke and Deane
26 Bassetlaw
27 Bath
28 Bearsden and Milngavie
29 Berwick-upon-Tweed
30 Berwickshire
31 East Yorkshire, Borough of Beverley (was Beverley)
32 Bexley
33 Birmingham
34 Blaby
35 Blackburn
36 Blackpool
37 Blaenau Gwent
38 Blyth Valley
39 Bolsover
40 Bolton
41 Boothferry
42 Boston
43 Bournemouth
44 Bracknell Forest
45 Bradford
46 Braintree
47 Breckland
48 Brecknock
49 Brent
50 Brentwood
51 Bridgnorth
52 Brighton
53 Bristol
54 Broadland
55 Bromley
56 Bromsgrove
57 Broxbourne
58 Broxtowe
59 Burnley
60 Bury
61 Caithness
62 Calderdale
63 Cambridge
64 Camden
65 Cannock Chase
66 Canterbury
67 Caradon
68 Cardiff
69 Carlisle
70 Carmarthen
71 Carrick
72 Castle Morpeth
73 Castle Point
74 Ceredigion
75 Charnwood
76 Chelmsford
77 Cheltenham
78 Cherwell
79 Chester
80 Chester-le-Street
81 Chesterfield
82 Chichester
83 Chiltern
84 Chorley
85 Christchurch
86 City Of London
87 Clackmannan
88 Cleethorpes
89 Clydebank
90 Colchester
91 Colwyn
92 Congleton
93 Copeland
94 Corby
95 Cotswold
96 Coventry
97 Craven
98 Crawley
99 Crewe and Nantwich
100 Croydon
101 Cumbernauld and Kilsyth
102 Cumnock and Doon Valley
103 Cunninghame
104 Cynon Valley
105 Dacorum
106 Darlington
107 Dartford
108 Daventry
109 Delyn
110 Derby
111 Derwentside
112 Dinefwr
113 Doncaster
114 Dover
115 Dudley
116 Dumbarton
117 Dundee City
118 Dunfermline
119 Durham
120 Dwyfor
121 Ealing
122 Easington
123 East Cambridgeshire
124 East Devon
125 East Hampshire
126 East Hertfordshire
127 East Kilbride
128 East Lindsey
129 East Lothian
130 East Northamptonshire
131 East Staffordshire
132 East Yorkshire
133 Eastbourne
134 Eastleigh
135 Eastwood
136 Eden
137 Edinburgh City
138 Ellesmere Port and Neston
139 Elmbridge
140 Enfield
141 Epping Forest
142 Epsom and Ewell
143 Erewash
144 Ettrick and Lauderdale
145 Exeter
146 Falkirk
147 Fareham
148 Fenland
149 Forest Heath
150 Forest of Dean
151 Fylde
152 Gateshead
153 Gedling
154 Gillingham
155 Glanford
156 Glasgow City
157 Gloucester
158 Glyndwr
159 Gordon
160 Gosport
161 Gravesham
162 Great Grimsby
163 Great Yarmouth
164 Greenwich
165 Guildford
166 Hackney
167 Halton
168 Hambleton
169 Hamilton
170 Hammersmith and Fulham
171 Harborough
172 Haringey
173 Harlow
174 Harrogate
175 Harrow
176 Hart
177 Hartlepool
178 Hastings
179 Havant
180 Havering
181 Hereford
182 Hertsmere
183 High Peak
184 Hillingdon
185 Hinckley and Bosworth
186 Holderness
187 Horsham
188 Hounslow
189 Hove
190 Huntingdonshire
191 Hyndburn
192 Inverclyde
193 Inverness
194 Ipswich
195 Isles of Scilly
196 Islington
197 Islwyn
198 Kennet
199 Kensington and Chelsea
200 Kerrier
201 Kettering
202 Kilmarnock and Loudoun
203 Kincardine and Deeside
204 Kingston Upon Hull
205 Kingston upon Thames
206 Kingswood
207 Kirkcaldy
208 Kirklees
209 Knowsley
210 Kyle and Carrick
211 Lambeth
212 Clydesdale (was Lanark)
213 Lancaster
214 Langbaurgh-On-Tees
215 Leeds
216 Leicester
217 Leominster
218 Lewes
219 Lewisham
220 Lichfield
221 Lincoln
222 Liverpool
223 Llanelli
224 Lliw Valley
225 Lochaber
226 Luton
227 Macclesfield
228 Maidstone
229 Maldon
230 Malvern Hills
231 Manchester
232 Mansfield
233 Medina
234 Meirionnydd
235 Melton
236 Mendip
237 Merthyr Tydfil
238 Merton
239 Mid Bedfordshire
240 Mid Devon
241 Mid Suffolk
242 Mid Sussex
243 Middlesbrough
244 Midlothian
245 Milton Keynes
246 Mole Valley
247 Monklands
248 Monmouth
249 Montgomeryshire
250 Moray
251 Motherwell
252 Nairn
253 Neath
254 New Forest
255 Newark and Sherwood
256 Newbury
257 Newcastle upon Tyne
258 Newcastle-under-Lyme
259 Newham
260 Newport
261 Nithsdale
262 North Bedfordshire
263 North Cornwall
264 North Devon
265 North Dorset
266 North East Derbyshire
267 North East Fife
268 North Hertfordshire
269 North Kesteven
270 North Norfolk
271 North Shropshire
272 North Tyneside
273 North Warwickshire
274 North West Leicestershire
275 North Wiltshire
276 Northampton
277 Northavon
278 Norwich
279 Nottingham
280 Nuneaton and Bedworth
281 Oadby and Wigston
282 Ogwr
283 Oldham
284 Orkney Islands
285 Oswestry
286 Oxford
287 Pendle
288 Penwith
289 Perth and Kinross
290 Peterborough
291 Plymouth
292 Poole
293 Portsmouth
294 Preseli Pembrokeshire
295 Preston
296 Purbeck
297 Radnor
298 Reading
299 Redbridge
300 Redditch
301 Reigate and Banstead
302 Renfrew
303 Restormel
304 Rhondda
305 Rhuddlan
306 Rhymney Valley
307 Ribble Valley
308 Richmond Upon Thames
309 Richmondshire
310 Rochdale
311 Rochester upon Medway
312 Rochford
313 Ross and Cromarty
314 Rossendale
315 Rother
316 Rotherham
317 Roxburgh
318 Rugby
319 Runnymede
320 Rushcliffe
321 Rushmoor
322 Rutland
323 Ryedale
324 Salford
325 Salisbury
326 Sandwell
327 Scarborough
328 Scunthorpe
329 Sedgefield
330 Sedgemoor
331 Sefton
332 Selby
333 Sevenoaks
334 Sheffield
335 Shepway
336 Shetland Islands
337 Shrewsbury and Atcham
338 Skye and Lochalsh
339 Slough
340 Solihull
341 South Bedfordshire
342 South Bucks
343 South Cambridgeshire
344 South Derbyshire
345 South Hams
346 South Herefordshire
347 South Holland
348 South Kesteven
349 South Lakeland
350 South Norfolk
351 South Northamptonshire
352 South Oxfordshire
353 South Pembrokeshire
354 South Ribble
355 South Shropshire
356 South Staffordshire
357 South Tyneside
358 South Wight
359 Southampton
360 Southend-on-Sea
361 Southwark
362 Spelthorne
363 St.Albans
364 St.Edmundsbury
365 St. Helens
366 Stafford
367 Staffordshire Moorlands
368 Stevenage
369 Stewartry
370 Stirling
371 Stockport
372 Stockton-on-Tees
373 Stoke-on-Trent
374 Stratford-on-Avon
375 Strathkelvin
376 Stroud
377 Suffolk Coastal
378 Sunderland
379 Surrey Heath
380 Sutherland
381 Sutton
382 Swale
383 Swansea
384 Taff-Ely
385 Tameside
386 Tamworth
387 Tandridge
388 Taunton Deane
389 Teesdale
390 Teignbridge
391 Tendring
392 Test Valley
393 Tewkesbury
394 Thamesdown
395 Thanet
396 The Wrekin
397 Three Rivers
398 Thurrock
399 Tonbridge and Malling
400 Torbay
401 Torfaen
402 Torridge
403 Tower Hamlets
404 Trafford
405 Tunbridge Wells
406 Tweeddale
407 Tynedale
408 Uttlesford
409 Vale of Glamorgan
410 Vale of White Horse
411 Vale Royal
412 Wakefield
413 Walsall
414 Waltham Forest
415 Wandsworth
416 Wansbeck
417 Wansdyke
418 Warrington
419 Warwick
420 Watford
421 Waveney
422 Waverley
423 Wealden
424 Wear Valley
425 Wellingborough
426 Welwyn Hatfield
427 Derbyshire Dales (was West Derbyshire)
428 West Devon
429 West Dorset
430 West Lancashire
431 West Lindsey
432 West Lothian
433 King's Lynn and West Norfolk (was West Norfolk)
434 West Oxfordshire
435 West Somerset
436 West Wiltshire
437 Western Isles Islands
438 Westminster, City of
439 Weymouth and Portland
440 Wigan
441 Wigtown
442 East Dorset (was Wimborne)
443 Winchester
444 Windsor and Maidenhead
445 Wirral
446 Woking
447 Wokingham
448 Wolverhampton
449 Woodspring
450 Worcester
451 Worthing
452 Wrexham Maelor
453 Wychavon
454 Wycombe
455 Wyre
456 Wyre Forest
457 South Somerset (was Yeovil)
458 Ynys Mon - Isle of Anglesey
459 York

constituency, but instead show which party came second. Seven thousand results are included in this one graphic. In summary the graphic shows how the Liberal party has risen from the ashes of its post-war low to be seriously contesting a majority of seats in the South of England by the end of this century. The rise of nationalism in Scotland and the exit of main stream parties from Northern Ireland are also clear messages from the figure (clear at least in its original colour form!). More importantly, by presenting this quantity of information the graphic can show to what extent these assertions are not universal. It is even possible to follow the fortunes of individual constituencies over time. A simple example is the Isle of Wight, the most Southern constituency on the cartograms. There, the second placed party has changed from Labour to Conservative to Liberal. These maps can also incorporate the effects of boundary changes. If, for instance, you examine the figure closely you can see that the number of constituencies alters over time in each region.

Conclusion

This paper has presented only twelve illustrations taken from a social atlas of Britain which contains over two hundred maps and two hundred graphs, most of which show patterns across ten thousand areas. Hopefully it has given a flavour of what can now be done given the wealth of information available and the ease with which that information can be manipulated. A home microcomputer was used to create all the maps and graphs shown here and to typeset the atlas. Apart from showing what is technically plausible this paper argues that care is needed if the maps and diagrams which social scientists produce are to represent society fairly. One guiding principle is that equal numbers of people are equally represented. Another principle is to present as much information as possible when as complex an object as society is being studied. What we need to see cannot be predetermined. Simplification conceals the complexity of reality.

Note

The census data used in this article is crown copyright-ESRC purchase.

Reference

Dorling, D. (1995) "Visualizing Changing Social Structure from a Census" *Environment and Planning A* 27 (2) pp.353–378.

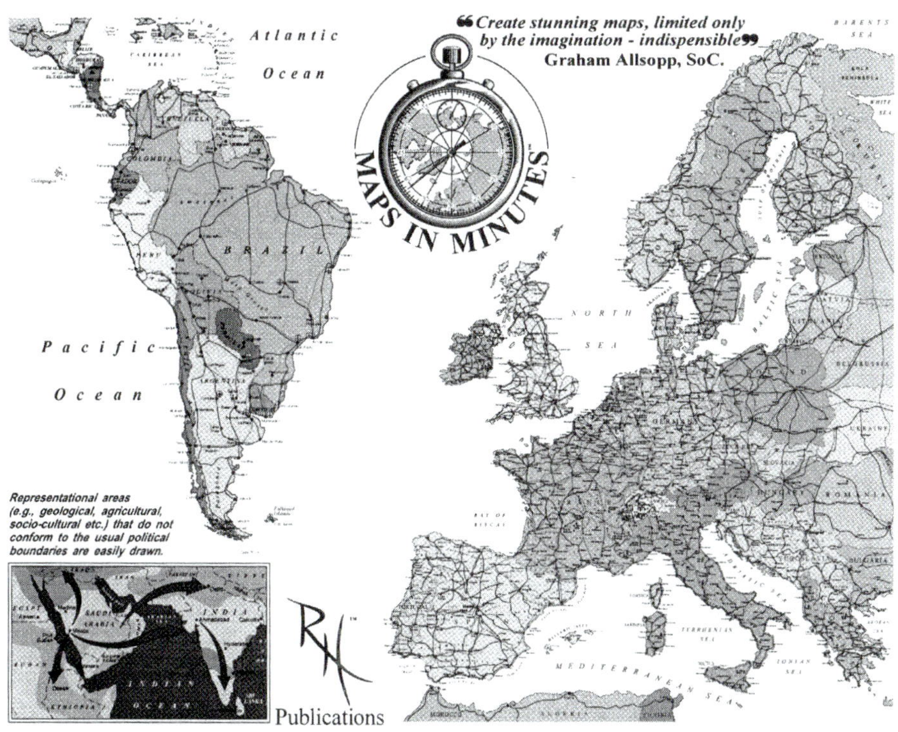

MAPS IN MINUTES™

"...the leader in PostScript map-making"

MacUser, UK

***MAPS IN MINUTES Set 1** (UK & Eire Map, Europe Map and World Map)* **.£165.00**
Total cost £199.75 (i.e. £165.00 plus £5.00 Special Delivery postage and packing plus £29.75 VAT @17.5%)

Special Offer to Members of the Society of Cartographers Set 1 £120.25
Total cost £147.17 (i.e. £120.25 plus £5.00 Special Delivery postage and packing plus £21.92 VAT @17.5%)

Single Map purchase - North America Map £79.00
Total cost £98.70 (i.e. £79.00 plus £5.00 Special Delivery postage and packing plus £14.70 VAT @17.5%)

SITE LICENCE RATES AVAILABLE ON REQUEST. Any European Community VAT Registered Company outside the United Kingdom is exempt from VAT charges subject to written confirmation of EC VAT Registration Number upon purchase order.

To itemise your order please fill in the form below and enter the appropriate number of units required:-

	Aldus FreeHand® 3.0	*Adobe Illustrator™ 3.0*	*Adobe Illustrator™ 5.0*	*Deneba Canvas™ 3.0*
MAPS IN MINUTES Set 1				
NORTH AMERICA MAP				

☐ Cheque enclosed for £ .... made payable to R.E. & S.E. Harrington; or

☐ Please charge my Visa / MasterCard (delete as appropriate) the sum of £ ....

Credit Card No. ☐☐☐☐☐☐☐☐☐☐☐☐☐☐☐☐ SoC M'ship No ....

Expiry Date .... Cardholder's signature ....

Co. Contact/Cardholder's Initials & Surname Mr/Mrs /Ms ....
(delete as appropriate)

Company Name .... VAT Reg. No. ....

Company/Cardholder's Address ....
(delete as appropriate)

....

.... Postcode ....

Daytime Telephone No. .... Fax. No. ....

Send to:- RH Publications, Driftwood, Treligga, Cornwall, PL33 9EE, United Kingdom.
Telephone: 0840 212135 Fax: 0840 213060 *N.B. Payment is accepted only in £ Sterling.*

A PERCEPTUAL APPROACH TO MAP DESIGN

Henry W. Castner

Originally published in Volume 30, No 1, pp.1–7

This paper reviews the perceptual characteristics of the four types of cartographic symbols in order to develop a practical approach to map design. For this, the distinction is made between activities and tasks in map use. The former relates to the information requirements for a map; the latter speaks to the ways that information should be processed visually by the map user. Making use of the two different but integral components of visual processing, map designers can encode information in symbology appropriate to matching the importance of the information placed on the map and the anticipated map use tasks. A matrix serves as an organizing tool for making these design decisions.

Introduction

It has been said that there is nothing more practical than a good theory! This paper reflects on some first principles of visual perception, which came out of the theoretical research into cartographic communication, in order to develop some practical approaches to map design.

In map design, we all recognize the need to determine what information users will need for the activities they wish to perform upon or with the map. Figure 1 shows two quite different map designs: a) is for general reference purposes, for many different map users engaged in many different activities; b) is for a few specific people to get to one particular event in an area which is probably familiar to them.

In between, we can envision a host of other designs in which we must consider: What is the map's purpose? Who are the intended users? What information should be kept? What information should be left out? What liberties can we take with the generalization, classification and symbolization of the included information?

For these possibilities, it will be helpful to know how users should visually process the image. What visual tasks will be involved? We assume that the designer already knows the area to be mapped, the intended activities for the map, and the constraints of money, format, landmarks, etc. But will the users be counting symbols, comparing densities or making estimates of distance, area, or numerousness? Will they be looking for a unique symbol or a particular set of relationships? Or what? Trying to determine these visual tasks may be the forgotten step in map design.

A perceptual approach to map design starts with an examination of the distribution of receptors in the retina, Figure 2. There we find a near reciprocal distribution of cones and rods. The some six million cones are clustered around the area we call the fovea. Because they have

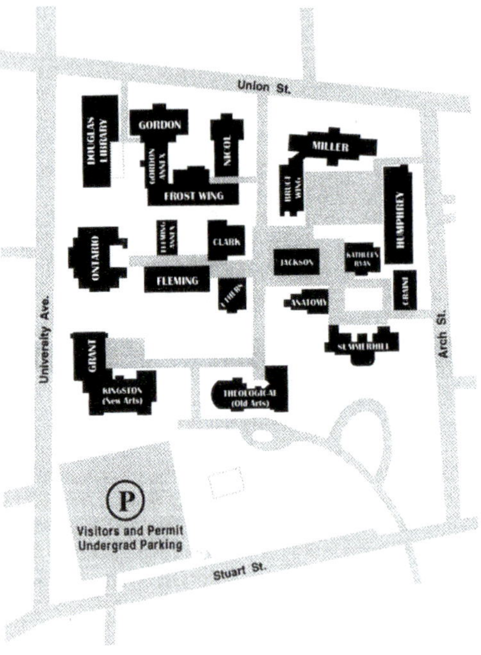

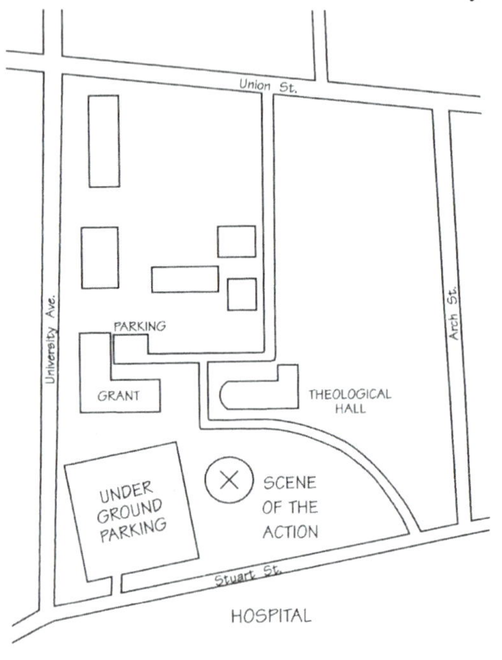

Figure 1
Two campus maps with very different goals, audiences, and thus designs.

individual linkages to the cortex, they provide us with this area of highest acuity. Foveal viewing is also associated with our conscious attention.

The some 120 million rods are distributed more evenly over the rest of the retina. Because they tend to be connected in groups, they better serve in the detection of motion or change and of edges, in the perception of textural gradients, and in night vision. They are involved in peripheral vision. As a result, in visual perception, we can talk about:[1]

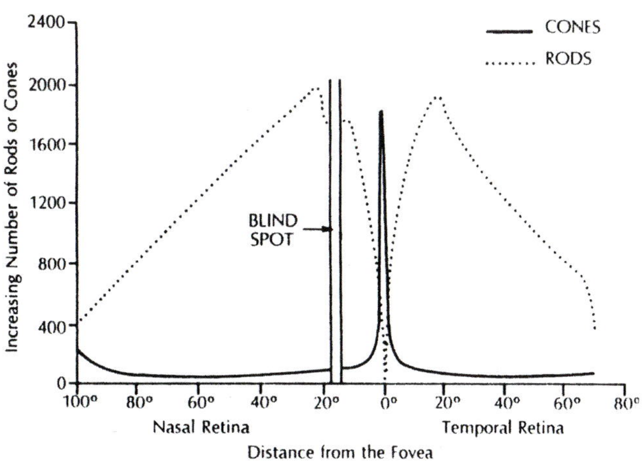

Figure 2
Cross section of the retina showing the relative distribution of rods and cones (after Schiffman, 1976: 164)

Discrimination: a process of peripheral vision where we monitor our environment and our position within it. It is also a filtering process in which potential targets are sorted, without scrutiny, in order to determine which should receive conscious attention in the solution of some problem.

Identification: the conscious attention to fine detail, or to a target isolated in peripheral vision, to determine its nature and meaning.

We can illustrate these two processes by examining Figure 3. Bela Julesz (1975: 34–35) uses two sets of figures to distinguish what he calls 'pure' and 'cognitive' perception. In each set of figures, one is made up of a single line; the other of two lines. In the case of the lower pair, the difference is obvious; we distinguish them immediately without any conscious attention. This is an example of pure perception. In the upper pair, however, it is impossible to make this distinction without following a line into the centre of the figures to see if it ends or turns upon itself. This determination is an example of cognitive perception for we must apply our conscious attention to solving this problem.

More relevant examples can be found in eye movement recordings of experimental subjects being asked different questions about the painting in the top left-hand corner of Figure 4. As the questions change, the patterns of viewing change. While this image is a painting, there is evidence[2] that the same changes in viewing occur in task-specific viewing of maps. Collectively, these records suggest that for every question we might ask of a map, there may be a different and optimum design which supports these viewing tasks.

In a study of nautical charts (Castner and McGrath, 1984), thirteen general chart activities and the visual tasks that support them were hypothesized. Figure 5 lists three of them. Each activity is reducible to a design problem. Our analysis led to a number of specific suggestions for chart design. Most of them related to reducing the amount of visual clutter in order to remove distractions so that more information could be discriminated in peripheral vision thus saving the application of conscious attention to more important tasks.

In the context of nautical charts, however, the kinds of design changes we suggested might not be possible to implement as charts are produced for a very special audience whose members have many specific expectations about the look of such maps. Thus any radical design changes might create problems of safety in operating vessels. But for most other designs, and especially the thematic and special purpose maps that a large number of

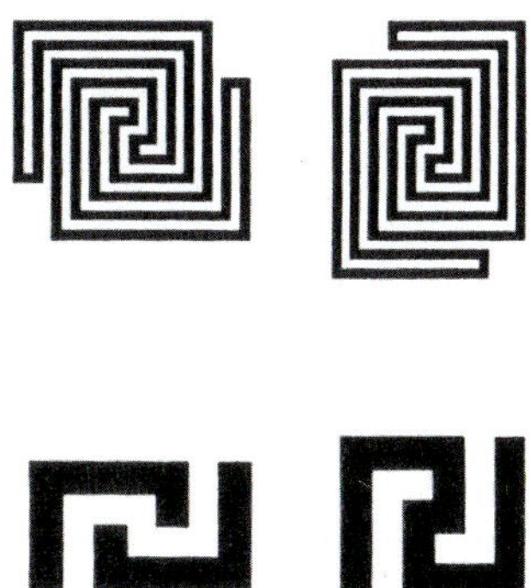

Figure 3
Two sets of figures, one of which is composed of a single line turning upon itself; the other is composed of two interwoven lines. In the bottom pair, this distinction is obvious. In the upper pair, one can only make this distinction by the application of conscious attention (see Julesz, 1975: 34–35).

map designers are involved, there are no such dire limitations. As a result, the analysis of visual tasks provides useful insights into how the map may best function and thus how it might best be designed. In short, knowing how a symbol should be processed visually will suggest how it should be designed.

For example, do we need to scrutinize a symbol or have only to discriminate it in peripheral vision thus isolating it from other similar elements? To discriminate a lighted buoy (Activity #7 in Figure 5), the symbol need only to be distinctly different from those around it. On Canadian charts, this is done with a magenta apostrophe. To identify it precisely, however, the symbol must carry information on the colour and flashing sequence of its light. But this latter information can be coded in very small letters and numbers for they are not needed to first discriminate the symbol. This suggests that in its most basic form, symbol design may be a matter of making sure symbols can first be discriminated. For this we must include in their design some attribute that stands out in peripheral vision. In a sense this means deciding whether a symbol should be obtrusive or recessive, and then how to make it so. In this context, we realize that our basic symbol types are not processed in the same way. While viewing Figure 6, consider the following.

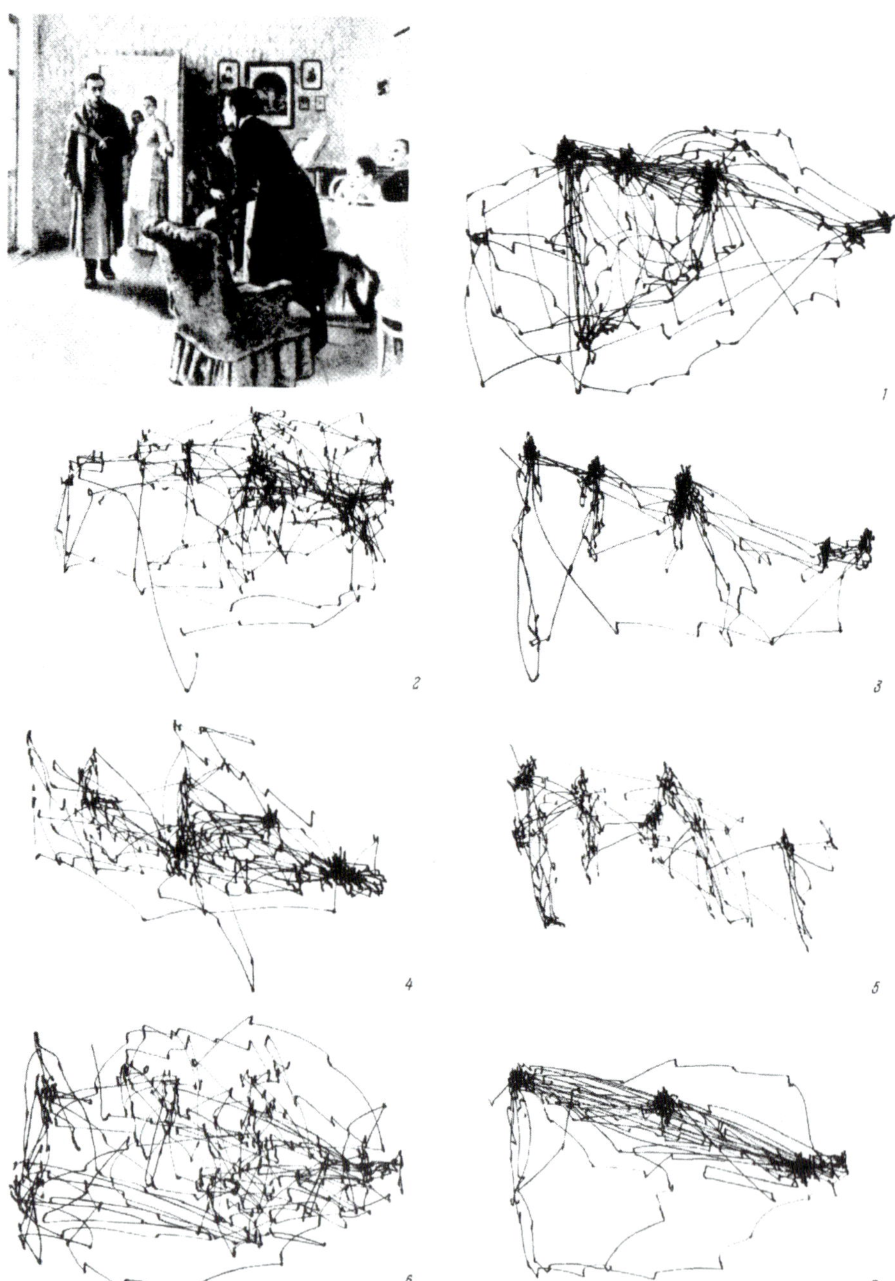

Figure 4 *Eye movement recordings of subjects viewing Repin's The Unexpected Visitor under the following conditions or questions:*

1. Free examination.

2. Estimate the material circumstances of the family.

3. Give the ages of the people shown.

4. What was the family doing when he arrived?

5. Remember the clothes worn by the people.

6. Remember the position of the people and objects in the room.

7. Estimate how long the visitor has been away.

(Figure reproduced courtesy of the Plenum Publishing Corp.)

Point Symbols:

By their very nature, point symbols attract foveal attention. They are attractive to the scanning eye. In order to make them more obtrusive, we can make them larger, bolder, or more complex. To make them recessive, we can make them smaller, lighter, or less complex. We also know that hue is better than shape as a sorting dimension in peripheral vision.[3] In other words, symbols of a particular hue are more efficiently discriminated from other hues than are symbols of one shape discriminated from among symbols of other shapes.

Line Symbols:

Inherently, line symbols are very attractive but scrutiny isn't necessary for their discrimination. The problem with lines is not in finding them but in their visual interference with and great attractiveness relative to other map symbols. For them to be obtrusive, we can make them wider, darker, more irregular, interrupted, or unpredictable. To make them recessive, we can make them narrower, lighter, smoother, more continuous, or more predictable, all in an effort to reduce visual noise and their attractiveness. Quite subtle but effective line boundaries can be suggested by changes in the quality of adjacent area symbols.

Letter Symbols:

Individual letters are like point symbols for which the same rules as above will apply. But chains of letters, i.e., names and labels, are more like line symbols and viewers can usually tell something about them without scrutiny. For example, the type of information symbolized (e.g., lakes, cities, etc.) and their importance can be built into the label by changing its type style and its size. For them to be

	ACTIVITY		VISUAL TASKS IN WHICH ELEMENTS ARE SEARCHED FOR & ISOLATED IN PERIPHERAL VISION	VISUAL TASKS IN WHICH ISOLATED ELEMENTS ARE FIXATED UPON
6.	Search for on-shore marks recorded on chart, e.g., radio antennae, flagstaff, stacks, water towers, prominent buildings, to assist in position fixing.		Discriminate marks and features within land area and general area of ship's position.	Identify precise mark or feature and position of vessel.
7.	Search for navigational aid, e.g., lighthouse, lighted buoy, unlighted buoy, to assist in position fixing and course setting.		Discriminate aids within water area within general area of ship's position and destination area.	Identify precise aid and vessel's position and destination.
13.	Familiarize one's self with (new) chart by locating:			
	13.3	nature of soundings and chart scale.	Discriminate large magenta letters in margin or discriminate chart title area; discriminate bar scale(s).	Read large word "metric" or find and read line stating soundings in meters; read words under scale bar, e.g., "meters."

Figure 5
Three activities in chart use and the supporting visual tasks in peripheral vision and foveal inspection (Castner and McGrath, 1984: 687).

obtrusive, we can make them larger, bolder, darker, or more complex as with open face or serifed styles. For them to be recessive, we can make them smaller, lighter, simpler or less complex.

Area Symbols:

Both the boundaries and the extent of area symbols are easily discriminated in peripheral vision and often their nature and identity as well. For them to be obtrusive, we can make them darker, coarser in spacing, and their component elements more complex. For them to be recessive, we can make them lighter, finer in spacing and with simpler component marks. Thus a pattern made up of small dots, closely spaced in a uniform arrangement is less obtrusive than one with large, irregular blotches in some random pattern.

All this suggests a design strategy of encoding as much information as possible in ways that allow their representative symbols to be discriminated and identified in peripheral vision. In so doing, more time is freed up for conscious attention to specific visual tasks involved in solving problems, learning geography, making new connections, or simply serendipitous discovery. [4]

There are two design practices which could help in this although, curiously, we don't make much use of them. The first is the practice of reversing symbols out of a tonal background, as in Figure 7. The degree to which such symbols are obtrusive or recessive can be modulated by changing the value of the background tint.

A second possibility, which is but a variation of the first, is the use of what Julesz (1975) calls secondary statistics in area symbols. His illustration of this phenomenon, Figure 8, shows how subtle changes in interior elements can be discriminated. While these fields of complex elements are too coarse to be used as area symbols, our practice of reversing small point symbols out of uniform tints suggests that we already make use of secondary statistics in order to map subcategories within some major information classification. For example, we might wish to distinguish areas of coniferous trees from broadleaf trees in areas of forest represented by a green tint. The reversed symbol is the secondary statistic!

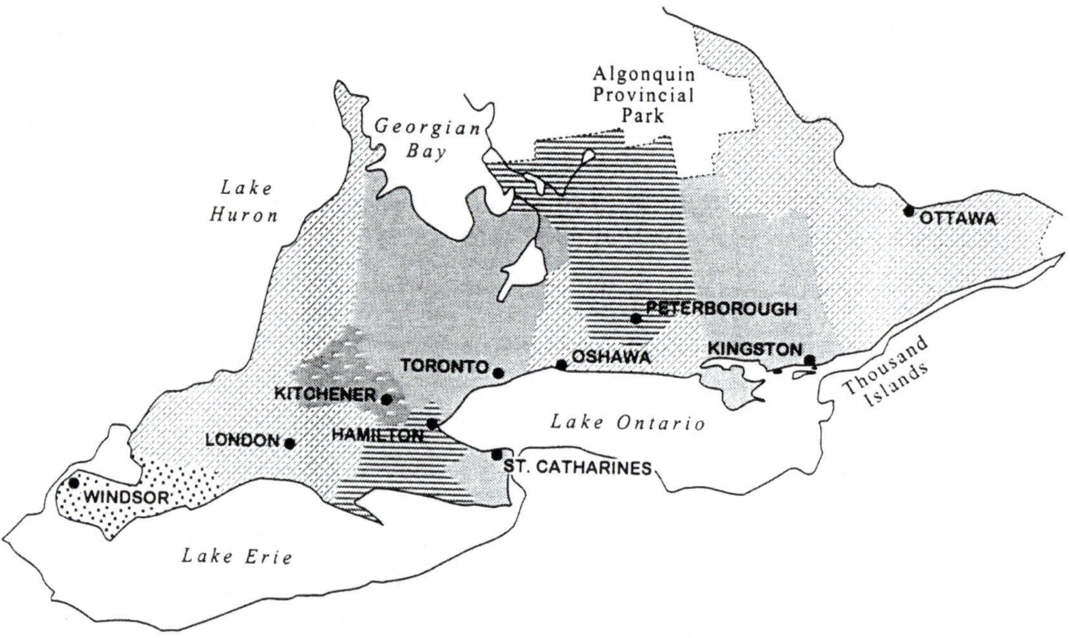

Figure 6
A map of zones of influence (by county) using point, line, area and label symbols.

Figure 7
A selection of point, line and letter symbols reversed out of: a) 90 per cent; b) 50 per cent; and c) 10 per cent background screens (McGregor, 1988).

Conclusion

Putting this all together suggests allocating information to our various symbol types so as not to overload any one way of discriminating and identifying the map's symbols. To do this, the further distinction should be made between two levels of importance: the subject information (which should be obtrusive) and the base or background in formation (which should be recessive) respectively in the design. These decisions can be represented in a design decision matrix as in Figure 9.

Several observations can be made about this diagram:

1. Obviously the important, subject information, the reason the map is being made, should be allocated to the top line, and the less important base or background information (which supports access to the subject information) should be allocated to the bottom line.
2. Care should be taken to make recessive any point and line symbols representing background information.
3. If more than one kind of point symbol is necessary to represent the subject information, they should be differentiated on the basis of hue. The same is probably true of lines if they are wide enough for the hues to be identified.
4. Perceptually, speaking, the exceptional symbol type is for areas where using more of them may be perceptually better as long as we can manage the potential problems of simultaneous contrast.[5]
5. An optimum of no more than four symbols at each of two levels of visual importance may be overstating the meaning of the diagram. But filling it in as one develops the design of a map may help (or perhaps force) the designer to identify the visual tasks necessary for the design to work, or are anticipated for the alleged activities, and to make use of as many symbol types as practical.

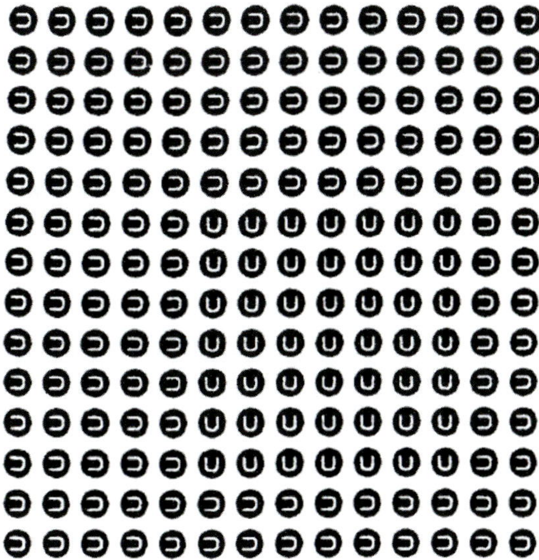

Figure 8
Two textured figures with embedded subareas: the one at left has a different secondary statistic from its surround; the one at right has the same secondary statistic. As a result, the embedded square at left is easier to discriminate and thus easier to use in symbolizing some secondary attribute of a distribution (from Julesz 1975: 37, used with his permission).

Some formal research may also reveal what symbol combinations are better. But the diagram might best be used as a diagnostic tool to analyze existing maps to see what combinations are associated with maps that we deem 'successful' given some prescribed design goals. Even such qualitative results will only be useful if map authors keep their designs within general perceptual limitations and try not to pack them with too much information. By this I refer to Figure 10 and its implicit warning that while increasing information content increases map accuracy, there is a perceptual limit in the amount of information that users can actually make use of or will be willing to respond to.

Type of Symbolization

	POINT	LINE	AREA	LETTER
SUBJECT INFORMATION				
BASE INFORMATION				

Figure 9
A matrix of design decisions concerning the allocation of subject and base information classes to the four basic types of cartographic symbols (Castner, 1990: 157).

Notes

1. For a thorough discussion of these terms and relationships see Castner (1990), Chapter 3.
2. The eye movement recordings of specific maps have been reported in a number of places. See, for example, Dobson (1979), Castner (1983), Castner and Eastman (1985) and Wood (1994).
3. One of the first experiments showing how different hues made for the most efficient sorting of point symbols was Williams (1971).
4. In a sense we do this anytime we work with one image and become familiar with it. By a process of verification we increasingly make use of peripheral vision to relocate ourselves within the structure of a viewed map after, for example, making reference to the landscape when way-finding.
5. The nature of simultaneous contrast and induction is mentioned in most basic cartographic texts. Campbell (1984: 189–190) does a better job than most in spelling out the implications and limitation for map design.

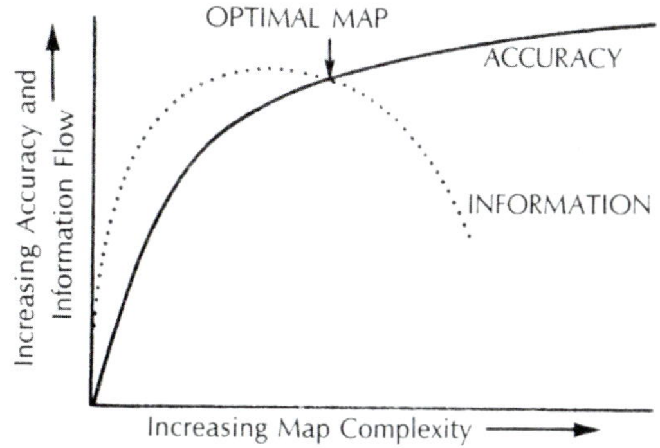

Figure 10
The relationship between information flow and map complexity as it affects accuracy and user performance (after Jenks and Caspall, 1971: 243).

Bibliography

Campbell, J. (1984) *Introductory Cartography* Englewood Cliffs: Prentice-Hall.

Castner, H.W. (1983) "Research questions and cartographic design" In Taylor, D.R.F. (Ed.) *Graphic Communication and Design in Contemporary Cartography* (pp.87–113) London: John Wiley.

Castner, H.W. (1990) *Seeking New Horizons: A Perceptual Approach to Geographic Education* Montreal: McGill-Queen's University Press.

Castner, H.W. and Eastman, J.R. (1985) "Eye-movement parameters and perceived complexity – II" *American Cartographer* 12 (1) pp.29–40.

Castner, H.W. and McGrath, G. (1984) "Educating map publishers: Evaluating changes in map design on the basis of map reading activities and visual tasks" *Paper presented at the 12th Conference of the International Cartographic Association, Perth, Australia, August* pp.680–690.

Dobson, M.W. (1979) "The influence of map information on fixation location" *American Cartographer* 6 (1) pp.51–65.

Jenks, G.F. and Caspall, F.C. (1971) "Error on choroplethic maps: Definition, measurement, reduction" *Annals of the Association of American Geographers* 61 (2) pp.217–244.

Julesz, B. (1975) "Experiments in the visual perception of texture" *Scientific American* 232 (4) pp.34–43.

McGregor, B.R. (1988) "The Perception of ReversedImage Line Symbols on Maps" *M.A. thesis* Queen's University, Kingston, Ontario.

Schiffman, H.R. (1976) *Sensation and Perception: An Integrated Approach* New York: John Wiley.

Williams, L.G. (1971) "Obtaining information from displays with discrete elements" *Cartographica* (Monograph) 2 pp.29–34.

Wood, C.H. (1994) "Effects of brightness difference on specific map-analysis tasks: An empirical analysis" *Cartography and Geographic Information Systems* 21 (1) pp.15–30.

Yarbus, A.L. (1967) *Eye Movements and Vision* New York: Plenum Press.

MAP MAKER Pro

The Map Processor for Windows

"Map Maker is the sort of useful program to look out for... Map Maker is a versatile tool..."

New Scientist, Sept. 28. 1996

"An excellent and inexpensive tool for dedicated map-making... Packs a lot of specialist equipment into a small footprint"

Personal Computer World, Feb. 1997

Until recently, computer based mapping has been expensive and complex. *Map Maker Pro* is inexpensive and easy to use and yet remarkably powerful.

You can make your own maps from scratch, use data from existing GIS, or create maps from scanned images such as aerial photos.

You can measure distances and areas on the screen as well as carry out more sophisticated analyses. 3D terrain models allow you to measure gradients and derive contours. A range of methods can be used to indicate data on a map.

Maps can be exported to *Windows* word processors. A map complete with data and annotation can be sent as an *e-mail map* to other *Map Maker Pro* users.

You can create maps in the field on your laptop using G.P.S. and more traditional survey techniques.

With the *poster print* option and a standard printer you can print your map as a wall map on multiple sheets.

Using one of the many map projections contained in *Map Maker Pro* (including OS and UTM) you can combine your field measurements with existing maps.

Compatible with OS NTF data, ArcInfo and much more.

148 page illustrated manual

Single-user licence...

£ **165** (+ VAT in E.U.)

Educational site licences available

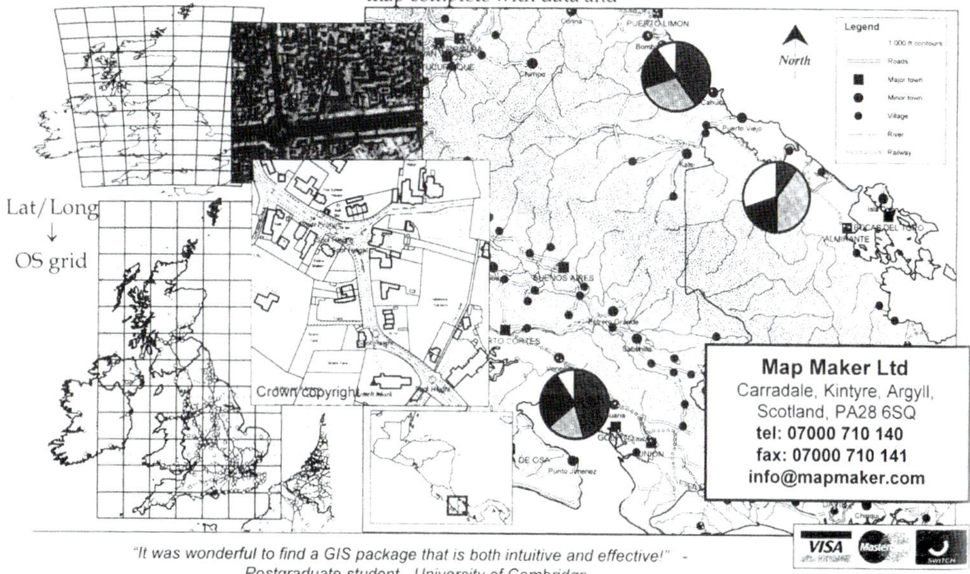

SITE DESIGN OR INTERNET CRIME?

Nick Tasker

Originally published in Volume 32, No 1, pp.1–4

This paper is my personal opinion and does not necessarily reflect the views of my employing establishment or the MOD. A version of this paper was first presented at the SoC Summer School in Edinburgh, 1998.

This is not an academic study of cartography on the Internet nor is it intended to suggest a quantum shift in the role of cartographers. However, as more and more cartographers are finding their job descriptions widening to include demands for all manner of graphic and visual interpretation skills it seems appropriate to investigate what we have to offer to the emerging discipline of the Internet. In particular, why we should use our skills to contribute to the World Wide Web.

I won't be covering how to publish maps on the Internet or how to design maps tailored for the Web in this paper. Instead, I want to look at how we understand the Web environment and the advantages cartographers can bring to the design of web sites.

Introduction

The traditional drawing office is disappearing fast, if not already a thing of the past in some areas. Certainly the tools of the trade have changed dramatically in recent years. Many see cartography becoming less distinct as a profession. Patterns of work and ultimately the structure of the cartographic workforce are also experiencing great change. True, there has been a definite convergence in the tools used by several 'graphic' professions. Scanning equipment, image processing and powerful digital manipulation are all part of the modem armoury of a good cartographic unit. Comparable equipment is also used in photographic, technical design and desktop publishing studios. As budgets tighten and cartographers increasingly find themselves working to several masters, the traditional boundaries between departments merge to form confused professional relationships.

It is in this climate that cartographers are looking to reaffirm their identity, whilst management forces attempt to fully utilize what are now expensive multipurpose production facilities. I have heard of demands for photographic and image reprocessing, general exhibition graphics, colour photocopying and scanning, even book-binding requests placed within the cartography section. This seems to apply equally in academic, public or private sector industries.

In this paper I have chosen to concentrate on the valuable assets we cartographers possess which transfer well into other disciplines. I do not suggest that a cartographer should become a Jack-of-all-trades or that we can rule the world (even though we may be able to draw it!).

It is my belief that it is not by accident that cartographers are now being asked to become involved in the latest business fashion – the design and construction of the company/department web site.

Qualification

At first, it may seem inappropriate to involve cartographers in web design, but as with any emerging science it is preferable to expand existing knowledge, rather than start with a blank skills sheet. Web site design requires a range of skills and as with most professions it is rare to find an individual competent at all aspects of the trade from the start. If one is trying to establish a web site from scratch there are a number of skills that will prove useful. Where can you find someone within an organisation with knowledge of digital graphic construction and manipulation, a good sense of balance and proportion, layout, fine visual communication skills and a smattering of programming experience? If your company or institution runs a cartographic unit then you need look no further.

However, web design is not only about graphic construction. It is also about maintaining disparate data sources and welding them all together into a coherent whole. It is about managing the process as much as designing the front end. Cartographers are born to do this. I put it to you, if the WWW came first, then cartographers would probably have become webmasters who naturally developed mapping as an extension of their web skills. I have no doubt that webmasters are, to an extent, cartographers already. They may not agree with the proposal, but they generally have an acute sense of schematic visualization. The most effective web sites are exactly like good maps. They communicate their information on an almost subliminal basis. When using a good map one almost looks right through the map, ignores it, and registers only the information that is designed to be communicated. So it is with good web sites. Visual communication is the common denominator here, but on closer examination the cartographic link is much stronger and I suggest much deeper. I am not just talking of the placing of map images on the Internet, how they look and how they may be utilized. I am talking of the fundamental design of web site architecture itself. The ability to conceptualize the whole from single elements and to literally map these into a new dimension.

Mapping Visualization

Mapping web site architecture is mapping at its most pure. Before the product, the concept, the visualization. Taking the complex and stripping it down into salient components portraying it in graphic form for others to understand. I hear shouts of dissent! If this is cartography, then where is the spatial relationship that separates a map from a complex graphic? Modern definitions of cartography seem to cling to this idea of attaching data to a spatial reality, the geographic element. Well, the World Wide Web is definitely secured within a geographic framework, Cyberspace does exist. And if you are looking for a graticule or network within web sites so that they may be referenced to each other, then the myriad of hyperlinks which can be formed between pages and beyond them, between sites, is structure sublime. Furthermore, cartographers have been some of the first to conceptualize the reality and actually map the concept and reality of Cyberspace.

Suspicion of the Unknown

In recent years I have noticed a backlash against computers in cartography. Or was that a backslash? Anyway... now we have the Internet upon us and I hear the same noises of discontent and suspicion. 'Never need to produce a map again, we're redundant, there won't be any need to produce new maps just find someone who has done it before on the Net and download it'. Or 'have you seen the things being passed off as maps on the Internet they are awful, firstly they take an age to download then you can't read them and finally you never see the whole image on a tiny screen anyway, so what's the point'. So, if they are so bad what are we worried about? What is the crime of maps on the Internet?

If there is any crime of maps on the Internet, apart from that which may require another paper on the very complex issues of Internet copyright, it is only the minor offence of poor reproductions or inappropriate use and display of data. But those problems (crimes) are still committed on paper! I think it is the Internet itself that really concerns us, not the maps upon it. The Internet and in particular the component of it known as the World Wide Web is growing faster than anything we have seen before. The technology is changing faster than even GIS technology before it. It's not only the environment that has changed, it is also the language. Just think about the language. Firstly, ask yourself where is the cartography, and secondly how many words weren't even invented three years ago. I know half the technology wasn't.

Mapping Cyberspace

As long as concepts and new environments develop, I suppose cartographers will rise to the challenge of mapping them. Jiang and Ormeling (1997) have recently been looking at how we can map the environment of the web. The virtual world of the Internet is known as Cyberspace. Not looking at the maps on web sites, but the mapping of the network itself. Actually representing a fluid dynamic mass of global communication and data flow. This is not an easy task as it is constantly changing, growing and developing. There are no rules to demand that everyone connects in the same way, or at the same time, or maintains the same provision they bad yesterday. However, it is a highly monitored environment and data logs exist for some very complex global statistics. As ever, statistics alone are next to meaningless, but visualized through a cartographers properly exercised mind they become understandable, at least in concept.

Examples of a range of cyber mapping can be found at the sites listed in the references.

I want to strip away the concept that the mapping of cyberspace stops at the front door, so to speak. I want to take it further, beyond the home page boundary, to look inwardly at the architecture of web sites themselves. Here, within web sites, I see a bidden cartography. Good web sites have a common set of principles that make them good for their audience. They may be based on a diverse range of subjects and presented in a variety of styles but all of the best web sites possess this hidden cartography. I came across a new Internet job title recently, that of Information Architect. What would you call yourself? I think information architect fits pretty well for cartographers and web designers alike.

An innate skill at representing a multidimensional, complex network in a clear understandable manner. The ability to take a mass of potential data and portray only the salient points at any one position. The ability to create hierarchies and lead the eye to focus points. The facility to navigate easily and quickly retrieve the information you require. The ability to rationalize space and connect similar concepts themes or values. Now does this start to sound like familiar territory to you?

Among the first people to successfully comprehend the complexity of the WWW have been cartographers, who have then designed new mapping techniques to portray an ethereal world. From these new maps of the web environment, others have been able to access, for the first time, an appreciation of the global importance of the Internet. Cartographers are used to portraying the complex, in an understandable manner and they routinely allow others to grasp information otherwise beyond them.

Web Design Rules

What then are the rules of this internal cartography of a web site? Let us look at the internal composition of the standard map product and find corresponding features in good web architecture.

Title

All maps should have a title, telling you what the map is about, who it is produced by and often little bits of important information on how to use it. The web site 'home' page should possess a title block as on a map, following the same rules. Kept simple, without overpowering the whole thing but clear as to who it is coming from and what will be found on the site.

Scope

When beginning a new map, one starts by defining the borders. With that in place the whole thing can be controlled, working inward from whole to part. Scheming what will be included and what will require another map to complete or have to be left out altogether. Starting a web site by defining its scope is an excellent idea. They have a habit of growing fast and can easily get out of control. It is equally vital that you know where the boundaries lie in a web site.

Scheme

When creating a series of maps it is taken for granted to maintain a common identity throughout, often common scales and projection dynamics, etc. Many people make a great mistake to create radically new, trendy looking web sites with sexy black backgrounds because they think that is what the web demands. Why create products which bear no relationship to the core identity of the firm or house style they represent? If planning series of web pages or bigger still a group of web sites, it is imperative not to lose your identity. Keep your message, keep your image or brand, keep your personality.

After all, you want to attract the sorts of people to your web site who already know and trust your traditional image. Make the site sexy by all means, but make it sexy for your users and their tastes. However, don't fall into the trap of designing your web site to reflect the company administrative hierarchy. Design is about products not principles. If the rules need to be bent then do so and use a little cartographic licence.

Planning New Editions

If you produce a new edition of an existing map you are going to get a bulge in sales, as people demand the latest and greatest. This is expected on the web; new editions are the norm. Regularly updated sites and attempts to provide something new encourage people to revisit. In this way you will attract more and more people to your web site. The simple rules of map marketing still apply on the web, only faster.

KISS Design

Nobody tries to design maps which are hard to follow or force the user to contact the cartographer for an explanation of what is going on or where to find things. Keep your web site symbols easy to understand and if necessary create alternative language options or segregate the site for different categories of user. Just as one may create general tourist maps and more specialist versions using the same data, good web sites are tailored to meet the needs of individual users. Make the direction clear – let people know where they are and where they can go. Give larger scales to expand a theme but make it clear how they fit into the big picture.

In my professional career, I have been told the quality of an Admiralty chart relies on four things:

- Quality of initial data
- The ability to navigate
- Facilitating interaction
- and reacting to feedback

These same attributes are the essence of web site design, get them right and you have cracked the successful formula of web site design. I thought it might be useful to provide the following advice for 'cartographic' consideration.

Navigation

- Where do you want them to go?
- What do you want them to see?
- What do you want them to learn?
- What do you want them to do?
- Don't lose them within the product.
- Provide easy choices and quick access (monitor how they move through your site and make it easy).
- Provide a key (or site map) or at least visual clues.
- Provide directional symbols.
- Give them a chance to search.
- No dead ends.
- Don't lose them to another product (or site).
- Put links to the outside in one place.
- Have partners implement cross links.
- Be creative (but not at the cost of readability or at the cost of usability).

Interaction

It is said that the designing of a web site is like the act of creating a pond in the middle of Cyberspace. You can create a big deep pond or a shallow little pond. Put up a sign telling people you have made such a pond and then let them come in for a swim. They can stay as long as they like, dive as deep as they wish or just paddle in the shallows. They can come and go as they please and come back whenever they wish. However, sensible folk who own ponds change the water frequently to stop it stagnating. They never push people in or hold their head under. They have safety features around the edge to help weak swimmers get out. Good web sites have natural hierarchy; they offer information on a range of levels and hold something for a wide range of users. These are all classic cartographic design requirements. I put it to you that by treating the web design task as a mapping project and structuring your web in a pseudo-cartographic manner you will not go far wrong.

The Quality Factor

I think 'A good map is like a good diamond', and a good diamond is rare, but worth looking for. The same applies to web sites. A good diamond can be measured in many ways but it must have quality. Quality of diamonds is measured by the four Cs:

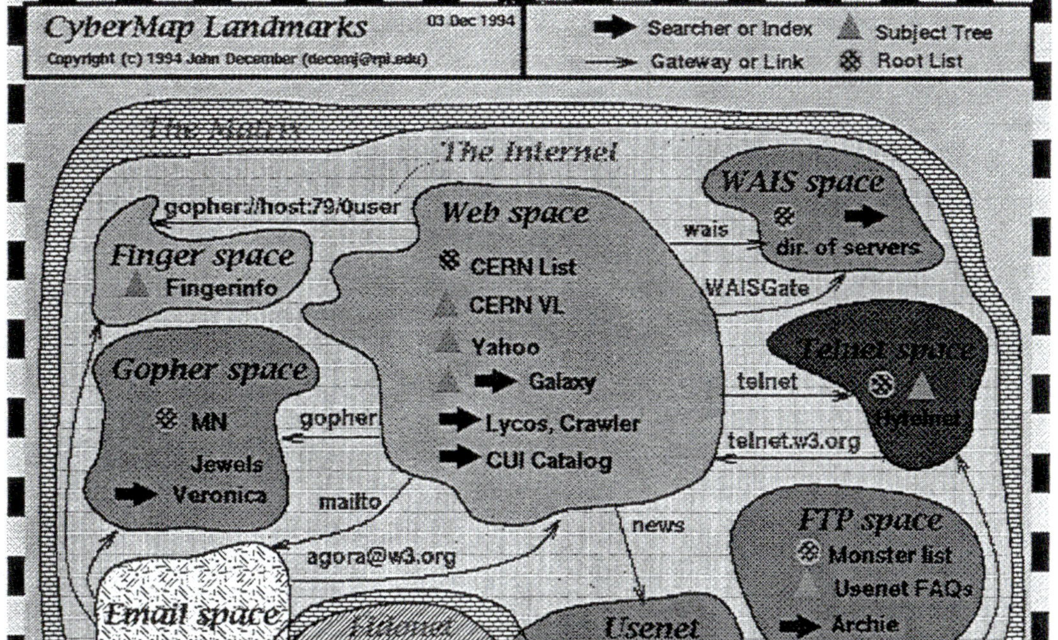

Figure 1 Cybermap Landmarks (TeleGeography, 1994)

Colour

What makes your web site special, a little different, out of the ordinary? If you can find a unique 'colour' for your site, your site will be priceless.

Carats

Size isn't everything but the bigger sites usually have more potential to appeal. As with maps a larger product usually can be sold to more users. 'Content is King' on the Internet. It doesn't have to look pretty if it satisfies the function. Make sure you include what browsers think they are going to get. Don't offer an online interactive graphic catalogue and then only provide a list. Don't oversell the site.

Clarity

Make the pages clear, use colour and graphics carefully, keep graphics small (under 35 KB), and design good strong navigation elements so browsers feel safe in your web site. They must feel they can leave if they wish to and know where they are. Provide a good site map.

Cut

Put the same effort into your web pages as you do into the rest of your designs. Then aim to provide at least the same levels of service or response as you would offer via traditional media.

Conclusion

Designing web sites can be common sense for special people called cartographers. I don't need to tell cartographers how to design web sites, they already know. They just need to recognize they are good designers and that their existing skills are very valuable to webmasters.

References

Cybergeography (1994) "Cybermaps of websites" Available at: *http://www.cybergeography.org/atlas/web_sites.html*.

Jiang, B. and Ormeling, F.J. (1997) "Cybermap: The Map for Cyberspace" *The Cartographic Journal* 34 (2) pp.111–116.

TeleGeography (1994) "Cybermap Landmarks" Available at: *http://www.telegeography.com/Publications/cyberland.gif*.

Wood, A., Drew, N., Beale, R., and Hendley, B. (1995) "Hyperspace Web Browsing with Visualisation" Available at: *http://www.cs.bham.ac.uk/~amw/hyperspace/www95/*.

LIVING IN A WORLD WITHOUT MAPS: SOME THOUGHTS ABOUT CARTOGRAPHY IN A SPATIALLY DISFUNCTIONAL WORLD

Gary Brannon

Originally published in Volume 32, No 2, pp.5–8

We remark that our world is shrinking, when what we really mean to say is that our spatial concept of the globe on which we live is undergoing a radical change. This article discusses the relevance of maps in an age where the immediacy of information is paramount, but where its geographical context is increasingly becoming irrelevant.

Dead Countries, Dead Maps

A colleague recently sent me a copy of a book that he has written, entitled *Dead Countries of the Nineteenth and Twentieth Centuries: From Aden to Zululand.*[1] It is an intriguing volume, full of evocative names that I still vaguely remember from the stamp albums of my schooldays but which have since vanished from the face of modem maps. Reading through the book with its references to exotic lands such as the Trucial States (now part of the United Arab Emirates), Straits Settlements (now variously administered by Singapore, Malaysia, and Australia), Ifni (now returned to Morocco by Spain), and Zanzibar (now part of Tanzania) was like taking a nostalgic journey back into the past. It also raised in my mind a provocative question: If, over a relatively short period of time, names and entire nations can disappear from maps – then why not the maps themselves?

An Endangered Species?

I have long contested that cartography, as a singular profession requiring definitive highly-developed technical skills, will ultimately disappear – or at least continue to exist only as a minor subset of some other allied profession. This evolution has, in fact, already been ongoing for decades, but it took the harsh economic realities of the 1990s and the obvious savings in time and manpower that the computer has to offer, to accelerate the pace of change ten-fold.

If, as I believe, map-makers are increasingly becoming an endangered species amongst technical professionals, then we should not be surprised that the maps they produce will inevitably suffer the same fate. Comfortably cocooned within a profession where maps are commonly held in almost mystical esteem, it is sometimes difficult to see the world as it really is (a particularly ironic situation for cartographers when one considers that cartography is – or at least was – a profession soundly rooted in the geography of terra firma). Unfortunately, this supposed importance of maps is sometimes elevated to a level that completely defies either rationality or common sense.

The Lives of Ordinary People

It did not require a carefully controlled scientific study on my part to discover a simple home truth: that maps are a largely unimportant factor in the everyday lives of most ordinary people. In my own circle of friends, family, and acquaintances – a diverse group that reflects wide-ranging interests, age, social demographics and education – casual observation and carefully directed conversation simply confirmed what I had already long suspected. Within this group, one or two individuals did admit rather sheepishly, perhaps conscious of my own cartographic background, to owning an atlas, though none could remember when they had last opened it or could even hazard a guess at how current it might be. Others admitted to having highway maps in the glove compartments of their cars, though the general opinion was that it was much easier to ask directions from a stranger when lost than to suffer the aggravations associated with consulting a folded map in a confined space (an opinion that I share). Few persons that I asked had ever actually consulted a map while on vacation, and fewer still ever bothered to look at maps in newspapers or in books or magazines. Interestingly, when I asked about maps they might have seen on television, I received mostly blank stares, though, when prompted, most admitted that they did sometimes look at television weather maps. Television weather maps, it seems, do not constitute real maps in the minds of ordinary people!

A Spatially Challenged Society

In a technological society that is inundated (overloaded?) with information, we are becoming accustomed to being told, especially as we turn the calendar on both a new century and a new millennium, that today's world is a very different place than it was even a few decades ago. There is no denying that this is a truism, the world has indeed changed radically in a vast number of ways and is likely to continue doing so perhaps at an even more rapid rate than we can even begin to imagine, but there is another truism to consider: we are also seeing our world in a very different way. We say that our world is shrinking, when what we really mean is that our spatial concept of the globe on which we live is rapidly being transformed and distorted. And the degree to which this transformation is happening is in direct proportion to the degree by which the technological age of instant information/communication/dissemination is burgeoning. As a society we crave information (aka news) but as the flow of information becomes more immediate and more accessible, what happens becomes of paramount importance and where it happens becomes increasingly irrelevant. When the geography of news becomes irrelevant to society, so too, I suggest, do the maps.

There was a time, not so long ago, when the map was our most reliable source for spatial information. If our Aunt Mabel sent us a postcard from her hotel on Fiji, we might first have looked at the postmark and found that it took three weeks to reach us, then pulled out a well-thumbed atlas and searched for the diminutive dot labelled Fiji in a sea of blue ink. The map might not have told us very much about the island of Fiji or about the people of Fiji, or indeed about Aunt Mabel's trip, but it would at least have satisfied our curiosity in terms of distance, and told us a great deal about the vastness of the Pacific Ocean and the minuscule nature of the land masses that lie within it. It would also have made it easier for us to understand the reason why a postcard mailed in such a place could take three weeks to find its way into our letterbox. From a simple dot on the face of a map we would have learned some fundamental geography and very much more.

Few of us will be old enough to remember first-hand either of the two world wars that have blighted the 20th century, but one need only watch some of the old newsreel war footage to understand the crucial importance of maps to the military during those dark periods of history. Maps were also crucially important in terms of boosting morale on the home front. Newspapers were full of them. They were spread out on kitchen tables and pored over or pinned on walls and stuck with coloured pins to show advances and retreats. At a time when the dissemination of news was exclusively the domain of the newspapers and radio, maps became inordinately important. They provided a very necessary spatial link between the battle front and the home front, despite the fact that much of the information they contained might be better described as propaganda than reliable news. Rulers would be taken out, distances would be measured, and the person at home could relate to battles in far flung places they had never set eyes upon and with whose names they were unfamiliar.

In a 1942 speech, American President Franklin D. Roosevelt, clearly conscious of the importance and value of maps in rallying support for the war effort at home, where isolationism was the prevailing attitude, said in part: 'I'm going to ask the American people to take out their maps. I'm going to speak about strange places that many of them never heard of – places that are now the battleground for civilization. I'm going to ask the newspapers to print maps of the whole world'.[2] In a radio address to the nation later that same year, Roosevelt kept his word, urging Americans to: 'Take out and spread before you a map of the whole Earth. Follow with me the references which I shall make to the world-encircling battle lines of this war'.[3] It was a clever, well-thought-out plan intended to bring into the homes of Americans a war that was perceived by many in the United States, at least up until the attack on Pearl Harbour, as 'Europe's war'.

Because of the major advances made in technology since the 1940s, the Vietnam War was a very different affair. From the Gulf of Tonkin incident, in 1964, to the fall of Saigon, in 1975, the American public had little need for maps to bring the horrors of war into their homes. The various television networks did their job in a way that would set the standard for all future conflicts. The map, it seemed, as a means of explaining a war, had now become passé. But not quite passé. Recently, listening to a televised speech, I was struck by the strong similarity between the words of Franklin Roosevelt, in 1942, and the following words spoken by American President William Clinton, in April 1999: 'I ask you, literally, get down an atlas and look at a map'.[4] The conflict this time involved Serbia and Kosovo and the Balkan nations surrounding them. Once again there was an urgent need to bring the American public onside. Again there was a need to explain why the United States, this time as part of a larger N.A.T.O. force, should get involved in a war far from home. Again there was a need for maps. Like the battlegrounds of Europe in the 1940s, like Korea in the 1950s, like Vietnam in the 1970s, like the Persian Gulf nations in the 1990s, the Balkan nations in 1999 presented a particular problem for military planners and governments alike: how to explain to the general public the importance of waging a war in a mysterious, little known, far-flung part of the world. And again, just as in 1942, an American president was suggesting that a nation dust off its atlases and delve into maps to discover the where, before any justification for the why could be adequately communicated.

Today, conceptually speaking, our world has shrunk to such an extent that distance has become meaningless. A typical 30-minute television newscast might contain segments from a dozen different countries in distant corners of the globe: tribal wars in Africa, ethnic conflicts in the Balkans, riots in South Korea, floods in China, an earthquake in Afghanistan, fires in California, a landslide in Argentina, etc., etc. The overwhelming effect of this live, in-your-face media coverage is that we have become generally desensitized by varying degrees, not only to disasters, but to geography itself. Maps do not matter anymore because seeing an event as it happens puts us right on the spot; we have travelled around the globe in an instant, the television or computer monitor has become our virtual transportation device. We are right there... wherever there might be. Turn off the sound, take away the captions, and we might as well be looking at generic, archival footage taken from the media vaults. Some American social psychologists have coined a term for this. They refer to it, somewhat derisively, as the 'CNN Effect'. President Clinton, in reference to the Balkan conflict, conceded as much by remarking that nobody in America had likely even heard of Kosovo until they saw it on CNN. It was a *de facto* admission that *Homo sapiens* are becoming increasingly spatially dysfunctional.

The Rare Case

Only in very rare cases, and with the noted exception of military conflicts, do maps and the spatial information they convey manage to interject themselves upon our consciousness. Such a case in point was the crash of Swissair flight 111, on 2nd September 1998, in Peggy's Cove, off the coast of Nova Scotia, Canada.

This particular disaster was remarkable, even in a world desensitized to such cataclysmic events, not only because of its picture postcard location in one of Canada's pre-eminent tourist areas, but because of the particular poignancy of the events that led up to it (an aircraft with smoke in the cockpit, the crucial yet laconic exchanges between crew and flight controllers, the choice of Halifax for an emergency landing rather than a closer airport, the decision whether or not to dump fuel, the complete disintegration of the aircraft at sea, and many other factors). Here was a news story that clearly had a strong spatial dynamic.

From the moment that it happened, the Swissair disaster was a significant media event. Canadian television crews were on the scene immediately (there is a major Canadian Broadcasting Corporation news bureau conveniently located in nearby Halifax) and media crews from around the world were not very far behind. Watching the daily briefings by representatives of the multitude of agencies that had become involved in the frantic recovery operation and subsequent disaster investigation, it was fascinating, at least from the particular point of view of a cartographer, to see that maps were being used as a front line tool in an attempt to provide accurate information to the assembled media from around the world.

The day before the final track of the aircraft was published, the Chief Investigator of the Transportation Safety Board of Canada had tried his best to explain verbally the complicated changes in course that had transpired over a period of 15 minutes from the time that the crew of the ill-fated aircraft had informed U.S. air traffic controllers that they had a problem to the moment of impact off the Canadian coast. It was quite obvious from the frustrated tone of the follow-up questions that were being asked by the media representatives, that they had little understanding of the concept of compass bearings and reciprocal courses, nor did they have very much knowledge of the geography of the Atlantic coast of Canada. The following day, when maps were made available showing the aircraft's course, superimposed on a topographic map, which also clearly showed the rocky intricacies of the coastline (and the degree of difficulty that this created for rescue and recovery), the information suddenly took on a new and very comprehensible quality. That map, or redrawn versions of it, appeared in many newspapers and on many newscasts around the world, and likely was the only map to be used on that newscast or in that newspaper on that particular day.In the days and weeks following the disaster it became apparent that some important early lessons had been learned, as many of the subsequent briefings had, as their central component, one or more maps showing such diverse topics as deployment of search vessels off the coast, location of wreckage, areas of coastline that had to be manually searched, the parameters of the commercial no-fly, no-shipping containment area, etc; all topics, it should be noted, that had strong spatial elements. The map had, in this rare case, reasserted itself not only as a primary source of clear, unambiguous information but as a valuable interpretive tool, confirming Andre Chastel's (1977) contention that 'the map is symbolic form. It possesses an almost hermeneutic function'.[5] Ironically, although some lessons were learned well, others still remain to be learned. One lesson, still not learned even as recently as the various military and humanitarian briefings on the Kosovo crisis, is that square or vertical maps/charts do not fit horizontally rectangular TV screens. Another, is that cluttered, busy maps or graphics fail to communicate anything. Basic stuff, but lessons still to be learned.

It is no coincidence, of course, in respect to the investigation of Swissair flight 111, that during the early days of the rescue and recovery, the operation was, for all intents and purposes, a Canadian military operation. And like the military operations noted earlier, maps frequently play a significant role in the dissemination of such information. As an anonymous Canadian reporter is said to have remarked during a Swissair disaster briefing, 'If this crash had happened in a farmer 's field in the middle of Saskatchewan, we wouldn't be looking at maps'.

Where Now?

Though there will likely continue to be special cases where maps interject themselves into our consciousness, it is clear that an inevitable trend is taking place and that maps are destined to play a much less significant role in our everyday lives than they have done previously. And this is true no matter whether we are part of the cloistered world of academe or not. Over the past several decades even the geographic sciences have been slowly yet relentlessly disassociating themselves from maps in favour of other forms of data dissemination. One has only to take a sampling of journals, magazines, and text books from the 1960s and 70s and compare them to those that have been more recently published, to fully comprehend just how far this trend has progressed.

Road maps, currently the likeliest reason for the average person to pick up a map, will eventually disappear from the scene (since 1996 the American Automobile Association reports that it has recorded a drop of 7 per cent in requests from the general public for paper route maps) as on-board global-positioning systems, linked to satellites for precise geographic positioning and roadside sensors for up-to-the-minute local traffic data become standard rather than expensive optional items in luxury automobiles. And as GPS systems for the boater, hiker, trekker and the average outdoors-type-person become increasingly more affordable, the necessity for many other types of map will also diminish.

To conclude on a note of dismay, a Canadian family magazine recently informed its readers about a new piece of computer graphics software that they should consider purchasing for their children. Your kids will never have to draw another map for their school projects, the reviewer espoused, perhaps without thinking about the disturbing ramifications of creating a population of young people that is spatially dysfunctional and neither knows nor cares about the shape of the world in which it lives.

References

1. Harding, L. (1998) *Dead Countries of the Nineteenth and Twentieth Centuries – From Aden to Zululand* Boston: Scarecrow Press.
2. Roosevelt, F.D. (1942) Speech.
3. Roosevelt, F.D. (1942) Radio address.
4. Clinton, W.J. (1999) Televised speech.
5. Chastel, A. (1977) "Preface" In Boudon, F., Chastel, A., Couzy, H. and Hamon, F. *Système de l'architecture urbaine: Le Quartier des Halles à Paris* Paris: Éditions du Centre national de la recherche scientifique.

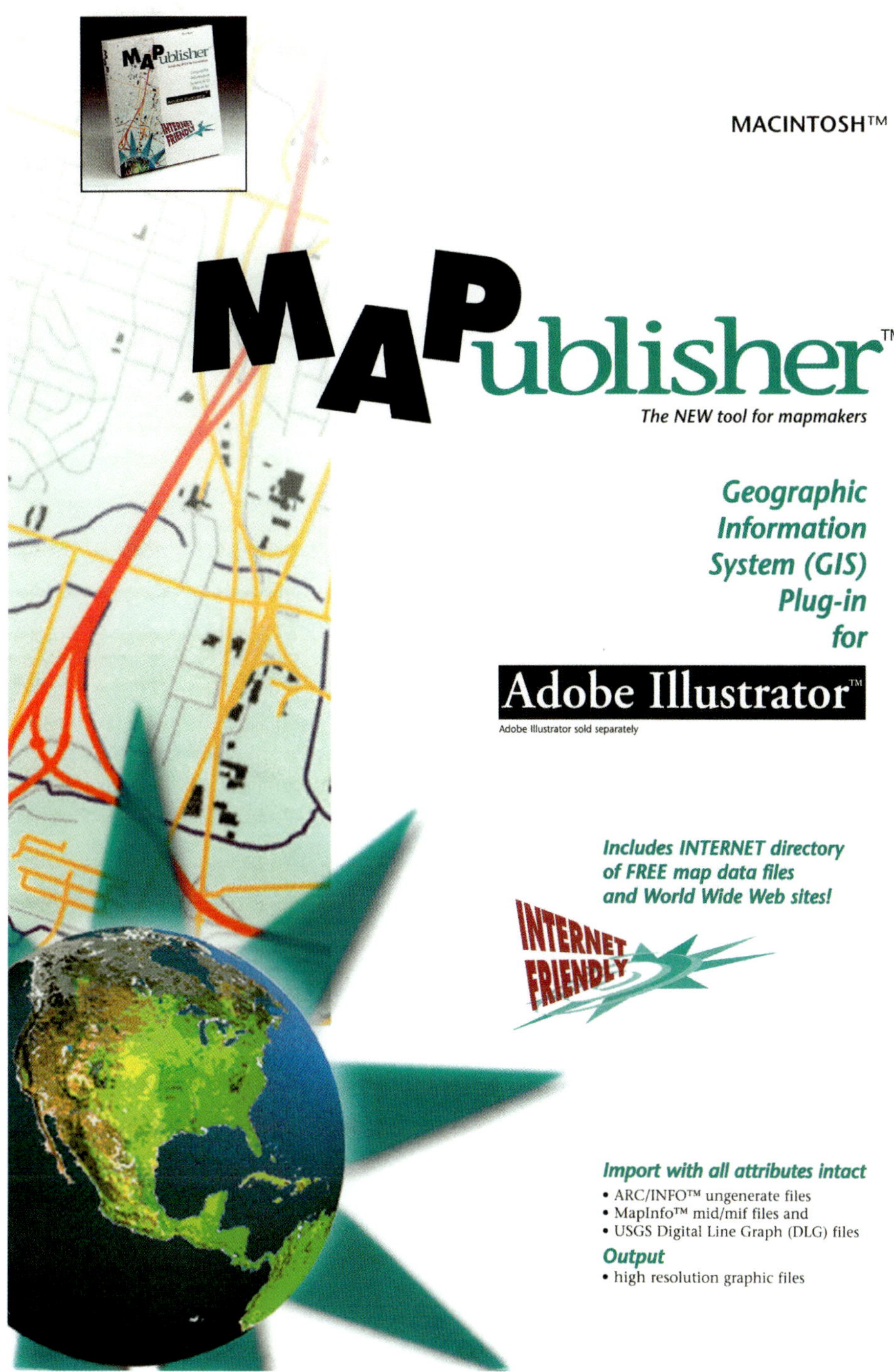

VIEWS OF THE WORLD: MAPS IN THE BRITISH PRESTIGE 'PRESS'

P. Vujakovic

Originally published in Volume 33, No 1, pp.1–14

Maps are an important element of the 'media's' representation of a range of important geopolitical and environmental issues. To date, few comprehensive surveys of news media map use and design have been undertaken. This paper provides an overview of the empirical results of a seven month survey (January to July 1999) of five UK prestige daily newspapers ('broadsheets'), *The Daily Telegraph*, *The Financial Times*, *The Guardian*, *The Independent*, and *The Times*, and compares these findings with earlier studies. The paper suggests several key questions for news media map research and sets the scene for further detailed analysis of the data set provided by the survey.

Introduction – Maps in the News Media

The 'news' provides an important conduit through which areas of geographic knowledge which lie beyond the direct experience of mass publics are reproduced and disseminated. This knowledge is an increasingly important factor in international affairs, especially as public opinion grows as a force influencing the decisions made by governments and supranational agencies (e.g. the United Nations). Many international issues are inherently geographical in nature, and require an understanding of spatial patterns, territorial divisions and boundaries, distributions and relationships, or dynamic geographical or demographic processes. Maps provide the main graphic medium by which this information is represented in the news media.

We live in an era of global communication, in which telecommunication technologies can provide mass publics, through the 'press', television, radio and the internet, with almost instantaneous reports on international events. Politicians, in states where the news media are relatively free of direct government control, are only too well aware that public support for their actions, for example humanitarian interventions or peacekeeping activities, can be affected by positive or negative 'news' (e.g. the loss of life amongst 'peace-keeping' troops). Dag Hammarskjold, former UN Secretary General, recognises and welcomes this development, which he sees as, 'the final, least tangible but perhaps the most important new factor in diplomacy: mass public opinion as a living force in international affairs' (cited in Bellamy, 1996: 33). Monmonier (1989), in his major study of news maps, claims a direct link between public understanding of these geopolitical issues and mass media cartography. He believes that 'the news media are society's *most significant* cartographic gatekeeper and its *most influential* geographic educator' (p.19; emphasis added), influencing perceptions of foreign, domestic and local problems.

Maps clearly play a significant role in the development of an individual's cognitive image of the world, and will inform and influence perceptions of geopolitical concepts and processes; the importance of world map projections in propaganda maps of WWII and the development of postwar conceptualizations of the globe is well documented (see for example, Hendrikson's (1974) discussion of the influence of world maps on the mind-set of post-war 'Air-Age Globalists', who 'pictured [the world] as a smooth, seamless ball, a monosphere.' (p.448)). Even today, the world maps used in a wide range of news publications, as well as atlases, school texts, advertisements, tend to reinforce and restate a 'eurocentric' conception of world affairs. Saarinen's (1988) study of nearly four thousand 'sketch maps' drawn by students from 49 different countries, has shown that eurocentric images remain a potent force shaping many people's cognitive geography (80 per cent of the sketch maps displayed a clear European bias, despite many being produced by students from all parts of the globe).

Denis Wood has coined the phrase 'to live map-immersed in the world', to describe the widespread exposure to cartographic products that characterises technologically advanced 'western' societies. By this he means that individuals are so surrounded by, and so frequently use maps that these become indistinct from other taken-for-granted consumer products. Maps are no longer special, the property of privileged elites or institutions, but are regarded by wider society as ubiquitous and common-place – 'apparently reproduced... without effort' (Wood, 1992: 34). This ubiquity probably means that much of the influence that maps have on our personal geographies is subliminal. During a century characterised by mass production and consumption, maps have become tools which are encountered, used or produced in every walk of life. They are commonly used in advertising, as route maps, at recreational sites, in schools, for weather information, and most significantly for the present discussion, in the news media and current affairs programmes. Maps are an integral part of the processes of symbolic communication embedded within an ever widening array of methods of news (re)production, transmission and consumption. As communication technologies develop, enabling 'information-graphics' to be more easily produced, it is likely that maps will be more commonly used to generate 'place knowledge' and aid (or obstruct!) understanding of spatial processes and geopolitical issues. Computer aided design has clearly enabled news producers to adapt and update maps of international events on an extremelyfrequent basis. At the height of Nato's intervention in Kosovo (March and April 1999), the five UK daily 'broadsheets' published no less than 166 maps of the region (44 per cent of all maps for the period) many in full colour, and often incorporating photographs or other 'information graphics'.

A deeper understanding of the impact of news maps on the geographic knowledge and understanding of mass publics, as well as decision-making elites, is clearly needed, especially as new technologies (e.g. internet editions of newspapers, 24 hour television news networks) provide greater scope for a range of cartographic representation (e.g. the animated and interactive ('virtual') election maps provided on BBC and *Electronic Telegraph* internet sites during the 1997 UK general election (Vujakovic, 1997)).

This area of research is still in its infancy and needs to address several related questions. First, in what circumstances are maps used in the news media, and what types of information do they contain? There has been some limited attempt to answer these questions, but comprehensive studies are still few in number (e.g. Monmonier, 1989; various studies in Scharfe (Ed.), 1997) and these generally lack comparability. Secondly, what is the quality of the information provided (including appropriateness of the cartographic techniques used)? Thirdly, what intended role do maps play within specific discourses disseminated by the news media? In other words, what 'work' does the authoring agency expect the map to do as part of its propagation of a specific 'message' about the world? And finally, how are maps understood by, and incorporated into the geographical understanding of their 'readers'? The last two questions demand that news media research moves on from its largely a theoretical and empirical position to take a more critical approach, grounded in appropriate theory. The author, for example, argues for an intertextual approach to news media maps, which builds on theoretical developments in cultural geography and the emerging field of 'critical geopolitics' (Vujakovic, 1997).

The longitudinal survey of UK elite newspapers discussed below is intended to provide a basis from which to answer some of these complex questions. This paper provides a summary of the empirical findings and compares these with previous studies. More detailed discussion is not possible here, and future papers will address map use within specific geopolitical discourses (e.g. the Kosovo crisis of early 1999), or specific cartographic design issues (e.g. use and abuse of world map projections, colour, symbolization, etc.).

The News Media Map Project: A Study of Maps in the UK Prestige 'Press'

A longitudinal survey of maps published in five daily UK elite newspapers ('broadsheets') was begun in early 1999. The newspapers included in the survey are *The Times*, *The Daily Telegraph, The Independent, The Guardian* and *The Financial Times*. At the time of writing data had been collated and classified for the months January to July 1991,[1] although map collection has now been extended to cover the whole of 1999, including map use by two Sunday 'broadsheets' (*The Sunday Times* and *The Observer*). All maps and their accompanying news stories published in the main news sections and the business news sections of the papers have been archived. Daily weather maps are not included (in common with other surveys, e.g. Monmonier 1989, Perkins and Parry, 1996), nor are maps from the sports or travel sections, or from any other review sections or magazines, as the survey is intended specifically as a study of 'breaking' news, with a strong emphasis on understanding the role of maps in current international political and environmental debates. Where sports or travel stories (accompanied by a map) are found in the main news sections, they are included in the survey. Previous surveys have included maps from sports, travel and other supplements. Inclusion of maps from these sources will inflate the 'count' of published maps, but offers little insight into news media cartography (although it is acknowledged that they may influence the readers' general geographic knowledge base, as might any published or broadcast map).

	Conservative	Labour	Liberal
Daily Telegraph	71	13	15
Times	61	18	17
Financial Times	50	27	18
Independent	31	40	26
Guardian	12	59	22

Table 1
Political affiliation of national newspaper readers (%) (adapted from MORI poll results quoted in McNair, 1994)

The survey concentrates on the UK prestige press for several reasons. First, these papers are more likely to be read by policy and decision makers, politicians, media professionals (who may be involved in re-making news or documentary material), and members of the public who are actively involved in 'political' interest groups. They are also frequently used as an information source and general learning resource in a range of educational settings. The five prestige papers, as a whole, represent a broad range of UK mainstream political affiliation (Table 1).

It is also worth noting that while only 25 per cent of the UK. public report newspapers as their main source of world news (with television topping the poll at 62 per cent), for readers of elite papers this figure rises to 57 per cent (and television drops to 32 per cent) (Negrine, 1994).

Secondly, the prestige newspapers are more likely to use maps to illustrate current news items.[2] Perkins and Parry (1996) report that the five UK daily 'broadsheets' published a total of 197 maps in June 1995, compared with only 29 for the five main 'tabloids' (ranging from only one map in *The Star* to ten in *The Daily Mirror*; the only tabloid to reach double figures). They also note that the numerically most important categories of mapped news items in the prestige press are military conflicts and geopolitical issues (ranging from 53 per cent of all maps published in *The Independent*, to only 17 per cent in *The FT*), while the environment, and political and social issues also form significant categories. These news maps may form key elements in competing contemporary geopolitical discourses; as King (1996) notes, 'the map itself became the battleground in the debate over how Bosnia-Herzegovina might be dismembered [...] [a] confusing array of maps accompanied the reporting of the process in the western media' (p.55). Vujakovic (1993, 1999) provides evidence for the importance of maps in the news media's coverage of the emerging post-Cold War European security landscape. The later study revealed a subtle change from 'triumphalist' cartographic and textual representations of the European political landscape in the UK press in the early 1990s (especially the idea of the eastward expansion of 'western' ideals, values and institutions as the route to regional security), to retrenchment into 'realist' images (including maps) and metaphors more redolent of the Cold-War period in more recent years.

The News Media Map Project: Survey Methods and Results

The survey of news maps is divided into three main sections dealing with thematic and geographic coverage, and basic design issues. Data was recorded for each and every map in the following categories.

A) Thematic coverage

- Major news themes (7) (see Table 2)
- Sub-themes (18) (see Table 2)
- 'Positive' or 'negative' nature of news item

B) Geographical issues

- Map location (based on major world regions (14)
- 'Ethnocentrism' – degree of association of news items to the reporting nation
- Elitism (based on geopolitical status of nations (3 categories)

C) Design

- Map size
- Use of Colour
- 'Involvement' (oblique/perspective versus perpendicular/frontal view)
- 'Composition' (4 categories; from single map to complex montage of map and other graphics)
- Map use (7 categories; eg simple locator maps, route maps, plans, etc)

Details of specific classifications are introduced in the following sections as the results are presented. All of the following discussion (and tabulated data) refers to the period 1 January 1999 to 31 July 1999 (inclusive), and includes results for all of the five UK daily 'broadsheets', unless otherwise stated.

i) Thematic coverage

Thematic coverage in this study is classified according to seven basic categories, further subdivided into eighteen sub-themes (Table 2). This allows for both broad and fine scales of analysis and overcomes some of the problems inherent in previous studies. For example, Monmonier's (1989) study (using only nine themes) lumps several distinct issues together (e.g. 'economic resources' (finance?) and 'transportation').

Table 2 News map themes – News Media Map Project 1999

Theme	Sub-theme
A Politics (internal)	1. Government, legislation, electoral, parties, non-violent protest/strikes
	2. Riots, terrorism, civil conflict/war
B Politics (international)	3. International relations, negotiations, agreements (non-trade)
	4. Military conflict/war, defence issues, territorial disputes
C Disasters/ Accidents	5. Large-scale disasters (earthquakes, floods, etc.), epidemics
	6. Accidents (transport, etc.), explosions & fires, industrial disasters, avalanches
D Environment and science	7. General science, natural science, engineering, medical
	8. Environmental problems/impacts, pollution
	9. Transport systems, development & planning
	10. Land use/resource planning & conservation, public works, neighbourhoods
E Society	11. Demography and social trends, housing, employment, education
	12. Crime, courts/judicial, police, missing persons
	13. Social disasters (famine, refugees)
F Cultural affairs	14. History & archaeology, heritage, the arts and 'media'
	15. Travel, tourism, recreation, sport
	16. Human interest/'odd events', VIPs/royals, scandals (non-political), religion
G Economics	17. Business & finance, industry
	18. Macro-economics, trade agreements, international finance, aid and economic development

For pragmatic reasons the maps are always allocated to one class only. There are clearly times when this can prove difficult, and in such cases the main thrust of the news story will determine the final choice; for example, a map illustrating a story concerning NATO intervention in Kosovo will be classified as 'military conflict/war' (A4) despite the fact that there may also be reference to 'environmental problems/impacts' (D8) or 'refugees' (E13). Too rigid a concern for quantification at the broad scale is not, in any case, pertinent to the study of news maps, rather, the statistics are used to provide an overview of the relative importance of themes, and act as a 'spring-board' for more detailed multi-thematic and intertextual analysis of the archived maps and news stories.

Despite some differences in approach, certain similarities are evident between the results for this study and other research. Both Monmonier (1989) and Perkins and Parry (1995) identify 'military conflict, defence, geopolitics, threats and riots' as the numerically most significant categories of mapped news. Gauthier's (1997) study of French Canadian papers identified armed conflicts as the largest category of non-domestic news maps, while Tremols (1997) study of the Spanish press recorded 'geopolitical conflicts' as the largest category. This general trend is confirmed in the present study, with 'military' maps accounting for 28 per cent of all maps for the period and 36 per cent of non-domestic news maps (see Table 4). This concern with military issues, international disputes and violence is in fact characteristic of most non-domestic news reporting (with or without maps) in Europe and north America. Wallis and Baran's (1990) study of US, UK and Swedish 'broadcast' news identifies these themes as accounting for anything between 31 per cent and 49 per cent of 'foreign' news items. In the purely domestic context other themes may dominate, Gauthier (1997) for example, identifies (non-violent) politics, economics and accidents/catastrophes as the main mapped themes in French Canadian domestic news, while in Spain 'social unrest' and 'ecology and environment' form major mapped themes (Tremols, 1997). Crime and nonviolent politics are the largest domestic themes identified in the 1999 UK study (see below).

The present survey also classified all maps according to the degree of 'ethnocentrism' of the accompanying news story (see key to Table 3).This allows thematic coverage in this study to be placed in the wider context of 'home' and 'international' news, and a spectrum from the purely 'domestic' to the entirely 'foreign'. The classes range from news stories concerned only with an issue related to the reporting country (a 'domestic' item – most ethnocentric), to one involving a number of states, but not the reporting country itself ('international' – least ethnocentric). The criteria are adapted from Wallis and Baran (1990), who used a similar classification for their study of news broadcasts. It is worth noting that 22.5 per cent of all maps published during the period deal with purely domestic (UK) news items. This is a slight decline when compared with Perkins and Parry's (1996) result for 1995, with domestic maps representing 29.6 per cent of all news maps (unfortunately, their data set only distinguishes between UK and non-UK maps, so no further comparisons of changes in 'ethnocentrism' can be made).

As well as the purely domestic ('home') news, the four other categories can be broadly categorized as 'parochial'

Table 3
News media maps classified according to the 'ethnocentrism' of the accompanying news story

'Ethnocentrism'	All maps	%
Domestic	254	22.5
National	221	19.6
International	236	20.9
Bilateral	31	2.7
Multilateral	386	34.5

Domestic (D) = Item dealing with reporting nation only;
National (N) = item dealing with one nation only (but excluding the reporting nation);
International (I) = More than one nation involved, but not including the reporting nation;
Bilateral (B) = involving the reporting nation and one other;
Multilateral (M) = involves the reporting nation and more than one other nation.

Adapted from Wallis and Baran (1990).

(37.2 per cent) (when the reporting nation has a specific interest or is directly involved) or 'foreign' (40.5 per cent) (where the reporting nation has no distinct involvement). This is clearly an important issue, as 'news worthiness' is often associated with both physical and cultural distance decay effects. This issue would, however, require detailed examination of the proportion of mapped and non-mapped news items; it can be assumed that less well known, distant and exotic places are more likely to be mapped than those which the news producers believe their audience to know (e.g. there is generally no need to provide a simple locator map with a story about London or Paris). A domestic reverse-distance decay effect is also likely to occur, with locator maps needed for (obscure) non-metropolitan locations.

The dominant theme, especially when considering nondomestic news, is the 'military conflicts' category, and this is itself dominated by the war in the Kosovo region of Yugoslavia. The Kosovo news stories for the period include both internal issues (e.g. Serb attacks on Albanians), classified as A2, and international/multinational events (e.g. NATO intervention) classified B4 (see above). The Kosovo crisis accounts for 25.3 per cent of all maps for the seven months of the survey, rising to a staggering 45.7 per cent in April, at the height of NATO intervention. This also accounts for the general month on month rise in the number of maps published in early 1999.

The series of crises in the Balkans, following the collapse of the wider Yugoslav federation during the 1990s, has resulted in much greater news reporting and map coverage of the region, and this ought to have altered people's geographic knowledge of this part of Europe if the news media is the gatekeeper to geographical knowledge suggested by Monmonier. However, anecdotal evidence from a pilot survey of undergraduate geographers suggests that even a basic understanding of the configuration of the new nation states of the region remains limited. First year Geography students (Single and Combined Hons.; n=47) were given an outline map of the 'Eastern European Gateway' region and asked to name the

* 'Parochial' = combined Multilateral and Bilateral news maps, 'Foreign' = combined International and National news maps

Table 4
Thematic coverage and 'ethnocentrism'

Themes	Sub-themes	All maps	Domestic	'Parochial'*	'Foreign'*
A politics (internal)	1. Government, etc.	84 (7.4%)	42	12	30
	2. Riots, etc.	144 (12.8%)	16	23	105
B Politics (international)	3. International relations, etc.	25 (2.2%)	0	14	11
	4. Military conflict, etc.	320 (28.4%)	1	213	106
C Disasters/ Accidents	5. Large-scale disasters, etc.	15 (1.3%)	0	3	12
	6. Accidents, etc.	74 (6.6%)	15	30	29
D Environment and science	7. General science, etc.	37 (3.3%)	9	10	18
	8. Environmental problems, etc.	44 (3.9%)	18	13	13
	9. Transport systems, etc.	26 (2.3%)	19	1	6
	10. Land use, etc.	41 (3.6%)	29	3	9
E Society	11. Demography etc.	30 (2.7%)	18	3	9
	12. Crime, etc.	84 (7.4%)	46	15	23
	13. Social disasters	8 (0.7%)	0	1	7
F Cultural affairs	14. History, media, etc.	28 (2.5%)	9	10	9
	15. Travel, etc.	35 (3.1%)	5	22	8
	16. Human interest, etc.	56 (5.0%)	12	17	27

Table 5
Maps of the Kosovo crisis as a proportion of all maps

Month	All maps	'Kosovo crisis' maps	
		number	as percentage (%) of all maps
January	112	12	10.7
February	142	6	4.2
March	174	56	32.2
April	203	111	45.7
May	156	42	26.9
June	182	49	26.9
July	159	9	5.7
TOTALS	**1128**	**285**	**25.3**

states and to locate six cities and the Kosovo region. Outlines of familiar states bordering the region (e.g. Germany, Greece, Turkey (to be named)); and Italy, Denmark (not to be named)) provided locational and orientational cues[3] (although no verbal information was provided). Germany achieved the highest recognition rating (87 per cent) and other states bordering the region fared reasonably well (Russia 68 per cent, Greece 62 per cent, Turkey 60 per cent), however, the region itself was very poorly known. The central European states were best known (highest recognition for Poland 51 per cent), while the Baltic or Balkan states, or ex-members of the USSR (excepting Russia) were rarely identified correctly (range; Estonia 17 per cent (n=8), to Moldova 0 per cent (n=0)). Despite the intense news coverage of the Balkan region during the Kosovo crisis and the large number of maps (both in the newspapers and on television) of the bombing of Belgrade and other Yugoslav (Serbian) locations, students displayed a very poor knowledge of the region; only four located Kosovo with any accuracy, only five placed Belgrade in Serbia (one with any real accuracy), and the territories of Serbia/Yugoslavia and Albania were correctly recognized by only four and three students respectively. Almost all of the students (79 per cent) identified the television as their main source of international news, while only six read a 'quality' paper with any regularity. It may be that people's perception of the ethno-politics of the regions has changed (although this itself may involve stereotyping of issues related to ethnic groups, histories and cultures), but despite increased map coverage, their geographic knowledge has been little affected. This raises important questions concerning the role and real impact of news maps, especially the ephemeral cartography of television.

Few other themes come close to dominating the data set as do these geopolitical themes. In terms of domestic (UK) news, two themes emerge as important; political (internal/non-violent (A1) and crime (E12), with land use planning (D10) trailing as the third largest (see Table 4). These form 16.5 per cent, 18.1 per cent and 11.4 per cent of domestic news maps respectively, and together represent just under half of all domestic maps (46 per cent). The mapped news is again domi-nated by specific events, for example, the killing of Jill Dando (the BBC television presenter), and the bombing of 'gay-bars' in London, form a high proportion of the 'crime' maps, while the elections to the Welsh Assembly and Scottish Parliament account for a considerable number of the political maps. These individual news events form a useful basis for further study of map use

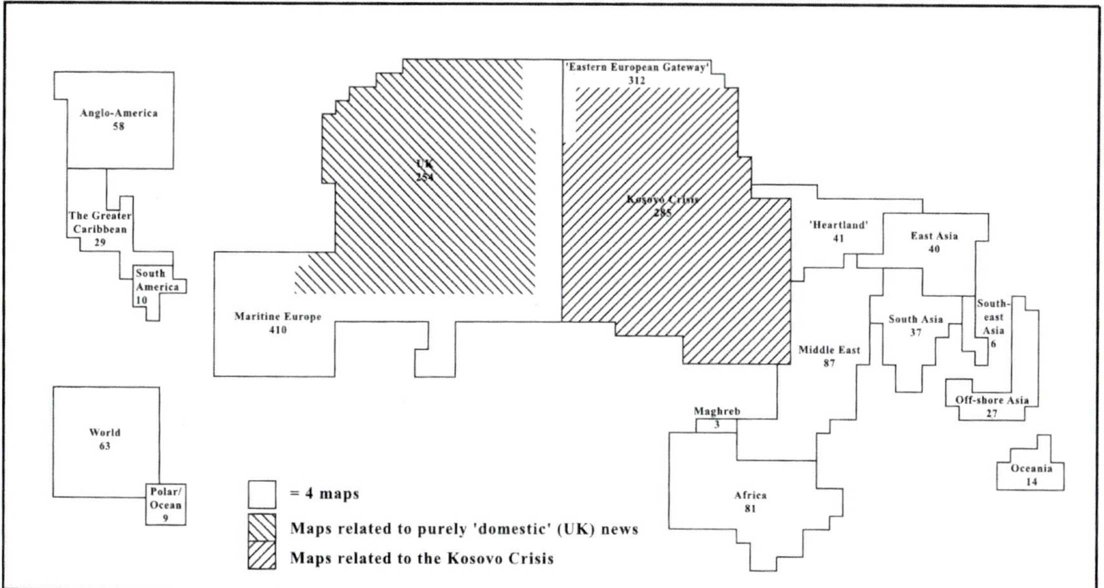

Figure 1
News map coverage (five UK prestige daily papers) according to major world region (regions based on geopolitical criteria, adapted in part from Cohen (1991) for the period 1 January–31 July 1999.

and design, and comparisons between the newspapers. The project archive also allows comparisons to be made with other coverage at the time (e.g. news magazines and reviews (*The Economist, New Scientist*, etc.) and retrievable news material in other media (e.g. the internet).

Maps accompanying environmental issues remain, however, a surprisingly small part of the total output from the papers. This may reflect the complex nature of some environmental issues in terms of the difficulty of graphic illustration (however, complex graphics of physical phenomena such as cyclones and tornadoes are relatively common in the press). It may also reflect the inherent conservatism of the elite press; 47.7 per cent of all maps illustrating 'environmental problems' are published in just one paper, *The Guardian* (although this theme represents only 7.8 per cent of all maps in that paper), while *The FT* has published none in the category.

ii) Geographical coverage

The geographical distribution of all news maps for the period is illustrated in a cartogram (Figure 1). The number of states within the regional classes adopted varies widely (from

Themes	Sub-themes	World Powers *	Regional Powers **	Others	TOTAL
A politics (internal)	1. Government, etc.	48	16	20	84
	2. Riots, etc.	26	32	86	144
B Politics (international)	3. International relations, etc.	17	5	3	25
	4. Military conflict, etc.	33	52	235	320
C Disasters/ Accidents	5. Large-scale disasters, etc.	8	2	5	15
	6. Accidents, etc.	49	3	22	74
D Environment and science	7. General science, etc.	26	5	6	37
	8. Environmental problems, etc.	34	7	3	44
	9. Transport systems, etc.	22	2	2	26
	10. Land use, etc.	36	3	2	41
E Society	11. Demography etc.	25	2	3	30
	12. Crime, etc.	69	12	3	84
	13. Social disasters	4	1	3	8
F Cultural affairs	14. History, media, etc.	15	8	5	28
	15. Travel, etc.	14	7	14	35
	16. Human interest, etc.	44	5	7	56
G Economics	17. Business, etc.	35	2	6	43
	18. Macro-economics, etc.	16	10	8	34
Grand totals		**521**	**174**	**433**	**1128**

* China, France, Germany, Japan, Russia, UK, US. ** twenty-four second order or regional powers adopted from Cohen (1991) including, for example, Brazil, Iran, Indonesia, Mexico, Nigeria, Poland.

Table 6
Thematic coverage and geopolitical status

Table 7 The function of newspaper graphics: June 1995 (adapted from Perkins and Parry (1996)

Newspaper	Locate	Route	Distribution	Explanatory
FT	25 (71%)	3 (9%)	5 (14%)	2 (6%)
Guardian	24 (71%)	3 (9%)	3 (9%)	4 (12%)
Independent	19 (59%)	4 (13%)	4 (13%)	5 (16%)
Telegraph	24 (75%)	2 (6%)	3 (9%)	3 (9%)
Times	51 (69%)	12 (16%)	3 (4%)	8 (11%)
TOTAL	143 (69%)	24 (12%)	18 (9%)	22 (11%)

'Anglo-America' represented by only two states, to 'Africa south of the Sahara' with over forty); the regions are chosen to reflect contemporary 'geopolitics', and are adapted, in part, from Cohen's (1991) classification of geopolitical realms. The classification reflects the author's approach to the study of news maps, grounded in the emerging field of 'critical geopolitics' (see Vujakovic, 1997 & 1999). Two world regions, 'Maritime Europe' and the 'Eastern European Gateway' dominate the data set; however, the first is swollen by the large number of purely domestic maps (62 per cent), while the second area reflects the current importance of central and eastern Europe as a region of geopolitical flux, a trend which has been evident throughout the 1990s. Similarly, other zones of instability in the Middle East, and parts of Asia and Africa, show up in the data as significant regions of mapped news. Examples of geopolitical and geostrategic issues include, continued western intervention in the Gulf region, capture of the Kurdish leader Abdullah Ocalan by the Turkish authorities in February 1999, India and Pakistan in conflict over Kashmir, and the Spratley Islands dispute in the South China seas.

Maps were also classified according to the geopolitical status ofthe states mapped. All states are classified into one of three categories, first order powers (global powers/elite nations), second order powers (regional powers, such as India or Brazil), or 'other'. The map is classified on the basis of the highest order state shown on the map which contributes directly to the news item.

Again, the importance of regions of geopolitical stress is evident. By far the largest category relates to military conflict in the global periphery and semi-periphery, with a significant number of maps related to conflicts involving second order powers (e.g. India in its dispute over Kashmir with Pakistan – itself a 'pretender' for regional power status, and a nuclear power). The relatively stable core regions centred on first order powers are less likely to be mapped in terms of conflict, except in situations such as the Kosovo crisis, were maps show the disposition of troops from the US and maritime Europe to this zone of tension. The only exception to this is Russia, which cannot be described as a stable geopolitical entity at the present; the war in Chechnia will certainly contribute to the internal conflict theme in the latter part of 1999. The global importance of the core is instead reflected in the large number of economic and financial mapped new items, with 66.2 per cent of all 'economic' maps involving a first order state (NB, only 19.5 per cent of all economic maps are entirely domestic (UK).

Table 8 Map use (function) in five elite UK newspapers

Newspaper	Logo	Locator	Route	Plans	Distribution	Dynamic	'Political'	TOTAL
FT	9 (6.1%)	61 (41.5%)	24 (16.3%)	2 (1.4%)	27 (18.4%)	6 (4.1%)	18 (12.2%)	147
Guardian	4 (1.5%)	175 (52.7%)	26 (9.5%)	6 (2.2%)	30 (10.9%)	22 (8.0%)	12 (4.4%)	275
Independent	3 (1.3%)	121 (54.0%)	18 (8.0%)	7 (3.1%)	33 (14.7%)	30 (13.3%)	12 (5.3%)	224
Telegraph	1 (0.6%)	81 (48.8%)	21 (12.7%)	11 (6.6%)	16 (9.6%)	26 (15.7%)	10 (6.0%)	166
Times	1 (0.3%)	178 (56.3%)	28 (12.0%)	9 (2.8%)	43 (13.6%)	29 (9.2%)	18 (5.7%)	316
TOTAL	18 (1.6%)	616 (54.2%)	127 (11.3%)	35 (3.1%)	149 (13.2%)	113 (10.0%)	70 (6.2%)	1128

Logo = Logo or decorative function (but does not include maps in cartoons); Locator = Simple locator function; Plans = Plans and layouts (generally large scale); Distribution = Distributions and spatial variations; Dynamic/spatial processes (e.g. troop movements); 'Political' = non-standard political and regional boundary, zones or areas (i.e. does *not* include established state or county boundary maps, but would include newly created boundaries, or areas such as the Iraq 'no-fly' zones)

Table 9 Map composition

Newspaper	Single map	With small locator	'Multiple-map'	Montage	TOTAL
FT	25	58	11	53	147
Guardian	49	183	12	31	275
Independent	59	114	19	32	224
Telegraph	66	35	13	52	166
Times	96	145	22	53	316
TOTAL	295 (26.2%)	535 (47.4)	77 (6.8%)	221 (19.6%)	1128

iii) Design issues

Perkins and Parry (1996) make the important point that past attempts to evaluate news map design have been undermined by using criteria more appropriate for conventional cartography. They point to Balchin's (1988) study which they claim epitomised the orthodox approach and tended to ignore the context and constraints of journalistic cartography (deadlines, print quality, opportunity costs between word and image space). They suggest that studies should focus on the functional complexity of the mass media maps – whether they succeed in meeting the needs of the news item and whether it tells the same story as the text – i.e. is the map 'fit for purpose' in its journalistic context. One way to examine this is, they suggest, to evaluate the functional role of maps. Perkins and Parry (1996) divide maps into four broad categories (see Table 7).

Their survey, limited to one month, suggests that simple 'locating devices' are the most common form of media map graphic used in the UK press, and this is confirmed by the present survey. Locator maps provide the basic spatial context for many news items. More complex maps showing routes or distribution are much less frequent, although still significant in numbers. Their final class is, however, less helpful. They term the remaining maps 'explanatory' – used 'to clarify military and geopolitical stories' (p.332), unfortunately this is not a exclusive class (many distribution or route maps, for example, could be classified as 'explanatory') and it has no specific functional criteria.

The present study seeks to overcome this and other potential problems by having a clear, and separate set of functional ('map use') and 'compositional' criteria. It also recognises other legitimate map functions, for example the simple, but politically important 'logo' function (also identified by Ormeling, 1997), in which a map outline provides the context for a news story, without providing more specific detail. With regard to classification of map use, any particular map is allocated to the highest level use it performs (i.e. most maps perform a locator function, but if one also shows the deployment ('distribution') of troops, the map would be classified accordingly). Table 8 provides a summary of 'map use' by the five newspapers surveyed.

Table 8 shows that locator maps predominate, but are less important than suggested in the 1995 survey. This may represent the larger and more representative sample of the 1999 survey, although it may also indicate a more diverse use of maps for other functions, perhaps due to increasing flexibility provided by use of computers. The Kosovo crisis provides an instance where a wide range of maps are used to show distributions (16.5 per cent of all Kosovo maps for the period), and dynamic/spatial processes (21.8 per cent). Many appear to be based on the same map, with adaptations for a particular news item. It may also reflect the inclusion of non-news maps in the 1995 survey. Such problems of direct comparison between studies are an important argument for a consistent and comprehensive system of news map research.

The dominance of locator maps is reflected in terms of 'composition'. Maps are here classified according to whether they are a single map, a main map with a small inset locator, a graphic with several maps ('multiple-map'), or map as part of a montage of photographs or other information-graphics (see Table 9).

Many of the second category are 'locator' maps themselves in functional terms, representing 63.7 per cent of alllocator maps in the survey.

Colour is an increasingly important design issue in the study of news media cartography. Perkins and Parry (1996) note that prior to the mid-1980s maps in UK newspapers were almost entirely monochrome, but that recent changes in print technology provide the potential to communicate complex geographic information in a concise and unambiguous fashion. They record, however, a very uneven use of colour amongst the papers. The Financial Times almost never used colour (only 3 per cent of its maps) in 1995, while nearly half of all *The Independent's* maps were in colour (47 per cent).

The 1999 survey shows that the trend to increasing use of colour has been maintained, with a general increase from 28 per cent to 41 per cent of all maps. Most papers have increased their use of colour, some dramatically, and only *The Daily Telegraph* appears to be bucking this trend. Colour has certainly been put to effective use to illustrate and explain complex geopolitical and environmental issues during 1999. It has been used extensively to illustrate the complexities of the Kosovan crisis, with regard to both military issues and the ethno-linguistic divisions within the province (and the wider Balkan region).

Table 10
Use of colour

Newspaper	Perkins & Parry (1996)		1999 survey	
	Colour	Mono-chrome	Colour	Mono-chrome
FT	1 (3%)	34 (97%)	21 (14%)	126 (86%)
Guardian	4 (12%)	30 (88%)	133 (48%)	142 (52%)
Independ-ent	15 (47%)	17 (53%)	116 (52%)	108 (48%)
Telegraph	13 (41%)	19 (59%)	45 (27%)	121 (73%)
Times	25 (34%)	49 (66%)	145 (46%)	171 (54%)
TOTALS	58 (28%)	149 (72%)	460 (41%)	668 (59%)

The use of colour is generally simple and effective. Colour is used to distinguish land from sea, or to provide topographic information. Political maps are generally treated with a subtle touch, with key territories being emphasised by use of deeper tones of a single colour. Very few of the choropleth maps in the survey use colour and those that do are generally well designed. An example of good practice (despite poor choice of projection (see discussion below) is *The Independent's* map illustrating 'Species in danger of extinction' (5.3.99; p.9), this uses a very effective tonal range (from pale yellow, through oranges and reds, to dark reddish-brown) as the basis for choropleth mapping of quantitative data. *The Independent* appears to be relatively consistent in its use of colour throughout the survey period. *The Times* provides another good example, entitled 'Serb withdrawal from Kosovo' (16.6.99; p.ll), in which the zones of withdrawal are differentiated by changes in tone from orange to pale yellow, providing a very clear indication of the temporal-geographical dynamic ofthe process. If a problem exists with the use of colour, it is the reproduction of maps which were clearly originally designed as colour maps as monochrome in the paper. The tonal variations in the original appear as murky greys in the final product. This appears to be a common problem.

Projecting Global Issues: World Maps in the News

World maps form a significant group within the survey. The seven months of the survey yielded 63 world maps, as well as a number of additional map which cover a significant proportion of the earth's surface. While the world maps represent only 5.6 per cent (nearly 1 in 18) of all news maps, they are functionally complex, and are probably of major significance in terms of the public's general understanding of geographic issues. If only those maps which accompany news ttems concerned with non-domestic issues are included, then the proportion of world maps rises to 10.1 per cent of the total.

There is now a considerable body of work concerning the cultural role played by world map projections in the formation of specific 'world-views', especially the creation and maintenance of a dominant, eurocentric conception of the globe. Widespread use of maps, such as the Mercator, or the Miller cylindrical, which distort area (massively exag-gerating the land mass of high latitudes) and which tend, by convention, to be centred on Europe, have been blamed for this (e.g. Black, 1997; Peters, 1983; Thomas and Crow, 1994). In essence, the discussion has centred on whether the frequent use of certain projections creates a biased and distorted geographic image of the world, which itself fosters or enhances other cultural and political biases. It has been argued, for example, that the exaggeration of high latitudes tends to diminish the importance (for people living in the rich 'North') of the 'developing world', and hence affect attitudes to economic and social assistance (Development Education Centre, 1985). It is argued that only by using equal-area maps (showing the continents in their true size relative to each other) to display thematic information about the world, can an equitable global image be created. Clearly this is in itself simplistic and may foster other biases; for example, a number of authors have argued for the (occasional) use of population cartograms to illustrate development issues, as this lays stress on people (the real concern of development) rather than abstract space (see Thomas and Crow, 1994; Vujakovic, 1989) .

Cartographers have long recognized the problems associated with the choice of appropriate base maps for thematic mapping, and have taken pains to encourage sensible usage (see for example Dent, 1990; Robinson *et al.*, 1995). As Kraak and Ormeling (1996) point out, 'there are no good or bad projections but, rather, there are good or bad applications of map projections' (p.86). However, it has to be acknowledged that cartographers and geographers have had only limited success in propagating this message to other practitioners in the field of graphic design, and to the general public (Snyder, 1993). In fact, a perceived 'outsider', the German historian Arno Peters has had more success,[4] although with unfortunate consequences. Peters' own equal-area map has been uncritically adopted by many development organisations to the exclusion of other, often better projections (Vujakovic, 1987; 1989). Despite the concerns of cartographers and geographers, projections which distort relative area continue to be widely used for general thematic maps. Inappropriate use of these projections is endemic; from their use in advertisements and other forms of 'persuasive' cartography, to their continued appearance in school and university level texts, including many authored by geographers. It is hardly surprising, therefore, that the erroneous use of map projections is also characteristic of the news media.[5]

The 63 world maps recorded in this survey are typically dominated by political themes (50.9 per cent) and those associ-ated with conflict (41.3 per cent) in particular. Two other themes also figure prominently, 'cultural affairs' (F) (28.6 per cent) and 'economics' (G) (20.6 per cent) (only 10.5 per cent and 6.9 per cent respectively for 'all maps'). These results are dependent, however, on particular stories covered during the survey period, and the proportion may eventually change. For example, a large number of world maps are cover the attempts to circumnavigate the globe in high-altitude hot air balloons (sub-theme 'sport' F15) (see discussion below). The use of world maps for 'environment and science' is a little above average (19 per cent, compared with 13 per cent for all maps), but interestingly, is dominated by the sub-theme of 'general science ... ' (9 maps), rather than 'environmental problems ... ' (only 2 maps), as might be expected. Economic maps are also important, and reflect both the growing importance of the 'global market' and processes of'globalisation', and the inclusion of the Financial Times within the survey.

Analysis of map function also provides some interesting, although not unexpected contrasts, between world maps and the full survey (Table 11).

Table 11
Map function

Map function	World maps number (and %)	All maps number (and %)
'Logo' function	4 (6.3%)	18 (1.6%)
Simple locator	18 (28.6%)	616 (54.6%)
Routes/networks	19 (30.1%)	127 (11.3%)
Plans/layouts	0 (0.0%)	35 (3.1%)
Distributions	16 (25.4%)	149 (13.0%)
Dynamic/spatial processes	6 (9.5%)	113 (10.0%)
Political/'regions'	0 (0.0%)	70 (6.2%)
TOTALS	**63**	**1128**

As a distinctive image, 'world maps' are used relatively frequently to perform a 'logo' function. No explicit information is contained within these maps, but the global context of the news item is denoted by the outlined continents. World maps represent over 1 in 5 of all 'logo' maps recorded in the survey. The use of world maps as a 'simple locator' is still significant (28.6 per cent), but much less common than for news maps as a whole (simple locators, are the most common form of all news map (54.6 per cent) – often showing the site of a single news event). Route maps instead form the largest class of world maps

(30.1 per cent). This figure may, however, be inflated by the large number of maps dedicated to the coverage of 'round-the-world' high-altitude ballooning during the survey period. The other major class (25.4 per cent) is distribution maps. Many of these maps show aggregate data at the national level (e.g. 'world population growth',

In some cases base maps have been distorted for graphic effect and it is not possible to identify the original projection with any certainty, this is indicated by (?-?).

Newspaper	Main projection(s) (frequent use)	Other projections (limited use)
The Financial Times	Times	n/a
The Independent	Gall stereographic	Oblique ortho-graphic
The Daily Tele-graph	Times Gall stereographic	Oblique ortho-graphic
The Times	Times Gall stereographic	Oblique ortho-graphic Cahill butterfly (?)
The Guardian	Eckert IV	Van der Grinten (?) Times Gall stereographic Oblique ortho-graphic Lambert azimithul equal-area (?-?)
The Sunday Times	n/a	Gall stereographic (?) Times (?) (?-?)
The Observer	n/a	Times (?) Oblique ortho-graphic (?-?)

Table 12 World map projections used in UK elite news-papers (January -July, 1999), including two Sunday 'Broadsheets'

The Independent (9.2.99: p7). Both of these categories are significantly larger for world maps than for news maps taken as a whole.

While these data provide a broad indication of the use of world maps by the press, and their general thematic coverage and function, the real significance lies in the manner in which individual world maps are used to illustrate specific issues. The survey shows that map design is in many cases inappropriate to the intended task, and may lead to a misunderstanding of the geographic context or the processes involved. This is particularly the case with regard to the choice of map projection.

The majority of world maps used in the UK press are based on a limited range of projections, although specialist projections are used on occasion. In some cases, world maps have been 'cropped' to produce a base map for a more limited region of the world. Table 12 provides a summary of the projections adopted by each newspaper. It is not always possible to be confident about the specific projection that has been adopted; news maps rarely include a graticule or acknowledge the projection used. In addition, the maps are often cluttered with other graphics or with 'boxes' of explanatory text, are 'cropped' to save space, or are too small or poorly printed to provide clear evidence of the exact projection. The inclusion of national boundaries is sometimes helpful in distinguishing visually similar projections; for example, the boundary between Alaska and Canada is drawn with a slight curve and at an oblique angle on the (eurocentred) Times projection, but is perpendicular (following a 'straight' rather than 'curved' meridian) on the Gall stereographic projection .

It is clear from Table 12 that the most popular projections are the Gall stereographic (or a variant) and the Times (itself a modified Gall projection). One, or both, are used extensively by four of the five daily 'broadsheets'. Neither, however, provide equivalency, the property of equal-area representation, which Dent (1990) claims to be the 'overriding concern' when choosing a world map for thematic purposes. Both projections create serious exaggeration of the area of higher latitude land masses.[6]

The only newspaper to consistently adopt an equal-area projection (the Eckert IV), as a base for its thematic maps, is *The Guardian*. This projection is generally regarded an appropriate choice for the display of thematic data, although there is considerable distortion of the shape of landmasses pole-wards. The Eckert IV is the preferred projection of the World Bank (used for its annual *World Development Reports*), and of several atlases concerned with development issues (e.g., the Open University's *Third Word Atlas* (Thomas and Crow, 1994), and the Developing Areas Research Group *Atlas of World Development* (Unwin (Ed.), 1994). However, despite its use of Eckert IV, the paper occasionally uses other, often inappropriate projections, and, as if to add insult to injury, applies 'fictional' graticules to these for dramatic effect. One example of this is the occasional use of (what appears to be a variant of) a van der Grinten projection (e.g. 16.1.99; p.26). The map grossly distorts the size of high latitude land masses and is drawn with a totally spurious graticule (in which the equator appears to run through Nicaragua!). This was also used during the late 1990s as the graphic accompanying a world news review page, identifying the location of items.

While the use of projections lacking equivalency as the basis for world thematic maps is the most frequent and highly visible of the 'errors' noted during the survey, other instances of poor practice exist. Two examples serve to illustrate this, one related to thoughtless adherence to 'convention' and the other to the erroneous use of projection. In the first instance, the ubiquity of eurocentric maps has created a situation where routes have been artificially interrupted because a map has been 'cut' across the Pacific, rather than the Atlantic. Maps illustrating the global circumnavigation attempt by the high-altitude balloon, Breitling Orbiter (e.g. *The Times*, 19.3.99; p.3, and *The Telegraph*, 19.3.99; p.13), show its progress from the launch site in Switzerland, eastward until it reaches the right-hand edge of the map (somewhere in mid-Pacific), it then reappears on the left-hand edge of the map, to

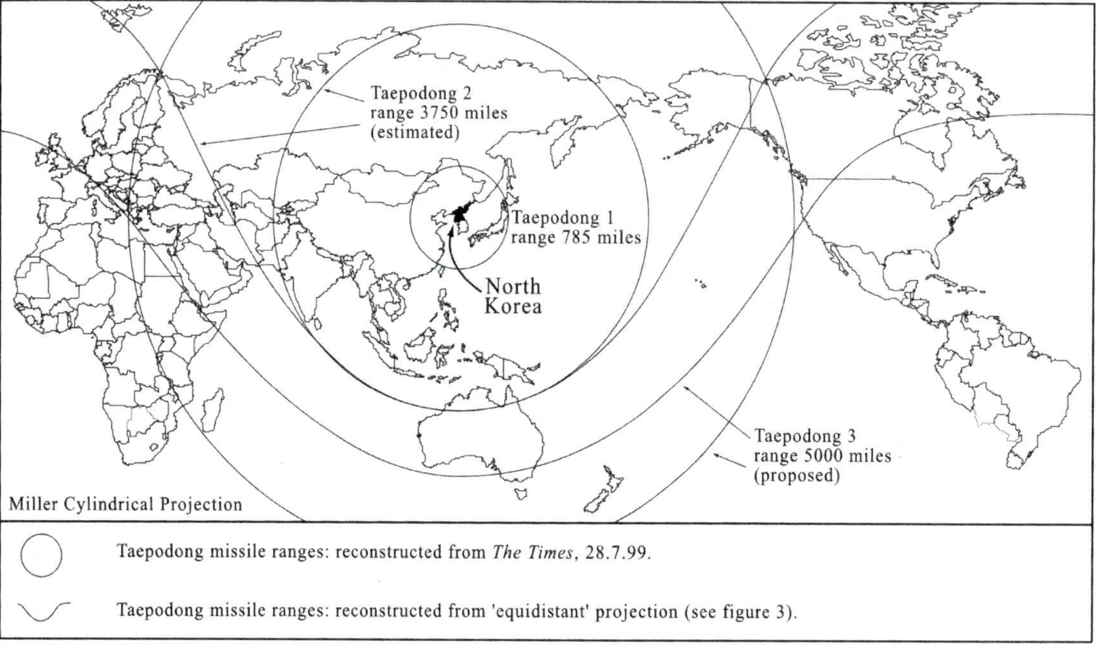

Figure 2
Ranges of the North Korean Taepodong missile series erroneously displayed as concentric circles on a cylindrical projection.

continue its journey as far as the Caribbean. It would be much more appropriate to use a sino-centric projection which maintained the integrity of the route from start to finish, presenting the Pacific component of its traverse as a whole. The second example is more serious in factual terms. Two of the 'broadsheets' published maps to accompany reports on a North Korean missile system (*The Times*, 18.6.99; p.17 and 28.7.99; p.13, and *The Guardian* 10.7.99; p.15). The story concerns the Taepodong missile series, which according to *The Times*, includes three missile systems (one operational and two planned) with ranges of 785 miles (1,300 km), 3,750 miles (6,000 km) and 5,000 miles (8,000 km) respectively. Unfortunately, both papers appear to use the Gall stereographic (or similar cylindrical projection) which does not allow missile ranges to be constructed as a series of concentric circles centred on the launch site. The cylindrical projections provide uniformity of scale along standard parallels only, hence a circle of constant distance (the missile's hypothetical range in this case) drawn on a globe cannot be reconstructed as a perfect circle on the map. The circular ranges used by both *The Times* and *The Guardian* are therefore spurious.[7] Figure 2 illustrates the effect of using the erroneous pattern of concentric circles adopted by *The Times* (here reconstructed on the Miller cylindrical) when compared to the true ranges interpolated from an appropriate equidistant projection (shown in Figure 3; oblique azimuthal equidistant projection; centred on Pyongyang, North Korea). The equidistance projection does allow the ranges to be constructed as concentric circles. Clearly the nature of the perceived threat (in

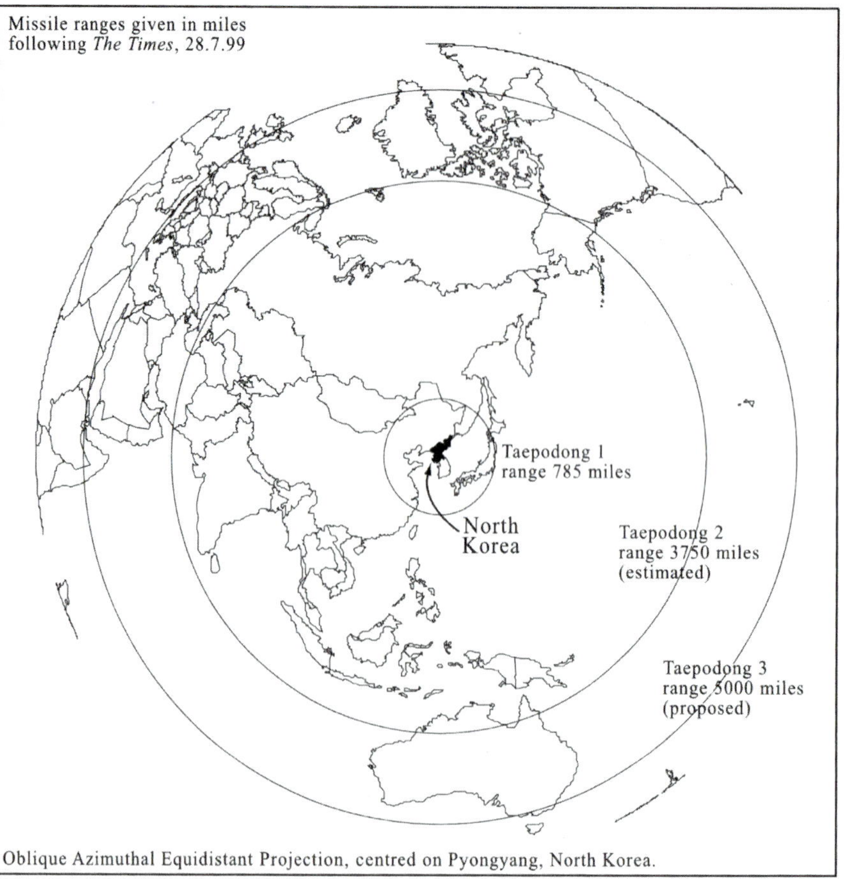

Figure 3
Ranges of the North Korean Taepodong missle series shown as concentric circles on an appropriate 'equidistant' projection (centred on Pyongyang).

geopolitical terms) of the missile systems will be quite different in terms of the images created by the two newpapers and an image of the real situation.

The issue of appropriate use of projections needs to be addressed by the news media. There is a danger, both in terms of factual information and in impacts on people's global concepts or 'world view' of the continued use of highly distorted images of the globe, especially where the projection is not 'fit for purpose'. It is worth pointing out that many western television news programmes still use the Mercator or other non-equal area map as a logo or as a back-drop to their newsdesks, an action which further reinforces false notions of world geography.

Conclusions

It has not been possible in this paper to do more than present some of the major empirical fmdings of the survey to date and begin to suggest areas for further discussion and research. The survey indicates that maps are an important component ofthe news media and are especially relevant to the study of military conflicts, border disputes and other geopolitical issues. This offers a rich vein of material for geographers interested in sites of popular geopolitical discourse and for cartographers concerned with map design and communication. The number of maps used by the press is shown to have increased dramatically during the Kosovo crisis, the key geopolitical event in 1999. This provides a very useful data set for the study of the role of maps in the construction of geopolitcal identities and formulation of notions of the 'other', especially the Balkans constructed as 'alien' or 'savage' lands, and certainly peripheral to concepts of a democratic, liberal European identity (see Vujakovic (1993) for discussion of news media maps, European identity and 'tribal war' in Bosnia in the early 1990s, see also Myers *et al.* (1996) for an alternative reading of the news media's 'inscription' ofthe Bosnian war).

The survey confirms the general trends identified by previous research concerning thematic coverage and map use, while adding to our knowledge with regard to the use of map projections and the geographical coverage of maps in the UK press. The survey also provides a comprehensive data base for more detailed studies of design issues and the use of maps in geopolitical and geo-environmental discourse. Future papers will address these issues. It is hoped that the present study will also provide the basis for further discussion and the development of common classification systems for mass media map research, a process initiated by the Mass Media Map conferences held in Berlin and Budapest during the late 1990s.[8]

Acknowledgements

Thanks are due to John Hills (Department of Geography and Tourism, Canterbury Christ Church University College) for producing the maps used in this paper, and for his fortitude in helping me to move vast piles of newspapers from the College's library to the third floor of a building with no lift!

References

Balchin, W.G.V. (1988) "Media Map Watch: A report" *Geography* 70 (4) pp.339–343.

Black, J. (1997) *Maps and politics* London: Reaktion Books.

Bellamy, C. (1996) *Knights in White Armour: The New Art of War and Peace* London: Hutchinson.

Cohen, S.B. (1991) "The emerging world map of peace" In Kliot, N. and Waterman, S. (Eds) (1991) *The Political Geography of Conflict and Peace* London: Belhaven.

Crow, B. and Thomas, A. (1983) *Third World Atlas* Milton Keynes: Open University Press.

Dent, B.D. (1990) *Cartography: Thematic Map Design* Dubuque: Wm C. Brown.

Development Education Centre (1985) *People before Places? Development Education and an Approach to Geography* Birmingham (UK): Development Education Centre.

Gauthier, M-J. (1997) "Cartography in the media: An overview of Canadian developments" In Scharfe, W. (Ed.) (1997) *Proceedings of the International Conference on Mass Media Maps* June 1997 (pp. 39–52) Berlin: Freie Universität Berlin.

German Cartographic Society (1985) "The so-called Peters projection" *The Cartographic Journal* 22 (2) pp.108–110.

King (1996) *Mapping Reality: An Exploration of Cultural Geographies* Basingstoke: Macmillan Press.

Kraak, & Ormeling, F. (1996) *Cartography: Visualization of Spatial Data* Harlow: Longman.

Loxton, J. (1985) "The Peters phenomenon" *The Cartographic Journal* 22 (2) pp.106–108.

McNair, B. (1994) *News and Journalism in the UK* London: Routledge.

Monmonier, M. (1989) *Maps with the News: The Development of American Journalistic Cartography* Chicago: University of Chicago Press.

Myers. G. Klak, T. and Koehl, T. (1996) "The inscription of difference: news coverage of the conflicts in Rwanda and Bosnia" *Political Geography* 15 (1) pp.21–46.

Negrine, R. (1994) *Politics and the Mass Media in Britain* (2nd ed.) London: Routledge.

Ormeling, F. (1997) "Mass media maps as objects of cartgraphic thought and theory" In Scharfe W. (Ed.) (1997) *Proceedings of the International Conference on Maps Media Maps, June 1997* (pp.22–33) Berlin: Freie Universität Berlin.

Perkins, C. and Parry, R. (1996) *Mapping the UK* London: Bowker Saur.

Peters, A. (1984) *Die Neue Kartographie – The New Cartography* New York: Universitätsverlag Carinthia: Klagenfurt/Friendship Press.

Robinson, A.H. (1985) “Arno Peters and his new cartography” *American Cartographer* 12 (2) pp.103–111.

Robinson A.H. (1990) “Rectangular world maps-No” *Professional Geographer* 42 (1) pp.101–104.

Robinson, A.H., Morrison, J. L., Muehrcke, P. C., Kimerling, A. J., and Guptill, S. C. (1995) *Elements of Cartography* (6th ed.) New York: Wiley.

Saarinen, T.F. (1988) “Centering of mental maps of the world” *National Geographical Research* 4 (1): pp. 112–127.

Scharfe, W. (Ed.) (1997) *Proceedings of the International Conference on Maps, Media Maps June 1997* Berlin: Freie Univerität Berlin.

Snyder, J.P. (1988) “Social consciousness and world maps The Christian Century” 24th February pp.190–192.

Snyder, J.P. (1993) *Flattening the Earth: Two Thousand Years of Map Projections* Chicago: University of Chicago Press. Thomas, A. and Crow, B. (1994) *Third World Atlas* (2nd ed.) Milton Keynes: Open University.

Tremols, M. (1997) “The social importance of geocartography in the Spanish media” In: Scharfe W. (Ed.) (1997) *Proceedings of the International Conference on Maps Media Maps June 1997* (pp.219–228) Berlin: Freie Univerität Berlin.

Unwin, T. (Ed.) (1994) *The Atlas of World Development* New York: Wiley.

Vujakovic P. (1987) “The extent of adoption of the Peters projection by 'Third World' organisations” In the UK *The Bulletin of the Society of University Cartographers* 21 (1) pp.11–15.

Vujakovic, P. (1989) *Mapping for world development Geography* 74 (2) pp.97–105.

Vujakovic, P. (1993) “Maps, myths and the media: cartography and the ‘new Europes’” In Mesenburg, P. (Ed.) (1993) *Proceedings of the 16th International Cartographic Conference* (pp.231–243) Beilefeld: German Society of Cartographers.

Vujakovic, P. (1997) “Toward the Millennium: Mass media map research in the United Kingdom” In: Scharfe W. (Ed.) (1997) *Proceedings of the International Conference on Maps Media Maps* June 1997 (pp.53–64) Berlin: Freie Univerität Berlin.

Vujakovic, P. (1999) “‘A new map is unrolling before us’: Cartography in news media representations of post-Cold War Europe” *The Cartographic Journal* 36 (1) pp.43–57.

Wallis, R. and Baran, S. (1990) *The Known World of Broadcast News: International News and the Electronic Media* London: Routledge.

Wood, D. (1992) *The Power of Maps* London: Routledge.

Notes

1. A small error in the relative figures given in this paper (and a slight underestimate of map totals) may exist because of missing editions in the library collection used (34 out of 1,060 editions (c.3 per cent), which were reasonably evenly distributed between the five papers). The results are, however, deemed to be representative and to reflect the current state of news media mapping in the UK's prestige press. It is hoped that the gaps will eventually be filled from other sources.
2. It is acknowledged that certain key international events with strong UK links (for example conflicts involving British forces, such as the Falkland and Gulf wars, or Nato's interventions in the Balkans) are covered extensively by the 'tabloids', and may be accompanied by complex information graphics. It is recognised that these might usefully form the basis for detailed studies in their own right, as sites of popular geopolitical discourse.
3. It is clear from examination of the individual maps that most students understood the geographical context of the map; the inclusion of Italy in particular provided an immediate cue to the scale, orientation and location of the region (most students (n = 42) correctly located at least two of the 14 key states which bound the region). However, a small but significant number displayed considerable problems (one named Turkey as 'Greenland', several mistook Germany for France (n = 3), and one student was so confused, despite locating 'USSR'(!), Germany and Poland correctly, that they identified Estonia as Norway, Sweden and Finland, Estonia's Hiiumaa and Saaremaa islands as the UK, and Crete as Japan!). Several students (n = 5) named the Ukraine, Romania or the Black Sea as 'Iran'. The same exercise was performed with the author's 12 year old son Alexander; he ranked 3rd out of 48 on his knowledge of the region (despite his father's own obsession with maps Alexander has not been 'coached') and when questioned, believed his knowledge of the Balkan region to be based on television news maps – he was one of only two people to correctly place Belgrade, and one of three who located Kosovo with any accuracy.
4. This is not the place to rehearse the 'Peters debate' in detail. The debate is polarised between Peters' supporters, who tend, in many cases unthinkingly, to adopt his ideas and projection (also known as the Gall-Peters; Peters claims that he invented the projection independently, and was unaware of the existence of the Gall Orthographic), and the cartographic community which condemns his work (e.g. The German Cartographic Society, 1985; Loxton, 1985; Robinson, 1985, 1990). It must be recognized, however, that Peters, and the adopters of his projection, have been very successful in raising awareness of map projections as an important political issue, and so the onus must now be on cartographers to provide alternatives, rather than simply 'wringing their hands [...] over the misuse of the Mercator Projection', as John Snyder (1988: 192) so eloquently puts it.

5. It would be extremely unfair to single out a few individual texts as exemplars of poor practice, but any scan of the shelves of a school or university library will provide ample evidence of poor use of world map projections in text books and academic journals.
6. Snyder (1993) notes several important modifications of the Gall stereographic and other similar cylindrical projections, especially Miller's cylindrical projection which has been used extensively in the USA. It is not possible, given the print quality of many maps in this survey, to distinguish between the standard Gall and these others. The Gall stereographic itself is the most likely candidate having been adopted and popularized in the UK by Bartholomews, although the Miller is a standard projection available in some computer mapping packages (the Miller used in this paper was generated using ArcView).
7. All distances are quoted directly from the newspaper, the range for each missile appears to have been converted from miles to kilometres and rounded to a convenient figure. *The Guardian* gives a figure for Taepodong 2 only, at 3,730 miles (shown as 6,000 km on its map).
8. The papers presented at the first of these events are published in Scharfe (Ed.) (1997), material from the second workshop in Budapest in 1999 can be found at *http:/naza-rus.elte.hu/hun/tantort/1999/mmmstart.htm*.

HIKING ON MARS? THE WHAT, THE HOW AND THE WHY OF PLANETARY MAPPING

R.B. Parry

Originally published in Volume 33, No 2, pp.5–12

Since the early moon shots of the 1960s, mapping of the other planets and moons of the solar system has formed a significant part of the American and Soviet space programmes. This article discusses the special characteristics of planetary maps, stresses the need to include them in the collection policies of map libraries, and gives advice on access to both printed maps and web-based image archives.

Many definitions of 'map' appear to overlook the possibility that this medium could be concerned with anything other than mother Earth, but as cartographers well know, the techniques of representing features in two or three, or indeed four-dimensional space, can have many non-geographical applications. One of the most obvious and dramatic is the mapping of the other planets in the solar system. Yet perhaps it is because of the common association of maps with geographical (i.e. Earth-bound) science, that while planetary mapping has proceeded apace, there has been relatively little literature about such maps. Over the last decade, this has begun to change. First, thanks to the work of Greeley and Batson (1990) there is now an excellent and substantial book on this subject, and more recently these authors have produced a superb atlas of the solar system (Greeley and Batson, 1997), which no map library should be without. Moreover, planetary mapping was a special theme at the International Cartographic Association's conference in Stockholm in 1997, and a Planetary Cartography Working Group has subsequently been established within the ICA.

Nevertheless, from the perspective of the map library, I suspect that planetary maps have been and still are relatively neglected, and it is this that prompted the inclusion of a new section on the genre in the second edition of *World Mapping Today* (Parry and Perkins, 2000), on which this paper partly draws. While there may not be much practical demand for them, Stoke (1989) has emphasized the desirability for libraries to develop historic archives of such maps.

A further incentive for writing this paper came from a chance occurrence. A number of map collections in the UK are fortunate recipients of boxes of 'weedings' distributed by the military survey at Tolworth. They usually contain quite a lot of unwanted material, but there are always some goodies too. A few years ago, I opened a couple of these boxes to find that we had been donated a virtual archive of the early lunar mapping, all the stuff that was done in the 1960s prior to the Apollo landings, and a fair sprinkling of Martian mapping as well (why the military survey required such mapping in the first place is a mystery!).

We already had a fair sampling of planetary maps in our collection, but now suddenly we had acquired an unusually extensive collection, especially of lunar maps (some of which were never widely available), and these give an insight into the development of the techniques and problems associated with making maps of these distant worlds.

This paper sets out to outline some of the characteristics of planetary mapping, and to draw attention to the availability of such maps to map collections. The main title – 'Hiking on Mars' – I will come to later; the subtitle is more straightforward, though in mapping it is often easier to explain 'what' and 'how' than 'why'. In the case of planetary mapping the what and the how are closely related, and will be dealt with together. The why I will mostly leave till the end.

It has been observed (for example, by Greeley and Batson, 1990: 2–3) that mapping of the home planet proceeded along with exploration and discovery, and progressed from the local to the global. We surveyed our immediate environments, slowly patched those surveys together and gradually came to know our planet as a whole. With mapping the other planets, it has been the reverse. For technological reasons we began by knowing them slightly, but globally, then as technology improved we got to know them better. From low resolution and rather hazy visions of a whole planet we progressed to increasing resolution and increasing accuracy. And rather than mapping what we have already discovered, as did the great explorers, in planetary mapping we discover through mapping. To quote Greeley and Batson again, we needed the maps in order to discover the planets.

It can also be said that historically there have been three, overlapping phases of planetary mapping (Shingareva, 1997). The first, which until 1959 was the only way, is the telescopic, Earth-bound phase, much improved of course, when photography was added to visual observation. The second is the so-called fly-by phase, as demonstrated by the Mariner 10 mission to Venus and Mars (though actually Mariner 10 is in solar orbit), and the Voyager missions to the outer solar system. Finally we have the orbiting phase, where satellites are put into orbits around particular planets and often with a specific mapping remit. For the Earth's moon, remarkably good maps can be produced from telescopic observation from Earth, but detailed mapping of the other planets and their satellites had to await the development of the technologies for sending spacecraft to those planets.

Another characteristic, and a truism of planetary mapping, at least in the post-telescopic phase, is the fact that the maps produced are closely tied in to the space missions on which they rely. You can usually match series of planetary maps to specific missions. (Although this is not always the case, as for instance the revision of Lunar astronautical charts (LACs) of the Moon by information sent back by Lunar Orbiters).

A further characteristic is that, since planetary mapping literally 'took off' in 1959, it has been almost entirely a product of 'new technology', i.e. from data captured digitally. But there is another curiosity here. Many of the digital mapping programmes for planet Earth have been undertaken by converting existing analogue maps into digital files. In the case of most planetary maps (the Moon is something of an exception), all the early stages of the mapping used state-of-the-art digital techniques.

Mapping the Moon

Detailed planetary mapping began in earnest with the race to set foot on the Moon. The American programme began with telescopic mapping and the main product was a series of 44 Lunar charts (LACs) published through the 1960s, which covered the Moon's near side at 1:1,000,000 scale. A series of 1:500,000 scale Lunar charts was also published in the 1960s. These early series had shaded relief but were mostly without contours.

Maps were refined with the data returned by the Ranger, Surveyor and Orbiter missions. The Orbiter carried vidicon cameras whose pictures were recorded pixel by pixel on tape on board the spacecraft, and the digits radioed back to Earth. More recently, other kinds of mapping sensors have also been introduced, and the vidicon camera has been superseded by the more accurate CCD (charge coupled device). The data had to be computer-processed to correct random errors, rectify geometric distortions, improve contrast, and transform the perspective into a vertical, rather than an oblique view. Finally, the imagery needed to be converted to a standard map projection. The favoured projections for planetary mapping are usually a Mercator for middle latitudes, Lambert conformal for intermediate latitudes and a stereographic projection for polar areas. Conformal projections are preferred because they preserve the shape of the landform features. The next stage was to produce cartographic quality products from these data. The corrected data were sent from the receiving station at the Jet Propulsion Laboratory (JPL) at Pasadena, California to a planetary cartography unit established in 1961 at Flagstaff, Arizona, and now operated by the United States Geological Survey (USGS). Here, the digital data were converted to conventionally printed maps which would be meaningful and useful, if not to hikers on Mars, at least to Apollo astronauts heading for the Moon. So although the solutions were high-tech, the thinking, at least in the 1960s and 70s, was still classical graphic cartography.

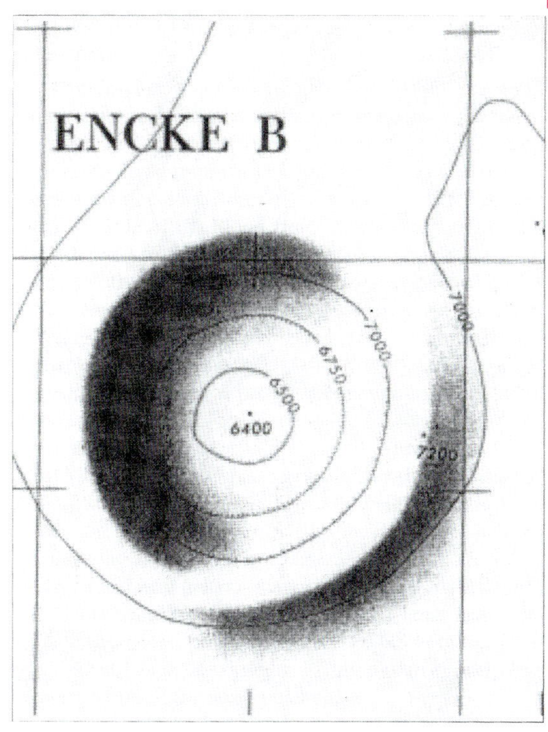

Figure 1
Part of 1:250,000 Topographic lunar map (Sheet 4 Encke B), published by Army Map Service, 1961

The principal products were shaded relief maps, and the technique used to produce the relief was the airbrush. Why was this apparently subjective method used? The problem was that no single set of images, taken under static illumination, provided the whole picture. To appreciate all the nuances of the relief required the cartographer to study many different images, sensed under different conditions and different viewing angles. Essentially, then, the air-brushed map was a summary, and an interpretation of the planet's relief forms.

Figure 1 is an extract of an early 1:250,000 scale air-brush map. The shading here gives the crater rims a rounded look, rather like rubber rings, but apart from the

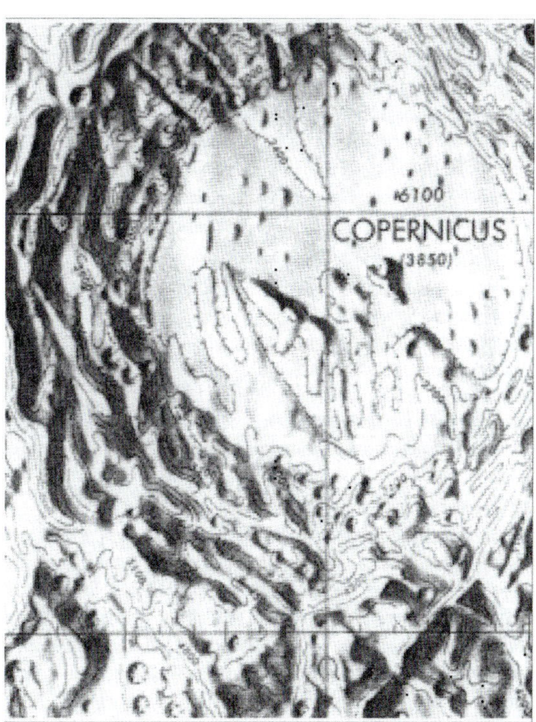

Figure 2
Part of 1:1,000,000 Lunar chart (LAC 58 Copernicus) with 300 m contours. Published for NASA by ACIC, 1964

craters little surface detail is shown. This map was published in 1961 and is an early attempt to produce contours, derived photogrammetrically from telescope photography, but the scale is really too large for the detail available. This map was published by the Army Map Service (AMS). AMS competed with the US Air Force Chart and Information Center (ACIC) at St Louis for the rights to lunar mapping. ACIC took precedence over the AMS, though the task was eventually transferred to the USGS (United States Geological Survey).

Figure 2 shows an extract of a sheet in the 1:1,000,000 scale series of Lunar charts (LACs). Published in 1964, this is derived from telescope photos and observations and 300 m contours have been added from calculations derived from a shadow measurement technique. The shaded relief assumes a western light source, while the background colour of the original depicts albedo variations under full moon conditions.

Successive sets of maps were made of the Moon through the 1960s. Although initially they were made from photos taken from Earth, as data came back from the Ranger, Surveyor, Orbiter and Apollo missions, it became possible to refine maps and produce very detailed maps of some areas. The Orbiter missions provided cover of 99 per cent of the lunar surface, and their coverage of the near side equals the accumulated telescopic/photographic coverage which preceded it.

Figure 3 is part of a second edition 1:500,000 scale Lunar chart published in 1966. It depicts the Mösting A crater which is close to the Moon's equator and is used as a fundamental point for geodetic control. On the reverse of this map is an airbrush rendition of part of the chart (not illustrated here) which also shows landform provinces, and is accompanied by written terrain descriptions.

Figure 4 is part of a 1967 Lunar map made from Lunar Orbiter images. To borrow Ordnance Survey terminology, we might call this an "Explorer" map, as the scale is an ambitious 1:25,000 with contours at 25 metre intervals. In fact it is one of 76 charts made of possible Apollo landing sites. This one was made by ACIC, while AMS also made a number of photomaps at the same scale.

Maps issued in the 1970s were able to benefit from Lunar Orbiter data and from the large format mapping cameras carried by the last three Apollo missions. A series of 1:1,000,000 scale Lunar maps (LMs) published in the 1970s have 300 m interval contours and supplementary contours at 100 m (Figure 5). In the 1970s sheets in a series of Lunar topographic orthophotomaps at 1:250,000 scale were also published. These sheets follow the orbit tracks of the Apollo missions and have contours at 100 metre intervals, but the shadows on the photo base are produced by illumination from the east which necessitates inverting the map to get the correct perceptual three-dimensional relief effect. Most (but not all) of the airbrush maps assume a western light source.

Concurrently with the detailed Moon mapping programmes, commercial and government mapping agencies published a number of general maps of the Moon

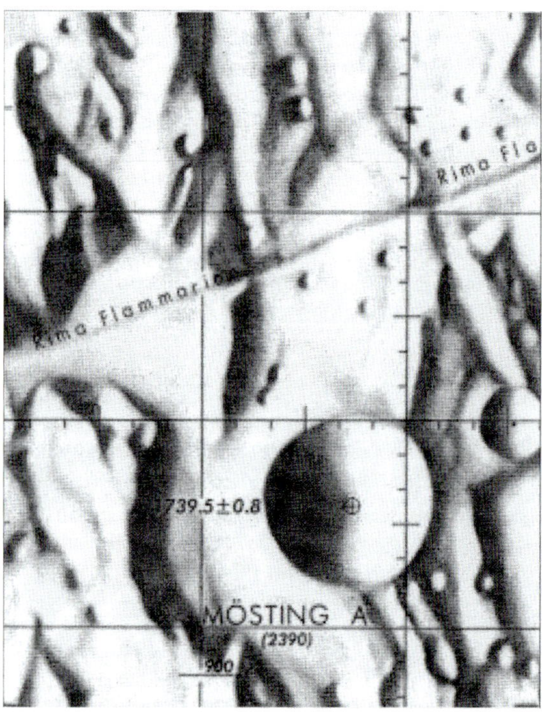

Figure 3 *Part of 1:500,000 Lunar chart (AIC 77A Flammarion), published by ACIC, 1966*

which were revised and improved as new data became available. The Soviets of course were first to acquire, in 1959, images of the far side of the Moon from their Luna 3 probe, and maps derived from these were published in a Soviet atlas. Subsequently, Lunar farside charts were published by the Americans, based on Lunar Orbiter photographs. The Hallwag Moon map, first published in 1978, shows the two hemispheres at the same scale (1:5,000,000), as do the maps published by the French Institut Géographique National in 1984, and by the National Geographic Society, while other maps, mainly of the near side were published *inter alia* by Rand McNally and Philip's. Some of these show the Apollo landing sites, and a map published by Falk Verlag in 1968 invites the

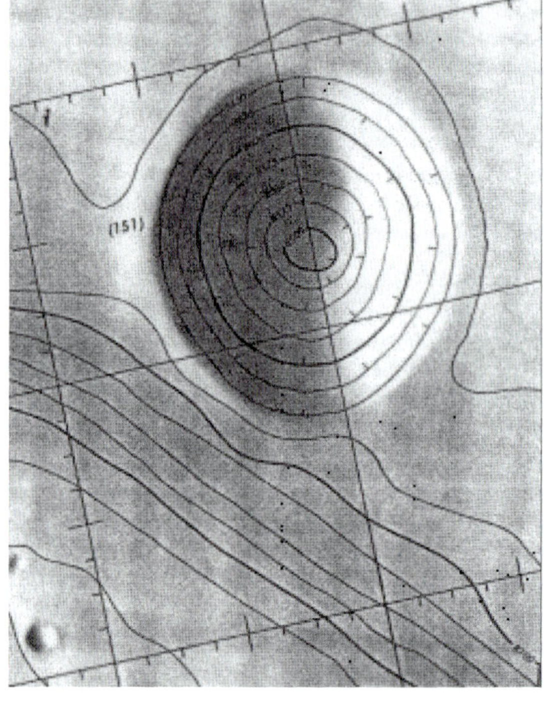

Figure 4 *Part of 1:25,000 Lunar map, prepared for possible Apollo landing sites. Published for NASA by ACIC, 1967*

Figure 5
Part of 1:1,000,000 Lunar map (LM 41 Montes Apenninus) showing Apollo 15 land site. Published by DMA Aerospace Center, St Louis, 1976

user to place a circle on the Apollo 17 landing site once the astronauts have landed. For map collections, the latest and best general map is a set of six shaded relief and surface markings sheets covering the lunar near side, far side and poles, published by the USGS between 1980 and 1992 at 1:5,000,000 scale. Figure 6 is an extract of an earlier version of the north polar sheet.

Mapping the Moon provides many interesting technical problems, for example how to provide geodetic control, a suitable coordinate system, and a vertical datum for a body which has no sea level. But in terms of representation, there is not a great deal to show, except that is, for a lot of craters, some large flat expanses (Mare), and occasional ranges of hills or curious linear features (lunar rays). There

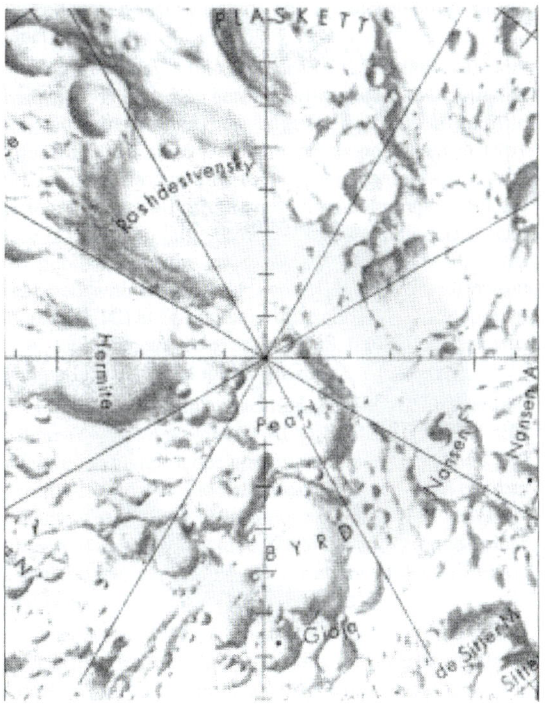

Figure 6
Part of 1:5,000,000 Lunar polar chart, north polar region, published for NASA by ACIC, 1970

is no vegetation, no human imprint to complicate the scene, and all the television cameras on the spacecraft can do is record the variations in reflected light which enable our interpretation. First there are the shadows, which help discern relief, and then there are more subtle variations in reflected light, albedo, which may indicate something about surface materials. Indeed albedo measurements combined with interpretation of landforms have made it possible to make a kind of geological map for the Moon and for some other planets and satellites. A photogeological map and a map of physiographic divisions of the Moon were published as early as 1960 by the Military Geology Branch of the USGS, and the whole of the Moon's near side was also mapped geologically at 1:1,000,000 scale during the 1960s and early 70s. But the topographic and relief shaded maps are essentially just maps of landform.

There is one other thing of course, and that is the names which have been given to features. Since 1919, the International Astronomical Union (IAU) has been the primary arbiter of planetary toponymy, and the first systematic listing of lunar names appeared in 1935. A new *Gazetteer of Planetary Nomenclature* was published by the USGS in 1995, superseding an earlier USGS Open File Report 84-692 published in 1986. It contains all the names of topographic and albedo features of the planets and satellites officially approved between 1919 and 1994. The names are those appearing on the USGS maps and also on maps of the Earth's Moon, Venus and Mars published by the former USSR. Names are decided and approved by a committee of the IAU, and when there are a lot of new features to be named, a suitable theme is selected and associated names are then used. It is a bit like new housing estates where you suddenly come across a lot of street names after culinary herbs or famous composers. Figure 6 reveals a fixation with terrestrial polar explorers.

Martian Landscapes

So if we search beyond the Moon do we find anything more interesting to map, or just more craters? The answer is 'yes, we do', and we find it on Mars. After the Moon, planetary mapping has focused most keenly on Mars. This is the planet most similar to Earth, of course, and which continues to give hopes of evidence of past life forms, though present ones can probably be discounted.

We owe our new understanding of Mars primarily to Mariner 9, launched successfully in 1971 after the failure of Mariner 8, and the images, mosaics and maps constructed from the 7,300 pictures returned to earth by its television cameras. Almost the whole planet was photographed with a wide-angle camera (1 km resolution) and areas of suspected scientific interest by a narrow angle camera (100 m resolution). The vidicon images were radioed digitally to the Jet Propulsion Laboratory, where they were reconstructed and image processed, then sent to Flagstaff where maps were created by mosaicing the images and fitting them to a geodetic framework which had been produced by aerotriangulation techniques. The first complete map was published in 1972 at a scale of

1:25,000,000. One result of this map was to eliminate forever the canals hypothesis of Lowell and his supporters, to the disappointment of those who had high hopes of intelligent life on Mars. However, it also showed that the surface of the planet had a physical geography of immense variety and interest, which had been created not only by tectonic and volcanic processes, but also by surface processes similar to those on Earth, which included the action of wind, water and ice. Once again the images were used to construct shaded relief maps. Relief was copied from rectified mosaics and then portrayed with a western illumination. These Martian maps have a reddish background: 'No attempt was made to precisely duplicate the color of the Martian surface, although the color used does approximate it'. Figure 7 is an extract of a 1991 1:25,000,000 map. This has additional information from the imagery returned by the Viking Orbiters which sent back 55,000 images beginning in 1976.

This imagery also facilitated the construction of 140 controlled photomosiac quads at 1:2,000,000 scale. These reveal dramatically the geomorphology of the Martian landscape, including dendritic channel systems and erosional evidence suggesting the former action of flowing water. Figure 8 shows part of the Coprates Chasma. This is part of a huge tectonically formed canyon system which dwarfs the Grand Canyon, being five times as wide and stretching a distance equivalent to that of Los Angeles to New York. These photomosaics of course are effective because of the exciting features they reveal, but are not strictly cartographic representations. However, contour maps have also been constructed at 1:2,000,000 scale, and a 1 km resolution DTM (Digital Terrain Model) was released in 1992 on CD-ROM, which can be used for three-dimensional reconstructions of the landscape. For the 1:5,000,000 scale sheets of the *Atlas of Mars* series, airbrush techniques were again used. Geological mapping of Mars has also been undertaken at scales as large as 1:5,000,000 and 1:2,000,000.

As with the Moon there are also commercially produced small-scale maps. These include a Haack Lambert equal area map of 1985 which stresses the red colour of the planet but also differentiates the variations in albedo, and a National Geographic Society map of 1973. Both show the biggest volcano in the solar system, now called Olympus Mons, at 25 km high with a caldera which alone would swallow the entire island of Hawaii.

The latest phase of Martian mapping is currently being undertaken by the Mars Global Surveyor, which is collecting terrestrial, atmospheric and magnetic data. There is a web site at *http://mars.jpl.nasa.gov/mgs* with some excellent images.

So now it is known that although Mars has no canals or vegetation, it does have a very interesting and varied landscape. Mapping Mars, as with most planets and their satellites, amounts to the construction of a landform map. We have noted the use of shaded relief techniques to give a qualitative rendering of relief. But many of the Moon maps were also contoured, and, as indicated above, the same is true of Mars. There have been many attempts to measure absolute variations in relief using principles of stereo photogrammetry (where overlapping cover is available) or radar altimetry or infrared spectrometry in other circumstances. In a lecture given to this society at its Summer School at Lancaster back in 1984, Lionel Wilson demonstrated the metrical principles of using brightness variations to calculate the angle of slope of a feature and thence its height variation (Wilson, 1984). Using these principles, contour maps have also been constructed for parts of the Moon, Mercury and some of the satellites of the outer planets.

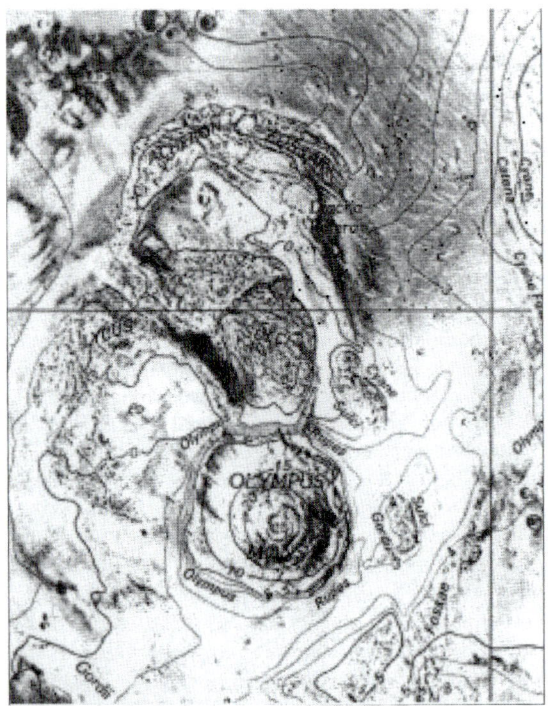

Figure 7
Part of Topographic map of Mars 1:25,000,000, published for NASA by USGS, 1991

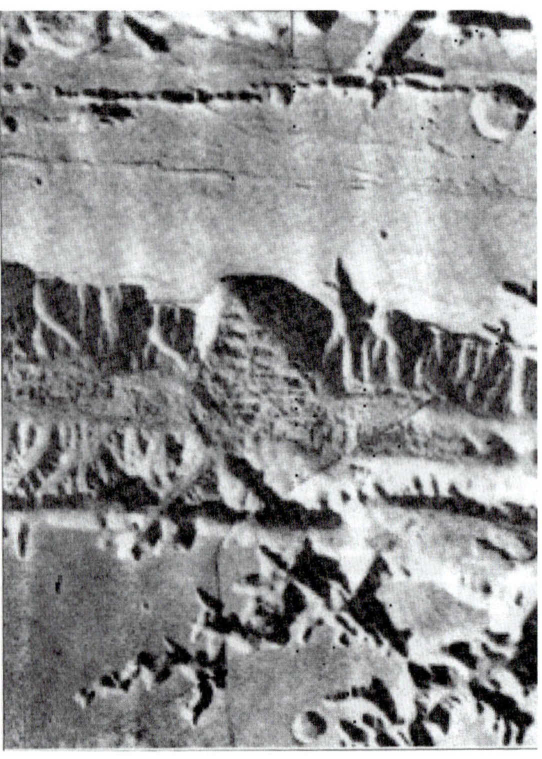

Figure 8
Part of Atlas of Mars 1:200,000 topographic series (Sheet MC-18 NW) Controlled photomosaic showing part of the Coprates Chasma. Published for NASA by USGS, 1979

Mapping Other Planets

This paper has concentrated on the Moon and Mars because these have been subjected to the most prolific mapping programmes. The other inner planets, Mercury and Venus, are more difficult to map and the outer planets are of course gaseous and not amenable to conventional cartography.

Mercury is the least known of the terrestrial planets, being difficult to observe from Earth, but it has had the benefit of three Mariner 10 fly-bys in 1974–75. The imagery returned from these visits is still being calibrated and digitally mosaiced. It covers only part of the planet. A semi-controlled reference mosaic was made from the images and a few 1:5,000,000 scale shaded relief maps have been issued.

Due to its thick, cloudy atmosphere, Venus cannot be mapped with conventional, visible wave-length television cameras except at a very low level, and so radar altimeters flown on orbiting spacecraft are needed to penetrate the atmosphere and provide a comprehensive picture. In addition to information returned by the early Russian Venera soft landings, the planet has been mapped with radar by a succession of orbiting spacecraft including the United States Pioneer-Venus mission, the later Russian Venera 15 and 16 spacecraft, and most recently the United States Magellan spacecraft.

The first comprehensive atlas of Venus – *Atlas Venery* – was published in Russia in 1989. It is a multicolour work which includes photo and altimetric maps, and geological and geomorphological maps at a principal scale of 1:10,000,000. Stereographic and conformal conical projections were used. The USGS has published five maps at the scale of 1:50,000,000, while three 1:15,000,000 scale maps were published in 1989 as a result of co-operation between scientists from the United States and the former Soviet Union.

The most recent Magellan Mission started mapping the surface with radar in 1990 and provided two more years of continuous mapping covering 98 per cent of the planet. New radar image, shaded relief and topographic maps were published in 1998 at scales of 1:50,000,000 and 1:10,000,000 and there are plans to produce new image and topographic maps of the Venusian surface at scales as large as 1:1,500,000. In the meantime, several hundred CD-ROMs have been produced containing the high resolution images, and data are also available on-line via the PDS Imaging Node at the Jet Propulsion Laboratory. Magellen data have also been used to create three-dimensional views of the planet's landscapes and computer-simulated flights across its surface. There is also currently a 1:5,000,000 scale geological mapping programme in progress. Sixty-two sheets will cover the planet and many of these are published or in an advanced stage of preparation.

As for the outer planets, it is their numerous rocky satellites which have received most cartographic attention, and airbrush maps have been made of many of these from Voyager images. Numerous photomosaic, shaded relief and geological maps are available for satellites of Jupiter, Saturn, Neptune and Uranus.

Hiking on Mars?

So now to the question posed in the title of this paper: are we really going to be able to go hiking on Mars? This title was adopted after hearing a paper in Stockholm in 1997 from a group of German researchers who proposed to map Mars in impressive detail (Lehmann *et al.*, 1997): perhaps not a hiking map, but at least, to use an Ordnance Survey analogy again, a 'Travelmaster'. Two German cameras were placed aboard the Russian mission Mars96 which was launched in November 1996. The cameras and mission had the potential to meet photogrammetric and cartographic requirements for the production of a 1:200,000 scale map of the entire planet, and for four preceding years, the team developed the software to process the data and designed a specification for the map. Two equal area projections were selected, one with an equatorial perspective, the other for the poles (respectively Sinusoidal and Lambert azimuthal). A graticule-based sheet index was developed, a sheet numbering system, sheet design, specification of content, and the map has an official name: *Topographic image map MARS 1:200,000.* Finally, a folding method was devised for the paper map 'resulting in a handy size... for the folded map'. (They did not say whether a Bender fold would be used – a good choice for hiking on Mars to discourage the map from blowing away in the strong Martian winds!) Apart from no one yet being in a position to hike on Mars, there is one other slight snag about this map: it does not exist. Not one of the 10,372 sheets has been produced. The Russian spacecraft launched on schedule but failed to enter the trajectory to reach Mars. Rocket engine trouble resulted in the craft crashing into the Pacific Ocean shortly after launch! Still we must not be discouraged at this misfortune, or the subsequent disappointment with the American Mars Polar Lander. If we can have OS 'Explorer'-scale maps of our Moon, there is every likelihood that we will soon have at least 'Travelmasters' of Mars, and indeed the Mars Global Surveyor, is now orbiting the planet and returning high resolution images. We need these maps, as scientists will confirm, to come to a better understanding of tectonic processes, and the nature of the solar system in general. The cutting-edge technology of mapping the planets has undoubtedly had spin-off in the development and use of remote sensing techniques for studying our own planet, and of course there is the inevitable desire to explore the solar system for evidence of extra-terrestrial life-forms.

While the idea of a hike on Mars might smack of science fiction, such ideas may only be a short step from reality. In his 1994 book *The Snows of Olympus*, Arthur C. Clarke mixed fact with vision and with virtual reality. His vision was the terraforming of the Martian landscape and the rendering of the planet's atmosphere fit for plant growth and for the support of human life. In 1990, Clarke was introduced to the landscape modelling program

VistaPro for the Amiga computer, and it so happened that Vista 1.0 came with demonstration data for the Martian volcano Olympus Mons. Clarke married the idea of being able to model virtual landscapes with ideas about creating a denser atmosphere for Mars, complete with a greenhouse effect, and raising its temperature sufficiently to support plant growth. To this end he used the program to show not what is or what has been, but what might be. The terraforming programme would be a long term one, measured in millennia, but with the aid of VistaPro its effects could be created in minutes. The book provides illustrations of what Olympus Mons might look like in wintertime around AD 3000, and of its huge caldera in about AD 4000 with a succession taking place from fractally-created pines to fractally-created oak trees! Other pictures in this richly illustrated book show what might happen to the greening of one of the chasms which run off the Noctis Labyrinthus. In AD 2400, the land surface is covered by lichens and the atmosphere is turning a deep blue as it becomes denser; two centuries later, lakes have been created which thaw during the summer, and trees have begun to spread from the lake shore.

This may sound very far-fetched, but it is a reminder that we now have a fully digital environment for planetary mapping. No longer are we restricted to printing paper maps, as one-time interpretations of the landscape. Instead, researchers and educationists alike are provided with digital data and programs for visualizing, exploring and even transforming the landscape in an infinity of different ways. Thus Magellan full-resolution radar mosaics of Venus, and imagery from many other missions, are now available on CD-ROM, and educational CDs such as Mars Explorer use DTM data to create fly-bys of the planet's surface.

Planetary Mapping and Imagery on the Web

This brings me finally to the web. Thanks especially to NASA, this has become an extremely good resource for finding information about planetary imagery, both current and archival, as well as space exploration in general. Numerous sites are run from the Jet Propulsion Laboratory at Pasadena, California, the Goddard Space Flight Centre, Greenbelt, Maryland and from the USGS at Flagstaff, Arizona, and there are hot links between these and many other sites: *http://www.nasa.gov/, http://www.jpl.nasa.gov or http://wwwflag.wr.usgs.gov/USGSFlag/Space/* are good starting points. NASA has established a Planetary Data System (PDS) whose function is to provide easy access to planetary data sets, with an on-line catalogue, and which has ordering facilities for CD-ROMs (*http://pds.jpl.nasa.gov/pds_home.html*). Each planet also has its own web page at JPL, which can be found by opening *http://www.jpl.nasa.gov/solarsystem/nameofplanet/.* The USGS site at Flagstaff allows you to 'browse the solar system'. The home page includes a poster of the planets, which is also published in paper form by the USGS. If you click on the planet of your choice, you enter into a file of images of that planet.

The new *Gazetteer of Planetary Nomenclature* published by USGS in 1995, and containing all the names of topographic and albedo features of the planets and satellites officially approved by the IAU between 1919 and 1994 is also now made available in its entirety at *http://wwwflag.wr.usgs.gov/USGSFlag/Space/nomen/nomen.html.*

The Clementine mission launched in 1994 to investigate the Moon's polar areas returned high resolution multispectral imagery of the whole of the Moon, together with laser altimetry data, A Lunar image browser of data from this mission is available to be viewed at *http://www.nrl.navy.mil/clemantine.*

October 1997 saw the launch of a new mission to Saturn and its mysterious satellite, Titan, and there is a special web site for this (*http://www.jpl.nasa.gov/cassini/*), which also announces the availability of CDs, videos and posters.

Finding and Acquiring Planetary Mapping

Several commercial publishers have produced maps of the Moon, of Mars, or of the solar system as a whole, including Philip's, Hallwag, Klett-Perthes Verlag (formerly Haack), Rand McNally and the National Geographic Society. There are also some excellent atlases, including the NASA Atlas of the Solar System, already mentioned, and atlases of individual planets published by Cambridge University Press.

Some mapping, including a new 1:35,000,000 scale map of Mars, has been published by the Institute for Cartography of the Dresden University of Technology in collaboration with the Moscow State University of Geodesy and Cartography (Buchroithner, 1999).

The most accessible detailed mapping is that produced by the United State Geological Survey (USGS) for NASA. These maps commonly conform to one of three styles: semi-controlled photomosaics, which are image maps patched from the scanned images captured by vidicon cameras or others sensors; shaded relief maps, conventional line maps with shading usually added by airbrush; and topographic maps, maps to which contours have been added. Additionally, geological maps have been published of all or parts of the Moon, the Jovian satellites, Mars, Mercury and Venus. The maps are published in the Geologic investigations series, and are listed in the quarterly USGS New Publications (also on the web at *http://pubs.usgs.gov/publications*). USGS also has an excellent poster describing and illustrating its planetary mapping programme and products. A web listing of published maps is provided at *http://wwwflag.wr.usgs.gov/USGSFlag/menu/FFCindex.html.*

Detailed Russian mapping does not appear to be readily available, although currently maps of Venus are being produced cooperatively with NASA, and are published by the USGS.

A CD-ROM entitled 3-D Tour of the Solar System (by P. Schenk, D. Gwynn and J. Tudor) has been issued by the Lunar and Planetary Institute (details at *http:cass.jsc.nasa.gov/lpi.html*). Planetary imagery is also packaged on numerous other CD-ROMs available from the National Space Science Data Center (NSSDC) at *http://nssdc.gsfc.nasa.gov/cd-rom/cd-rom.html.*

Coda

Whether we will ever be able to take a hike on Mars, I leave to the reader's imagination, or failing that to Arthur C. Clarke's who postulates that, due to the difficulties of the terrain, rock climbing will become an important feature of Martian life and livelihood. In any event, I believe mapping of other planets, just as much as mapping of our own, can be justified in the words of J.N. Wilford (1981) in his book *The Mapmakers*: 'Maps embody a perspective of that which is known and a perception of that which may be worth knowing'.

Note

All URLs (Uniform Resource Locators) were accessed on 07/07/00.

References

Buchroithner, M.F. (1999) "Multilingual Mars map – first in a new series of planetary maps" *Touch the past: visualize the future: Proceedings of the 19th International Cartographic Association Conference* Ottawa: Canadian Institute of Geomatics pp.1775–1778.

Clarke, A.C. (1994) *The Snows of Olympus* London: Victor Gollancz.

Greeley, R. and Batson, R.M. (1990) *Planetary Mapping* Cambridge: Cambridge University Press.

Greeley, R. and Batson, R.M. (1997) *The NASA Atlas of the Solar System* Cambridge: Cambridge University Press.

Lehmann, H., Scholten, F. and Albertz, J. (1997) Mapping a whole planet – The new topographic image map series 1:200,000 for Planet Mars *Proceedings of the 18th ICA International Cartographic Conference, Stockholm Gävle:* Swedish Cartographic Society, pp.1471–1478.

Parry, R.B. and Perkins, C.R. (2000) *World Mapping Today* (2nd ed.) East Grinstead: Bowker Saur.

Rodionova, Z.F. (1991) "Space achievements and maps and globes of planets" *Geodesy and Cartography* N 7 pp.15–22.

Shingareva, K.B. (1997) "Planetary cartography: Results of the Russian program and prospects for the development" *Proceedings of the 18th ICA International Cartographic Conference, Stockholm* Gävle: Swedish Cartographic Society pp.521–525.

Stooke, P. (1989) "Maps of other worlds" *ACML Bulletin* 71 pp.9–15.

Wilford, J.N. (1981) *The Mapmakers* New York: Alfred A. Knopf.

Wilson, L. (1984) "Cartographic and topographic measurements from spacecraft" *The Bulletin of the Society of University Cartographers* 18 (2) pp.99–102.

The 2000s

In the year 2000, Malcolm Gladwell's book *The Tipping Point* was published. Described as 'a rip-roaring account of how cultural events happen', it is in some ways a metaphor for what happened in cartography during this decade, as will be seen.

So, what has the mapping landscape? By the start of the decade the SoC *Bulletin's* 'Computers in Cartography' column had become a summary of Carto-SoC (the society mailing list) contributions on the subject. The first article in the *Bulletin* of the decade was on mobile mapping on demand. At the same time, as a mirror of society, there were two legal cases involving mapping/software organizations. Centrica and the Ordnance Survey had just settled the AA copyright of mapping case. Then there was news of Macromedia filing a patent infringement counterclaim against Adobe Systems over aspects of their graphics software. On a brighter note, rivals Corel had released version 10 of the CorelDraw software package, and there was soon a four-page review of the ever-popular FreeHand MX package in the *Bulletin*.

Early in the decade Menno Jan Kraak's *Web Cartography* book was published, as was Dodge and Kitchen's *Mapping Cyberspace*. In 2001, Simon Winchester's book *The Map that Changed the World: The Tale of William Smith and the Birth of Science* was released. This was one of the first books to popularize cartography and mapping for a wider audience, and paved the way for later books by the likes of Mike Parker and Simon Garfield.

But the real change that led to what may be considered a cartographic tipping point came from 2003 onwards. The *Bulletin* published an article by Ed MacGillavray on collaborative mapping, showing it had a finger on the pulse. Esri's mapping packages were penetrating the market, as their virtual campus website (training modules and seminars for products such as ArcGIS) came online in 2004. A significant game-changer happened quietly that year, when Google announced that they were buying Keyhole (predominantly a 3D imagery database), which led to the eventual deployment of Google Earth.

By dint of good networking, and a bit of serendipity, the 2005 SoC Summer School at Cambridge was a ground-breaking occasion (which may well not have been realized at the time). Papers were presented on Open Geodata, Open Access Journals, next-generation web mapping, community mapping and OpenStreetMap (OSM). Together, these were all indicators of a major shift in cartography – the beginning of a crowd-sourced opendata future. At the same time the *Bulletin* listed a series of significant blogs related to cartography that were carrying this same message.

In 2007, Alex Kent took over from me as Editor of the *Bulletin*, and shortly he had introduced an effective re-design of the publication. In the next Issue, Nick Black claimed that OSM was aiming to complete its map of the UK in 2008. Whilst OSM maps, and more crucially data, have had a huge impact on cartography, it is a moot point as to whether that completion has happened – due to variations in data granularity achieved across the nation by this vast volunteer project. But even more was to happen in 2009, when the UK government announced that the national mapping agency (Ordnance Survey) were going to be asked to release a significant proportion of its data as 'open data' – allowing other cartographers, agencies, and so on access to map data formerly tightly controlled by the Crown Copyright licence model.

Four papers celebrating this new direction for cartography are included in this volume. Ed MacGillavray's article runs through some collaborative mapping projects that are moving mapping away from the national mapping agencies and the GIS vendors. Richard Fairhurst's paper gives a good résumé of the first three 'generations' of web-mapping and envisions a role for cartographers who understand 'the art of maps and the science of GIS'. He pinpoints the influence of Google Maps, but also posits more open data alternatives being based on open source ideals. OpenStreetMap founder Steve Coast outlines the early days of the project, ending with a rallying call to cartographers to get involved in order to move it on. Nick Black and Steve Coast points out the community and data aspects of the OpenStreetMap project, and predicts that consumer demand would shake-up the map data market.

There is also a reflective piece by the then SoC President Mike Wood, who argues cogently that there should be improvements in context-based software options, and that there is also very much a role for cartographers in this design process. Respected geographer Peter Haggett then takes a look back over 50 years and gives his thoughts on changes in university cartography, taking the opportunity to thank cartographers for their work of 'beauty and utility'.

A further paper is included by Ifan Shepherd (Middlesex University) on interactive graphics, whereby he argues that the software tools being developed for interactive games are leading the way. He uses a case study of the poverty data collected by Charles Booth to show how this technique can aid 3D exploration interactively. Bernhard Jenny and Nathaniel Kelso give some good background on designing maps for the colour-vision impaired, also emphasizing how the digital environment should make it easier for cartographers to experiment with colour, shape, size and contrast to produce better results for those with the differing colour impairments. Finally, Alan Collinson poses the question 'Why can't a map have sky?', suggesting there are very radical ways to look at mapping our environment more artistically, and Alex Kent discusses what he calls 'cartographic blandscapes', arguing for map-makers to 'learn the language' but also to use the new methods to experiment so that their vocabulary describes the landscape in an effective and attractive way.

Steve Chilton

References

Dodge, M. and Kitchen, R. (2001) *Mapping Cyberspace* London: Routledge.
Gladwell, M. (2000) *The Tipping Point* New York: Little Brown.
Kraak, M.J. and Brown, A. (Eds) (2001) *Web Cartography: Developments and Prospects* London: Taylor and Francis.
Winchester, S. (2002) *The Map That Changed The World* London: Penguin.

THE HEART OF THE MATTER — A PERSONAL VIEW

Michael Wood

Originally published in Volume 34, No 1, pp.1–6

Early last century cartography was recognised as a specialized profession of the few. Today everyone may become a cartographer. What are the implications of this transformation? The talk examines the past and possible futures of the subject, changes in its character and in the nature, tasks and responsibilities of cartographers. Cartography has been evolving for millennia, but recent changes to digital (+DIY) mode, the dominating roles of the Internet, governments and business in map production and the availability of high resolution imagery, have especially confused people on the periphery of the subject and beyond who still believe that cartography is stuck in a paper map time-warp. Many even believe that the subject is dead, surviving only as GIS output from spatial databases, or in isolated niche companies. This of course is wrong. Cartography may have begun as lines in the sand but has now matured into a dynamic, versatile and powerful facility. The paper states, simply, what the author believes is the heart of the matter and re-introduces the reader to Cartography which has now grown up.

Introduction

One of my earliest recollections of using cartography was for the illustration of Geography notes at school. We had not been taught map-making nor had we been told to include maps, but it seemed natural to do so! Just as writing and mathematics help us understand and express particular ideas, cartographics have always given power to the analysis and presentation of geospatial information. Using maps/cartography in an integrated, instinctive way is fundamental to the theme of my paper. At a time when cartography seems to be losing its clear definition and character in some communities, it is useful to reflect and re-focus on its core, the natural instinct on which it is based. This is a personal interpretation which may satisfy some of the complexities which appear to surround a subject (cartography) which seemed more clearly defined in the past. I review some of cartography's origins and the stages of its evolution to the present day. More importantly, however, I consider cartography as, fundamentally, a powerful facility, the exploitation of which has been limited, even in the recent past. Today, with the aid of new technology, cartography, in all its dimensions, is available to everyone.

When I began to study cartography in its own right I soon discovered that there was more to it than I had first appreciated, and certainly more than just using a technique to help analyse or explain a spatial problem. Indeed in the 1960s and '70s much professional and academic attention was directed less at using the facility than at the improvement and modernization of production procedures and the technical processes of image-making and reprographics. The 1960s also saw the beginnings of accessible computing technology, which has increasingly dominated our subject over the last 40 years. Major issues such as the capture, quality and timely availability of spatial data, and the methods of manipulating them and making these manipulations visible in graphic form – a kind of cartographic engineering – have always been of concern to cartographers. But there is more. When composing written text, once we have had the satisfaction of finding the necessary source material, from the Web or local library, and when we have accepted the technical wonders of the word-processor, we simply get down to creative and personal activity of writing itself. Perhaps, in cartography, I have been distracted, too often, from the latter by the former, and this echoes part of the theme of my presentation.

Definitions and Changing Times

As the subject had been developing through the centuries it became tempting for observers to identify a boundary which would include all those who might be sufficiently involved with maps to be called 'cartographers'. But what was a cartographer – author, producer or both? Perhaps the International Cartographic Association (ICA) has already resolved this matter, defining cartography as 'a discipline dealing with conception, production, dissemination and study of maps', and 'map' as a symbolized image of geographical reality representing selected features or characteristics, resulting from the creative effect of its author's execution of choices – and designed for use when spatial relationships are of primary relevance'. But what of non-realities such as the land of the Hobbits, medieval maps of Paradise or modern simulations in virtual reality?

Since the mid-twentieth century maps have changed from solely static printed hardcopies to visualizations produced from a range of software and viewed on a variety of media. Traditional products still persist but much of the 'new' cartography is denied to most of the World's population, as only about .001 per cent have access to a computer. Digital cartography is still for the very few! My travels for the ICA have taken me to regions still largely without new technology, although the people know of its existence and potential. They, and others across the World, see traditional methods disappearing and hear of the success of GIS – of which Cartography is now assumed, by many, to be merely a subset! Even in Greece, as late as the early 1990s, there was little popular awareness of day-to-day map-making and use, until a change in government policy opened up possibilities for a more cartographic future. The president of the newly-founded Hellenic Cartographic Society recognized the potential dangers of sudden and uncontrolled acquisition of GIS by cartographically naïve local government departments. His strategy was to launch an educational/awareness programme which I was invited to join.

There have always been geographical information systems. Unfortunately, in the past they were chaotic, weakly structured and often quite unreliable, comprising map sheets of different scales, ages, projections, levels of accuracy, and tables of spatially-related information, etc., often stored in different physical locations. And, of course, they were manipulated primarily by human hands and brain! Fortunately the new digital era has transformed the ways in which geographic information is acquired, stored, structured, explored, manipulated and retrieved in the process of problem-solving. It was claimed, recently that the bulk of 'cartography' is now produced as spin-off from geo-databases rather than through conventional cartographic processes, but the truth is that cartography has always been produced from GIS and geo-databases, it is just that in the past they were not digital. It is a matter of perspective.

The Heart of the Matter: Seeking Origins

Cartography is now more widely used than ever before, but seems even more difficult to define. Modern cartographic visualizations can seem so different from the traditional map (which also functioned as a pre-digital geographic database) that they may even be classified by some people as part of a different discipline, such as applied computing. Much of the cutting-edge discussion today is also more concerned with the provision and quality of geo-data, than how they can be explored, analysed and communicated. What has caused this apparent change of focus? Is it possible to return to cartography and find an idea which expresses its common core – the true heart of the matter?

We may have to begin by looking at what might be called fundamental human instincts. These seem to include 'language', 'mapping' and, possibly, 'music'. The origins of language are unknown, but it may be 200,000 years since sounds from the mouth truly began to represent structured communication. Steven Pinker, in his book *The Language Instinct* (Pinker, 1994), identifies this ability of our species as remarkable in that with language 'we can shape events in each others brains with exquisite precision'. It is primarily a mental process translated into external sounds. The more recent ability to write and read 'makes communication more impressive by bridging the gaps of space and time…' (Pinker, 1994). Pinker regards writing as an 'optional' extra, the real engine of communication being the spoken word. However, my concerns relate to this 'optional' extra.

The growth of the alphabet and how we write is relevant to our subject. Important languages began as pictures, especially Chinese and Japanese. But even parts of modern English have pictographic origins: letter 'A' is from Semitic aleph meaning OX, originally a rough drawing of an ox's head. Expressive drawing, therefore, may predate the development of language.

Mapping, like language, is also basically a mental process common in humans and animals and it too has been expressed externally for centuries, even millennia. As with language, its origins may also post-date graphic drawings, which are certainly known to date back at least 30,000 years (in the Caves of Lascaux, France) and possibly even earlier, from more recent discoveries, 15,000 years older, in Northern Italy. Clear prehistoric physical evidence of the 'mapping' instinct is more difficult to find, although early historical examples, such as the work of aboriginal peoples in North America, Australia, Oceania and the Arctic suggest a tradition which must go back many generations. During his sixteenth-century travels to Tahiti, Captain Cook reported having received guidance from a native through use of a simple map. Such experiences were also reported by European explorers in 19th century Africa. It is believed that the combination of Alexandrian Science and Hindu number lore, important in that early pre-Christian period, ultimately led to developments in serious mapping in the form of sea charting.

Where early maps have been found, their detailed purpose cannot always be known, although maps discovered in China, dating back about 2,200 years, were clearly military and had a planning function. Although the survival of very early maps is rare, absence of evidence need not be evidence of absence and we can assume a continuous, if not steady, growth in the development of this graphic means of spatial expression. The history of cartography is rich with maps and map-like images of all types and has evolved into the period, in recent centuries, when maps were to become familiar and commonplace.

Maps and Cartography as Aids to Spatial Thinking

Once a cartographic representation exists its influence and impact will grow. It is primarily a valuable external aid to the internal, mental processes of 'mapping', an element of what has been called 'visual thinking'. And this is perhaps our first approach to what could be called a core element in the argument. A map is a representation, but it is not only that. It becomes a facility to stimulate, inform and even accelerate the spatial thinking/problem-solving process – an external work space on which one can plan and navigate one's thoughts. Such thinking may be professionally focused for route selection (as in early ocean navigation), seeking patterns, or exploring and matching what is discovered with what is already known.

On the other hand the process may be quite casual, the eye and mind gently roaming over a cartographic image.

Once people have experienced how maps can help them explore spatial patterns and investigate spatial problems, they have the opportunity to respond more specifically to the mapping instinct in themselves, to think visually but also to try and create new external maps to take their investigation forward. Before reaching this point they were merely using existing maps (the products of the cartographic industry) passively, as visual, virtual resources. The next stage, (when 'map use' begins to include the making of additional maps as part of the research process) launches them into active employment of 'full-blown' cartography as an investigative tool.

The Legacy of Map-Makers

There is now a huge legacy of map production, covering many themes; a resource which continues to yield information to prepared minds. Also, contradicting the more simplistic theory that a map is merely 'communication', much of the information obtained by passive map-users was never intended by the originator. It derives from a combination of prior user-knowledge and the visual thinking processes interacting with the visual evidence in the map. The 19th and 20th centuries brought new map genres, especially specific themes such as geology, vegetation and land use. Field survey information was passed to plotting specialists (cartographers) to compile, specify, design and produce the cartographic end product. Although these maps covered specific topics they could also yield other in- formation to enquiring, motivated minds focused on particular problems.

Who makes maps? Obviously anyone can attempt a casual (sketch) map by using readily available tools such as sticks for sand drawing, or pencil and paper. But, although many tasks can be supported by maps which are only relatively accurate, the demand or desire for greater accuracy and completeness grew. Special survey instruments and techniques were employed to identify features (field boundaries, geological boundaries, etc.), measure and locate them and finally plot them on paper. Although still incomplete, this stage involved a kind of preliminary cartography, where the map acted as a primary data store. Land surface measurement techniques became increasingly sophisticated, with air photo survey and satellite remote sensing (photography, scanner images and GPS). Inevitably specialists in these new and initially distinctive fields came together in their separate groups, shared their knowledge, and (as the need and opportunity arose) devised ways of developing and adding value to their data and data-collecting methodology. Specialist national and international societies were one of the results.

Gradually those who were more involved in the assembly and production of maps, developed an identity (and title) of their own, i.e. 'cartographer'. In some cases, especially in national mapping agencies, these people formed part of a production process, each with quite restricted remit and skills. The organization was the 'author' but the final map output resulted from the combined efforts of specialist cartographers. Commercial companies became established to supply smaller-scale maps, atlases and other special products. This was where further subdivision of labour emerged, including author-editors (who made and checked the compilations) and cartographic producers who transformed the basic compilations into printed designs.

Yet more groups emerged, within military agencies, the media and academia, each with increasing needs for the products of cartography. The result was greater convergence between author, producer and user and this stimulated academics with cartographic experience to write textbooks and teach the subject. (e.g. A.H. Robinson, J.S. Keates). Many students, educated in the complete process of map making, from data collection to conception, design and production, were fed back, as cartographic employees, into national and commercial mapping organizations. Other graduates chose to follow academic research and examine fundamental issues related to aspects of theoretical and practical cartography. Although some research results were acknowledged and partly adopted by map producers, most remained in the academic archives. However the latter were to contribute to a growing knowledge base for what would become a kind of cartographic 'engineering', in which the processes of cartography were analysed, industrialized and eventually automated through digital computing.

Cartography in the 1960s

Government and commercial mapping had increased by the 1960s but most 'users' of cartography were still primarily map-readers. Since the 1940s many in the military, government, utilities, industry, resource management and elsewhere, had become increasingly frustrated with the unsuitability of standardized products for specific tasks. They frequently had to depend on the wrong cartographic tools for the job because the right tools did not exist. The obvious solution (heralded, in the 1950s, by the book, *The Look of Maps*, Robinson, 1952) was for these investigators to develop their own cartographic 'tools', which they did by establishing in-house drawing offices to help deal with problems as they arose. In such situations employment of the cartographic process was becoming less passive. Problems were being investigated (with the support of cartographers) and specially-designed maps were created to help analyse, clarify, explore, but still mainly to present spatial themes. As cartography moved beyond mid-century, 'cartographers' had become established as a kind of profession. Although many were still employed in the compilation and production of standard sheet maps in conventional map-making organizations, some were more directly involved in the fuller cartographic process, offering a response-based map service, beyond mere representation. The functions of these cartographers could be likened to a human equivalent of the data manipulation and output software in a digital GIS. As there was no uniform standard education structure for every cartographer in the UK, those in employment had backgrounds ranging from on-the-job training to university education. But as this new profession gained experience and knowledge, some of its members began to act as professional consultants (like an architect) examining clients' problems and helping guide them towards solutions. In many cases they were working with people, such as environmental professionals, who already had a well-developed mapping instinct (in the context of their work) and just needed help to refine it and create the appropriate visualization tools. These included more complex cartographic techniques (easy to visualize mentally but not to create manually) such as 3D modelling, multi-layering of information, and animation, all of which would become much simpler with computer assistance. However, despite this positive localized activity, most maps being used professionally, despite their shortcomings, were still from national topographic or thematic series. Scientists and other professionals regularly encountered spatially-based problems, but obtained little advanced help from the conventional cartography of the time. The 1960s may have seen quite high levels of map production (of sheet maps and atlases), but it could best be described as a period of limited or even impoverished use of what could be called the full potential of the cartographic process, methods being restricted to map use, rather than fully-integrated thinking and map creation.

Many professionals, whose work regularly involved spatial problems, became aware of the potential of more creative mapping and felt the urge to develop and apply their own mapping instinct as an investigative tool. Unfortunately most lacked basic graphic/cartographic skills and, when special maps were needed, these professionals had to turn to in-house cartographers or external map/graphic agencies. As the graphic/mapping instinct can be strong in such active environmental scientific researchers, it is possible that they may always have created sketch maps, etc., as investigative aids. Inevitably such ephemeral 'working' maps or graphics tend to be abandoned and disappear from the record. The only survivors are those which have been produced to present the results of research. A classic example from history is John Snow's map (1855) showing deaths from cholera in the Broad Street area of London.

The 1960s, therefore, saw a rapidly growing demand for more effective 'cartography'. And then it happened. Just as early mapmakers became aware of the printing process and saw its potential, so also did those who knew the potential of computers. Examples included the work of Ian McHarg (a landscape planner), and the Canadian GIS. Early computer graphics were coarse and ugly, but during the last 30 years, technology has developed rapidly. Now it is possible to create realistic, fully-animated, virtual images and environments (of all kinds), linked to the internet and acting as interface for international databases. Initially cartographic organizations (such as mapping agencies) restricted computer-assistance to the replication of conventional mapping processes but with the development of effective GIS, graphics software, etc., almost any graphic image is now possible. Cartographers are not just using GIS, they are using these modern tools and many more, to develop what might be called a 'new' and expanded cartography. But it is not really 'new'. Many of the ideas and images now possible had been attempted, manually or, at least, mentally, in the past. Now inventive cartographers are able to do what they had always done in the past (and much more), but better and faster. Information technology had enabled the more restricted cartography of earlier years to blossom out into a fully mature and powerful facility for data analysis, exploration and display.

Despite the expanded possibilities from computer graphics, multimedia, etc., much of the digital mapping of the last decade has been quite traditional in nature. The more important ambition, especially of national mapping agencies, has been the creation of topographic databases - new resources with huge value and potential. It has been observed that these and similar organizations, using various GIS, CAD and graphics software (e.g. from Autodesk, ESRI, Intergraph) are producing most of the maps in the world today... without the involvement of traditional cartographers (although their legacy lives on in much of the graphic output software). The number of traditional cartographic employees may have reduced, but the above reference to increased map creation is good news for car- tography, and highlights the expanding importance of visual mapping in so many fields. The revolution is also not necessarily bad for cartographers. Much of the pre-digital work of those employed within mapping organizations was restricted and quite repetitive. New generation cartographers have much more scope to apply, not only their technical skills, but also more fully to exploit their creative abilities. It is not surprising, therefore, that a niche business, i.e. cartographers as consultant producers, is expanding rapidly in many parts of a world in which the exploitation of geospatial information is increasingly regarded as an essential ingredient in most environmental activities.

Visualization – Scientific, Cartographic and Geographic

Another important development, which paralleled that in digital cartography and GIS, was Scientific Visualization (or Visualization in Scientific Computing, ViSC). This evolved in the late 1980s from collaboration between advanced computing, science and engineering. Some of the biggest problems being addressed by such groups involved very large volumes of multidimensional data, varying in time and space. The only effective solutions seemed to be through the development of innovative visualization software to permit exploitation of the combined analytical processing power of the eye and brain to facilitate understanding and insight. The common characteristics of many of the tasks and goals of pioneering geographers/cartographers and other scientists led to the forging of important links in the early 1990s between workers in ViSC and in the field of mapping/GIS. This interdisciplinary theme was taken up, in the mid 90s, by a pioneering group within the International Cartographic

Association (now called 'The Commission on Visualization and Virtual Environments'). They established an historically significant joint project with the Special Interest Group on Graphics (SIGGRAPH) of the American Association of Computing Machinery (ACM). The result has been one of the most fundamental surges of research into the basics of our subject – cartographic visualization.

Having evolved from the fundamental mapping instinct, cartography (new cartography or cartographic visualization) has developed and matured, and its potential for use and application has both widened and deepened. At one extreme it can offer simple and effective images, indistinguishable from a pre-digital conventional map; at the other it is a powerful research tool (integrating techniques from conventional cartography, ViSC, image analysis, Exploratory Data Analysis [EDA], and GIS) now being referred to by the wider term, Geovisualization. The ability to employ the full potential of scientific Geovisualization may, initially, be restricted to a narrower group of scientists facing complex problems. However, 'new cartography' tools of all kinds, available locally or on the World Wide Web, are becoming increasingly user-friendly and will soon be as accessible as wordprocessors, databases and spreadsheets. In the 'youthful' stage of its development, cartography seemed to have few advantages beyond representation, and was produced by (and accessible to) only an elite class of artisans and users. Today the flexibility, power and relative ease-of-use of specialized software tools have democratized cartography, offering it as a facility to everyone. Current users include:

- Government and businesses, employing cartographic engineering tools linked to GIS,
- Lay people, in increasing numbers creating their own computer maps, even if only from packages like Encarta, Excel, or from the Internet,
- Spatial scientists, using all of the facilities of Geovisualization,
- A new breed of consultant 'super-cartographers' developing cartographic visualization products of all types for their clients.

The Future – Looking Good for Cartography and its Users

From these words it should be obvious that I have few worries about the long-term future of what I have chosen to call the mature cartographic process. Now that it is easier to use, its application will continue to spread across science and society. However, so rapid has been the change from the commonplace paper map to, for example, a 'sensitive', animated-3D-image-flythrough on a website, retaining the word cartography may confuse people for a time. But whichever overall name is finally adopted I will always believe strongly in the unity of the basic concept. I also believe that like language, it has its fundamental origins in human instinct but, in a sense, it is even more powerful. The analogue pictorial/graphic structure of cartographic representations (especially traditional maps) means that they can be transmitted between different individuals, groups and cultures much more readily than abstract language representations. Ironically, while standard analogue maps created thousands of years ago are still interpretable by the modern eye, today's digital equivalents will become unreadable and obsolete in a matter of decades as the technology that created them is left behind.

The Heart of the Matter

This is the Heart of the Matter. These geospatial aids, from lines drawn in the sand, through the so-called conventional paper map, right down to the most advanced multimedia, holographic VR tools accessed on the Web, are all part of the Cartography/Carto visualization/New cartography family. That is why I do not agree that Cartography is dead. Today, more than ever, we are not just using maps, but using cartography in all its dimensions. People used to be divided into map-makers (e.g a cartographic editor working for Bartholomew) and map users ('Dad' reading a road map on a holiday journey). But now the boundary between those categories has, effectively, been removed. Anyone can open Microsoft's Encarta World Atlas or access Canada's National Atlas Information System on the web to create maps. This range of passive and interactive involvement must now be referred to not as map use, but as 'using cartography' through hand, eye and brain.

But that is not the end of the matter. There is still another issue close to the heart of things which, after many years in the subject, I believe to be significant. We are not just automatons carrying out actions such as map-making. We are conscious beings and the instincts which I have mentioned (which include music and language) have important emotional dimensions: music, writing (consider poetry, great writers such as Shakespeare, and even modern authors who can move a reader to laughter or tears by their literary skills), paintings … and, of course, maps. Some people are fascinated by the graphic or historical quality of historical cartographic images (i.e. they 'love' maps!), but amongst cartographic artisans, past and present, we know that map-making can also bring pleasure (from a job that has gone right and looks good) and pain (when things have gone the other way!). I identified this through an international map design questionnaire a few years ago.

The famous American neurosurgeon, Antonio Damasio, who has written about the meaning of human consciousness (*Descartes' Error*, Damasio, 1995), believes in the importance of what he calls 'somatic markers'. These are fundamental reactions to things which we seem to store in our bodies as coded feelings, consolidated by a wealth of similar, repeated experiences of what is right or wrong in a particular circumstance. For example, cartographers can produce many maps over the years, which involve, not just the overall design of each product, but also application of myriads of tiny graphic decisions which seem to be instinctive to the practised professional. It is believed that we store, subconsciously, the feeling/sensation related to each 'right' and 'wrong' decision made. When we apply a series of fine adjustments to improve the appearance or legibility of part of a map, the reasons for these decisions are hard to explain to a casual observer. The actions just 'feel right'. In such working circumstances we can carry and apply many of our skills somatically – and these instinctive and immediately-available facilities prove to be the most effective controls in the work of an experienced creator.

Conclusion

Many GIS-based maps are satisfactory because the software has been engineered to help solve some of the main cartographic problems semi-automatically. Such maps may not come directly from the hands of a fully-trained cartographer, but the guidance software has been informed by the combined knowledge they have helped to establish. The quality of mass-produced maps will continue to improve as context-based guidance software becomes more intelligent and sensitive. But as working tools, designed to support the efficient management of the environment and its sustainable development, such mapping software should be operated directly by the responsible scientists and other professional managers, exploiting their mapping instincts without being disadvantaged by lack of design knowledge or drafting skills. This might also apply to some leisure-based tools such as web atlases used at home. But there is still scope for original custom designs by the professional niche-market cartographer.

In conclusion, I believe the heart of the matter lies in the fundamental process of cartography which has never been more important to our world, and, as human creators we will continue to contribute to that success. Back in the early centuries, even from the time of Ptolemy, chorographers used the mapping process when it suited their need to describe a region or landscape. The map was not so precious, but a tool to be used when required. In the late 18th and 19th centuries, when the subject was beginning to define itself more rigorously and acquired the term 'cartography' it may have lost some of its workman-like character and became more of an object of desire. However, with the latest stages of computer-related democratization of the cartographic process, we may now be returning to its pre-18th century worktool character. The subject continues to evolve. I believe that a river of distinctive cartographic endeavour has flowed through history and, with continuing adaptation to changing technology and circumstances, that river continues in strength.

References

Damasio, A. (1955) *Descartes' Error: Emotion, Reason and the Human Brain* London: Picador.

Pinker, S. (1994) *The Language Instinct* London: Penguin Books.

Robinson, A.H. (1952) *The Look of Maps* Madison: University of Wisconsin Press.

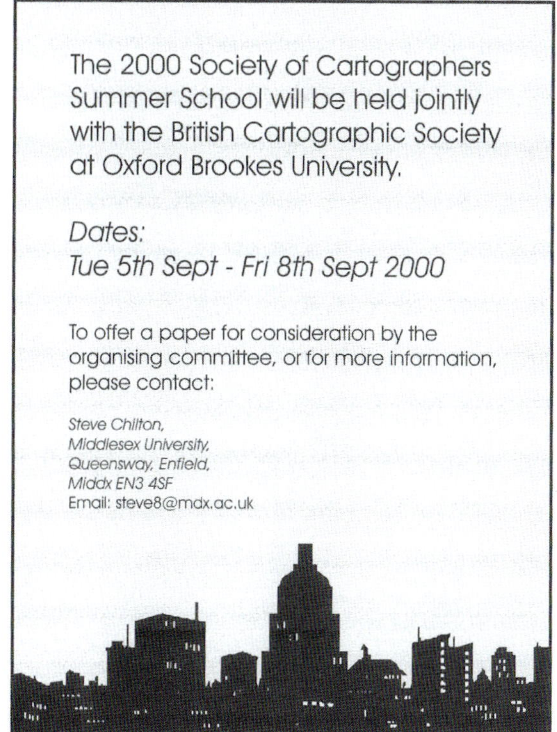

IT'S ONLY A GAME: USING INTERACTIVE GRAPHIC MIDDLEWARE TO VISUALIZE HISTORICAL DATA

Ifan D.H. Shepherd

Originally published in Volume 36, No 2, pp.51–55

This paper reports on the author's recent research into family-level mapping using data from manuscript notebooks compiled in the late nineteenth century by Charles Booth, author of the ground-breaking study of poverty in London: 'Labour and Life of the People'. Using Booth's original data, the paper suggests ways in which recent visualization technologies can be used to provide insights into historical data. In particular, it reports on the use of state-of-the-art videogame technology to develop an interactive 3-D visualization engine to display point-located social data. The paper describes the software, its advantages and disadvantages for visualising historical data, and reports how it resolves a widespread and vexing problem in thematic mapping.

Introduction

Digital mapping software is beginning to come of age. However, during the past decade, a variety of alternative technologies have emerged that offer effective visualization of both 2-D and 3-D data, either in desktop environments or over the Internet. Alongside modern desktop mapping and GIS software are to be found an increasing range of software that can visualize spatial data, including: scientific data visualization software; games engines; virtual reality software (including the now defunct Superscape product and its companion Web engine, Viscape); VRML (and GeoVRML) interpreters; and Java applets.

This paper reports on an experiment to create interactive 3-D visualizations of data for a late-Victorian survey of families in London. Rather than adopt conventional approaches rooted in geographic or cartographic technology, the project turned its back on existing approaches and instead developed software using a combination of the high-level programming language C and powerful videogames middleware (described below). The interactive performance of this software, even when visualizing large datasets, is ensured by optimising the software for the modern, high-level 3D graphics adapters available on current desktop and laptop PCs.

Mapping and Re-mapping Booth

The sample data used for this project come from a ground-breaking social survey carried out in the 1880s by Charles Booth (Booth, 1889, 1889–1891). Booth's poverty maps of late-Victorian London are well known in the map- ping community, and have achieved almost iconic status in the history of cartography (Hyde, 1975; Topalov, 1993). It is hardly surprising, then, that several attempts have been made over the past decade to re-map Booth's original printed cartography: on paper (e.g. Reeder, 1984), on CD-ROM (e.g. O'Day, 1996–7) and over the Internet (e.g. Ahmad, 1999). The author's contribution to research in this area has been the creation of the first-ever family maps, based on data recorded in Booth's original notebooks (Shepherd, 2000), and it is these data that provided the stimulus for the current visualization experiments.[1]

Data have been extracted from Booth's manuscript notebooks for a sample district located within the East End that includes well-known London landmarks such as Spitalfields Market and Middlesex Street (better known as Petticoat Lane). This district contains 79 streets, over 1,700 dwellings and some 2,400 families. The entire family data for these streets have been transcribed into a spreadsheet and imported in a geocoded form into a desktop GIS (MapInfo). (The geocoding was made possible by locating the street addresses recorded in Booth's notebooks for this area to precise locations on contemporary large-scale plans.) In order to a provide a suitable base map, the second-edition OS 25-inch base map of the area was scanned and stored as a raster image, and various boundary files were also created, including polygons for each of the coloured patches which were displayed on Booth's famous map to indicate the social condition of individual streets. Using this database, numerous family maps have been created (see Shepherd 2000 and 2003a for examples), largely using conventional point symbol mapping.

Enter the Dragon: An Insidious Mapping Problem

When the original family maps were created, using standard desktop mapping software (MapInfo), an unwelcome problem emerged: on all of the maps displayed, only a proportion of the point symbols representing individual families were visible. The cause of the problem was simple, though not immediately obvious: where several individuals, families or households lived at the same residential address, they shared identical geocodes. As a result, the point symbol maps created with standard mapping software concealed from view all but one of those individuals who live at the same address. (Because mapping software displays coincident point symbols in a 'stack', only the topmost symbol is visible to the map analyst. All others are concealed from view.)

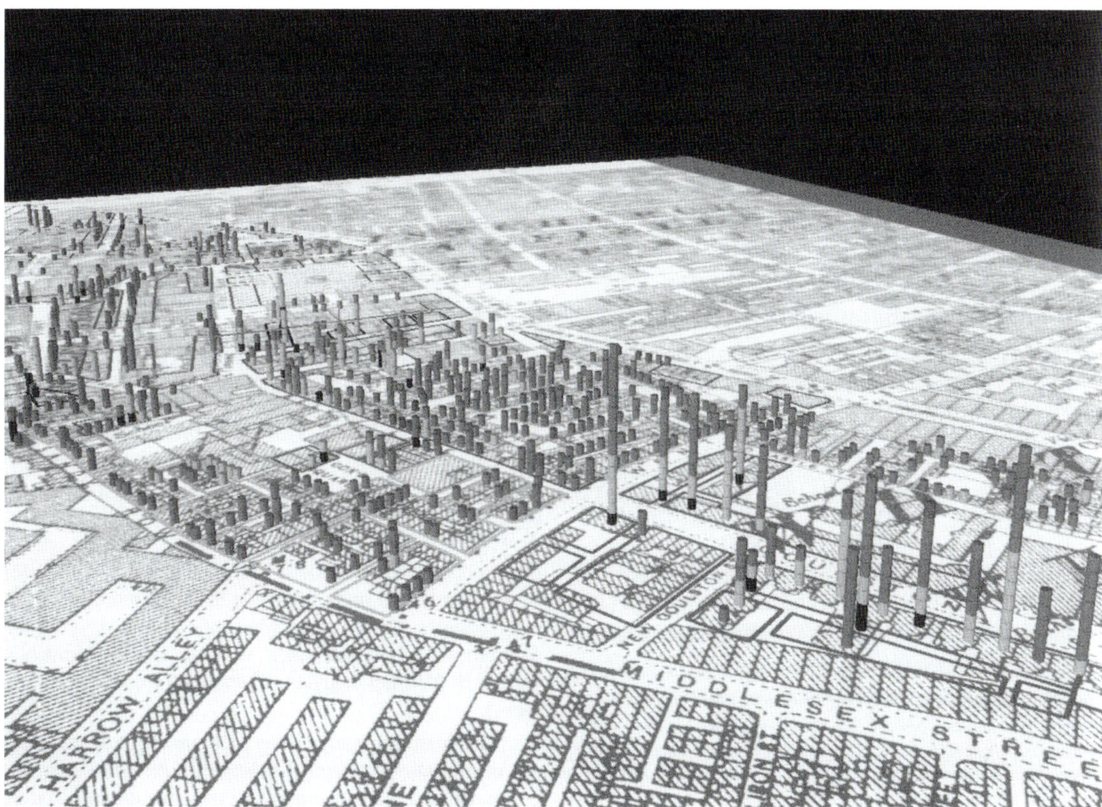

Figure 1

This is an extremely thorny problem which does not admit of an entirely effective solution in the context of the 2-dimensional map, whether it be paper or digital. Feature coincidence is a potentially widespread problem in point symbol mapping, and gives rise to what the author has termed the 'hidden symbol' problem. A full review of this problem, its extent, causes and solutions, together with a discussion of the related problems of symbol overlap and symbol occlusion, is provided elsewhere (Shepherd, 2003a).

The solution to the hidden symbol problem devised by the author involves a shift from 2-D mapping to 3-D visualization. The use of 3-D symbolism and 3-D navigation within an interactive data visualization environment has been demonstrated to provide a satisfactory resolution of the problem. In addition, the purpose-built interactive 3-D visualization software provides a number of broader advantages for the visual exploration of historical data and, by extension, contemporary spatial data. This software is described in the following section.

Using Games Technology to Display Point-located Data

Rather than continue with conventional mapping, GIS or scientific visualization software, it was decided to develop real-time 3-D visualization and navigation software based on videogame development technology to map the Booth family data. Most desktop mapping and GIS soft- ware is still designed for 2-D worlds and the static display of data, though extensions, add-ons and plug-ins are now available for most of the market leader programs to enable them to handle 3-D data and/or to provide some semblance of real-time interaction with spatial data through visualization. The attractiveness of videogame technology lies in its ability to facilitate super-fast navigation through very large and often complex 3-dimensional worlds. Moreover, within the past few years, this technology has matured in ways that make it ever more capable of handling the types of spatial data (both vector and raster) that were once the exclusive preserve of digital mapping and GIS software. Increasingly, the spatial data handling capabilities of videogame engines are meeting and, in the 3-D arena, surpassing those of mainstream digital geotechnologies.

The obvious approach to harnessing this mature technology is to adapt an off-the-shelf games engine to handle real rather than imaginary world data. Games engines are becoming more and more sophisticated, and an increasing number of games (e.g. Grand Theft Auto 3 and The Getaway) involve detailed, large-scale representations of real (especially urban) landscapes. Some pseudo-realistic strategy games (e.g. SimCity) have been used for educational purposes, flight simulator games have been used for flight training of non-commercial pilots, and at least one games engine has been adapted for use as an educational tool (Foster, 1997). It is possible to licence a games engine for commercial exploitation, or to reconfigure the spatial content of an existing game's database (e.g. Doom's WAD file) to handle real-world data.

For this research project, a different approach has been taken. Rather than use a standard games engine (the approach taken in the author's earlier research into the cognitive spatial strategies of games players), the technology underlying 3-D games engines has been used to write a specialist data visualization program. In taking this decision, the author has stepped down a level in the games software hierarchy (which broadly consists of games engines, games development middleware, and low-level graphics libraries or APIs – i.e. application programming interfaces), and has selected one of the leading graphics middleware systems available for commercial games development: Renderware.[2] This system permits games software to be rapidly written in C or C++, which is the standard programming language used to create videogames with Renderware.

A major advantage of games middleware is that it combines a rich graphics library (i.e. a graphics API) with full access to a mainstream programming language (in the case of Renderware, this is C or C++). In addition, games middleware usually provides high-level graphics modelling and games-playing facilities. In the case of Renderware, these include: texture mapping (including complex surface shaders), polygon modelling, light mapping (including real-time dynamic lighting), LOD (level of detail) facilities, object animation and 3-D sound. Renderware also enables applications to be written for a range of platforms, thus ensuring the widespread usability of resulting software. Currently, supported platforms include modern games consoles (including Sony's Playstation 2, Ninetendo's Gamecube, and Microsoft's Xbox) and PCs with suitable 3-D graphics adapters. The challenge for those wishing to develop visualizations using this approach is the expense of purchasing Renderware, and the learning curve involved in using C or C++ and the Renderware API. (It should be noted that the Renderware licence cost is typically a fraction of the cost of licensing a games engine, and programming skills in C or C++ are widely available among computer graduates.)

Using videogames middleware and a high-level programming language, it takes far less time and effort to write visualization software than it did in the days of FORTRAN and lower-level graphics libraries. (Readers will doubtless have fond memories of writing applications software with such low-level graphics libraries as HPGL or Plot-10, higher-level graphics libraries such as GKS, PHIGS or PICASSO, or even dedicated mapping and spatial analysis libraries such as GAG.)

The Renderware system provides more than enough facilities for the development of interactive software to enable fly-abouts through 3-D space populated with 3-D point symbols displayed against a raster image base map. The completed Booth visualization software includes functions to:

- import various data files from ASCII export files generated by the MapInfo and ArcView desktop mapping programs
- organize the imported data for speedy search and selection
- build and export efficient binary file versions of the various data components
- display 3-D symbols representing family data located on the base map image
- handle the mapping of multiple coincident families (using vertical displacement)
- display family attribute data thematically on the displaced 3-D point symbols
- facilitate 'fly-about' user navigation
- change display options and datasets during the course of the visualization
- grab a screen dump at any point during the interative map navigation and store it on file
- These facilities were rapidly developed over the course of a few evenings and weekends.[3]

Using Renderware to Visualize Booth Data

In order to import the MapInfo database into the Booth visualization software, a number of modifications were made to the GIS files. The main focus was on pre-processing the large scanned base map image, which measured 2,400 pixels horizontally by 3,150 pixels vertically, to ensure effective screen rendering. First, the single GIF image was diced into standard square patches (or 'textures'), 256 pixels on each side, because standard-sized small texture tiles can be handled far more efficiently by modern graphics hardware. Textures also adapt more easily to LOD (level of detail) algorithms which will be needed in a future stage of the project when larger areas of London are to be visualized in real time. Next, the image was mip-mapped, in order to allow the image to be viewed from any distance without introducing aliasing artifacts, and bi-/tri-linear interpolation (a process similar to a 2-dimensional moving average filter) was applied to smooth the transitions between pixels. (The filter was clamped to all edges of each texture to avoid the visual artifacts caused by the default wraparound filter.) Because mip-mapping and linear interpolation operations are provided by most modern graphics adapters, their use considerably improves image quality during real-time fly-throughs with no appreciable performance penalty in terms of reduced frame rates.

The boundary and family point datasets were exported from MapInfo using the standard MID/MIF data formats, and the resulting text files were read by utility routines included in the visualization software. Following initial tests, various modifications and improvements were made to the software. For example, a routine was added to display the boundaries of the street polygons in varied widths, with rounded inflexions, and another was written to reduce the greyscale intensity of the rasterized base map so that it did not visually overpower the point symbols used to represent the family locations.

Discussion of Results

The Booth visualization software enables the effective interpretation of data displayed in three dimensions, because of its smooth 'fly-about' facilities. In order to interpret the spatial distribution of family characteristics, the navigator moves effortlessly around the 3-D map, using a number of navigation methods, until salient characteristics of the family distribution patterns have been identified and under- stood. The navigation is completely jitter-free, despite the detailed scanned image and the numerous point symbols displayed on screen. (For example, the map of cigar makers consists of 199 points, the map of family economic position consists of 1,365 points, the map of family social class consists of 1,478 points, and the map of all families includes 2,128 points.) Thanks to the power of the graphics adapter, the software can run at high resolutions (the author's laptop runs the software at 1,400 x 1,050 pixels with 16-bit colour depth in full-screen mode and 32-bit colour depth in a window), without degrading the navigational experience.

The visualization software also provides a simple but highly effective 3-D solution to the seemingly intractable hidden symbol problem that frequently occurs in 2-D space. By plotting coincident families as vertical pillars of 3-D point symbols, the program enables all points in the data- base to be seen. In order to avoid point-of-view occlusion, the user has only to move around the scene to disambiguate the patterns on display. (See Shepherd 2003a for a detailed evaluation.)

The software offers various facilities to enable it to be scaled up to much larger databases. Tests with databases of varying sizes suggest that up to about 10,000 data points can be handled with the software running in its current mode of operation, in which every geometrical item appearing on screen is recreated afresh for each frame of the visualization. Using a number of the technical 'tricks' used by games developers and available within the Renderware system (e.g. persistent mode rendering, LOD enabling, incremental data loading and unloading, world subdivision), it will be feasible to navigate in real time through databases which are one or two orders of magnitude larger than this. By using a simple project file mechanism, the user can switch rapidly from one database to another. (Project files are similar to the workspace mechanism in MapInfo, and permit the import of combinations of data files that make up the spatial database for a particular study area.)

Conclusions

This paper has revealed how a commercial software development library, the Renderware product for videogames programming, has been used to visualize extensive spatial data in three dimensions. In addition to enabling easy navigation through 3-dimensional spaces in ways that permit effective interpretation of historical (and contemporary) spatial data, the software also provides a solution to the difficult hidden symbol problem frequently encountered in point symbol mapping. Further work is in progress to scale up the software to handle very large datasets, and to extend the functionality of the software to handle data exported from other desktop mapping and GIS programs. Experiments to date suggest that games development software provides a highly effective alternative to conventional GIS and mapping software for the interactive visualization of large point feature datasets that need to be explored in 3-D space.

Acknowledgements

I am delighted to acknowledge the help of my son, Iestyn Bleasdale-Shepherd, in building the Booth visualization software. A senior software engineer at Criterion Software Ltd, he is a member of the Technical Team responsible for developing the core architecture of the Renderware software. I am also grateful to the organizers of the 2002 Annual SoC Technical Symposium for inviting me to present a paper entitled: '(Re-)Creating Charles Booth's Poverty Mapping of Late-Nineteenth Century London', from which the current paper was developed.

Notes

1. Background information on Booth's maps and the family mapping project can be found on the author's Web site (*http://mubs.mdx.ac.uk/~Boothmaps/*).

2. Renderware is a product of Criterion Software, a Canon company based in Guildford, UK. The software is used in a number of leading interactive videogames, including: Grand Theft Auto III and Vice City, Tony Hawk's Pro Skater 3, Pro Evolution Soccer, Airblade, and Burnout 1 and 2. It is also used in a number of more specialist titles such as: MX2002, City Crisis, Driven, and Super Bombad Racing.

3. Needless to say, this productivity owes a great deal to the experience of my expert programmer.

References

Ahmad, S. (1999) "Charles Booth's 1889 descriptive map of London poverty" Available at: *http://www.umich.edu/~risotto/home.html* (accessed 17/11/02).

Booth, C. (Ed.) (1889) *Labour and Life of the People* (Volume 1, including map) London: Macmillan and Co.

Booth, C. (Ed.) (1889–1891) *Labour and Life of the People* (2 vols, plus maps) London: Macmillan and Co.

Foster, J. (1997) "Doom and education" Available at: *http://ci.cs.vt.edu/~mm/s97/projects/doom.txt* (accessed 17/11/02). (The original report on the use of Doom in an information systems course in the US military, at: *http://www.eecs.usma.edu/cs383/doom/default.htm,* appears to have been withdrawn from this Web site).

Hyde, R. (1975) *Printed Maps of Victorian London 1851–1900* Folkestone: Dawson.

O'Day, R. (1996-7) "Charles Booth and social investigation in nineteenth century Britain" *Craft* 16 Available at: *http://www.arts.gla.ac.uk/www/ctich/Publications/craft16_1.htm* (accessed 17/11/02).

Reeder, D.A. (1984) *Charles Booth's Descriptive Map of London Poverty 1889* Publication 130 London: Topographical Society.

Shepherd, I.D.H. (2000) *Mapping the poor in late-Victorian London: A multi-scale approach* In Bradshaw, J. and Sainsbury, R. (Eds) *Getting the Measure of Poverty: The early legacy of Seebohm Rowntree* (pp.148–176) Aldershot: Ashgate.

Shepherd, I.D.H. (2003a) "Point symbol maps considered harmful: The hidden symbol problem in thematic mapping and spatial data visualization" Paper in preparation.

Shepherd, I.D.H. (2003b) "Three dimensional visualization for personal marketing and geogemographic data" Paper in preparation.

Topalov, C. (1993) "The city as *terra incognita*: Charles Booth's poverty survey and the people of London, 1886–1891" *Planning Perspectives* 8 (4) pp.395–425. (Originally published in *Geneses: Sciences Sociales et Histoire* (1991) 5 pp.5–34 English translation by A. Sutcliffe.)

Mapping with Options™

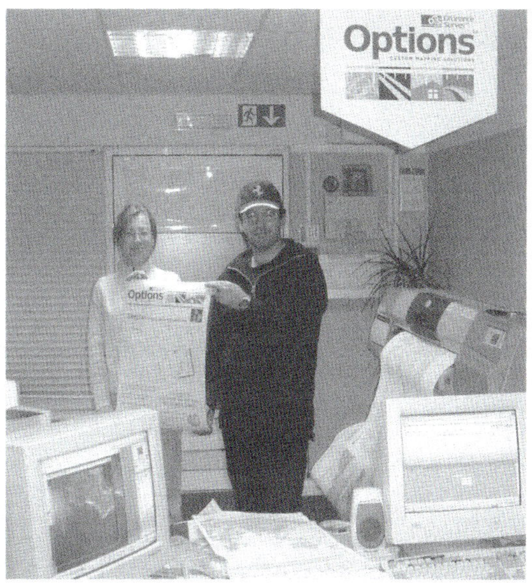

The *XYZ* Digital Map Company has added a further addition to their extensive range of map, airphoto, and GIS services. *XYZ* has recently become a supplier of the Options™ map range of products. Options™ is produced using Ordnance Survey data. The data provided are the most up-to-date, locationally accurate data that the OS can provide. All maps are plotted to order using our live connection to the OS master database.

Options™ has a variety of benefits. The map is centred on your chosen site location and excludes extraneous coverage so you only purchase what you need. There is prompt service and delivery as products can be ordered and e-mailed the same day or be delivered by Royal Mail. Options™ can prove to be of particular use in planning applications submissions, surveying, for farming or for site referencing.

Aerial photography can be used to compliment this data. It can provide detail to further reference a site location and its surroundings. *XYZ*'s high resolution images are rectified to fit Options™ data. *XYZ* carries both contemporary and historical air photos. 1946/47 and 1988 images are available for all of Scotland while coverage from 2001 to 2003 is available for England, Wales and most arable / urban areas in Scotland.

Contact Point:
Ms Marina Smith, Sales / Marketing Executive, The *XYZ* Digital Map Company Ltd
Email: marina.smith@xyzmaps.com
Tel: 0131 454 0426 Fax: 0131 454 0443

Further information on products and services, browse the *XYZ* web site at *http://www.xyzmaps.com*.

WHY CAN'T A MAP HAVE SKY?

Alan Collinson

Originally published in Volume 36, No 2, pp.57–61

This paper describes the adoption of 3D virtual reality mapping and other graphic techniques in the general publishing market whilst the cartographic world remains stoically conservative. In working for a number of clients in the publishing world we were encouraged to develop more interesting ways of portraying the Earth's surface. In contrast there is little or no encouragement from the cartographic world to develop these techniques. Why is this so? I would like to raise a few of the issues involved, and present some of the work we have done for your comment and consideration.

Introduction

IBM were not interested in Windows when it was offered to them, they couldn't see any sales potential for just prettying up an operating system. And so Bill Gates was left to muddle on, and we all know what a disaster that was.

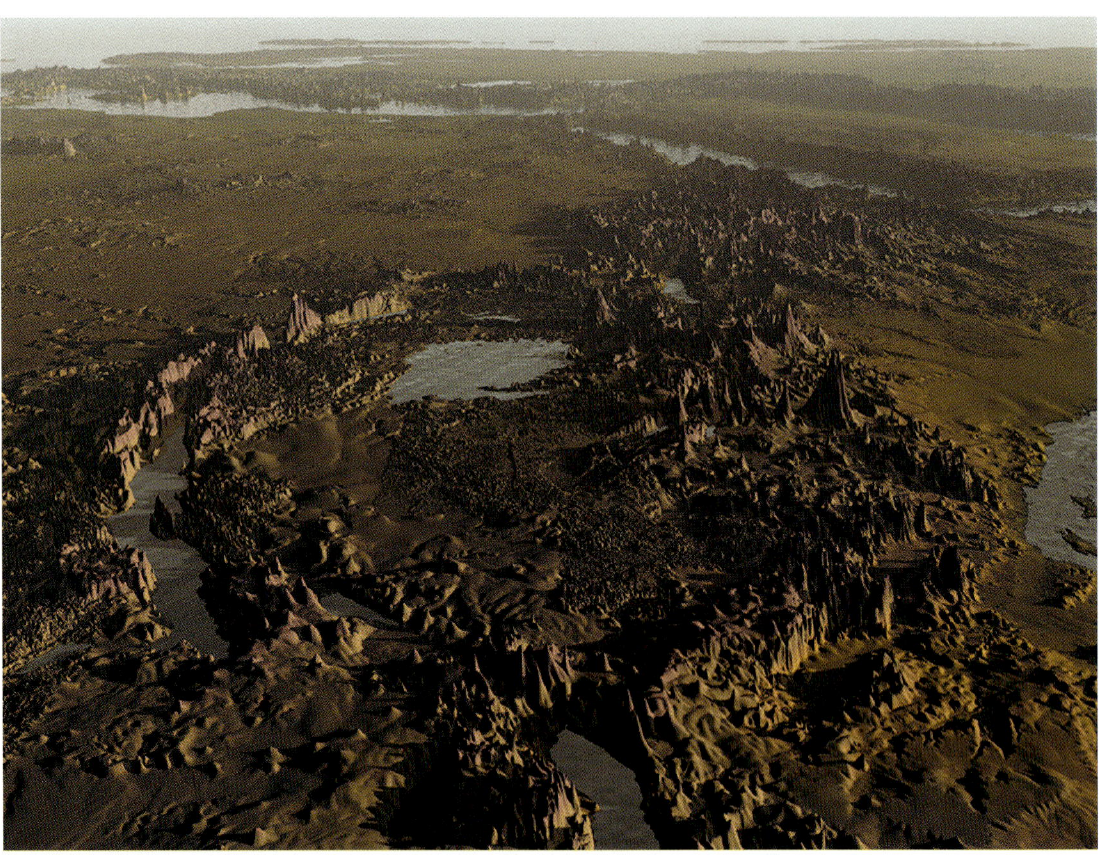

Xerox made the first OS system but couldn't grasp the context of using a 'Mouse' instead of co-ordinates. So they gave the idea to Apple.

In the early 1960s I remember an executive of Triumph Motor Cycle Company showing absolute disdain at the new little Honda bikes coming onto the market. His very words were 'They may compete a little in the small bike market, but they'll never affect our big bike sales'.

In each case, control of vast potential benefits slipped through the sleeping giant's fingers, in the arrogant self-assurance that forces that governed the past would continue to govern the future.

The rest is history, but also, how many times have I heard those exact sentiments expressed by our cartographic community when it comes to the fledgling efforts of our graphic and media colleagues at drawing maps. And yet the threat to the cartographic world is far greater than Honda ever were to Triumph.

'Trust me, I'm a cartographer' doesn't work any more, (if it ever did).

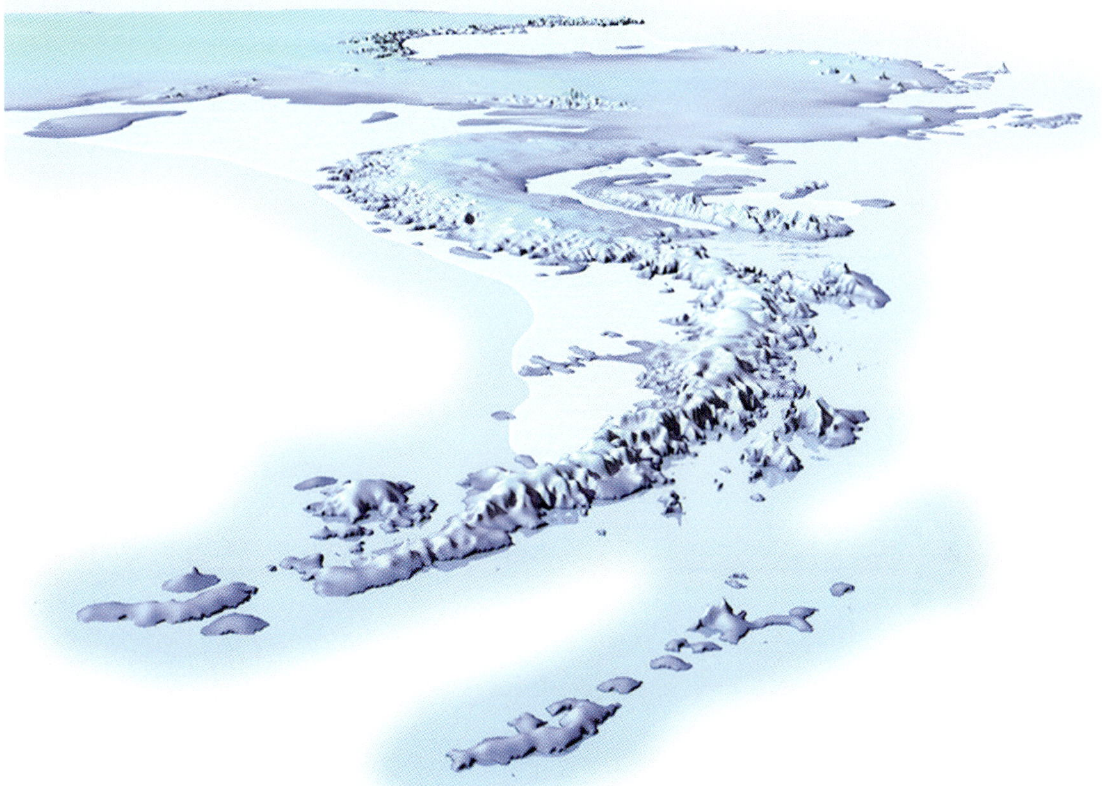

Two questions arise.

WHY IS THE CARTOGRAPHIC WORLD SO CONSERVATIVE?

and

WHY DOES THE PUBLISHING INDUSTRY APPEAR SO MUCH MORE RADICAL?

Before I start I would like to explain where I'm coming from. For the purposes of this article I would like to focus on the main subject of cartography, the topographic map, and especially the mapping of relief. Additionally I would like to talk about the visual product rather than the digital systems, which may underlay it or even in some cases obviate any visual media.

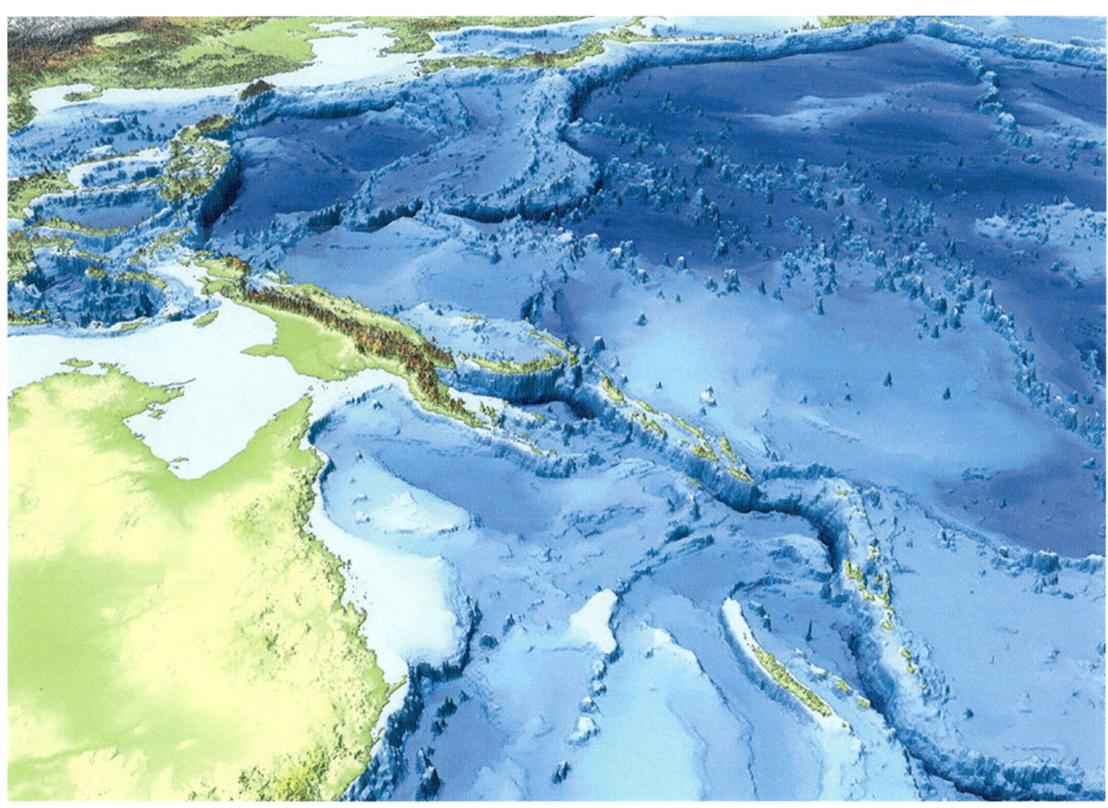

Perhaps, before I begin, a third question needs to be asked: 'Why have topographic map products of the computer age still not produced anything radically different to the hand drawn products of 50 years ago?'

For an example of this conservatism in specifically British cartography you only have to look at our Road atlases, and to ask why the various commercial and public products look so similar. And why haven't they changed over time. (Actually I'd like to show them to you but because most of them owe a debt to the OS that would entail charging each of you £2.50 to cover the scanning costs, the £200 license fee, plus a media conversion charge.)

SEVEN REASONS

I think that there are SEVEN reasons for this conservatism, some of which apply to cartography worldwide and some are peculiar to the British scene. However, what is puzzling is that most of these conservative restrictions do not seem to apply to other disciplines when they draw maps. Perhaps there is good reason, perhaps not.

Reason 1. Restrictions Inherent in Map Making

The first reason applies to cartography in general. It is in the nature of the profession that there will be restrictions on the freedom to order and distribute the content of maps. We accept this constraint in the same way that a car designer accepts that a car must have wheels, but somehow this constraint weighs more heavily on cartography than it does in other disciplines.

Reason 2. Restrictions in the Current Ethos of Cartography.

Without doubt there is an incredible technological change is taking place in cartography. We are now at a time in history when 'up to the minute' data of 100 per cent accuracy is almost within our grasp. But what happens when we achieve it? Does cartography as we know it come to an end? Or is there more to cartography than some glorified database?

In our rush to create a greater store of knowledge we are in danger of overlooking many of the purposes for which maps are designed to serve. Especially the grasp and understanding of multi-layered spatial information with the minimum of effort. It in not just knowledge but the communication of that knowledge which is the crux of the map-making process.

Reason 3. The Public

It is claimed by most cartographic companies that the public themselves demand the products they produce. They blame the conservatism of cartography onto the conservatism of the map-using public. It is certainly true that the public perception of maps is limited to very few products. When you tell someone that you are a cartographer, the inevitable retort in the UK is 'Oh! do you draw Ordnance Survey maps then', as if there were no other kind.

But ask the same question in the United States and the reply is unlikely to feature any of the national mapping agencies, but rather Michelin or Rand McNally. Perhaps our British public has been conditioned. By only offering one thing, perhaps they feel that there is only one thing.

Reason 4. The Cartographer

What is lacking in cartography today is an education and understanding which sets cartography in a modern media setting. When you look at the professional training experienced in other professions, it would appear that most cartographers have little formal training above GCSE woodwork. I don't know of a single cartographer with a qualification in design, yet what media house in its right mind would create a publication without a qualified and experienced design team at its helm. Most of our map design involves merely altering the road widths. In fact it is getting worse, the day is coming when there will be no

formal training of cartographers whatsoever in the UK and when the existing pool of qualified people dries up, who will be the cartographers of the future. Perhaps there is a group of fledgling graphic artists waiting in the wings to take over when cartography finally slips through our fingers, unless we can come up with something new that others can only admire. If we do not have something unique to offer, then we should not be in the business. And probably won't be if we don't buck our ideas up.

Reason 5. The Cartographic Organizations

If we look around, almost all our commercial and public organizations are now run by men and women who began their careers in an age when maps were drawn by hand. The age-old conservative traditions of map making are part

of their psyche. It's like those old people who still go to the seaside in Britain to sit in their deck chairs on the beach, because it brings back memories of happy days gone by, spent in tranquillity and repose, where the sun shone every day and you could get an ice cream cone, a bag of crisps, a charabang ride and a knickerbockerglory and still have change out of sixpence. Of course the old farts are on the beach on their own these days because everyone else has buggered off the Ibiza.

Reason 6. The Cost

The cost of representing cartographic information coupled with the fear that there might not be a financial return, tends towards a conservative attitude to map making. Also acquiring new data to create maps has always constrained cartography. The time, effort, and costs involved are prohibitive of experimentation and development. And yet, step outside the narrow confines of cartography and there is a vibrant (and profitable) mapping market out there.

Reason 7. Copyright Restrictions

If a fashion designer had to pay royalties on the fabric they used, then fashion design would come to an abrupt halt. It is well documented that copyright restrictions affect British cartography more than anywhere else in the world. Whatever the justifications may be, the interpretations put upon copyright law in the UK are restrictive to the

wellbeing of the cartography of Britain.

Conclusion

The restrictions inherent in the product, the perceived reaction of the general public to cartographic innovation, the ethos and inherent conservatism of cartographers and the cartographic industry, the cost of change and copyright restrictions, all contribute to an austere climate of conservatism and lack of innovation in cartography. Or so it would seem.

If this is indeed true of the cartographic industry, then why is the publishing world not so constrained? It isn't good enough to raise a haughty sneer at these other products, and claim that they are in some way inferior and therefore don't count. It's the Triumph syndrome all over again.

Quite simply I think that the reason why the publishing industry welcomes cartographic innovations is that they are in a white hot design cauldron, needing new solutions

every day and at the drop of a hat. It hardly describes cartography does it!!

The publishing world doesn't want two or three ideas to choose from, it wants twenty or thirty, and even then it won't choose any of them, but will go away and create new design products based them. Our cartographic methodology is too slow and too rigid to find a place in this active environment where the graphic designer and media creator have already largely replaced the cartographer. Cartographers say 'can't'. Publishers say 'why not'.

Sadly, as cartographers, it is hard for us to say it but, at Geo-Innovations we have been impressed with the flair and freshness of media driven cartography for some time. So, we tried to see through their eyes to see what it was that they saw, what they wanted from cartographers and where their future was taking them. We ventured into the lion's den.

It became obvious immediately that there was a basic problem. Firstly, our basic mapping tools FreeHand and

Illustrator were little used in the media world. Photoshop, Painter and Lightwave were much more in evidence. Also whilst the publishers we worked for were happy to commission work from us and would continue to do so for special contracts, they really wanted to do most of it themselves. So here is the rub. There are graphic artists and media agencies out there looking for cartographers to help them draw their own maps.

So, why not!!! Why shouldn't we give the graphics world the means to draw their own maps. If we haven't impressed them by now of the God like status of the cartographic profession then we're never going to do so. We have what they want but seem to have some reticence in giving it to them in the way that they want.

So following discussions with a number of media publishers, we developed a series of ideas which will allow graphic artists, media houses (and cartographers) to have access to our product, world relief data, at the press of a button. We felt that what the publishing world wanted was a choice of solutions that could be accessed easily and economically. Also, they would still prefer to construct their own maps rather than get a cartographer to do so.

A management and product assessment of our company identified a market for off the shelf products which did not rely on working in the purely commission environment we were in. Perhaps a few cartographers might be interested too.

WE CALL OUR APPROACH GEO-INNOVATIONS, and it comes in four parts.

1. The GEO-ATLAS

Print Resolution World Relief Images in Various Projections for Atlases and Multi-Media. Worldwide print resolution coverage at scales of up to 1:5,000,000.

The GEO-ATLAS is ideally suited to designers and cartographers who require accurate relief images for their work with a thousand and one variations at their fingertips in order to customize them to their own specification.

The images are fully editable in Photoshop format and

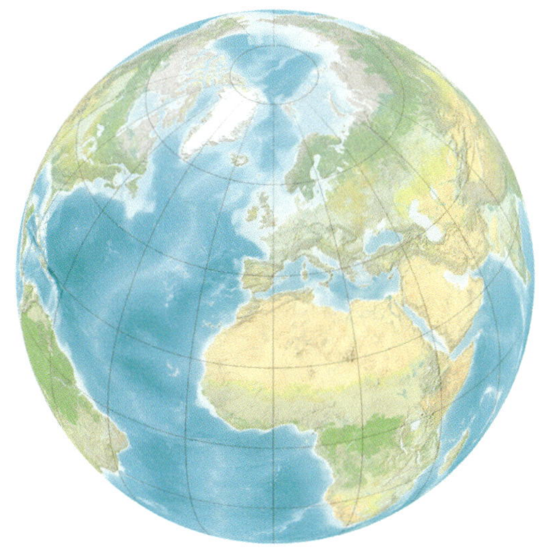

are capable of adding layers of additional data, rivers, lakes, graticule, oceans etc. and there are seven basic colour schemes, two height enhancements and greyscale as standard. Also there are a range of colour choices:

1 With tundra and deserts for general media and cartographic use.
2 Generated to highlight lowland areas.
3 For general atlas use. Three height variations.
4 A monochrome cartographic variant, with three height variations.
5 A distinctive colour variation for cartographic or media use.
6 Greyscale version for customization.
7 A lightly coloured cartographic version.

2. GEO-GLOBES

A unique collection of globe images of the earth taken directly from our master series. With 78 viewpoints, these 11 inch print resolution globes represent not only the most complete set of cartographic images of the earth ever published but also the highest resolution images available in such a format. These images too are fully editable in Photoshop format and are capable of adding layers of additional data, rivers, lakes, graticule, oceans etc. and there are six basic colour schemes as standard. In all there are millions of combinations of image possible with this 13 CD or 2 DVD set.

3. GEO-WORLDS

GeoWorlds is a series of panoramic images of the earth from many different angles and lighting conditions. They supply the needs of media and publishing houses to illustrate the earth's surface in other than merely plan views, and yet still remain cartographically credible. AND ANYWAY, why shouldn't a map have sky?

Why shouldn't the land reflect the atmosphere which shrouds it and encompasses it. Why can't a map have shadows, rain, mist and sunsets, and still retain its cartographic credibility? The real world has all of these things and it does not detract in any way from our understanding or interpretation of it. On the contrary our understanding is enhanced by these very elements which cartographers in the past have been unwilling or unable to portray.

4. GEO-PROFESSIONAL

But real cartographers have not been forgotten in all this after all. We have kept something back for them.

Following the launch of the SRTM satellite, (Shuttle Radar Topography Mission) two years ago we have been preparing for the introduction of new large-scale terrain data to come onto the market. This is planned to be complete in January 2003.

This will provide us with data capable of going up to 1:250,000 for 80 per cent of the world. Created by professional cartographers for professional cartographers. Whether they want it is another matter.

Our Unique Contribution

Much of the media we use is purchased off the shelf with only minor, but important, modifications from ourselves. However, whilst the original source data is available from a number of sources there are many restrictions to its optimal use by the commercial sector. The technical skills necessary to manipulate these images are not readily available, and also most graphics studios do not have equipment or the time free to manipulate images of this size. If we were to print our database out at full size it would be wall map 68 feet long and 30 feet high and there is no single computer that could work with it.

What we have contributed is to create WYSIWYG products that require no complicated mathematics or particular skill to operate. A wide standard versatility but with open opportunities for the user to create unique customized variations using only the average computer powers. BUT THERE IS ONE MORE STEP TO TAKE.

So finally, the last of our Geo-innovations products, the Geo-Creator.

5. GEO-CREATIONS

No matter how comprehensive and versatile our catalogue of images has become, there are still be millions of variations we could not cover. So, for those with imagination, flair and just a little graphics skill we have GEO-CREATIONS, for the designer and publisher or media house to create their own custom images in a Bryce landscape generator.

With Geo-Creations a designer or cartographer may prepare custom images for any media. Colour at will; rotate to any angle; add sky, reflections, sunsets, and mists; enhance or subdue the relief, and do a thousand and one things until the image finally matches exactly what is required.

Here at last is the world at your fingertips.

WE CALL THE WHOLE THING: GEO-INNOVATIONS

Spare a moment to take a look at our developing website: *http://www.geoinnovations.co.uk*

COLLABORATIVE MAPPING: BY THE PEOPLE, FOR THE PEOPLE

Edward MacGillavry

Originally published in Volume 37, No 2, pp.43–45

Anyone with a location-aware device could potentially create his or her own personal map. Collaborative mapping is an initiative to collectively produce models of real-world locations online that people can then access and use to virtually annotate locations in space. This paper explains why this trend has recently emerged and presents some projects that are currently being undertaken. It highlights the pros and cons and discusses some barriers that hinder the progress of collaborative mapping. Is the virtual world really becoming another means to navigate the physical world?

Introduction

Since the Enlightenment, mapping and the production of geographic information have been institutionalized: the map is the power. At home, maps were used as an instrument for nation building as nation states emerged. People learned about their country and administrations needed a tool to govern the territory. Away from home, maps were an instrument for colonization when Africa and Asia were split among the European nation states.

During the last few decades there has been rapid democratization of geographic information and maps. Geographic information systems moved from mainframes and UNIX operating systems onto personal computers and the Windows operating system. From research and government, GIS spread into the business sector. The PARC Xerox Map Server and Virtual Tourist brought maps to everyone's PC, followed by Mapquest, Maporama, Multimap, and Streetmap. Soon enough, everyone will carry around location-aware portable devices such as PDAs and mobile phones offering highly customized and sophisticated location-based services.

Although maps are more widely used than ever, the production of geographic information, and especially mapping, is still highly concentrated among national mapping agencies and the GI industry. But this oligarchy is very likely to be dissolved sooner rather than later.

Earlier this year, the online high-tech magazine 'Need to Know' heralded 2003 to be the 'year of the geospatial hype'. Indeed, several online applications that revolve around the importance of location have been launched this year. Particularly noteworthy were the launch of UpMyStreet Conversations and the overhaul of OpenGuides in the UK. These services enable people to discuss various issues and make suggestions about their area. It differs from common notice boards on the Web, as postings are not only order chronologically, but also by distance. Therefore, people can quickly access and contribute information and opinions pertaining to a specific area. Whereas UpMyStreet Conversations is a commercial service, OpenGuides is a network of community-maintained city guides running on open source software.

The Amsterdam RealTime project in the Netherlands invited visitors to participate in an exhibition of historical maps of Amsterdam by carrying around a PDA with a GPS unit for a week. Their movements in the city were instantly projected in the exhibition room for other visitors to follow the traces on screen. Gradually, individual traces joined up and started to overlap, thus creating a self-made map of Amsterdam.

These are just three examples of an emerging trend called 'collaborative mapping'. Other terms being used to describe the current trend are 'localized social software' or 'grassroots GIS'. Whereas these terms emphasize the underlying technology, 'collaborative mapping' focuses on the importance of individuals or groups working together.

Collaborative Mapping

Collaborative mapping is an initiative to collectively create models of real-world locations online, that anyone can access and use to virtually annotate locations in space. The value of the annotations is determined by physical and social proximity (expressed in distance and 'degrees of separation'). Thus, the information is not only filtered based on proximity, but also ranked according to the trust relationship between individuals or groups of people through social networks: the 'Web of Trust' (Espinoza *et al.*, 2001).

The notion of 'social proximity' as a measure of relevancy of geographic information sets collaborative mapping apart from corporate providers of location-based services (LBS). When sending a request for the nearest Italian restaurant, the service not only takes into the account the distance to the restaurant from the current position, but it also navigates the social network to find a suitable result. People are willing to travel further because the particular restaurant is a favourite haunt of one of their relations. Furthermore, geographic information is not merely broadcast to users, but users can actively contribute to the service: geographic information flows back to the service. A collaborative mapping service then works like a spatially enabled notebook or message board, depending on the privacy settings people attach to their postings.

As geographic information is created collaboratively,

people can store their contributions on their networked computer, PDA or mobile phone, just as they share digital music files over the Web. Ed Parsons, CTO of Ordnance Survey, talked in an interview about the 'napsterization' of geographic information, claiming that 'the experience of the music industry may also apply to geographic information' (Westell, 2003). Napster and Apple iTunes Music Store have certainly changed consumer attitudes towards acquiring digital content, but online distributed file sharing networks such as Kazaa and Gnutella are still hugely popular. Collaborative mapping enables the 'gnutellasation' of geographic information. Furthermore, it is a vehicle for critical cartography. Collaborative mapping does not provide one authoritative view of world, but it is a platform for people to publish and discuss their opinions about locations and areas.

Most collaborative mapping projects are Web-based, whereas corporate LBS providers mainly target the PDA and mobile phone market. Since web standards such as XML, RSS, and RDF are the building blocks of many collaborative mapping projects, other platforms can be catered for in the future. At this stage, projects are primarily developed as 'proof of concept'.

Geospatial Hype or Reality?

Since the beginning of this year, there has been a growing interest for cartography and GIS in online communities, notably the blogging community and the open source community. Bloggers maintain online diaries, so-called 'weblogs' or 'blogs'. Diary entries are ordered by time and date. There are various portals that provide long listings of blogs. Some of these portals have a geographic scope, only listing blogs in a particular country, or town, for example Brighton Bloggers (*http://www.brightonbloggers.co.uk*). As such listings can become rather long, some portals structure blogs by tube stop, such as NYCBlogger (*http://www.nycbloggers.com/*) and London Bloggers (*http://www.londonbloggers.co.uk/*). The portal GeoURL (*http://www.geourl.org*) lists thousands of blogs that pro- vide their location using HTML meta tags:

<meta name="ICBM" content="52.09022,5.08159">

<meta name="geo.position" content="52.09022;5.08159">

<meta name="DC.title" content="webmapper.net: what the map can be.">

Some bloggers travel regularly and blogs can be updated using mobile devices: moblogging. Location has thus become another means to structure the diary entries. Also, bloggers frequently write about locations they are visiting. Movable Type, a popular tool for creating weblogs, automatically detects town names in each diary entry and inserts a map of that location. Furthermore, it keeps track of geographical location, travels, and other nearby blogs.

Online vector-based mapping used to be the domain of proprietary plug-ins from leading GIS vendors. As the XML-based Scalable Vector Graphics (SVG) file format gains popularity, vector-based mapping is not anymore the monopoly of the GIS industry and government agencies. Many individual web developers have started to create compelling vector-based web mapping applications. Not only general web standards such as SVG, but also various standards approved by the OpenGIS Consortium (OGC) contribute to the wider adoption of web mapping outside the traditional fields. Another important event in online mapping is the release of a new version of the open source MapServer system this year. These developments make collaborative mapping a feasible exercise. Finally, geography-related games are becoming more and more popular, for example geo-caching (*http://www.geocaching .com/*). Caches are set up all over the world and their locations are shared on the internet. GPS users use the location coordinates to find the caches. Once found, a reward for the visitor may be found inside the cache.

Barriers to Innovation

Obtaining mapping data is one of the challenges facing collaborative mapping. Whereas geographic information created by US government agencies such as NIMA and USGS is available for free, in most European countries, geographic information has to be paid for. As the market position of National Mapping Agencies (NMAs) is changing rapidly, their geographic databases are their main assets. To survive in the market, they have to exploit these assets on which they also hold a monopoly: no other company or organization has the technical and financial resources to create accurate, large-scale geographic data sets. Especially when it comes to online transactions, pricing structures do not reflect general practice. Recently, the Geographic Industry Forum (GIF) was been created to lobby in Parliament for a fairer position of Ordnance Survey in the GI industry.

Not only pricing and copyright put up barriers. The AGI and the OGC recently highlighted the issue of patents. As the GI industry is evolving rapidly, patents have been granted to various key players in the market. This practice is a delicate trade-off between encouraging innovation and encouraging monopolies. Privacy legislation or rather the lack thereof, is another impeding aspect. Social implications are being investigated by the Urban Tapestries project (*http://www.proboscis.org.uk/urbantapestries/*).

Maps and Interfaces

Several projects are underway to collect geographic data to create base mapping. The RealTime project has covered large parts of Amsterdam and Riga, Latvia. As the need for base mapping is particularly pertinent in the United Kingdom, both Geowiki (*http://www.geowiki.com/*) and GPSdrawing (*http://www.gpsdrawing.com/*) are creating maps derived from GPS data.

To view, comment, and edit the data, there are various interface being developed. The Blogosphere is a Java applet that reads an XML file of locations to present a global view (*http://blogosphere.headmap.org/*). The SpaceNameSpace project uses an instant messenger paradigm to navigate a digital model of streets and locations in London (*http://space.frot.org*). The Swedish GeoNotes project (*http://geonotes.sics.se/*) and the

Canadian GPSter project (*http://www.gpster.net/*) have built applications for location-aware PDAs.

The Web: a Window to the Physical World

Collaborative mapping is the latest development in the democratization of geographic information. The technology of location-aware devices, blogging tools, standards approved by the W3C and OGC, and open source web mapping systems have paved the way for people being able to contribute and access localized information. Various projects are being undertaken to investigate the potentials and pitfalls of these services. Certainly, these initiatives are contributing to recognize online communities as real world communities: they are putting a sense of place back into cyberspace. Hopefully, collaborative mapping will become a complimentary input for corporate location-based services.

References

Espinoza, F., Persson, P., Sandin, A., Nyström, H., Cacciatore. E. and Bylund, M. (2001) "GeoNotes: Social and Navigational Aspects of Location-Based Information Systems" In Abowd, G.D., Brumitt, B. and Shafer, S.A.N. (Eds) *Ubicomp 2001: Ubiquitous Computing, International Conference Atlanta, Georgia, 30th September–2nd October* (pp.2–17) Berlin: Springer.

McClellan, J. (2003) "Get Caught Mapping" *The Guardian Online* 27th March 2003.

Need To Know (2003) "Tracking" Available at: *http://www.ntk.net/2003/01/17/#TRACKING* (accessed 21/11/03).

Westell, S. (2003) "The Napsterization of geographic information" *GI News* July/August 2003.

Acknowledgements

I thank various people of the Locative Media Lab for the discussions and their suggestions. I am also grateful to the organizers of Cartography 2003 for inviting me to present the paper entitled "Collaborative Cartography: Grassroots LBS" from which the current paper was developed.

CHANGES IN UNIVERSITY CARTOGRAPHY OVER THE LAST HALF CENTURY: A PERSONAL VIEW

Peter Haggett

Originally published in Volume 38, Nos 1 & 2, pp.31–35

Professor Haggett looks back over the half century since he graduated from Cambridge in the Geographical Tripos in July 1954. He reflects on changes in both the technical basis of university cartography and in its institutional organization. Despite these changes, certain constants in the practice of cartography remain. The original lecture was given at the Society's Summer School on Monday September 6th 2004 and was illustrated by fifty coloured slides (two of which are reproduced here in monochrome).

Introduction: Cartography at Bristol

It gives me very great pleasure to be invited to give a paper to the Society of Cartographers on its first visit to Bristol. A University College was established here in 1876 and in 1909 this duly acquired full University status. Geography was taught as part of teacher training from the turn of the century but it was to be only after World War I that geography became established as a formal discipline. A young lecturer (W.W. Jervis) was appointed to teach geography within the Geology Department in 1920. A separate geography department was established five years later and the first geography graduates (all six of them) duly graduated in July 1929. Jervis was appointed to the first chair of geography in 1933.

From the beginning, cartography and geography were closely intertwined (Haggett, 1995). Jervis himself had a research interest in early maps and his *World of Maps* (Jervis, 1936) drew heavily on the valuable Bristol collection. O.D. Kendall, a young Cambridge surveyor, was appointed in 1925 to teach geomorphology and topographic survey, thereby establishing the degree courses in surveying which (with those at Cambridge and Durham) provided the bulk of the young Colonial Service surveyors for the next three decades. The 1920s were also important in seeing the appointment of a young teenager as the first cartography technician. Harold Freke had a fine hand and sense of cartographic style and in a period when wall maps (particularly the prized Hermann Haack series from Germany) were expensive, Freke developed his own set of canvas-backed classroom maps to serve as the essential backdrop for regional courses. Until the appointment of Simon Godden in 1965, Freke was the sole cartographer in the Department. Godden's retirement after over thirty years service saw the appointment of Drew Ellis in 1998 being only the third cartographer in the Department's 84-year history.

When Professor R.F. Peel (after whom the lecture theatre in which we meet today is named) succeeded Jervis in 1957, he also had a strong cartographic background. He had served on Saharan expeditions as surveyor and map maker in 1938 and when called into war service with the Royal Engineers he worked on the maps needed for the D-day campaigns and later served with the VIIIth Army in Italy producing daily operational maps for Alexander's H.Q. Graduates of the department over the years have included a full quota of surveyors and cartographers They include David Rhind, Director General of the Ordnance Survey, and David McGuire, chief scientist at ESRI.

Personal Perspectives on Cartography

I came to Bristol from Cambridge in 1966 as a young professor to join Peel. My own love of cartography seems to have been developed early and what family archives survive include a 12-year old's diary with constant reference of 'making map of the village' with a mass of early maps mostly copied from those in the end pages of Arthur Ransome books. Topographical surveying and cartography were important parts of the Geographical Tripos at Cambridge which I read from 1951 to 1954. Plane tabling was a special love since it was one of the few surveying methods where the 'map' itself emerges as an integral part of the surveying process rather than having to be reconstructed later from angles, distances and elevations in field notebooks.

It served me well during all three years and never more so than in my second where it lead to my first published map. January 31st and February 1st 1953 saw the great tragedy of the North Sea floods leading to hundreds of deaths in England and thousands in Holland with massive inundations. But the impact of the surge was also marked on the detailed geomorphology of the low island chains off the north Norfolk coast. With a fellow undergraduate, John Small, we surveyed the 18 mile stretch between Hunstanton and Wells-Next-The-Sea by plane table using resection to mark changes on the existing large-scale maps. A corner of the sheet from Scolt Head Island is shown in Figure 1. It records a typical partnership between the geomorphologists (Professor J.A. Steers and A.T. Grove), the field surveyors (Small and myself) and the university cartographer (Les Thurston who produced the finished map).

Similar partnerships accrued in work in Portugal the following summer with maps for *The Geographical Journal* (Hayes, 1956) being beautifully drawn by George S. Holland at the Royal Geographical Society. What is striking today about the cartography of a half-century ago is its bespoke nature. Every map was designed for the job in hand, with the lettering located and executed with a style and flourish that spoke of years of craft. Holland must be one of the few cartographers then who for delicate work set aside his steel pens and had recourse to the quill pen. When coloured slides for the RGS lecture were required, he used an eye glass to paint with consummate care the 'new' and smaller 35 mm glass plates then used. Such maps linked back to a style which dated back to the high points of Dutch cartography a half-century ago.[1]

Cambridge for me was the start of a university career in which geography and cartography were to be closely linked. Looking back now on the paper trail of that career – 300 scientific papers, 20 books, 3 atlases – what is striking is that there are less than a dozen items (all small papers) in which I did not turn to a university cartographer for help, advice, and usually the finished product. For the literally thousands of maps drawn for me over the last half century I'm deeply indebted to several generations of craftsmen and craftswomen (most of whom, in the years since 1964, were members of this Society) and include those who did the photographic and reproductive work. At Cambridge[2] where I was undergraduate and research student (1951–55) and demonstrator and university lecturer (1957–66) the names include Tim Cliff, Ian Gulley, Stella Gutteridge, Pamela Lucas, Gillian Seymour, Philip Stickler, Les Thurston, Roy Versey, Lois Wright, Jenny Wyatt and Make Young. At University College London where I was an Assistant Lecturer (1955–57), Ken Wass and Roy Versey. At Bristol where I was Professor (1966–present), to Robert Bignall, Simon Godden, Drew Ellis, Harold Freke, Tom Phillips, Tony Philpott, and Jon Tooby. Spells overseas increased my debt to university cartographers there and I'm grateful to Kevin Cowan at the Australian National University, Canberra, for help with all the maps in a monograph written there.

Technical Changes in Cartography

When I was an undergraduate, the two books on cartography which we were expected to read and understand were the newly-published *Maps and Diagrams* (Monkhouse and Wilkinson, 1952) and the latest edition of *Map Projections* (Steers, 1949). To them was soon added the latest import from the United States, Arthur Robinson's *Elements of Cartography* (1953). Robinson was to make his Cartographic Laboratory attached to the Geography Department at the University of Wisconsin at Madison the Mecca for university cartographers. As edition succeeded edition, so the originally slim volume increased in both weight and number of authors with the current edition (Robinson *et al.*, 1995) having no less than five authors. Turning the pages of successive editions of Robinson is to view a half-century of rapid change in the technical changes within cartography.

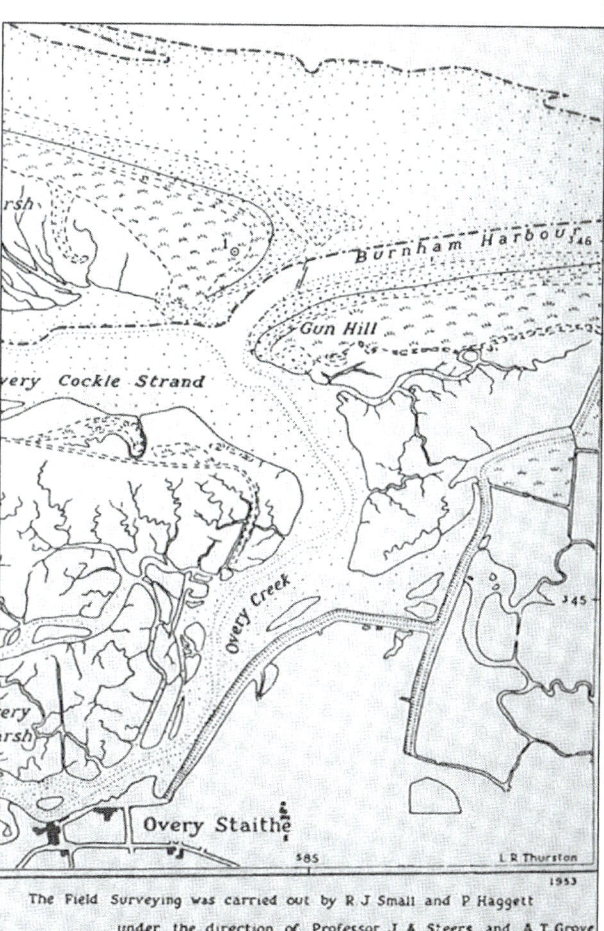

Figure 1
Map partnership in the style of a half-century ago. A fragment of a map of Scolt Head Island, Norfolk, surveyed by the author and R.J.Small (later Professor of Geography at Southampton University) in the summer of 1953. under the direction of Professor J.A.Steers and A.T.Grove. The final map was drawn from the plane-table field sheets by Leslie Thurston in Cambridge Geography Department's Cartography Laboratory and published in 1954 by the Nature Conservancy.

Before this professional audience who have lived through more recent changes and live with them day by day, it would be invidious to list the changes: they have affected all aspects of map design and delivery. But for me four stand out: (1) the analogue to digital revolution in which maps are now conceived and handled as immense numerical arrays or matrices and in which mappable data are flexibly handled in vector or raster form; (2) the switch from ROM ('read-only' maps) to RAM ('random access' maps) in which maps are retained in computer memory and can be interrogated to give an infinite range of locations, components, scale, colours, etc. to meet the particular needs of a particular user at a particular time; (3) the accompanying GEO-data compression which allows truly mind-blowing amounts of geographical data to be held on file and interrogated in so many different combinatorial forms; and (4) the static-dynamic switch which allows maps of geographic change to now be shown as short film sequences, blurring the boundary between cartography and movies. One geographer who made effective use of the latter was the late Peter Gould who converted his monthly maps of reported HIV-AIDS in the 3,000 counties of the United States into a striking colour film of the unrelenting progress of the epidemic through regions, states, towns and countryside (Gould, 1999).

The technical revolution, much of which has come in the last decade of the half-century has many popular images of which 'the end of the map corner problem' (map centring by the user) or 'maps on your mobile' (downloading GPS material on your mobile phone) must serve. For me as an active geographer, I measure the contrast in delivering map material to my publisher. Only 16-years ago it was still a hazardous operation in which large folios of black-and-white map drawings (for the *Atlas of Diseases Distributions*, Cliff and Haggett, 1988) would be taken in the boot of a car, hoping no vital piece of Letraset or shading would be dislodged on the journey, and that eventually some would survive and be returned to the huge map chest in which all work resided. Today the latest atlas in full colour (Cliff *et al.*, 2004) is transportable as a set of CDs backed up on a portable hard disk.

With these changes, the working partnerships illustrated in Figure 1 have changed. As an example I take a typical working manual produced by the Department of Geography at the Memorial University of Newfoundland. Map Design and Production using CORELDraw 7 and CORELDraw 8 (Conway *et al.*, 1998) involved an academic Professor Cliff Wood and the two cartographers Charles Conway and Gary McManus but also a third component, the COREL Corporation whose software engineers designed the original software. Increasingly, feedback from geographers and cartographers is allowing the software companies to ensure that general-purpose drawing software is adaptable for specialized cartographic use.

Institutional Changes in University Cartography

Alongside the technical changes in the practice of cartography have come a number of institutional changes. Some have been driven by those technical changes described above, others by wider changes within the university world in which cartography units are embedded.

Changes in the size of the university sector. In 1950 the proportion of cohort of young people in the 18–22 years cohort (adjusted for late entry for males undergoing National Service) who were going to university was around 1:40. A half century later the official government target is to reach 1:2. This overall increase in the size of the university sector has important implications for geography departments and the cartography units which have normally served as their homes. Going up to university in 1950 I had a choice of around a score of geography courses offered by universities in England and Wales from which to choose; today, for my grandchildren, that figure would be five times greater. In very rough terms, cartography units have mirrored that overall growth in geography departments.

Changes in the size of departments. Very few geography departments in 1950 had an academic staff in double figures. Today (if research staff, then a largely unknown category, are included) several would now have academic staff levels of 40+. Cartography units have also tended to grow in size but at a lower rate. This is in part due to much of the growth coming in academics in human geography which, to judge from journal papers in the last couple of decades, has had a smaller need for cartographic support than in physical geography and environmental studies.

Convergence of cartography with associated disciplines. The digital revolution which we have noted in cartography has been part of a general IT revolution. Over the last decade this has particularly impacted on general photography and on the specialized areas of aerial photography and remote sensing. Rather than maintaining separate small units with often a single staff member, a number of departments (including Bristol) have chosen to combine these into larger units where IT skills can be effectively shared. Such larger units also offer better promotion prospects for staff onto IT-related grades.

Merging of cartographic skills into university-wide units. The digital revolution in cartography means than many colleagues across the university are now sharing rather similar technology. Conventionally, mapping facilities were provided for geography departments as part of their research and teaching facilities, but these needs were not exclusive to geography. Other academic departments (notably geology, but also archaeology, other environmental subjects, and civil engineering) had mapping needs as did the estates department of university administration.

In generous times for universities with rapid growth as in the 1960s and 1970s, different departments tended to be left to make their own priorities. But as investment per student dropped in the 1980s and 1990s and as departments were encouraged to adopt competitive financial models so duplication across a university began to be exposed. The old 'grace and favour' regime by which maps were drawn within geography units for other departments as needed and as time allowed were replaced by costed time-sharing. Several different responses have been adopted within a range which runs from (a) cartography laboratories within geography departments being asked to take on a wider university role to (b) cartography laboratories being made into independent university institutes with their own budgets and with geography as only one of several departments using and being billed for their services. Each model (and several intermediate variations) has its advantages and drawbacks for the cartographers involved.

Beyond the university. There is no reason to think that the spatial integration of cartography should stop at the university limit. The ability to shift increasing large data files via the internet means that the potential to move maps around is unbounded (in theory, if not yet in practice). One example of this is provided by the 17 volumes of the *National Atlas of Sweden* (Wastenson, 1990–96) published over a seven-year period. This massive cartographic work was designed in seven different departments of geography, four government agencies and three specialist firms. A similar multi-department project is illustrated by the several fine volumes of the *Historical Atlas of Canada* (e.g. Gentilcore, 1993).

The Continuing Challenge of University Cartography

Despite the half-century of rapid change, in some ways the role of the cartographer remains surprisingly unchanged.

I'm concerned that the digital revolution in cartography, for all its huge advantages, may see it merely as part of some CAD-CAM continuum and put at risk the very special skills which are imbedded within map making.

So my final comments to the Society are threefold:

First, be mindful of the great history of your discipline. I'm not sure if the early Sumerian empire had universities in 500 B.C. but it certainly had cartographers and the British Museum has today clay tiles which are among the earliest surviving examples of the cartographer's art. I urge all of today practitioners to turn the pages of at least some of the volumes of the great History of Cartography (Harley and Woodward, 1987–), another fine legacy of the Wisconsin cartographic tradition.

Second, recall that map-making is only one half of the full mapping cycle. Figure 2 shows that cycle as a two-stage process. In the first, map encoding, the cartographers task is to convert information about real world into a map-like form. In the second, map decoding, the user of that map has to use it to make inferences from the map about what the real world may be like. In an ideal world (left) the cycle would be perfect and complete with the inferred world a perfect replica of the real world. But in practice (right) we know that maps are full of errors both of encoding and decoding. As Monmonier (1996) has argued, the cartographers age-old task is to reduce the closing errors.

Third, remember that maps should have beauty as well as utility. Do, within the framework of the new digital revolution, make your own maps in your own image and develop a distinctive and recognizable cartographic style.

Thank you, again, for inviting me to address this fortieth Summer meeting of the Society. I thank you and your colleagues for supporting so much of the work we academics do and for providing me with this personal opportunity to say thank you for providing me with an essential service without which my own research and teaching within academic geography would have been utterly impossible.

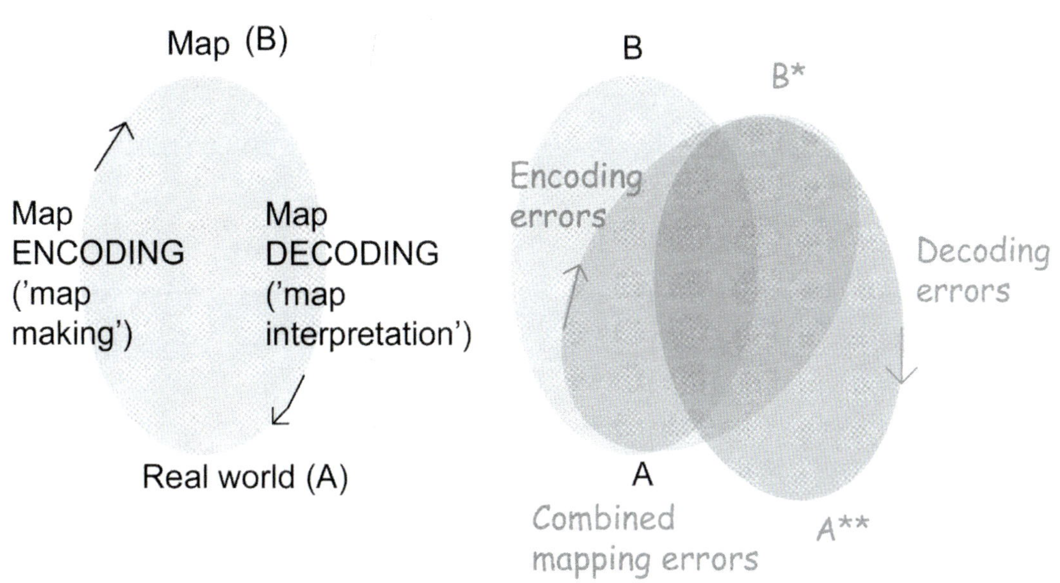

Figure 2 *The mapping cycle. (left) Idealized cycle showing the twin processes of map encoding and map decoding. (right) Realistic cycle with encoding errors (B to B*) and compounded decoding errors (A to A*).*

Notes

1. I have in mind one of the maps hanging on the wall of my study Peter Plancius's map of the Molluccas and the East Indies, printed in Holland around 1598.
2. The Cambridge list is longest because I continued to do joint research work there with Professor Andrew Cliff in the period 1972 to present.

References

Cliff, A.D., and Haggett, P. (1988) *Atlas of Disease Distributions* Oxford: Blackwell Reference.

Cliff, A.D., Haggett, P., and Smallman-Raynor, M. (2004) *World Atlas of Epidemic Diseases* London: Edward Arnold.

Conway, C.M., McManus, G.E. and Wood, C.H. (1998) *A Guide to Map Design and Production using CorelDRAW 7 and CorelDRAW 8* St John's: Memorial University of Newfoundland Cartographic Laboratory.

Gentilcore, R.L. (Ed.) (1993) *Historical Atlas of Canada, Volume II: The Land Transformed, 1800–1891* Toronto: University of Toronto Press.

Gould, P. (1999) *Becoming a Geographer* Syracuse: Syracuse University Press.

Haggett, P. (1995) *75 Years of Bristol Geography, 1920–1995* Bristol: Department of Geography, University of Bristol.

Harley, J.B., and Woodward, D. (Eds) (1987–) *The History of Cartography, Volume 1 –* Chicago: Chicago University Press.

Hayes, R.D. (1956) "A peasant economy in north-west Portugal" *The Geographical Journal* 122 (1) pp.54–70.

Jervis, W.W. (1936) *The World in Maps: A Study in Map Evolution* London: George Philip.
Monkhouse, F.J. and Wilkinson, H.H. (1952) *Maps and Diagrams: Their Compilation and Construction* London: Methuen.
Monmonier, M. (1996) *How to Lie with Maps* (2nd ed.) Chicago: The University of Chicago Press.
Robinson, A.H. (1953) *Elements of Cartography* New York: Wiley.
Robinson, A.H., Morrison, J.L., Muehrcke, P.C., Kimerling, A.J. and Guptill, S.C. (1995) *Elements of Cartography* (6th ed.) New York: Wiley.
Steers, J.A. (1949) *An Introduction to the Study of Map Projections* (7th ed.) London: University of London Press.
Wastenson, L. (Ed.) (1990–1996) *The National Atlas of Sweden* (17 vols) Stockholm: Almqvist & Wiksell.

NEXT-GENERATION WEBMAPPING

Richard Fairhurst

Originally published in Volume 39, Nos 1 & 2, pp.57–61

The last year has seen rapid advances in standards of map presentation over the Internet. In February 2005, Google – the market-leading search engine and perhaps the web's No. 1 brand – introduced a mapping service that was significantly easier and more enjoyable to use than the market leaders. (In the UK, these were principally *Multimap.com* and *Streetmap.co.uk*; MapQuest is popular in the US, ViaMichelin in Europe, and so on.) With its fluid movements, intuitive user experience and competent cartography, Google Maps has dramatically raised users' expectations. The high cost of industry- standard GIS software might lead webmappers – whether they be cartographers, programmers, or graphic designers – to think that the only way to match Google's achievement is by matching Google's bank balance. But this is far from the case. This article recaps the current state of the webmapping market, and describes how a system roughly comparable with Google Maps was developed on a very limited budget.

First-generation Webmapping

The first webmapping services to gain prominence in the UK were those offered by *Streetmap.co.uk* and *Multimap.com*. These are still the most popular mapping sites with the general consumer.

In essence, these 'first-generation' sites reproduce paper mapping on-screen, using the very simplest web browser technology. Even today, *Streetmap.co.uk* uses very little technology introduced after the Netscape 2 browser, ten years ago.

Typing a placename or postcode into either site will bring up a page of mapping centred on that page. To move around the map, or view a different scale, you click a navigation or scale control: this loads a whole new page with the desired map. This method of control is simple, but slow.

By using familiar map imagery – such as the Ordnance Survey's 'Landranger' 1:50,000 range, Collins Bartholomew's road atlas products, and various street plans – these sites instantly appealed to people who had previously only used traditional paper maps.

Their business models were also strong enough to survive the dot-com crash: essentially, they balance advertising revenue (all map pages are liberally festooned with banner ads) against the licence fees charged by the map suppliers. *Multimap.com* has also generated significant revenues from business services – selling its webmapping expertise to other sites. A well-known example of this is its 'store locator' service for High Street chains.

Of the two principal UK sites, *Streetmap.co.uk* offers the simpler user experience, whereas *Multimap.com* is considerably more refined and includes some extra features (such as superimposed icons). Ordnance Survey has a similar site, Get-a-Map, which benefits from the OS brand name and the most individual graphic design of the three. As befits 'the best mapped country in the world', Britain's first-generation mapping websites are arguably a cut above the equivalents in other countries – notably the US MapQuest, owned by Internet behemoth AOL but with maps that look remarkably crude to those brought up on OS paper mapping.

Second-generation Webmapping

Nonetheless, a market-leader in 2000 could not naturally expect to retain its position five years later. Personal computing is a fast-moving industry, and web technology, perhaps, the fastest-moving part of it.

Even for those applications (such as word processing) where people's demands have not changed significantly, today's versions offer much more 'polish' than those of ten years ago – compare, for example, Microsoft Word 2003 with a version from the mid–1990s. Expectations rise, and like any other application, webmapping sites cannot afford to fall behind.

The first site to significantly break away from the first-generation user experience was *Map24.com*. Map24 uses Sun's Java technology, a whole computing platform which can run within the confines of your browser window. This gives the webmap developer much more flexibility than was offered by the simple HTML used by first-generation sites, which – in its original incarnation – could do little more than display static images and text.

With this flexibility, Map24 became the first site to present a truly web-only experience, rather than simply reproducing paper maps. When you navigate around Europe on Map24, the map literally does zoom in, with the scale gradually increasing as you zoom from an all-Europe map to a close-up of your local area. Panning around the map no longer requires a new page for every single new view: instead, the map is redrawn seamlessly as you move.

This impressive, all-European site nonetheless has failed to gain much headway in Britain. This can partly be attributed to unfamiliar cartography. There are some stability and installation issues with the Java platform, and Map24's own user interface is not as simple as it could be. But the main reason the site has not caught on widely is that the brand is not familiar to UK consumers.

That was never going to be a problem for another second-generation mapping site: Google Maps (*maps.google.co.uk*).

In the web business, Google is the archetypal 800 lb gorilla – a roughly equivalent position to that held by Microsoft in the desktop market. Any start-up web company seeking to compete on Google's turf would be lucky to win funding from even the bravest venture capitalist. Most computer industry observers, however, believe that Google's dominance has been achieved by the quality of its product – not just by its marketing muscle and business savvy. Any Google mapping product would be met with high expectations. For the most part, Google Maps has fulfilled them.

For the consumer, Google's webmaps have three main distinguishing features. The first is their convincing integration with search data and (as with some other mapping sites) a motorists' route-planner; this is outside the scope of this article. The second is their class-leading ease-of-use. And the third is their cartography – perhaps best described as 'GIS with a human face'.

Ease-of-use is achieved through an uncluttered screen layout, as it is with many other Google sites. The user can move around the map simply by holding down the mouse button and dragging: new sections of the map are loaded when needed, without having to reload the entire page. A simple zoom control enables the user to switch between different scales, though the single-click 'recentre and zoom' action popularized by Multimap is sorely missed.

The maps themselves are drawn by Google, based on mapping data sourced from such suppliers as Navteq and Tele-Atlas. Clearly, they are maps generated by a GIS with minimal human intervention: if nothing else, this was forced by the vast reach of the mapping (a US launch of Google Maps was followed swiftly by the UK site, and more countries are being added).

But that said, Google's designers have achieved a much more attractive result than that shown in previous GIS-derived webmaps such as MapQuest's. Stateside webmappers have even bestowed the word 'cartography' on Google Maps – though cartographers may have a different opinion. By producing its own maps, Google has avoided the patchwork quilt of styles that characterized first-generation sites, where you might switch from an OS map at one scale to a Bartholomew at another.

The maps, however, are defined by the restrictions of the GIS data and by the rules that Google has applied to it to create maps. The style is clearly tilted more to street mapping than countryside leisure: there is no option to choose between a street atlas and something more akin to OS Explorer/Harveys, though the addition of aerial photography has ameliorated this slightly. There are also a few questionable artistic decisions, such as the rounded line caps which can give the illusion of street intersections where none exist.

For the technically inclined, there are two further reasons to commend Google Maps. One is the innovative way in which it stretches standard web browsers to achieve such a seamless, easy-to-use user experience. No Java plugin is required: instead, the site uses the standard-issue JavaScript language (unrelated, despite the similar name). In effect, when you use Google Maps, a small program is running in your web browser, sending requests to Google's servers for the mapping you request, and changing the display accordingly. The success of Google Maps has popularized this principle (which programmers call 'AJAX') for all sorts of websites.

Google Maps have also endeared themselves to web developers by allowing connections from other sites. These are not just simple links ('Click here to see where we are'): rather, Google actually allows you to use its maps as backdrops for your own location information. If you had created a website listing (say) the locations of historic buildings at risk throughout Britain, an hour's programming would enable you to add Google's maps to your own site, with clickable pins in the map to show each building. (This aspect, nicknamed 'Google Maps mash-ups', was further explored in other talks at the SoC Summer School.)

And yet... although Google Maps are now the webmaps of choice for the technologically literate, anecdotal evidence suggests that casual users, in the UK at least, are proving slower to migrate from the first-generation sites.

Building a Second-generation Webmap

Waterscape.com was launched as British Waterways' leisure website in June 2003. I worked as its editor from just before site launch until September 2005.

The first version of the site was built by a London-based IT consultancy using industry-standard content management software, plus ESRI's market-leading ArcIMS software to supply the webmapping. But in the time-honoured way of public sector IT projects, there was significant 'disconnect' between the software, the site design, the subject matter – Britain's canals and rivers – and, not least, the mapping. Though the maps had been specified by an immensely talented team with a track record of great waterway cartography, the disconnect with the web technology meant that good intentions failed to translate into a winning site.

The ESRI software was expensive to purchase and costly to run, requiring a separate server from the main website. Significantly, this meant that a promised tie-up between the map pages and the rest of the site didn't happen. Our waterside pubs and recommended walks never appeared on the maps, which were (badly-scaled) Ordnance Survey basemaps with an added blue line to show navigable waterways.

Similar disenchantment with the rest of the site resulted in a decision in late 2004 to scrap the software used in the original build. It would be replaced by a system designed and built entirely in-house. We took the decision that the new site would have geography at its heart – every single attraction, every waterway, and every facility would be mapped. But more than this, the site had to be enjoyable to use… – something that would encourage and inspire users to visit their local waterway. We couldn't persuade people to discover their local canal if it took half an hour to find it on the website, never mind on the ground.

In short, it had to be a second-generation design. High-quality countryside mapping – which, for UK-wide coverage, essentially means Ordnance Survey's 1:50,000 and 1:25,000 products – would be augmented with highlighted waterway routes, clickable icons, locks and bridges, and presented in an attractive, intuitive format.

The decision that transformed the project was when we decided to use Flash to display the maps. Flash, a web animation system developed by Macromedia, has a slightly unfortunate reputation. Widely associated with the phrase 'Skip intro', it has long provided an opportunity for graphic designers to indulge their worst excesses at the cost of usability.

But when used properly, it can deliver smooth, reliable web applications quickly and cheaply. It is almost universal (over 98 per cent of Internet users have Flash), reliable (as a mature technology developed by one company alone, it is harder to crash than Java), and – unlike the JavaScript so skilfully exploited by Google Maps' expert coders – will work consistently on all browsers and systems without hours of fine-tuning. Unlike standard HTML, it can cope with both vectors (scalable lines, such as you would draw in Illustrator) and raster images (a collection of pixels to form a picture, as in Photoshop). It was this that allowed us to superimpose waterway lines, drawn as vectors, onto the OS raster base-mapping.

Waterscape's mapping works on the same principle as other second-generation sites. The 'client' program runs on the browser: Flash for Waterscape, JavaScript for Google Maps, Java for Map24. This provides a series of buttons and other controls to the user. When one is clicked, the client sends a message to the webserver, saying what button has been clicked. The server returns the new map data to the client, which the client then displays on-screen.

The Waterscape maps provide the same click-and- drag panning capability as Google Maps. Like Map24, they offer 'live' zooming – when you move from one scale to another, there is a gentle transition rather than an abrupt redrawing. Check-boxes enable users to toggle layers such as visitor moorings and angling information.

The new site launched, on time and to (minimal) budget, in June 2005. The end result was comparable to other second-generation webmaps, and a great advance over the original site. (You can see an example at *www.waterscape.com/Oxford_Canal/map.*)

The Open-source Difference

But in contrast to the seven-figure cost of the original, externally-developed Waterscape site, the new mapping system was achieved with virtually no outlay. Only the base maps (from the Ordnance Survey and Bartholomew) were paid-for, with a combined cost of £5,500 p.a. The technology was all free.

Of course, that's an over-simplification. We didn't have to buy in any third-party software, but we did have to spend time developing the new system. (As I was employed as 'editor', rather than cartographer or developer, I actually carried out most of the work in the evenings over a three-month period.) Nevertheless, the total cost was a fraction of the original.

To achieve this, we used free software – programs developed by enthusiasts for no financial reward. The main website was built using a database called MySQL, a webserver called Apache, and a web programming language called PHP. The names might not mean much to cartographers, but they are the most common of their type: millions of developers around the world are skilled in this combination.

The piece of software magic used to deliver the Flash maps is called Ming (readers of a certain age will recognize the Flash Gordon reference). Completely free, Ming lets you develop Flash software without the involvement of any commercial Macromedia product. Used in conjunction with PHP and MySQL (or similar), it can draw information from a database, so all of our waterways and icons could instantly be superimposed on the basemaps.

With these building blocks, we were able to create a Google Maps-like site using just 600 lines of ActionScript (Flash's in-built programming language) and 300 lines of PHP, which for a hardened programmer is very little indeed. And there was nothing special about what we did – these tools are available to anyone. You just need a programmer for a few weeks, either as a volunteer for a community project, or paid as part of a commercial project.

Compared with standard GIS packages, the cost remains low, and the results impressive. The success of this approach can be measured by the fact that one of Britain's very biggest websites is in discussions with Waterscape about the possibility of using this technology. But, to reiterate an earlier point, you don't need to be as big as Google or even Multimap. Such a site is within the range of a hobbyist or small business.

The Base Mapping Problem

Actually, when I said 'you just need a programmer for a few weeks', I wasn't being completely honest. Unless you're creating a completely new map from scratch, you need a base map, or at least a set of geodata to feed into your own GIS. At the present time, in Britain, such maps and data are only available from commercial suppliers. This principally means the Ordnance Survey, but other suppliers have carved out niches at particular scales.

This is the single remaining barrier to increased innovation in UK webmapping. Much technical innovation on the web is now fuelled by the open source movement, which runs purely on people's enthusiasm and ingenuity, rather than financial imperatives. But these 'no- budget' websites cannot afford to buy mapping from commercial providers, even at the comparatively low rates which (say) Collins Bartholomew charges for Internet use of their large-scale raster maps.

This is perhaps why programmers have seized on the ability to connect their sites to Google Maps. Google is offering free base maps, within certain confines, to web developers. This is currently the only way to get UK base mapping for web use without paying a lot of money.

Though this provides a quick-and-easy mapping solution for simple websites, it has many limitations – legal, technical and artistic. You can do little more than add pins to Google's own mapping. Google's Terms of Service rule out many possible uses. Recent discussion on the mailing list for the OpenStreetMap project has highlighted a further drawback: any geographical data which your users create by reference to Google's maps ('click on the map to tell us where you are') is, legally, a 'derived work' from the original Navteq/Tele-Atlas geodata, severely restricting future uses.

A number of community projects, principally Steve Coast's OpenStreetMap, are beginning to create free geodata – data which could be used to create this much-needed free base map. GPS tracklogs can be combined with large-scale satellite imagery, traces from out-of-copyright maps, and personal research to slowly build national coverage. Within two or three years, such projects are likely to have achieved a free equivalent of Bartholomew's 1:400,000 raster data; a gazetteer of place-names; streetmaps of central London, Bristol, Cambridge, and a few other cities; and the beginnings of a postcode database. This is good news – but too far in the future.

Few would expect or want the Ordnance Survey's whole Mastermap database, or the entire 1:25,000 'Explorer' map set, to be made freely available tomorrow. But if Britain's reputation as the best-mapped country in the world is to be carried through into the web age, we need a 'starter kit' of basic map data to be made available free. At the least, this should include large-scale coverage of the whole nation, and a low-resolution postcode database.

Third-generation Webmapping

The pace of innovation on the web is such that, even today, mapping technologies are being created which are light-years ahead of the second-generation sites. Webmaps are moving further away from reproductions of paper maps, and into completely new representations of world geography.

The best example is another Google product: Google Earth. This combines 'fly-throughs' with a Digital Elevation Model and heavy use of aerial photography to produce a unique experience, blurring the boundaries between webmapping and computer gaming. You can fly around a relief map of the world, diving into Himalayan crannies one minute, zooming in on New York city centre the next.

Once again, Google Earth has thus far been celebrated by the technorati but relatively little noticed by the general public. We can, however, expect it to slowly gain greater currency, and for its concepts and approaches to be adopted by other webmaps.

Already, Microsoft is frantically playing catch-up with Google with its own product, MSN Virtual Earth (*virtualearth.msn.com*). ESRI has announced a software solution (ArcGIS Explorer, *www.esri.com*) 'for download in 2006'; NASA, no less, has released and open-sourced a similar program called World Wind (*worldwind.arc.nasa.gov*); another open source program, WW2D (*ww2d.berlios.de*), currently offers only 2D views but can doubtless be expected to spawn a 3D version before long. As this article was finalized, long-established Internet portal Yahoo! stepped into the second generation with a service using exactly the same Flash/PHP combination as Waterscape.com (*maps.yahoo.com/beta*); perhaps they, too, will launch a third-generation product.

The pace of development is dizzying given that such real world visualization tools were barely available to the consumer a year ago. It remains to be seen whether corporate-focused developers like ESRI and Microsoft, who speak through the medium of product roadmaps and like to devise 'solutions', can convincingly ape the 'agile' ethos of Google and the open source movement.

Meanwhile, electronic mapping techniques are starting to find a home outside the web browser. The impressive graphics of Google Earth have perhaps pointed the way for the use of geographic information in computer games – imagine car racing games or flight simulations with a limitless playing area. Away from the desktop, the past year has also seen an explosion in the popularity of in-car GPS navigation systems, with maps which are updated in real time and which are centred around the user's real-world location.

Location-aware mobile phones can be expected to continue the trend – the mobile phone mapping market is still young and undeveloped. As for updates in real time, few webmappers have yet explored the possibilities of animated maps. Consider, for example, how the real-time train location information offered by the nationalrail.co.uk website could be combined with a base map to show trains moving around the country.

It would be a mistake to think that innovation in webmapping is restricted to producing ever faster, more colourful mapping products. One unusual aspect of the new Waterscape mapping system is the 'dynamic PDF' – a canal guidebook in PDF format, assembled and typeset on the user's request and to their specification. Using the same data that was shown on the Flash webmaps, these guides contain regularly updated navigational information, safety advice, and listings of facilities.

Because they are produced 'on the fly', a lock closure announced at 10am one day will be included in a guide downloaded just a minute later. Reassuringly, the greatest challenge in developing these guides was not technical but cartographical – the old chestnut of label placement.

So where does the cartographer fit into all of this?

Without doubt, webmapping services mean that maps are now being consulted more often than ever before – which must be good news for cartographers. The positive reaction to Google Maps demonstrates that well-drawn maps are still valued, and that cartographers who can combine an understanding of the 'art of maps' with the 'science of GIS' will become increasingly prized. Though the methods of delivery are changing, an attractive and accurate map is still as important as ever.

OPENSTREETMAP

Steve Coast

Originally published in Volume 39, Nos 1 & 2, p.62

As all cartographers know, mapping data in the U.K. is produced and maintained by the government-funded Ordnance Survey, and is subject to Crown Copyright. This means it is not only expensive, but that derivative works and reproductions are restricted. The same is true for maps produced by mapping agencies throughout most of Europe and the rest of the world. One notable exception to this is in the U.S. where the government mapping agency must make its data available for anyone to use, free of charge. This leaves individuals and organizations in the U.S. free to make use of maps in their own work without the express permission of the producers of those maps.

OpenStreetMap is an online collaborative mapping project initiated in London by Steve Coast and run by a collection of volunteers from the mapping and open source software communities. Using a model similar to that of Wikipedia (an encyclopedia which can be contributed to and edited by anyone on the internet), OpenStreetMap allows anyone with an internet connection to produce and edit maps. Rather than use traditional and expensive survey methods, the source data for OpenStreetMap is user-submitted GPS tracks (collected from inexpensive personal and in-car satellite navigation systems), out-of-copyright maps and aerial photography. Users trace paths and annotate street names using an online interface, and the resulting maps are free for use by anyone.

The quality of commercial maps is undeniably high (though some maps contain deliberate errors [trap streets] designed to catch people who copy them without paying), but through the bottom-up process employed by OpenStreetMap several advantages are gained. For example, using GPS data it is possible to annotate roads with information about average speeds for different times of day or year.

From just one generous partner, a London-based online courier firm, OpenStreetMap volunteers have demonstrated just how rich aggregate GPS data can be. All of the courier's staff carry GPS devices to enable their customers to accurately track the progress of their packages, and as a by product of this data from just one week, a picture of streets in London has been built reflecting the most popular routes taken by the couriers. Clearly, this information can be used to better inform the courier's automatic routing algorithms – a key component of their business – whilst simultaneously providing London with free and timely mapping data.

Recently, OpenStreetMap put together a large poster using it's London data to raise money for the project. This is shown in the figure. It can clearly be seen that more travelled routes are thicker than those seldom taken. OpenStreetMap is currently scheduled for a version 1.0 release by February 2006, and is seeking further contributors of raw data and software expertise. Potential consumers and producers of free maps and location-based data are also invited to get in touch, especially cartographers!

GEODATA COLLECTION IN THE 21ST CENTURY

Nick Black and Steve Coast

Originally published in Volume 41, Nos 1 & 2, pp.3–8

OpenStreetMap is a community-driven project that makes maps and gives them away for free. Using cheap, consumer grade GPS units and software tools written by the community, contributors across the world are creating high-quality geo-datasets. The project and its associated dataset would not exist if not for the community of volunteers who write software, design maps, and collect data. There are numerous similarities between the Open Mapping community and the Open Software community, with members of both communities sharing motivations. 'Community' is a favourite buzzword of Web2.0 and OpenStreetMap is a leading example of a Web2.0 community – one that collaboratively produces a useful, valuable dataset. The OpenStreetMap community is an extremely social one, holding regular meet-ups and mapping parties, during which volunteers meet to intensively map a single town, city or region in one weekend. The community is underpinned by a sense of collective ownership in the dataset and by a passion for creating and using geodata.

1 Introduction

OpenStreetMap (OSM)[1] is a community-driven project that is mapping the world. Founded in August 2004, OSM now has over 20,000 members spread across the world, who walk, drive, sail and cycle the roads, railroads, rivers, and footpaths of the world. Armed with consumer-grade GPS units, OSM volunteers collect geodata in the form of GPS traces, photographs, and notes, which they use to create maps that are then given away for free. This paper looks briefly at some facets of OpenStreetMap; how the data is collected; what the economic implications of freely available geodata are; and then looks at the OSM community itself – we look at communities; how OSM galvanizes its community; and explore the challenges of community-building. Finally, we look briefly at some future directions of OpenStreetMap and open geodata.

2 OpenStreetMap for Beginners

Creating geodata with OpenStreetMap is not difficult. Anyone who has access to a computer connected to the Internet can contribute in some way. The most common way people contribute data to OSM is by collecting GPS traces. GPS receivers can be bought for as little as £60 and most can be used to record traces. Traces are formed from a series of points, which have a latitude, a longitude and a time stamp. If a person were to set their GPS unit to record traces and then cycle round a city, their trace might look the trace shown in Figure 1. If two people were to jointly cycle round a city, concentrating on different areas, their traces might look like those shown in Figure 2. If several people were to record their movements around a city, their traces would probably look like those shown in Figure 3, in which you can see the beginning of a map which shows

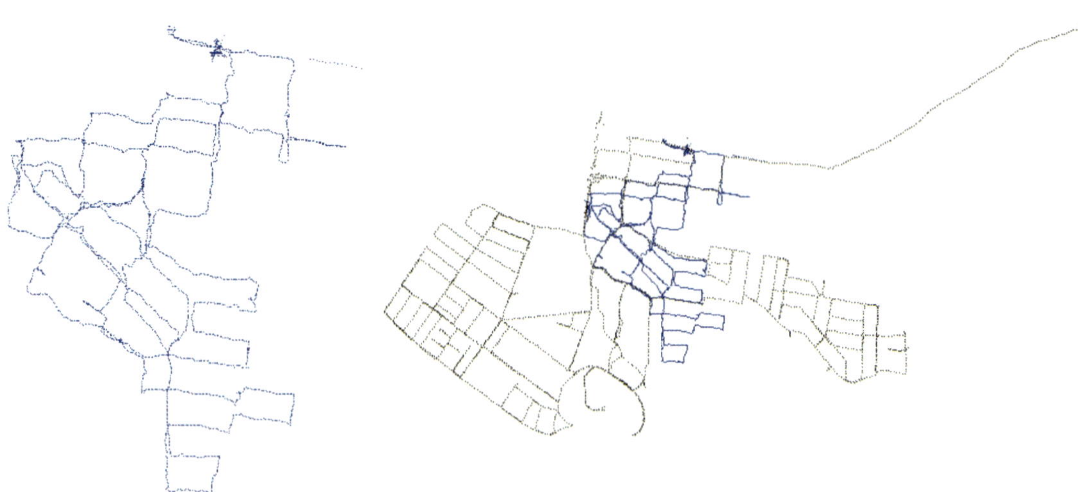

Figure 1
One person's GPS Trace

Figure 2
A Second person's GPS Trace

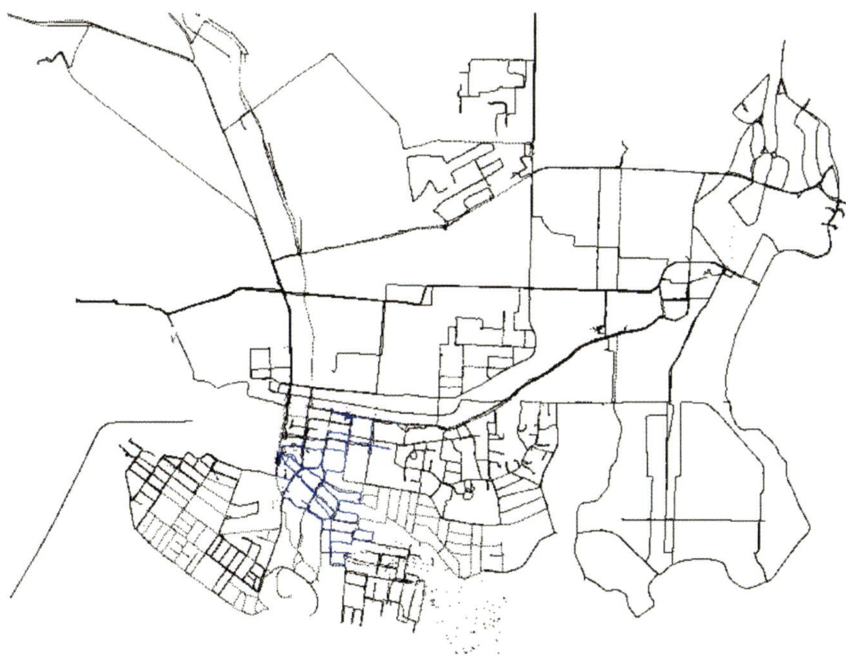

Figure 3
Combined traces from OpenStreetMap

roads and city blocks. You can also see that some of the lines are thicker than others. GPS receivers are not perfectly accurate and if you were to travel down the same section of road with a GPS receiver every day for a year, you would see a slightly different trace on each day. Lines appear to be thicker on the more travelled routes in Figure 3 because of these inaccuracies. Random errors like these are not a major problem for OpenStreetMap, whose volunteers attempt to draw roads along the thickest section of the line, thereby estimating the average of the combined GPS traces of several people.

Whilst the map in Figure 3 is a good start, it is far from complete. There are no attributes, so it isn't possible to tell which street is a primary road and which is a service road, or the name of a street. In short, most of the features that we associate with a useful map are missing in from our early iteration. In the same way that several people can jointly collect GPS traces, they could also add useful information, such as street vectors with attributes, or points of interest or land-use areas to a map. Figure 4 shows the same set of GPS traces inside an editing program called JOSM. JOSM is free software, licensed under the terms of the GNU General Public License, which anyone can download and use to create and edit OpenStreetMap data. Using JOSM or another editing tool, you can start to draw lines over the top of GPS traces.

Lines are known as 'ways' in OpenStreetMap parlance. To make ways useful they can be given attributes, such as 'motorway' or 'one way' or 'Douglas Street'. Because OpenStreetMap is a global project that seeks to collect any

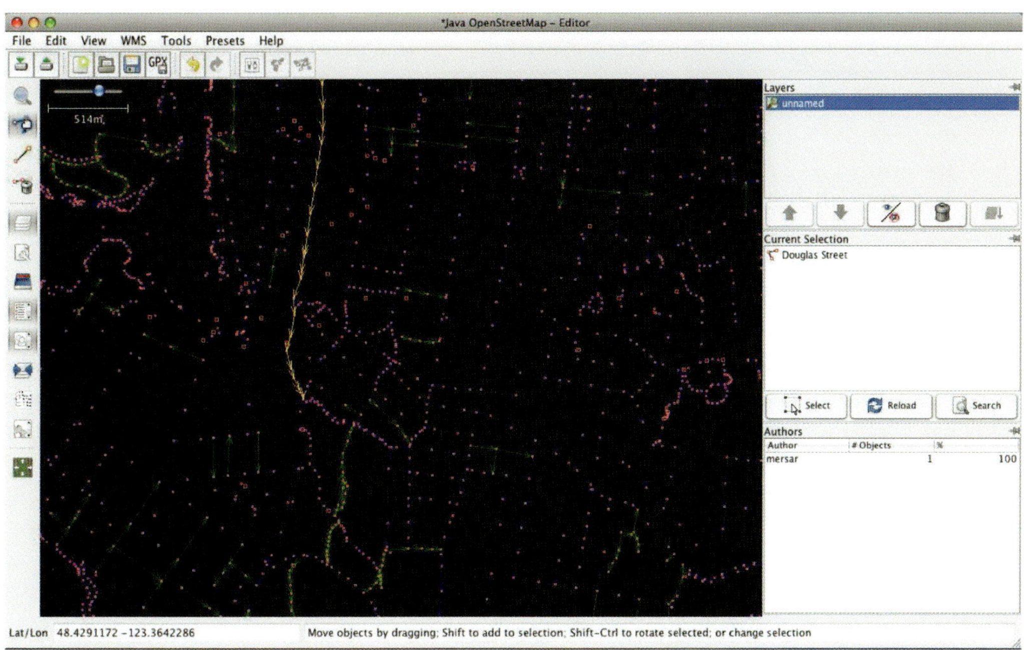

Figure 4
The city of Victoria, BC, Canada viewed in JOSM

form of geodata (not just streets), and because it would be impossible for one person to create an ontology that categorizes every type of geographic feature that the world will ever see, OpenStreetMap uses a system of user-defined tags which have a key and a value. A key might be a word like ‘highway’ and a value might be a word like ‘primary’. Putting the two together allows you to describe the properties of a primary road. Adding the key ‘name’ and the value ‘Douglas Street’ adds further meaning to the feature you are adding. Whilst any OpenStreetMap contributor can add any tag, there are conventions, some local and some global, that are adopted by most contributors. This means that there is a reasonably high level of consistency throughout the OpenStreetMap dataset whilst allowing new contributors to get started very quickly (without first having to define their own set of tags). Significantly, the key/value tagging system allows for new or previously unknown features to be quickly integrated into the tagging system.

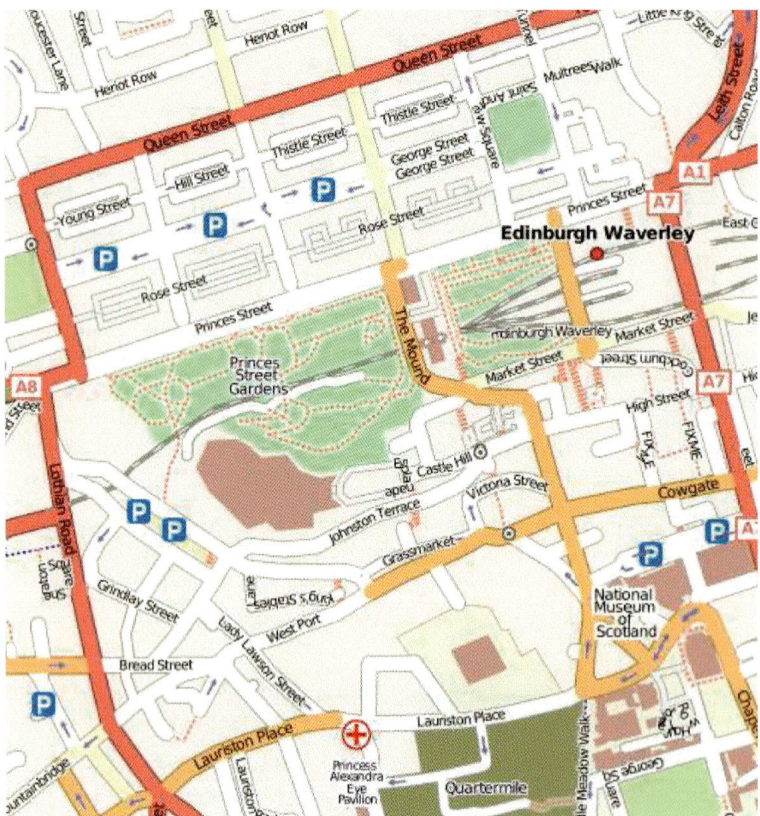

Figure 5
Central Edinburgh, Scotland

After a few hours of careful editing, a set of GPS traces can be turned into useful, attribute-rich geodata sitting in the OpenStreetMap database. Anyone with a connection to the Internet can get hold of OpenStreetMap data through the project’s Application Programming Interface (API). The OpenStreetMap API allows other computer programs to access and modify OSM data and by using the API it is possible to create maps like the one seen on the project’s homepage, or that shown in Figure 5 above. Using cheap consumer-grade GPS units combined with free software, it is possible to collaboratively produce a high quality dataset that can then be used to create aesthetically pleasing maps. All of this can be achieved at low monetary cost because the software is developed, and the data collected, by volunteers.

3 Economics of Commons-Based Peer-Production

3.1 Coase’s Penguin

The production methods used by OpenStreetMap, whereby a group of volunteers collaboratively produce a work which they then give away for free, are analogous to those employed in the development of free software, such as the Linux operating system. Yale economist, Yochai Benkler investigated such production methods (which he terms ‘commons-based peer-production’) in his paper Coase’s Penguin.[2] Benkler examines the motivations of the people who contribute to projects like Linux or OpenStreetMap and concludes that there are three key elements to such projects that facilitate commons based peer production:

- Non-monetary motivation;
- Discrete multi-size pieces; and
- Low-cost integration.

In the context of OpenStreetMap, Benkler’s observations require that contributing geodata be:

- Fun;
- Quick; and
- Easy.

These are the goals that OpenStreetMap strives to attain, and the more fun, quicker, and easier it becomes to add data to OpenStreetMap, the more contributors will join the project. The data collection processes described above illustrate these three points and many of the activities surrounding OpenStreetMap function to make contributing to the project more fun, faster or easier; a shining example of such activity being mapping parties.

4 The Free Mapping Community

4.1 Mapping Parties (Fun with a Map)

The first mapping party was held in May 2006 on the Isle of Wight. Around 30 OpenStreetMap volunteers travelled to the small island which lies off the south coast of England from across Europe, with the goal of mapping as many of the island’s roads and footpaths as possible in one weekend. The exercise was an enormous success; not only was a comprehensive geographic dataset (containing roads, footpaths, points of interest and land use) created and made available to anyone, but the concept of a ‘mapping party’ was born. More mapping parties followed the Isle of Wight and now there are parties every few weeks in the UK and Europe, as well as further afield. To date there have been mapping parties in Germany, The Netherlands, Spain, Italy, England, Scotland, Republic of Ireland, South Africa, Canada, Sweden, Isle of Man, Slovenia, and Norway.

Mapping parties play several important roles for the OpenStreetMap community. First, they offer an opportunity to map a single area intensively, facilitating the collection of geodata using the methods described above. They also act to publicize OpenStreetMap's efforts. The first mapping party attracted the attention of the local press and the Isle of Wight's Member of Parliament and subsequent mapping parties have often attracted similar media attention. Because contributing data to OpenStreetMap is easy and open to anyone, mapping parties offer the perfect opportunity for a newcomer to the project to meet some more experienced mappers and learn the tricks of the trade. A goal of a mapping party will be to convert at least a few local people; to get them hooked enough on OpenStreetMap that they will finish-off bits of mapping that may have been missed during the party and also to maintain the dataset. Again, the Isle of Wight party set the tone by attracting the support of Isle of Wight locals like David Groom. After helping with the initial mapping effort, David went on to be instrumental in completing the Isle of Wight dataset and has since mapped vast areas of Baghdad.

As anyone who works with geodata knows, currency is a major issue. Out-of-date geodata will usually have less utility than current geodata. For what other mapping companies describe as 'change intelligence', OpenStreetMap relies on local people. After all, no one knows more about a place than the people who live there. Many online communities try to bridge the divide between the virtual and real world with meet-ups and other social events. The nature of OpenStreetMap's activities means that real-world community activities like mapping parties are a core part of the project, rather than an activity undertaken solely to galvanize a community. The community-building effects of mapping parties, however, are key to the success of OpenStreetMap. Again, the Isle of Wight weekend set the tone and mapping parties are as much about allowing contributors to meet, talk, argue and debate as they are about collecting geodata.

4.2 What is a Community Anyway?

The buzz of Web2.0 has a lot to do with communities and more and more commercial organizations are starting to understand the benefits a community might bring them. But what is the OpenStreetMap community? On the most simplistic level, it is a group of people who want to make a free map. If you look in more detail at their motivations, the situation becomes more complex. The OpenStreetMap community consists of:

- People who want everything to be free;
- People who love maps;
- People who dislike Ordnance Survey;
- People who want to make maps;
- People who want to use data; and
- People who want to write software, and so on.

If the goal is making a free map – an activity that requires a certain level of cooperation – how can you get all of these people to do something coherent? Benkler's 'non- monetary motivation', which we paraphrase as 'fun', holds the key to successfully galvanizing a community, along with three other factors that we have identified as being paramount in the success of OpenStreetMap:

- Giving people ownership;
- Creating a short feedback loop; and
- Allowing especially talented people to do amazing things.

4.3 It Really is Your Map

Getting you to care about my project, my family or my map is quite difficult. Governments typically enforce such social cohesion through taxation and coercion. Other forms of motivation (financial reward, coercive threat, or ideological) are all extremely powerful forces, but each requires a large infrastructure to administer. Creating a free mapping cult may well work, but could take hundreds of years to gather enough momentum to actually make a map and would isolate so many people in justifying its existence that it would be highly ineffective. Coercive force is largely outside of the realms of possibility for all but governments, and as for money, if you pay peanuts, you get monkeys. Getting people to care about their project, their family or their map is far more powerful. The reason that we, as OpenStreetMap contributors, will go out mapping on a rainy Sunday afternoon in December is because it is our project. If I want to add a new road, I can. If I want to define a new category of monument, I can. I then get the satisfaction of seeing my efforts translate into a beautiful map and a freely available data set, which other people can do even more interesting things with. By contributing to OpenStreetMap, my efforts become part of a much larger resource and it is precisely because I feel such ownership over that resource that I continue to contribute to it and maintain it.

In order to maintain this situation, OpenStreetMap exists alongside the OpenStreetMap Foundation (OSMF): an international non-profit organization dedicated to encouraging the growth, development, and distribution of free geo-spatial data and to providing geo-spatial data for anybody to use and share.[3] The OSMF is a trusted, democratically elected, accountable body who own OSM's physical assets (such as servers) and act as conduit for financial transactions. As a UK Limited Company, the OSMF also provides a degree of protection against the threat of legal actions. Before the Foundation was registered, any legal action against OpenStreetMap, such as any alleged infringement of copyright, would have been the liability of the project's founder. The Foundation also serves as a vehicle for fundraising. It has a bank account and achieves a certain level of transparency by publishing its accounts and the minutes of its meetings each year. Finally, membership of the Foundation is open to all. The OSMF adds to the legitimacy of OpenStreetMap by placing important decisions in the hands of elected, accountable members of the community. Similarly, the Creative Commons license,[4] under which all of OpenStreetMap's data is released, enhances both the project's legitimacy and the trust of OpenStreetMap contributors.

4.4 I Want to See My Changes Now

With these guarantees in place, OpenStreetMap contributors feel an enormous sense of ownership in the project. This is reflected in the demands for a short feedback loop. The lag time between making an edit to the map data and the change being reflected on the map must be as short as possible. In November 2006, openstreetmap.org first featured Mapnik[5] tiles. The main map had previously consisted of white lines with road names drawn over Landsat tiles. The new tiles were undoubtedly better than the old ones, but they were less current. Whereas the old, Landsat tiles would be updated a few hours after editing, the Mapnik tiles could take more than a week to be updated. This is not acceptable to many contributors who wanted to see the effects of their hard work immediately. Whilst this might be slightly irritating for the people who write the code to put rendering and the slippy map in place, it shows that the community is working. If contributors didn't really care about the project and their contributions to the project, they wouldn't make a fuss. The pressure that the community creates to cut the feedback loop from edit to map tile means that the lag time is constantly dropping.

5 OSM08

In 2008 OpenStreetMap is aiming to complete its map of the UK, providing a dataset which contains all of the motorways, roads and railways as well as a large amount of point of interest and land use data. OSM will continue to grow outside of the UK, with well-mapped areas like Germany and the Netherlands developing even richer datasets. Areas such as The Netherlands, in which OSM has a complete road network, will see the switch from 'production mode' to 'maintenance mode' as the focus of mappers activities switches from creating a map to maintaining one. As the numbers of contributors in currently under-mapped areas reaches critical mass, we will see an explosion in mapping activities. South Africa is a country to watch in 2008, being home to an active and fast-growing OSM community and a rapidly developing, geodata-hungry economy.

As of January 2008, mainstream adoption of OSM data is still limited. The lack of completeness of the dataset is a key factor behind the reluctance of many companies to adopt OSM data. An equal – if not greater – factor is a reluctance to break with the status quo. In the same way that no one every got fired for buying IBM, no one ever gets fired for buying Navteq. Over the next few years, further consolidation in the geodata market, privatization of previously state-owned mapping companies, and increasing consumer demand for geodata will shake up the market. New players are bound to emerge at every level of the supply chain. OpenStreetMap's role in the new mapping economy is far from certain, but the signs look good. With more active contributors by the day, OpenStreetMap has a wealth of skilled systems administrators, software developers, cartographers, and surveyors who are dedicated to providing free, high-quality worldwide geodata for anyone to use and share.

References

1. *http://www.openstreetmap.org*
2. Benkler, Y. (2002) "Coase's Penguin, or, Linux and The Nature of the Firm" *The Yale Law Journal* 112 pp.369–446.
3. *http://wiki.openstreetmap.org/index.php/ Foundation*
4. *http://creativecommons.org/licenses/by-sa/2.0/*
5. *http://mapnik.org*

DESIGNING MAPS FOR THE COLOUR-VISION IMPAIRED

Bernhard Jenny and Nathaniel Vaughn Kelso

Originally published in Volume 41, Nos 1 & 2, pp.9–12

To design maps that are readable by the colour-vision impaired but are also appealing to those with normal colour vision successfully, cartographers need to know how the colour-vision impaired person perceives colour and which colour combinations become confused. In this article, we concentrate on red-green colour confusion, which is by far the most common form of colour-impaired vision, and suggest how maps can be designed to consider this user group. We also introduce Color Oracle (see *http://colororacle.cartography.ch*), a free application that allows the designer to see colours on the monitor as people with colour-impaired vision see them.

Introduction

Colour-impaired vision, where certain colours cannot be accurately distinguished, is typically inherited through a sex-linked gene and predominantly affects about 8 per cent of the male population. Although this may seem a small proportion, when publishing in a mass market (e.g. for a major newspaper), the number of affected readers may reach tens of thousands. Naturally, maps with a smaller circulation will affect fewer readers, but these may be critical members of the audience. Designers of maps and information graphics therefore cannot disregard the needs of this relatively large group of media consumers. Furthermore, barrier-free, 'universal' design becomes especially important when readers have very limited time to read maps and information graphics, as, for example, when reading evacuation plans in emergency situations. Not only can universal design be required by law, but cartographers may also consider barrier-free design as part of their professional ethics, since colour-impaired vision is one of the most widespread physiological conditions to hamper map reading.

Figure 1 *The visible spectrum as perceived by the normal viewer (top) and by those with red-green vision impairment (bottom)*

Red-Green Vision Impairment

A wide range of colour-vision anomalies exist. Some of these are genetic and some are caused by degenerative diseases, by poisoning, or by physical injury. The commonly called 'red-green blindness' is by far the most frequent form and affects about 8 per cent of all males, mainly causing difficulty when distinguishing colours within the red-green portion of the visible spectrum. Figure 1 shows the spectrum as perceived by people with normal vision and by people who have problems distinguishing between red and green. The degree of impairment varies from one person to another from almost full colour vision to 'pure' red-green blindness. Indeed, the measurable variation among individuals with 'normal' vision is so large that the boundary between normal and colour-impaired vision is arbitrary. Women are much less likely to be affected by red-green confusion than men, with only 0.4 per cent women impaired (Birch, 1993). Other rare forms of colour- impaired vision exist, which affect less than approximately 0.3 per cent of all men and women.

The number of colours that readers with red-green vision impairment can distinguish without ambiguity is rather small. Besides confusion over the well-known red-green combination, other colour pairs may be confused, as illustrated by Figure 2. Swatches grouped at the left show those colours which readers with normal colour vision can easily distinguish, while those swatches on the right illustrate how readers with red-green vision impairment perceive these colours. Dark green, brown, orange, and dark red in the first row appear as almost indistinguishable olive-green tones to the red-green vision impaired. The second row contains less saturated blue, turquoise, and purple, which are all seen as indistinguishable pale violet-blue. The saturated purple and various blue tones of the third row manifest as almost identical bluish tones. Cartographers should be wary of pairing these colours, especially when colour is used as the sole means for quantitatively or qualitatively distinguishing items.

Software for the Map Designer

Specialized software, such as ColorBrewer – a popular online tool that suggests colour schemes (Harrower and Brewer, 2003; *http://www.colorbrewer.org*) – can help the designer to choose between universally-legible colour combinations. Complementary tools help to verify the legibility of a design by simulating colour-impaired vision. One example is Color Oracle, an application developed by the authors of this article.[1]

Color Oracle allows the designer to see colours on the screen as people with colour-impaired vision see them. It is accessible via the Mac OS X menu bar or the Windows system tray. The Color Oracle user selects the type of impairment in a drop down menu (Figure 3). The program then filters whatever is displayed on the computer monitor to simulate colour-impaired vision. The filtered image disappears when the user presses a key or clicks a mouse button and it is possible to toggle between normal colour vision and three types of simulated colour-impaired vision to identify problematic colour combinations. This approach does not interfere with the user's usual workflow and works with any graphics or mapping software.

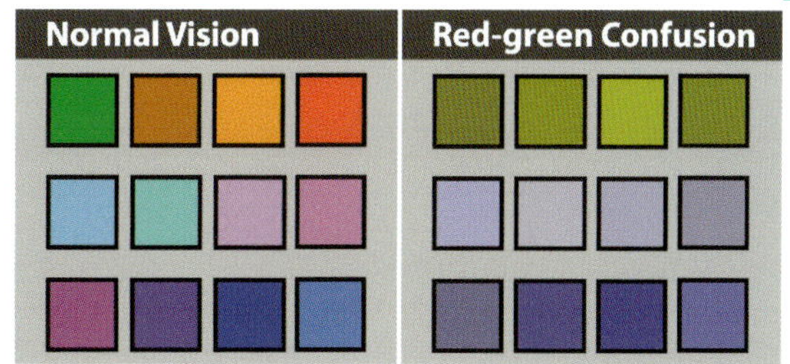

Figure 2
Colours as they appear to readers with normal vision and to those with red-green vision impairment

Color Oracle simulates 'pure' forms of colour- impaired vision, which are not as common as the milder forms with partial or shifted sensitivity. It can, however, be assumed that if a colour scheme is legible for someone with extreme colour-impaired vision, it will also be legible for those with a minor impairment.

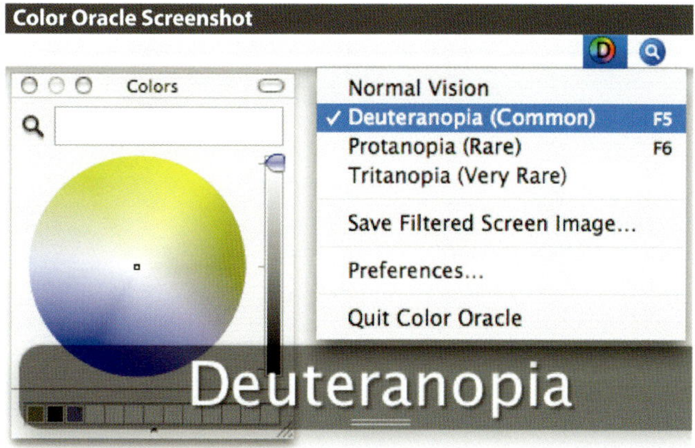

Figure 3
Screenshot of Color Oracle on Mac OS X simulating red-green vision impairment on a colour picker wheel

Color Oracle uses a well-established algorithm based on confusion lines (Brettel *et al.*, 1997 and Viénot *et al.*, 1999). Feedback from users with colour-impaired vision confirms that the simulations generated by Color Oracle are accurate except for very saturated colours, which might slightly deviate from the values seen by persons with 'pure' forms of colour-impaired vision

Designing Maps to Accommodate the Colour-Vision Impaired

The consequence of colour-impaired vision is that afflicted people are slower and considerably less successful in search tasks where colour is the primary attribute of the target object or if colour is used to organize visual displays (Cole, 2004). A low level of illumination impedes media consumers' successful reading of colour-coded information further. Investigations have shown that under reduced illumination, subjects with impaired colour vision make considerably more errors when identifying colour.[2] Colour coding should therefore not be used as the sole means of conveying information.

Greater clarity can be brought to maps by: (1) choosing unambiguous colour combinations; (2) using supplemental visual variables; and (3) annotating features directly. These techniques will not only improve maps for those with full colour vision, but will establish a good level of distinction for those afflicted with colour-impaired vision.

Distinguishing Point Classes

Dot maps often use hue as the only differentiating variable between classes of points. This hue coding can be difficult to interpret for readers with colour-impaired vision. Figure 4 illustrates how point symbols can be redesigned to increase legibility. Varying the saturation increases contrast and differentiates the dots only slightly for red-green impaired readers ('poor' column); shifting hue from green to blue improves legibility ('better' column); while the best solution is achieved by combining the use of different geometric shapes with varying hue and saturation. The last column shows that colour could even be discarded and the map would still be legible using the different geometric shapes. Well-designed symbols are easy for the reader to decode without consulting a legend.

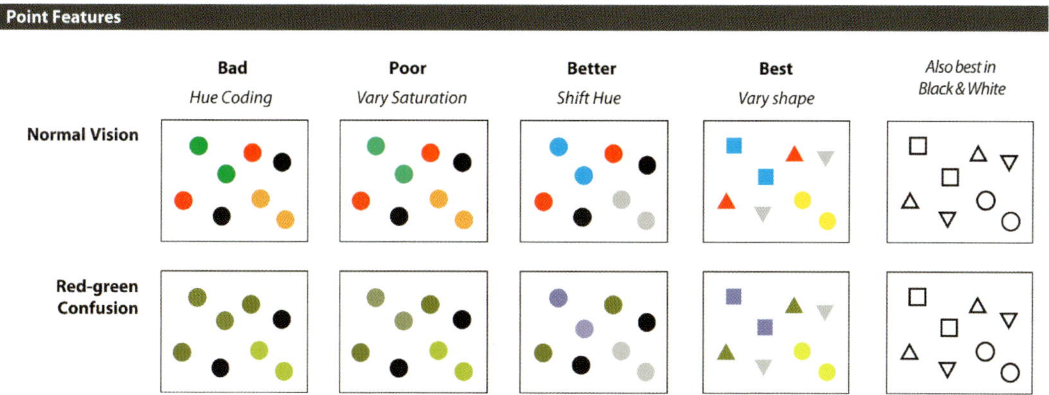

Figure 4
Point classes typical of a dot map distinguished by saturation, hue, and shape

Distinguishing Line Classes

To minimize confusion, colour-coded lines on maps can be redesigned in a manner similar to colour-coded dots (Figure 5). Changes to line width must be applied with care, however, since different stroke widths can imply varying quantities ('poor' column). Directly annotating the lines with labels is a better solution that clarifies ambiguous colours and reduces the need to refer to a legend. Altering colour hue is another way to improve legibility ('better' column). A combination of modified hue and saturation with varying line patterns and annotations is our preferred solution, because it is legible to everyone ('best' column). While line patterns (e.g. dash, dot) can imply unwanted qualitative or quantitative meaning or create undesirable visual noise, for complex maps that are also easy to read and interpret. This challenges the opinion of many cartographers, who advise against the use of spectral schemes for ordered data. To accommodate readers with red-green vision impairment, Brewer makes the following suggestions: (1) vary the lightness on the red-orange-yellow end of the rainbow; (2) omit yellow-green to avoid confusion with orange; and (3) for bipolar data, omit green and use a scheme with red, orange, yellow, light blue, and dark blue, and align the yellow-blue transition with the pivot point of the diverging data range.

Figure 5 *Line classes distinguished by width and saturation, annotation, hue, and line pattern*

utilize several line classes the distinction in texture can become essential.

Distinguishing Area Classes

It is possible to devise schemes for qualitative maps using hues that are potentially problematic, provided that these are differentiated by both saturation and value (e.g. dark red, bright green), while the use of overlay hatching can sometimes present a viable alternative for choropleth maps.

For continuous-tone raster data (where colours merge into one another), scientists often apply spectral rainbow colour ramps, which typically include red, orange, yellow, green, blue and purple. Brewer (1997) found that many readers prefer such spectral colour schemes, and that they

The precipitation map in the first row of Figure 6, for example, shows low quantities of rainfall in red and intermediate values in green. Hence low and intermediate values would appear identical for readers with red-green vision impairment. The map in the second row uses an alternative spectral ramp that omits yellow-green, uses a darker red, and places the transition between yellow and blue at the mean of all values. To bring further clarity to the map, selected high and low values could be labelled when the map is printed or for a digital map, the user could query values by moving the mouse over map locations.

Conclusion

Colour-impaired vision affects a significant portion of the population and therefore must be taken into account by the cartographer. When adjusting a colour scheme, the cartographer has to find a balance: on one side, 8 per cent of men who are colour-vision impaired have the right of equal access to information; while on the other side, the 96 per cent of the population with normal colour vision would welcome pleasing maps that are easy to read. It is the

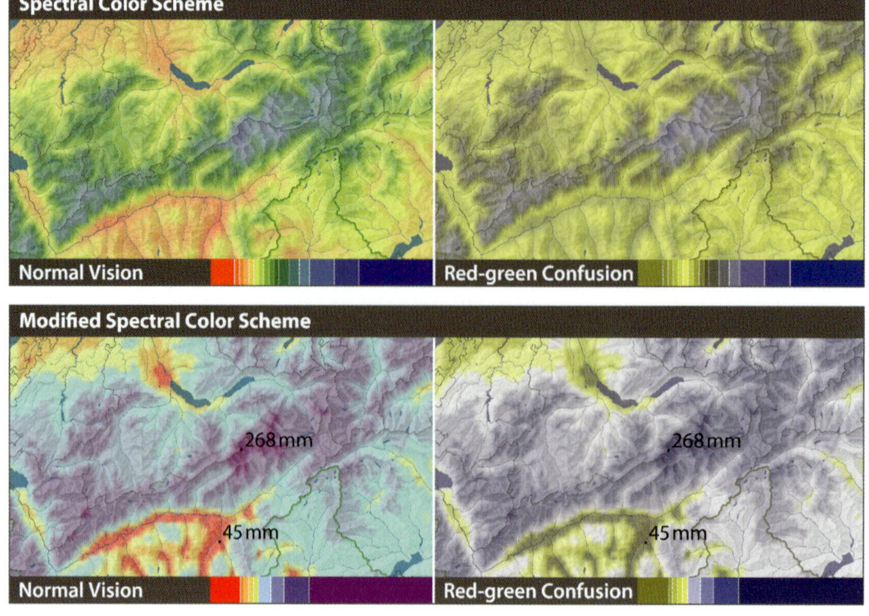

Figure 6 *Spectral colour schemes for precipitation maps, with rainbow colours (top row) and with an improved spectral scheme (bottom row). Colour ramps are depicted below the maps ('Mean monthly precipitation in January', ©Atlas of Switzerland 2, 2004)*

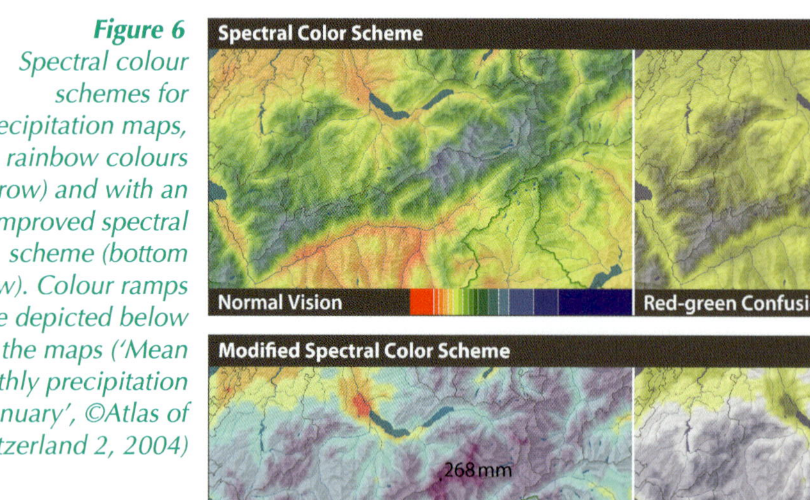

cartographer's responsibility to decide when colours should be adjusted.

To avoid problematic colour combinations, the cartographer should use colours with strong contrast and supplemental visual variables, such as shape, size, and pattern variations to allow all readers to discern and directly interpret a symbol without consulting a legend. Additional techniques include simplifying the map's design and annotating the map directly wherever the reader might become confused.

Interactive digital maps can further support readers with colour-impaired vision by providing tooltips or labels that are displayed on-demand. Digital environments can allow the user to customize colour schemes to suit their needs and provide methods to query individual values.

Color Oracle provides a convenient method to verify that colours on a map are legible to everyone. Indeed, we have discovered many problematic colour combinations in our daily mapmaking work in using this software. Adjusting colour schemes is not always simple and forces the cartographer to reassess well-established conventions – for example, rainbow colour ramps for precipitation maps or red-green colour schemes for voting maps (Jenny and Kelso, 2007). Color Oracle provides a convenient tool for seeing maps in the same way as readers with colour-impaired vision. It is now an integral part of our workflow and it is hoped that others may find it a valuable tool for designing universally accessible maps.

References

Birch, J. (1993) *Diagnosis of Defective Colour Vision* Oxford University Press.

Brettel, H., Viénot, F. and Mollon, J. D. (1997) "Computerized Simulation of Color Appearance for Dichromats" *Journal of the Optical Society of America* 14 (10) pp.2647–2655.

Brewer, C. A. (1997) "Spectral Schemes: Controversial Color Use on Maps" *Cartography and Geographic Information Systems* 24 (4) pp.203–220.

Cole, B. L. (2004) "The Handicap of Abnormal Colour Vision" *Clinical and Experimental Optometry* 87 (4/5) pp.258–275.

Harrower, M. and Brewer, C. A. (2003) "ColorBrewer.org: An Online Tool for Selecting Colour Schemes for Maps" *The Cartographic Journal* 40 (1) 27–37.

Jenny, B. and N. V. Kelso (2007) "Color design for the Color Vision Impaired" *Cartographic Perspectives* 58 pp.61–67.

Viénot, F., Brettel, H. and Mollon, J. D. (1999) "Digital Video Colourmaps for Checking the Legibility of Displays by Dichromats" *Color Research and Application* 24 (4) pp.243–251.

Notes

1. Freely available at *http://colororacle.cartography.ch* for Macintosh, Linux and Windows.
2. For more details, refer to Cole, 2004.

CARTOGRAPHIC BLANDSCAPES AND THE NEW NOISE: FINDING THE GOOD VIEW IN A TOPOGRAPHIC MASHUP

Alexander J. Kent

Originally published in Volume 42, Nos 1 & 2, pp.29–37

Should we expect web map service (WMS) applications to present a faithful description of the landscape? The number of providers continues to grow and concerns about their cartographic quality continue to be raised, generating debate on whether such maps are fit for purpose. This paper explores why cartographers consider the design of these maps to fall short. It suggests how judgments and expectations are drawn from cartographic communication theory and state topographic map design. The paper then proceeds to discuss the meaning of cartographic quality in an age of democratized mapping.

Introduction

Since MapQuest was launched in 1996, web map services have grown to become the face of ubiquitous mapping at the beginning of the 21st century. Put very simply, these online applications allow users to retrieve a seamless, multi-scale map centred on a location of their choice and perform route-planning queries. They occasionally provide the option of vertical and oblique aerial imagery as an independent or hybrid layer (e.g. Google Maps), as opposed to virtual globes (e.g. Microsoft Virtual Earth 3D), which are digital models comprising satellite and aerial images as their base. With collaborative mapping applications, such as OpenStreetMap, users generate the geographical base themselves by uploading GPS (global positioning system) tracks to create an editable map of the world. Some of the above permit users to create a 'mashup' by adding their own data (which may be drawn from a variety of sources) to the base data through an application programming interface (API). The combination of this explosion of data with the ability of the user to interact with them presents unprecedented possibilities for the exploration of geographical information.

Whether they allow the user to add their own data or not, WMS applications usually present an initial view of the desired location that is not intended to be highly detailed. However, such maps (as shown in Figure 1, for example) seem to be judged by cartographers as if they are meant to offer something definitive, perhaps even as an evolutionary development of the state topographic map (Figure 2). This is curious, because their route-planning functionality and lack of detailed relief information might render a more plausible evaluation – at least in terms of initial appearance and purpose – as a development of the road atlas.

Rather than attempting to compare and contrast the

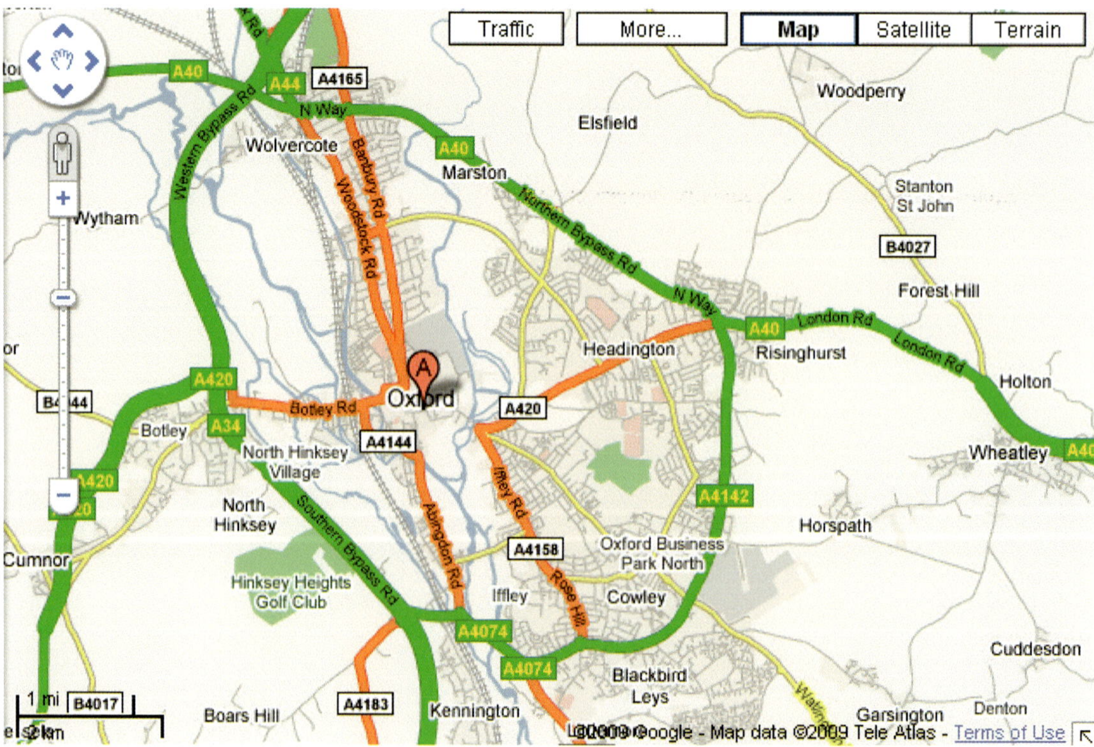

Figure 1 Excerpt from the initial view of Oxford, United Kingdom, as presented by web map server Google Maps (2009)

cartographic representations employed by various web map services, this paper aims to explore why cartographers call their fitness for purpose into question. It deconstructs the subject of the topographic map in order to the address this issue and discusses the relevance of cartographic quality in an age of democratized mapping.

A Sense of Place in the Global Blandscape

'the more good maps which are available to the public in any country, the better the chance that country has to progress and develop as a nation' (Thackwell, 1969: 7).

Do web map services supply us with 'good maps'? A recent series of debates[1] has shown this to be a highly contentious issue, reaching beyond the paper-versus-digital canon and addressing the deeper question of whether such maps are fit for purpose. Criticisms of their cartographic design have tended to focus upon the representation of their subject as a whole. As Spence (2008) asserts, 'We're in real danger of losing what makes maps so unique, giving us a feel for a place even if we've never been there'.[2] The non-communication of a 'feel for a place' may be explained in part by the selective omission of some features in particular (e.g. local landmarks) and a higher degree of abstraction in the map's symbology which those familiar with reading state topographic maps at comparable scales would expect to see. Resistance to the creation of a global 'blandscape' that pervades a sense of artificiality and unauthoredness though a lack of detail and homogenization of its subject is perhaps reminiscent of Collinson (1997: 119), who created his Virtual Worlds because traditional cartographic symbology did not convey enough 'geographical reality'. Both views suggest maps which demonstrate a lack of responsibility, a neglected duty; that a map showing part of the world should be more faithful to its subject in some way. More importantly, it also suggests that without providing a 'feel for a place' (rather than an accurate set of directions), WMS render their mapping less fit for purpose.

Others, particularly those who are involved in the proliferation and development of WMS, take a different view:

Modern online maps and satnavs don't display as much detail, it is argued by the BCS [British Cartographic Society], *missing out features like churches, village greens, etc., of course this is rubbish! Most online maps contain more detail than any traditionally designed map could ever do, but that detail is hidden behind an interactive interface, features are displayed dependent upon the level of zoom (scale) or the purpose of the map itself* (Parsons, 2008).

The essential task of cartographic generalization, i.e., deciding what to show, is influenced by the nature of the medium through which it is to be shown, e.g. paper or computer screen. Certainly, the drive for more in the amount and type of information to be shown on maps has outgrown the limits of static media. Moreover, advances in the development and accessibility of desktop publishing technology, and the rise of geographical information systems (GIS), on-demand mapping, and post-Fordist just-in-time (JIT) means of production, have revolutionized the availability of geographical data. This perhaps resembles the first streams of printed maps that were made available to the public hundreds of years earlier; the woodcut editions of Ptolemy's *Geographia* (such as that by Sebastian Münster) for example, and subsequently the copper-engraved atlases compiled by Ortelius and Mercator. In an age where the democratization of map-making is gathering immense pace and the availability of WMS, virtual globes, user-customization, and APIs have provided the user with more control over geographic

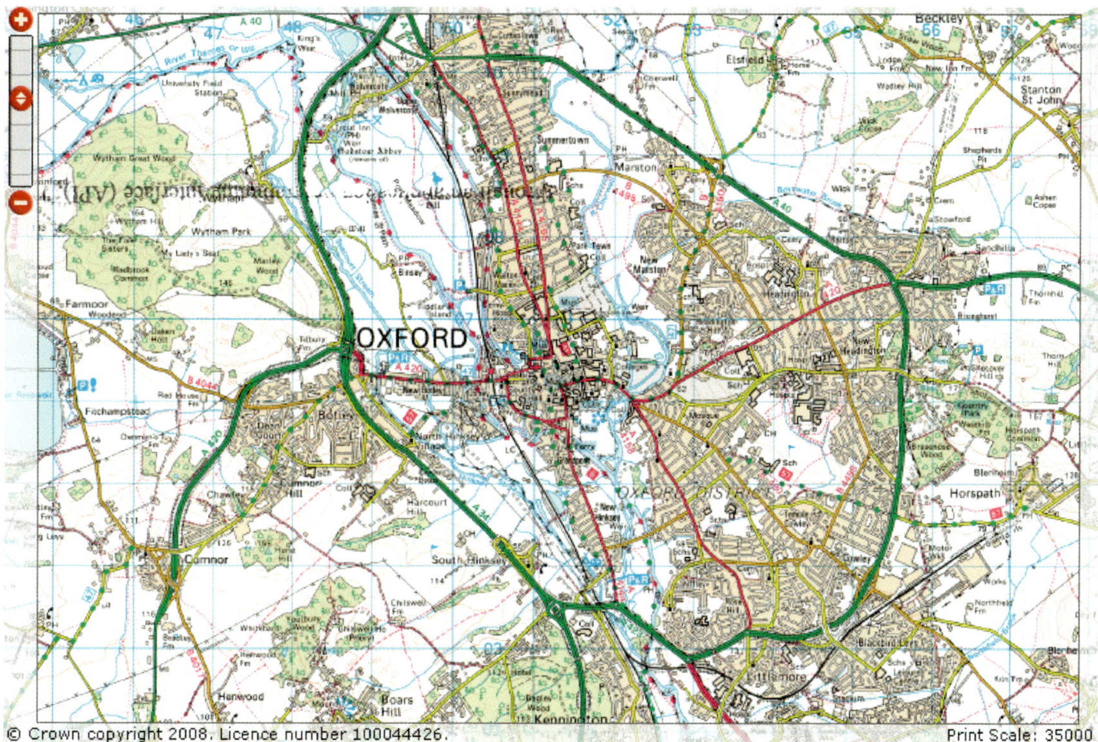

Figure 2
Excerpt from the initial view of Oxford, United Kingdom, as presented by Ordnance Survey via a web map server (Cassini Historical Maps, 2009) © Crown Copyright Ordnance Survey. All rights reserved.

visualization than ever before, we are also witnessing a colossal redistribution of power and knowledge in the relationship between the cartographer and the map user.

As a consumer, the user wields power in determining the success or failure of the map and by freeing them from the limits of scale, content, and appearance to some extent, web map services seem to offer an evolutionary leap – a move away from the alienation of imposed datasets and towards choice – that invites the user to decide what information to show and (with mashups in particular) how to show it. At first sight this seems a far cry from the situation reported by Monmonier (1982: 99) well over twenty years ago:

Most of our maps are made by organizations, principally governments and large companies, but mostly governments. Cartographers and cartographic technicians might be involved in various stages of planning and producing these maps, but the important decisions are institutional – federal, political or corporate, rather than individual.

However, Dorling and Fairbairn (1997: 80) warn that the recent advances offer nothing particularly new: 'It is easy to see the arrival of technologies such as the Internet as heralding a new age of openness and equality, and it is easy to forget who pays for maps to be made and who is allowed and able to make them.' So while the availability of geographical data and user-customization is likely to increase, the choices available to the user remain limited and are fixed by the mapping agency – state or commercial – responsible for making these data accessible online. Unless the user has complete control over what is shown as well as its graphic expression, they are only left with an illusion of choice.

Moreover, despite the power to do so, there is perhaps no reason to suggest why users should venture from the default settings to change the design and visualization of data. The initial view is the first (and possibly) last impression of a place represented on the map and therefore carries a particular relevance as a whole, regardless of any hidden potential held within innumerable layers. As Monmonier (1991: 5) suggests, 'Map viewers ought to condition themselves to questioning whether an ulterior motive might have led to a biased view of reality favouring the map author's philosophical, political or economic goals, or whether a lazy map author simply failed to explore designs offering a more coherent or complete picture of reality'. If the selection of features shown has the greatest impact on the potential use of the map; it is what maps describe that determines their function and, hence, influences judgments over their fitness for purpose. Is the ultimate function of WMS applications to provide a sense of place, which, in essence, is derived from the sum of its parts? Should providers of these applications change the default view?

Communicating Landscape

Practising cartographers today continue to align to the idea that the essence of cartography is communication (Lilley, 2007: 208), a position infused by the first theory of cartography to gather momentum in the twentieth century, i.e., that the whole gamut from making to reading maps is one of information transfer, as summarized by the models of cartographic communication (e.g., Board, 1967; Koláčný, 1969). Under this hegemony of optimization, cartographic design is regarded as a process whereby graphical variables (such as colour, size, and shape) of various map elements are manipulated and coordinated to improve efficiency in the flow of information from the cartographer via the map to the user. As Robinson and Petchenik (1975) explain, in order to achieve this, the degree of 'noise' inherent within the system (e.g. badly designed symbols) should be reduced.

It is not difficult to see how this theory has encouraged cartographers to suppress certain information and attract the user's attention elsewhere (e.g. through the use of stronger colours) in order to communicate a particular theme. Road maps for example, are generally designed to show roads with more potency and in higher detail than other features (these may exhibit a visual hierarchy according to how useful they are deemed to be to the user). Similarly, satellite navigation devices either suppress or omit other features to provide an effective display in situations where fixations of users' eyes on the map are likely to be fewer and shorter; the cognitive load is reduced so as to not interfere with the act of driving, for example.

But what about WMS applications: do their maps have a central 'theme', are they intended to present information 'equally', or are all themes suppressed to let the user see their own data in a mashup more clearly? Generally, users first interact with a WMS by defining a location, which results in a map – the scale of which follows a number of pre-defined increments and depends (initially) upon the 'resolution' of the data entered by the user (e.g. street, town/city, postcode, country). The interaction is therefore driven by location, by place. If topography is defined by *The Oxford English Dictionary* primarily as 'the science or practice of and description of any locality' (Simpson and Weiner, 1989: XVIII, 257), then WMS applications are, in essence, topographic maps. The capacity for some to offer mashups comprising data sourced by the user – underpinned by the geographical base – means that they operate on the premise that they are not as concerned with the specific feature as they are with the wider milieu. Improving the communication of one theme can inhibit the communication of others and this paradox not only makes it difficult to make reasoned judgments about the cartographic success of these maps, but has also led to the neglect of topographic maps in cartographic theory. As complex, polythematic cartographic products, the communication models were too narrow in their drive for the pursuance of a single message to incorporate their essential nature – topographic maps have no obvious point to make. It is therefore necessary to explore what cartographers possibly aim to achieve when designing such maps, and whether, at a time when making and disseminating maps is easier than ever before, understanding how they present their subject can contribute to the development of WMS applications.

Maps aiming to provide detailed and accurate observations of topography were amongst the first maps to be made, and according to Piket (1972: 267), they form the ancestral line in the cartographic family tree from the origins of map-making to the present. Unlike aerial photographs, the role of the topographic map is not simply to record the Earth's features in whatever shape or form they may be; maps deliberately offer an abstract rather than a mimetic representation in order to provide a manageable interpretation. As Robinson *et al.* (1995: 450) suggest, 'This quality of maps to typify and simplify is one of the fundamental reasons maps are needed by society. Otherwise we could operate with photographs alone'. For a subject to be mapped, it must undergo selection and generalization, usually through classification, simplification, exaggeration, symbolization and induction (*ibid.*: 451). As they are ultimately derived from detailed survey of terrain, topographic maps have usually been the preserve of national mapping organizations, putting them beyond the reach of most commercial cartographic enterprises.

In an overview of the varying content and appearance in contemporary topographic maps produced by different national mapping organizations, Hodgkiss (1981: 174) draws on an interesting bias towards aesthetic appreciation:

'the mountainous terrain of Switzerland [...] *lends itself to the making of visually attractive maps.* [...] *The Swiss National Survey* [...] *maintains a long tradition of beautiful mapmaking in which great care is exercised in the choice of type faces [sic], symbols and colour schemes to ensure maximum clarity and effectiveness of communication.*

His observation of Norwegian maps, that *'The importance of outdoor recreation is evident in the inclusion of hotels, fishing and shooting huts, tourist centres, ski-tows and a network of well-marked footpaths and tracks'* (Hodgkiss, 1981: 176) suggests the inclusion of a particular selection of features because of their significance to society. This leans away from the idea of the map as an objective, value-free record and towards a subjective, value-based selection of features which, if symbolized in a particular way, might together help construct and communicate a sense of place.

Whether paper or electronic, made by institutions or individuals, all maps offer selective views that reflect and serve the interests of their makers. The choice of features exhibited on state topographic maps serves the interests of the state and their success lies in presenting these choices as both neutral and objective. As Wood (1992: 73) explains, '[it] will be seen to serve so many purposes that none can predominate, or its means will be so widely spread in so many social institutions that it can be claimed by none'. While the availability of topographic base data and GPS technology has since facilitated the creation of WMS, a comprehensive representation of relief is not provided. Arguably, this does not detract from the air of 'objectivity' that these maps inherit from the legacy of topographical mapping. The selection of features presented in state topographic maps may not always be relevant or directly useful and the potential of web map services, APIs, and mashups to broaden the range of themes and purposes – and interests of users served – provides them with an advantage. Moreover, with some control over the selection and appearance of features, they offer the user a hand in defining a sense of place that appeals to them. But achieving a sense of place continues to be elusive and it is difficult for abstract symbology, as is perhaps illustrated by Figure 1, to be perceived as faithful and authentic, even if it helps construct open texts that stimulate imaginative interpretation of the subject. The subject of all maps dealing with topography – whether state, commercial, paper, WMS, API, or mashup – therefore requires more investigation.

According to Nielsen *et al.* (2002: 2), 'The purpose of topographical mapping is to describe the landscape on a map'. A landscape, according to art historian Malcolm Andrews, is 'what the viewer has selected from the land, edited and modified in accordance with certain conventional ideas about what constitutes a 'good view'' (Andrews, 1999: 4). Could this description be applied to the process of topographical mapping? If so, what would these 'conventional ideas' be, and more importantly, what would constitute a 'good view'? In the case below, it is easy to point to cartographic tradition as determining what constitutes a 'good view' in both making and reading maps, despite the apparent dedication of this particular design in seeking a faithful representation and perception of its subject:

To flout well-established conventions, such as the depiction of the sea in blue can be a way of introducing artificial noise into the map. The Reader's Digest Atlas of the British Isles shows the sea in shades of sea-green on the grounds that the British seas do not appear blue. It takes a little time to adjust to the unfamiliar colour (Board, 1967: 701).

The processes of map-making and landscape painting (Figure 3) share similarities in the fundamental treatment of their subject, although the hegemony of scientific objectivity and institutional authorship ensures that these are masked in the state topographic map to a far higher degree. Koeman (1971: 171), for example, states that 'In landscape-painting the modes of expression of the creative artist and the cartographer are approaching each other'. This is clearly visible in the work of Eduard Imhof (1895–1986); one of the most influential and respected figures in state cartography of the twentieth century and himself a landscape painter. The following description of the processes involved in creating topographic maps given by the 'Father of European Cartography' (Piket, 1972: 268) carries considerable potency in how it corresponds to landscape painting:

Due to scale restrictions, the cartographer makes a selection, classifies, standardizes; he undertakes intellectual and graphical simplifications and combinations; he emphasizes, enlarges, subdues or suppresses visual phenomena according to their significance on the map. In short, he generalizes, standardizes, and makes selections and he recognizes the many elements which interfere with one another, lie in opposition and overlap, thus coordinating the content to clarify the geographical patterns of the region (Imhof, 1982: 357).

Notwithstanding the creative restrictions implied by scale, this nevertheless suggests that topographic maps are structured towards the creation of a 'good view' and thus More commonly, perhaps, places are omitted due to their military sensitivity. Such features – in addition to a particular colour of the sea – do not conform to ruling ideologies about what constitutes a 'good view' either. But without the state imperative of providing a good view of the national landscape, web map services enjoy fewer limitations on what can be shown and how. By contrast, if state topographic maps offer a portrait – a good view – of the land, they do so without the abandon of Oliver Crowell's 'warts and all' epithet.

National mapping organizations have therefore developed a 'vocabulary' of cartographic symbols, which

Figure 3 *Frederic Edwin Church (1826–1900) Heart of the Andes (1859), oil on canvas, 167.9 × 302.9 cm, housed in the Metropolitan Museum of Art, New York (Wikipedia, 2009)*

parallels can easily be made with the selection of features from the land that is associated with landscape art:

In judging what is a 'good view' we are preferring one aspect of the countryside to another. We are selecting and editing, suppressing or subordinating some visual information in favour of promoting other features. We are constructing a hierarchical arrangement of the components within a simple view so that it becomes a complex mix of visual facts and imaginative construction (Andrews, 1999: 3).

The cartographic representation of landscape therefore depends upon the application of certain preferences and judgments that would appear to be based upon aesthetic values. However, while the pursuit of some cartographic beau ideal is likely to be more palpable in reactions to the appearance of certain symbols (such as that demonstrated by Board's response to the colour of the sea above), it should not be forgotten that the design of a topographic map conforms to the ideology of those exercising its power. The inclusion of particular features may potentially jeopardize the 'good view' of the state, as 'framed' by its international borders. As Taylor (1992: 128) observes (if somewhat overzealously), 'Throughout the cities of the world, maps omit the shantytowns, bidonvilles, black townships, public-housing projects and council estates as nonplaces, revealing their irrelevance to the public eye'.

has evolved to be more 'articulate' in expressing the state landscape for that particular society. Initiatives to devise an internationally standardized set of topographic map symbols, such as the International Map of the World suggested by Albrecht Penck in 1891, have not tended to be successful because the expression of these symbologies is inadequate for the purposes of a state wishing to present its own landscape. As with the ill-fated and artificial language of Esperanto, they do not provide the same refinement that hundreds of years have given to the development of a rich and varied (yet homogenizing) language of cartography. The challenge for all cartographers, therefore, is to continue to express their subject in a way that is acceptable to others.

Rage Against the Default: The Renaissance of Cartography

If maps are made because of the needs of particular social situations and to fulfil a particular function (Pickles, 2004: 66), like any product of society, their immediate value is defined by their relevance at a given place and time. Commercial cartographic institutions may serve their own interests in the pursuit of an agenda for exercising internal power, but if they become alienated from the needs of society they will only offer impersonal products that fail to appeal to the user community. This is crucial in a time

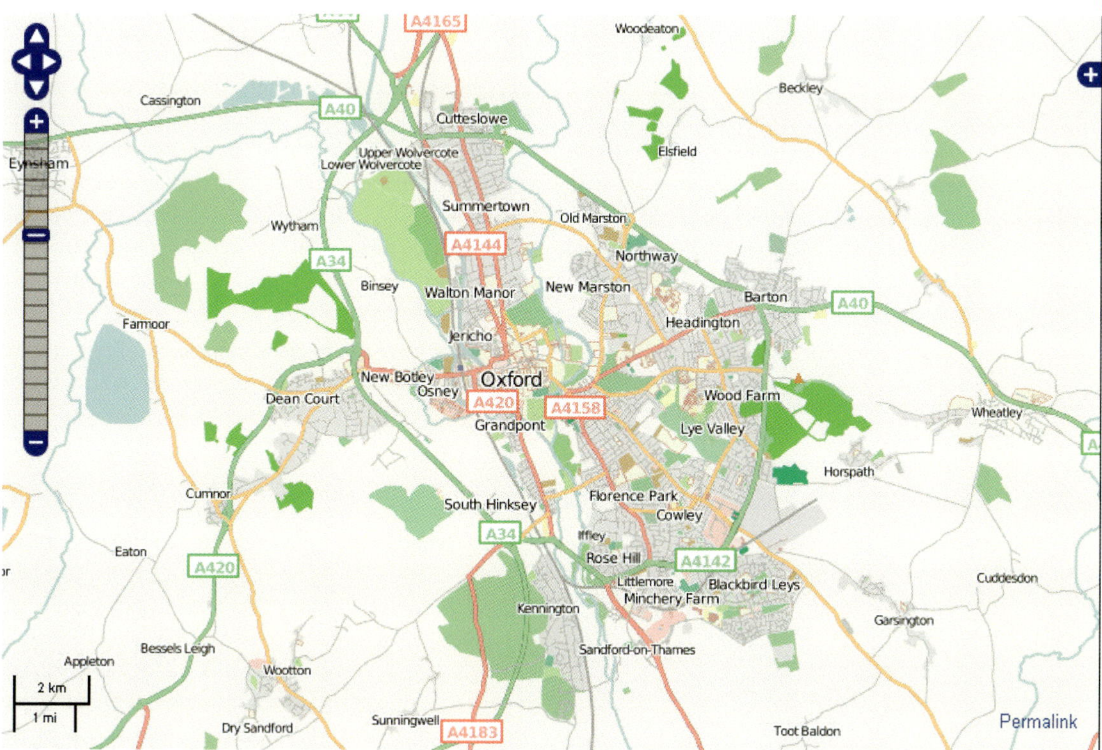

Figure 4
Excerpt from the initial view of Oxford, United Kingdom, as presented by OpenStreetMap (2009)

when more people are engaged in the creative exploration of geographical information than ever before. Users are consequently discovering that the presentation of geographical data actually matters and realizing that there is more to mapping than plotting data: 'Simply put, good maps come from good data combined with the application of sound cartographic and geographic analysis principles' Exler (2008).

But while the simultaneous presentation of spatial information requires a higher degree of cartographic expertise to coordinate, with the advent of electronic maps and geovisualization, some believe that the drive for creating an optimal cartographic solution is no longer relevant:

Harley and GVis [geographic visualizations] *indicate that mapping should proceed through multiple, competing visualizations which are not created by a cartographer and communicated to the user but made on the spot by the user acting as his or her own cartographer. In other words, the search established by Robinson for the single optimal map through ever-clearer methods of map communication is over* (Crampton, 2001: 236–237).

Similarly, Monmonier (1991: 6) envisaged a 'new cartography' based on dynamic displays, stating that the experiential map is making the traditional one-map solution less and less defensible.

Where does this leave the future of cartography in an era of democratized map-making? If Koláčný (1969: 47) was right in that a map has to satisfy the consumer's needs and interests, it could be argued that current geovisualizations follow a natural progression in the path to fulfilling that goal; the impetus behind this technology being the empowering of the user. However, while technology has now developed to allow users to make their own maps for themselves using a compilation of various data, these new map-makers rarely seem to make decisions on how best to meet the needs of the user. Does the 'good view' of the WMS or mashup therefore remain 'good' as long as it serves the interests of those behind its creation – the user-cartographer?

The empowering of users through GPS technology to create the building blocks of geographical information perhaps provides that degree of connection between maker and map which is necessary to generate concern about the way these data – and landscapes – are represented. The concept behind OpenStreetMap, for example, presents a distinctively different and radical contribution to other web map applications, not least because the geographical database is compiled by users, but also because there is considerable attention paid to the way the maps represent their subject. The development of rendering toolkits such as Mapnik to 'make beautiful maps' (Pavlenko, 2007: 13) and the subsequent integration of these within OpenStreetMap demonstrates the importance of aesthetics in capturing a sense of place (Figure 4). Here, the capital that is geographical data is reinvested to serve the interests of the mapping community more closely.

So if users were left feeling disillusioned by the homogenization of landscape imposed by the symbology of state topographic maps, they are reclaiming the land and choosing to present it as they want to see it. This initiative represents a kind of 'folk cartography', which expresses something altogether more authentic in the relationship between people and their landscape, with the raw data (GPS tracks) acting as signatures of movement – perhaps life itself – through it. The difference between this initiative and so many others is the absence of alienation between map-maker and map provider – this is a product of a community, with a growing identity, of people who want to define the landscape – the good view – as they want. Here, cartography is progressing beyond mere 'mapping'. As with any representation, if this definition of the landscape becomes rigid and alienated from the community it serves, it will be no more authentic than the state homogenizations it took such pains to avoid. We should therefore become less accepting of cartographic silence as more people make maps and the freedom and equality to express this landscape above others will therefore resemble the dilemma faced by any social liberty.

Conclusion

The symbology of state topographic maps has evolved to become an articulate way of describing the national landscape, so for those familiar with this rich language, in terms of cartographic quality (as constructed by society), WMS applications have a hard act to follow. They may be (perhaps wrongly) compared all to too easily with state topographic maps and criticized for their representation of what is essentially a socially constructed landscape, but all maps require a cartographic vocabulary to express their subject and this grows to meet a need for a wider and more precise expression. The way we use maps is therefore undergoing a change, away from reading maps as static objects and accepting the simultaneous presentation of information in the initial view to encompass their potential as dynamic and interactive representations.

We are at an early stage of internet mapping and the power of good cartographic design is beginning to make an impact, especially through collaborative mapping initiatives. But as Field (2008) states:

> [...] *surely what is of most value is the rich mapping landscape that is rapidly increasing and the sheer number of people and organizations that place value in geographical information and the need to map it in some form.*

Experience in making and using maps leads to a greater familiarity with the way maps represent space and place. Making maps is great fun, after all. As more and more users become accustomed to the language used to present geographical information, the vocabulary which constructs that language will be refined. The revolution in mapping, led by the democratization of map-making technology, is therefore leading to the renaissance of cartography, where users demand maps that are designed with clarity of expression. That refinement will be the cartographic achievement of our time.

The challenge for the global community of map-makers is therefore to continue to experiment, to learn the language of cartography, to discover and implement what makes their maps work better. The challenge for WMS providers is to encourage this and make it easier to experiment. If the relevance and meaning of maps is to lie beyond that of the individual user-cartographer, it is crucial that map-makers – both new and old – consider the user beyond themselves. Otherwise, our maps may bear all the familiarity of a signature to us, but be illegible to everyone else.[3] As the development of folk cartography shows, people seek an authentic representation of the land that derives its relevance from expressing the interests of the people. This is, in part, derived from the emotive associations with the sense of place, which the homogenization of landscape and narrowing of cartographic vocabulary can overlook. Whether national preferences in the symbolization of landscapes will emerge (as is the case with state topographical mapping) is yet to be seen. As people are now discovering the freedom of cartographic speech, the goal should therefore be a harmonization of language and landscape, achieved through a vocabulary of symbols that commands articulation. As their professional ancestors knew, a proficiency in cartographic language lends maps with a valuable quality for earning trust.

Notes

1. The issue has been raised in the national press, television, radio, and several personal blogs, particularly since a panel discussion entitled 'The Future of Mapping' between Mary Spence (President of the British Cartographic Society), Ed Parsons (Chief Technology Officer, Google), and Denis Wood (author of *The Power of Maps*) took place at the annual RGS-IBG conference in London on 28th August 2008.
2. Some have questioned this approach. As Treves (2008) points out: 'there are some much better cartographic criticisms you could make of online mapping e.g.: red dot disease: web developers using large red splodges as icons. When there are lots of red splodges it overwhelms the view and looks like map measles.'
3. Arguably, this dissonance contributed to the need for the social reconstruction of cartographic language that we are now witnessing.

References

Andrews, M. (1999) *Landscape and Western Art* Oxford: Oxford University Press.

Board, C. (1967) *Maps as Models* In Chorley, R.J. and Haggett, P. (Eds) *Models in Geography* London: Methuen.

Cassini Historical Maps (2009) Available at: *http://www.cassinimaps.co.uk/* (accessed 07/01/2009).

Clement, B. (2006) "Forget the Map and Just Pass Me that Flat Screen" *The Independent* 31st August 2006.

Collinson, A. (1997) "Virtual Worlds" *The Cartographic Journal* 34 (2) pp.117–124.

Crampton, J.W. (2001) "Maps as Social Constructions: Power, Communication and Visualization" *Progress in Human Geography* 25 (2) pp.235–252.

Dorling, D. and Fairbairn, D. (1997) *Mapping: Ways of Representing the World* Harlow: Longman.

Exler, R. (2008) "Should You Be Making Maps" Available at: *http://www.thegeofactor.com/* (accessed 07/01/2008).

Field, K. (2008) Editorial: "Maps, Mashups, and Smashups" *The Cartographic Journal* 45 (4) pp.241–245.

Google Maps (2009) Available at: *http://maps.google.co.uk/* (accessed 07/01/2009).

Hodgkiss, A.G. (1981) *Understanding Maps: A Systematic History of Their Use and Development* Folkestone: Wm Dawson & Son.

Imhof, E. (1982) *Cartographic Relief Presentation* (trans. Steward, H.J.) Berlin: Walter de Gruyter.

Keates, J.S. (1972) "Symbols and Meaning in Topographic Maps" *International Yearbook of Cartography* 12 pp.168–181.

Koeman, C. (1971) "The Principle of Communication in Cartography" *International Yearbook of Cartography* 11 pp.169–176.

Koláčný, A. (1969) "Cartographic Information: A Fundamental Concept and Term in Modern Cartography" *The Cartographic Journal* 6 (1).

Lilley, R.J. (2007) "Who Needs Cartographers?" *The Cartographic Journal* 44 (3) pp.202–208.

Monmonier, M. (1982) "Cartography, Geographic Information, and Public Policy" *Journal of Geography in Higher Education* 6 (2) pp.99–107.

Monmonier, M. (1991) "Ethics and Map Design" *Cartographic Perspectives* 10 pp.3–8.

Nielsen, S.R., Christensen, S.F. and Michaelsen, P.B. (2002) "Topographic Mapping in Denmark" *The Danish Way* 10 pp.1–15.

Parsons, E. (2008) "Cartography is Dead, Long Live the Map-Makers" Available at: *http://www.edparsons.com/2008/09/cartography-is-dead-long-live-the-map-makers/* (accessed 07/01/2008).

Pavlenko, A. (2007) "Open Source Renders the World" *The Bulletin of the Society of Cartographers* 41 (1,2) pp.13–16.

Pickles, J. (2004) *Cartographic Reason: Mapping and the Geo-Coded World* London: Routledge.

Piket, J.J.C. (1972) "Five European Topographic Maps: A Contribution to the Classification of Topographic Maps and Their Relation to Other Map Types" *Geografisch Tijdschrift* 6 (3) pp.266–276.

Robinson, A.H. and Petchenik, B.B. (1975) "The Map as a Communication System" *The Cartographic Journal* 12 (1) pp.7–15.

Robinson, A.H., Morrison, J.L., Muehrcke, P.C., Kimerling, A.J. and Guptill, S.C. (1995) *Elements of Cartography* (6th ed.) New York: John Wiley & Sons.

Spence, M. (2008) Quoted in: BBC (2008) "Online Maps 'Wiping Out History'' Available at: *http://news.bbc.co.uk/1/hi/uk/7586789.stm* (accessed 30/08/2008).

Simpson, J.A. and Weiner, E.S.C. (Eds.) (1989) *The Oxford English Dictionary* (2nd ed.) Oxford: Oxford University Press.

Taylor, P.J. (1992) "Politics in Maps, Maps in Politics: A Tribute to Brian Harley" *Political Geography* 11 (2) pp.127–129.

Thackwell, D.E.O. (1969) "The Importance of Cartography to Modern States" *The Cartographic Journal* 6 (1) p.7.

Treves, R. (2008) "Ed Parsons calls for a 'New Cartography'" Available at: *http://googleearthdesign.blogspot.com/2008/09/gis.html* (accessed 07/01/2009).

Wikipedia (2009) "Frederic Edwin Church" Available at: *http://en.wikipedia.org/wiki/Frederic_Edwin_Church* (accessed 07/01/2009).

Wood, D. (1992) *The Power of Maps* New York: Guilford Press.

The 2010s

The papers we have chosen to include for this final decade – even though we have not even reached the first half of it – together encounter some of the recurring themes witnessed on this journey over the last 50 years, the introduction of new technology, the role of maps in society, the maintenance of high cartographic standards, but also introduce some new ones. Consequently, the papers selected for this decade highlight the degree of innovation taking place in cartography and the many levels it can be found – from mapping Haiti to help aid organizations to creating artworks that encourage us to think about maps and the places they represent in new ways.

Many would agree that Harry Beck's map of the London Underground is a truly iconic design, but in the first paper, Andrew Smithers argues that the success of the map has stifled innovation in the mapping of other railway networks and persuades us that these complex entities require a creative approach to solve the problem of their representation more effectively. By illustrating a series of new local and national designs, Andrew demonstrates how an appreciation and use of the existing geography and shape can provide some answers.

The following paper was written by three Geography students at the University of Manchester, which describes and illustrates their project for the 'Maps and Society' module led by Chris Perkins. Naomi Hurrell, Emma Kerry and Helen Parsons chose to make and perform a feminist 'embodied map' to help 'reclaim' a public park following an incidence of rape. The truly innovative wielding of the aesthetic language and power of maps to make this bold and lasting statement is complemented by the beautiful execution of their artwork.

In situations where the capability of the national mapping agency has been neutralized, such as that in Haiti following the devastating earthquake of 2010 for example, counter-mapping initiatives take on a value far beyond what could originally have been envisaged. One of the challenges international aid organizations face in mitigating the effects of a disaster is a duplication of effort in acquiring geographical information arising from a lack of overall coordination. While working at the University of Southampton as Head of the Cartographic Unit in the School of Geography, I took part in a coordinated initiative to assess the damage caused by the earthquake using imagery made available via Google Earth. At the same time, OpenStreetMap was providing invaluable geographical intelligence to teams on the ground in Port-au-Prince. Using an initiative from the 2008 Wenchuan earthquake, these datasets and many more were pooled through the web-based Virtual Disaster Viewer (VDV), which made this crucial information accessible to many agencies thus helping them to better focus their resources.

The following paper by Oliver O'Brien reflects the growing maturity of the OpenStreetMap project which had been documented in the Bulletin in its very early days by founder Steve Coast (see previous section). The article explains that after six years of operation, OSM covers some locations around Great Britain in high detail and accuracy, but argues that significant gaps in its coverage remain. Nevertheless, with the continuing enthusiasm of OSM's growing community of followers, Oliver has confidence that they can address the situation.

Another application of volunteered geographic information – this time to combat poaching in Congo-Brazzaville. Doctoral student Michalis Vitos, together with Matthias Stevens, Jerome Lewis and Muki Haklay, describe the introduction of smartphones to the Mbendjele as a means of collecting anti-poaching data and some of the challenges which were faced, from devising methods for describing space to a non-literate society to the use of saucepans adapted for recharging phones once they had been (mis)used as torches and for recording music!

It is unlikely that authors of the earlier papers in this book did not envisage the many ways in which maps are being used and/or created as pure works of art that we are currently witnessing. Christian Nold's emotional maps and Kate McLean's smell maps were then a far cry from the communication models and unimaginable to those who felt uneasy about the art/science dualism of cartography (let alone anything more artistic!). Demonstrating that maps can communicate so much truth about a place, whether they subscribe to formal methods of topographical mapping or not, Stephen Walter's stunning 'London Subterranea' exposes what lies beneath the surface of the city and invites you to explore more – as any map should.

So as we enter the final decade on this journey through the first 50 years of the Society of Cartographers, it is worth reflecting on its original aims and their relevance today. The Society was formed in 1964 with specific aims, one of which was to promote standards of cartographic illustration, but just as important was to address the sense of isolation felt by cartographers working in university departments. Although the majority of Society members do not work in university departments today and the profession itself is finding its way through new challenges (there will always be new technology and new skills to learn) there is more potential for the application of cartography than ever before. The introduction of GIS in the 1960s, video games in the 1970s, desktop mapping in the 1980s, web map servers in the 1990s, Google Earth and satellite navigation systems in the 2000s, and whether we use a smartphone, tablet, smart TV, or wristwatch in the 2010s, is that all these technologies – and many more – rely on the effective presentation of spatial information. Cartography, whether or not by name, and cartographers, whether or not by profession, have been, and will always be, needed – whatever shape or form our maps may take.

I end with an extract from Greenough's Presidential Address to the Royal Geographical Society from 1840, which John Robertson mentioned in the 20th anniversary of the formation of the Society of Cartographers. It seems fitting to repeat it on the occasion of its 50th year:

In the extent and variety of its resources, in rapidity of utterance, in the copiousness and completeness of information it communicates, in precision, conciseness, perspicuity in the hold it has upon the memory, in vividness of imagery, in convenience of reference, in portability, in the happy combination of so many and such useful qualities, a Map has no rival.

Alexander J. Kent

References

Greenough, G.B. (1840) "Presidential Address to the Royal Geographical Society" (25th May 1840) *The Geographical Journal* 10 pp.xliii–lxxxiv.

McLean, K. (2014) "Sensory Maps" Available at: *sensorymaps.com* (accessed 31/07/14).

Nold, C. (ed.) (2009) "Emotional Cartography: Technologies of the Self" (Available at: *www.emotionalcartography.net*) (accessed 31/07/14).

NEW IDEAS IN THE INTERPRETATION OF COMPLEX RAIL NETWORKS

Andrew Smithers

Originally published in Volume 43, Nos 1 & 2, pp.13–21

The iconic status of Harry Beck's Tube map has prevented innovation in the graphic visualization of local travel systems as they have become more complex. The map designer should endeavour to show the individuality and personality of the network being depicted, rather than force it to follow traditional abstract rules. A design of a map for the railways of Great Britain gradually evolved, discovering that maps at national, regional and local level can share a common theme. Online access now presents opportunities for interaction and access to detail not possible with the traditional poster map. Arguably, the map is the brand. It is core to the identity of train operating companies and passenger transport organizations, yet is invariably low profile. Much more investment should be made, not only to provide improved understanding and legibility but beautifully designed and crafted maps, full of personality and even fun.

Background

Like many I was fascinated from an early age by the London Tube map (or diagram, as we should more properly call it). I had a go at my own designs – I still have them – made with felt-tip pens and poster paint. But as the underground system grew I became aware of the map's many shortcomings. Harry Beck's original idea of enlarging a central area where stations are closer together and reducing outlying branches on a relatively simple system worked in the '30s. But as the system expanded and became more complex, the distortions of reality have got more problematic. For example, when the creation of the new financial sector on the Isle of Dogs gave birth to the Docklands Light Railway, a new centre was created to the east of London's traditional central focus (the Circle Line). The distortion of the connections between the two centres has caused the lines around London Bridge to be stretched.

Other problems continue as the Tube map is developed by other designers. Lines that are straight often have to be made crooked (an enemy of clarity) because of the limitation caused by the 45° maximum angle. Interchange stations suffer by having multiple interconnecting nodes that bear no relationship to the ease or difficulty of a change – purely for the cartographer's convenience – and the dominance of the interchange symbol (which is too big for its boots) over the tick distorts the importance of stations, which is questionable. The map also leaves unsolved the issue of different lines sharing the same route, shown inconsistently as either touching or separated. Furthermore, the Beck formula does not work when applied to many other systems. For example, using his formula to the Paris map gives a result that is no more helpful in planning a journey than the geographic approach.

Innovation, Craft, and Geography

The simplicity of using just horizontal, vertical and 45° lines had much to be admired, but over time Harry Beck's design has prevented innovation and its iconic status has made imitation endemic. Now, too much map design is based slavishly on the principles of the London Underground map and not enough new thinking goes into the interpretation of complex modern travel systems. The aim should be to make what is unclear on the ground easier to understand, yet what is straightforward on the ground is often made to look less practical. Frequently, maps are just badly crafted with no effort to improve readability or legibility and time and again they are tucked on the back of a small leaflet as a token gesture, with their type rendered at an unreadable point size.

The map designer should not force a network to follow someone else's set of abstract rules that may have worked then, but should look for clues on the ground to show the individuality and the personality of the area being depicted – which can even contain humour.

For example, London has the Thames and the droopy cola-bottle shaped Circle Line that defines the shape of the West End and City; Europe has radial routes from Paris and the Rhone Valley with its distinctive arrow shape having the point at Frankfurt; Manhattan has its characteristic tilt; and there's a tilted parallelogram that links Liverpool and Manchester together. The axis of a major thoroughfare, river or coastline forms the shapes and angles that help the user to identify with a map.

Towards a New National Railway Map

At the privatization of British Rail in 1994, I realized that there were no satisfactory diagrams of the British rail network. There was a geographic map issued with the National Rail timetable which had schematic maps on the back showing service patterns relating to the timetable page; there was an Inter-City map with limited detail trotted out in diaries; and maps by individual passenger transport organizations (including the Tube map, for example). When one considers the high profile of the Tube map, why does one never see a national rail map at British

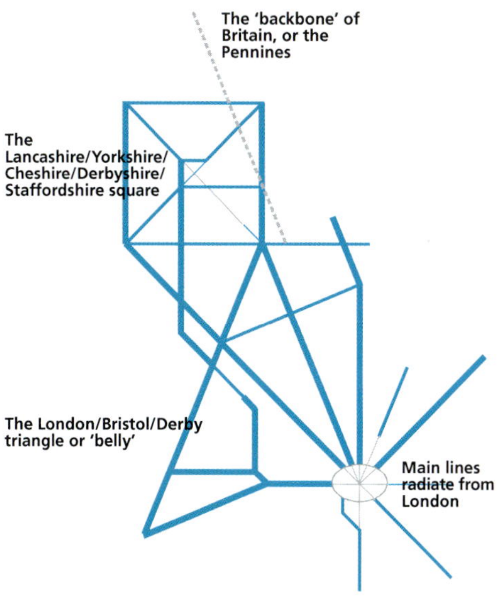

Figure 1
The basic structure for creating a new national railway map

railway stations? Why are the railways so different to the Tube? Was it too difficult? Had anyone ever tried?

So one day I had a go at designing one. The aims and principles were to:

- Create a map that would benefit public transport with an integrated image, overview and icon;
- Simplify routes as much as possible into straight lines, removing the effects of topography where these are a handicap, but retaining alignments that reflect the character of Britain;
- Show better the system as a network, improving the appearance of cross-country routes in particular;
- Reflect the geographical relationships between countries, major conurbations and regions more closely; and
- Pay particular attention to interpreting complex networks (e.g. Cheshire–Lancashire, Yorkshire, and Strathclyde).

The key to the solution was the establishment of a grid to simplify the complex Lancashire–Yorkshire network and a triangle for the critical London–Derby–Bristol 'belly'. Adding a new 22.5° angle (even 11.25° occasionally) enabled all mainlines to radiate from London and the East Coast and Midland mainlines to flow with the slanted shape of Great Britain (which also reflects the Pennines, the backbone of Britain), as shown in Figure 1.

The priority in the design was to keep the mainlines as straight as possible – the 'bones' of the network – so that when viewed from a distance one could clearly understand the underlying structure of the network, with regional lines linking at a secondary level and local lines filling the gaps or radiating from their respective cities.

The map uses a method of showing locations with multiple stations on different lines to avoid station name repetition, the coloured ovals also giving focus to major centres (Figure 2).

The map also uses curves with a large radius, for example, for the Cumbrian coast and north-east Scotland (why force gently curving routes into jerky lines just to retain a fixed radius?). This has become an important aspect of the map's design, with softer curves leading to a more pleasing appearance, particularly on the outer edges. Stations are shown as nodes without the correct bifurcation of routes, as modern multiple-unit trains often reverse to complete their journeys.

Interestingly, I discovered that over time, all design problems can eventually be resolved – what I considered impossible became possible. First, I was able, after all, to label all London termini in their right positions. Later, I

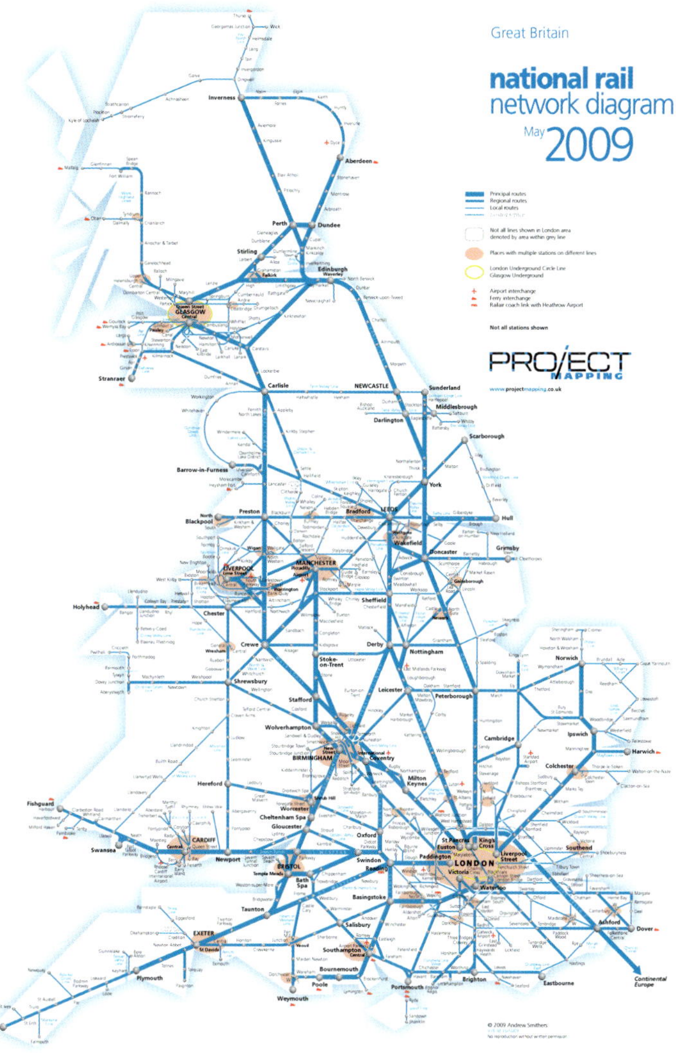

Figure 2
The national rail network diagram

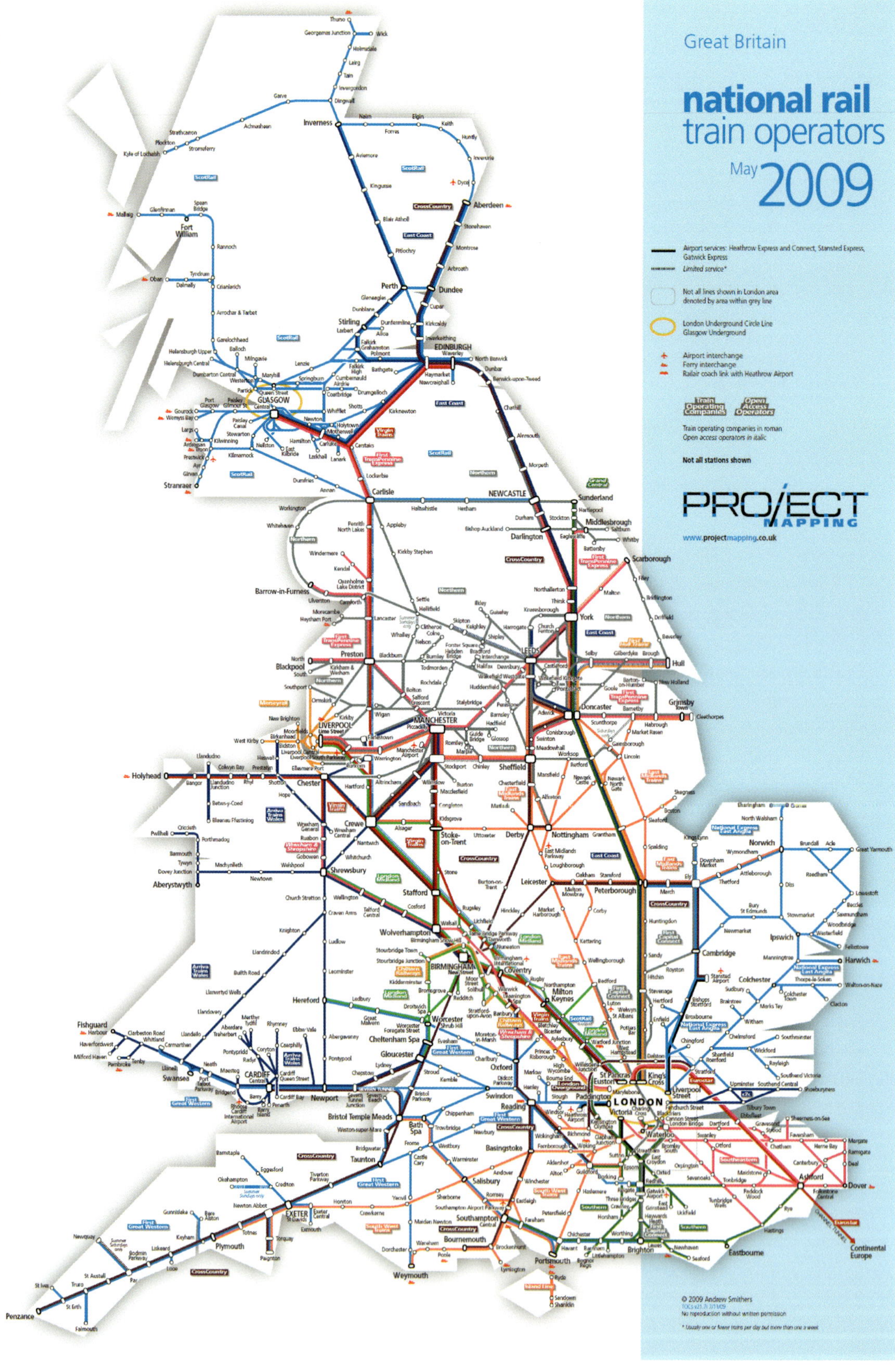

Figure 3
Map showing the national rail train operators

was able to show all the TOCs (Train Operating Companies) in colour.

But at no time did I allow this structure (the 'bones' of the network) to be compromised by local detail. This map was developed into a smaller, diary-size map showing only main routes. Both sizes showed only a selection of stations as it was impossible to include all of these on an A2 sheet or A6 page.

Figure 4
The Merseyrail network map

Other Innovations

Many of my new maps have been designed to improve on existing maps that I consider to be badly crafted pieces of cartography, often in response to releases in the rail press: 'x passenger transport executive has published a Tube-style map to make it easier for customers to understand our services' (or perhaps not!).

With maps based on a major conurbation, an important but self-evident idea is to put the centre of the city in the middle of the map – but this is surprisingly uncommon. Some show services, which is better for local area networks, and some show routes, which is better for national networks (detailed examples can be seen at *www.projectmapping.co.uk*).

The Merseyrail map uses 30º and 60º angles which help to shrink the size of the map to a square and reflect how the network looks. The balloon loop under Liverpool's city centre is shown as such and not forced into squares with the corners rounded off as depicted by Merseytravel. Ticket areas are described better.

The Manchester map indicates the city centre by the use of a large capital M (as seen on roundabouts approaching the city), shows the platform layout at the divided Piccadilly station and also represents the GMPTE (Greater Manchester Passenger Transport Executive) ticketing area using a simple oval.

The Railteam map shows all stations in Paris, a new way of showing the routes and a much more effective use of space.

The London Overground map interprets the orbital nature of the four disparate lines that formed this politically created network by the use of very shallow curves; a new way to show the central London focus and a different way to show the Thames.

Revealing the underlying structure shows the repetition of shapes and that London is a fried egg.

The Valleys map reflects the nature of the South Wales valleys and has English and Welsh versions to avoid the pitfalls of a bilingual map.

The Great Britain 'All Stations' Map

I received some criticism from people who couldn't find their own station on the main rail maps, which do not show all stations because the maps were designed to be a summary of the system.

But as PDFs on the Internet, without the restrictions of

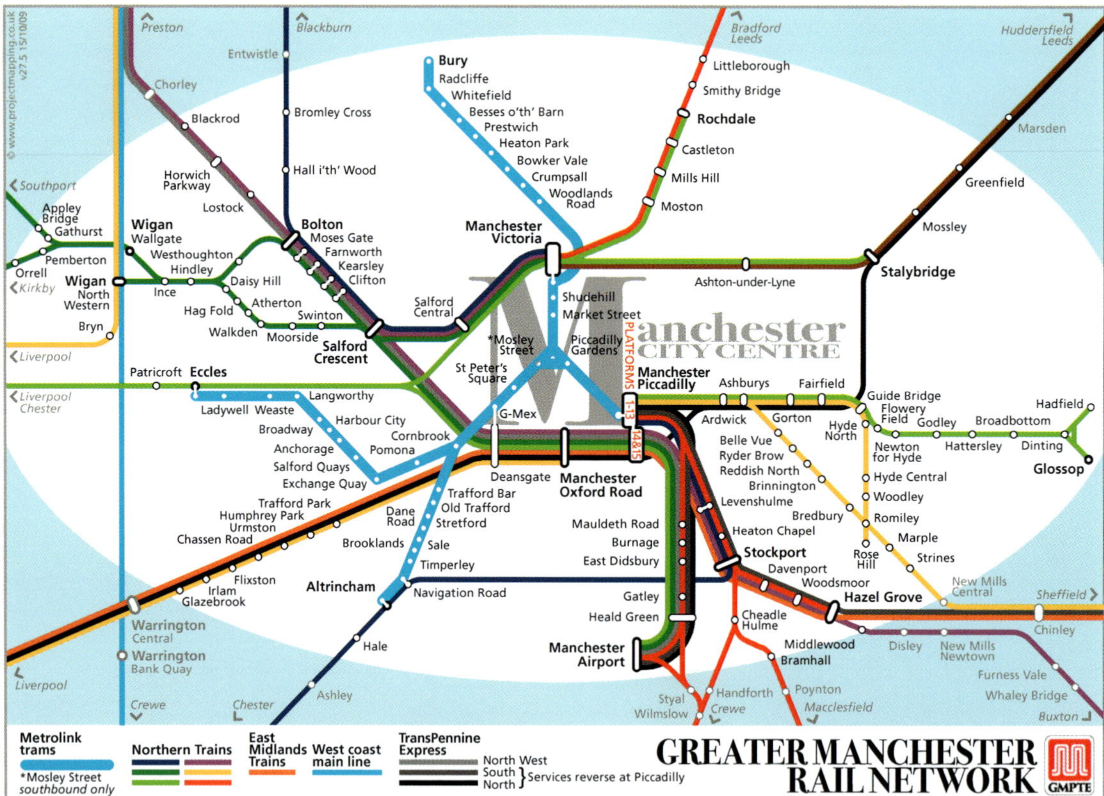

Figure 5
The Manchester network map

printed material, size is no longer an issue as you can zoom-in on the maps. I have just completed an 'all stations' map (Figure 10 shows just the London section). This demonstrates that national, regional or conurbation maps can use the same design. While not wishing to deny designers creative freedom, the utilization of different designs cannot be helpful to the user. For example, the Dublin tourist leaflet

Maps on the Internet

My skills don't extend to creating interactive maps or automatic maps from geographical datasets. But surely there is so much that can be done; the surface of the possibilities has hardly been scratched. For a start, when searching for a journey on a website, the results could be presented as a graphic route instead of (or as well as) the

Figure 6
The Railteam network map

Figure 7
The London Overground network map

long lists of text currently provided. Graphic route cards could be given with tickets. Lines and stations could be clicked-on to reach timetables or station information.

Website

As a graphic designer, not a professional map-maker, I started to develop my ideas for maps and established my website in 2008. The website was intended as a resource portal (including over 650 diagrams) for education, to stimulate debate, present new ideas, criticize, and congratulate. All the examples described in this article can be seen (as well as other designs mentioned by the train operating companies and passenger transport executives) at *http://www.projectmapping.co.uk*. Traffic to my website has climbed to around 300 hits a day and this year, I finally got two maps onto the National Rail Enquiries website,

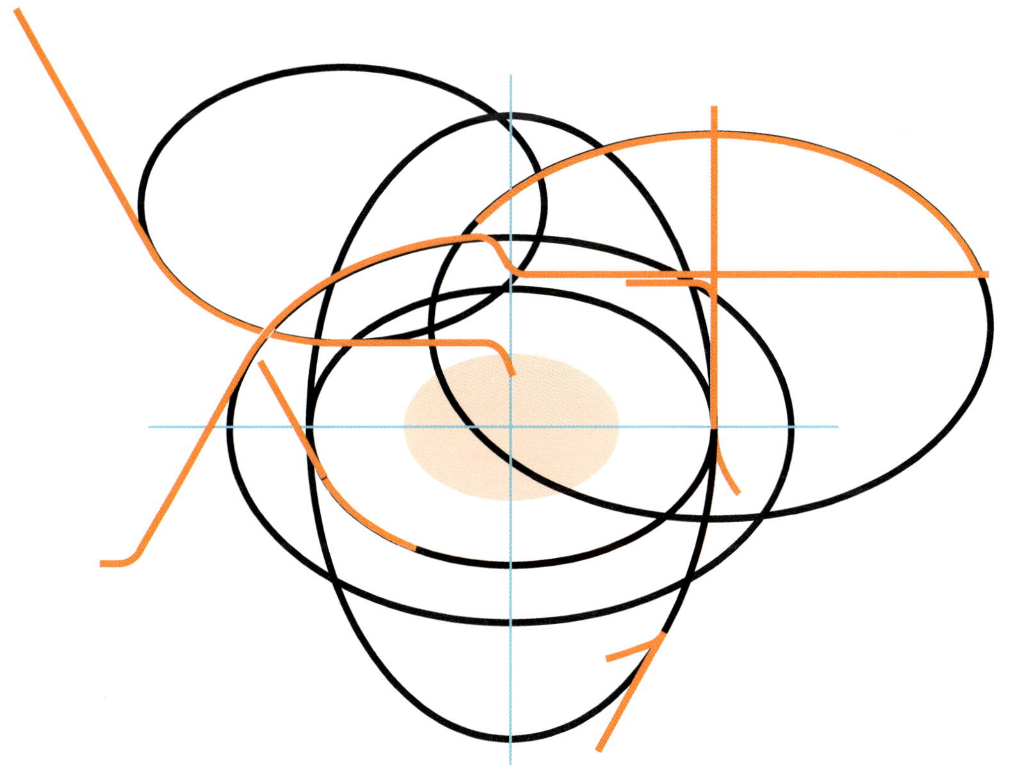

Figure 8
The London Overground network map (design framework)

Figure 9
The Valleys network map

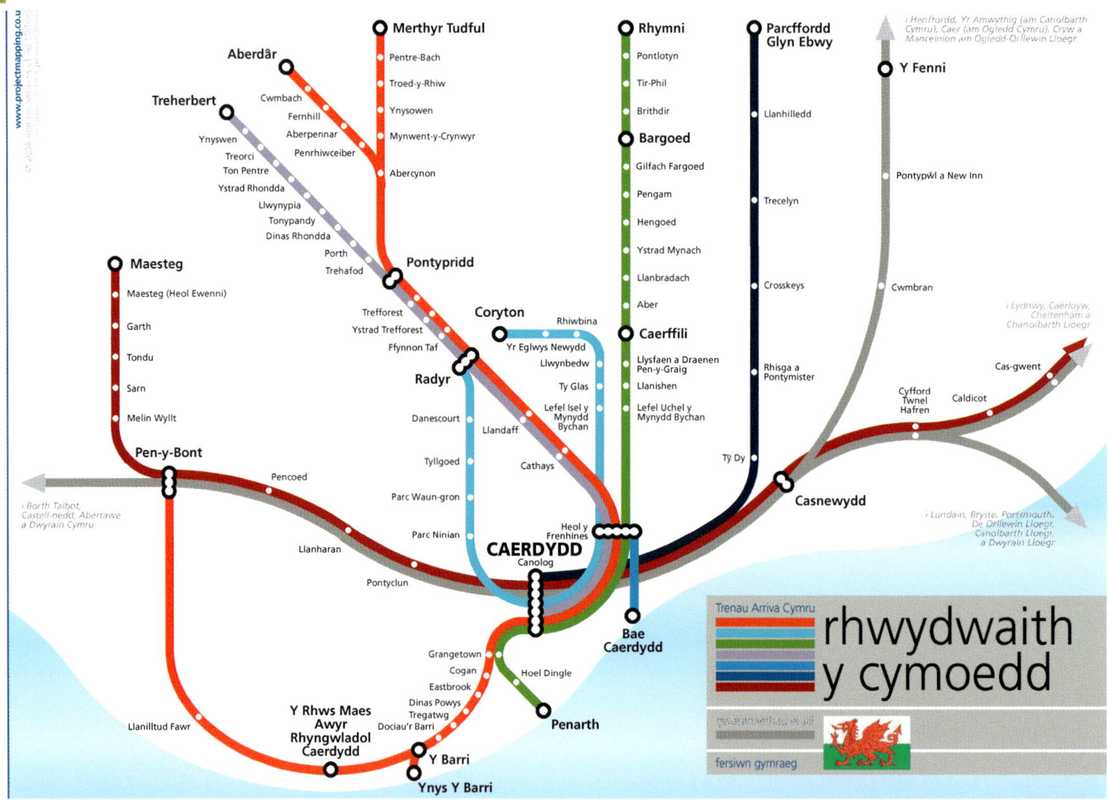

Figure 10
Excerpt from 'all stations' map

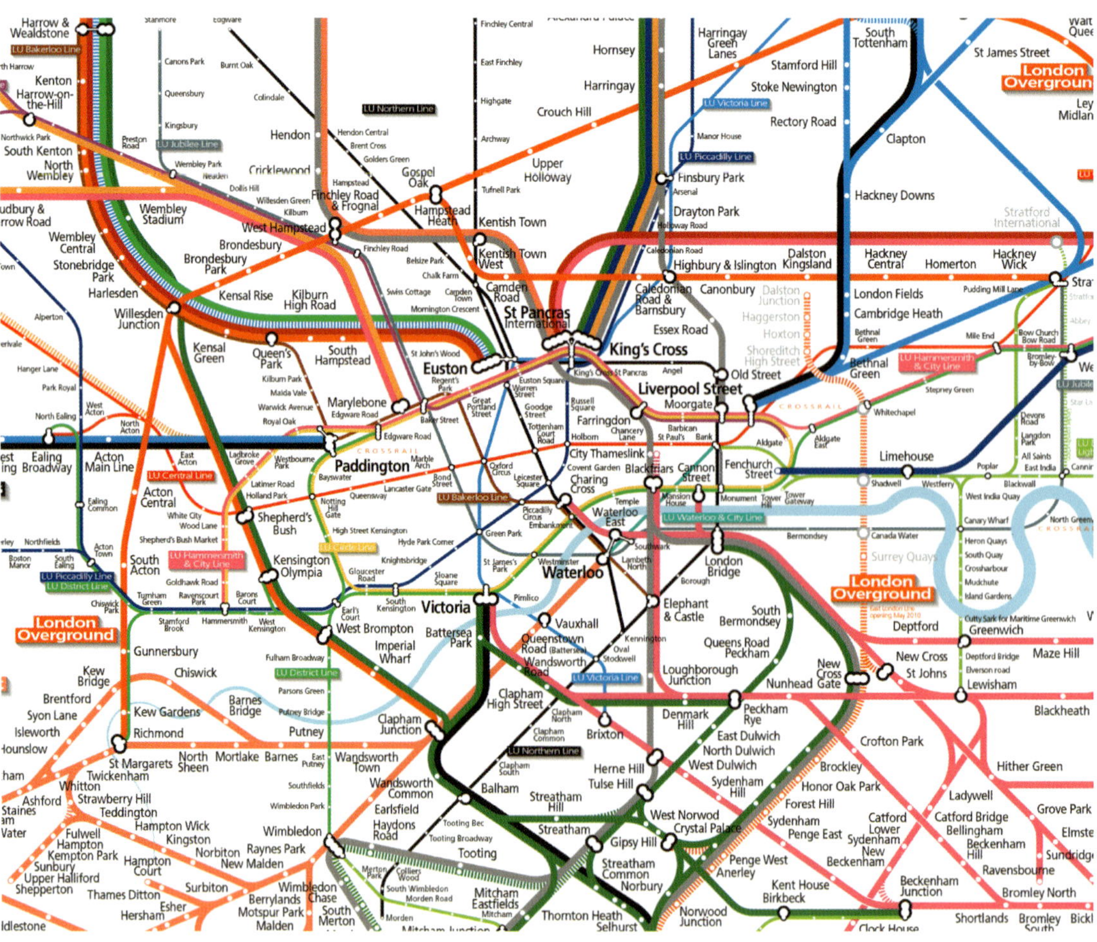

where they currently enjoy 1,000 hits a day.

The Map is the Brand

The map is core to the identity of train operating companies and passenger transport organizations. It's what makes one operator distinct from another. Those corporate identities – symbols, logos, and colour schemes – may provide identity and recognition, but are they just wallpaper? Only the map can ever present the individuality of the operator in a graphic form and so should be promoted positively as a major component in the corporate identity – instead of being hidden away. The map needs deep thought and craft to develop into an icon, so that customers and industry alike can rally around them.

register now

home about support forum faq contact

Develop a great idea with Ordnance Survey

Got a great idea? Make it work with the best mapping in the world.

Experiment with OS OpenSpace to incorporate Ordnance Survey data into your web application.

Develop your idea and see where it can take you.

Find out more

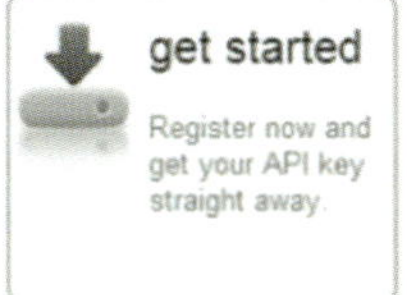

get started
Register now and get your API key straight away.

create
Get the coffee on; it's time to start coding. Experiment with your ideas!

collaborate
What did you think? Become a member of our forum and share your thoughts.

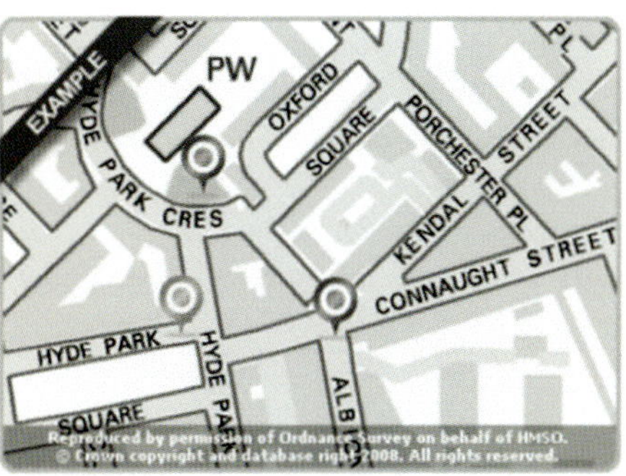

This is an example of what can be achieved with the OS OpenSpace API.

© Crown copyright 2008 Ordnance Survey

Developer agreement | Website terms and conditions | Privacy notice

"BY DAY, BUT NOT BY NIGHT": COUNTER-MAPPING PLATT FIELDS PARK

Naomi L. Hurrel, Emma Kerry and Helen E. Parsons

Originally published in Volume 44, Nos 1 & 2, pp.17–24

Mapping is a medium that is particularly well suited to challenging taken-for-granted assumptions about the world. Most mapping practice denies subjectivity and purports to offer an authoritative and neutral representation. This paper tells the story of an artistic mapping encounter which challenges these assumptions. It explores the strongly situated practices involved in the making and performing of an embodied map of part of Platt Fields Park Manchester. The park can be a dangerous space and mapart can be a mechanism for contesting existing power relations. The mapping was drawn *in situ*, photographed and documented, and subsequently deployed as part of a political process to reclaim the park. Emanating Ordnance Survey styles, this strongly personal artwork subverts established representational cartographic practice and celebrates the potential for a feminist reworking of mapping that embodies aesthetics and political challenge.

Introduction

On Friday 18th of December 2009 a woman was dragged into Whitworth Park, close to the University of Manchester, and raped as she walked home from a Christmas party (BBC, 2009). We walked past the crime scene the next day. For us, this brought to light the stark reality of sexual violence and highlighted our own emotions about the incident, where apparently 'public' spaces, such as parks, are perceived as spaces of danger. Reading the NUS Hidden Marks survey (2010), we discovered that other women students shared a perception of fear of public spaces, despite the stark reality that crime is more frequent in private than in public space (Pain, 1999). Something needed to be said, both about the crime and people's perceptions of spaces of fear.

As Geography Undergraduate students, we knew that maps could act as a powerful tool for expressing spatial relationships and decided to adhere to stereotype by indulging in mapmaking. Our 'Maps in Society' lecturer, Chris Perkins of the University of Manchester, asked us to complete a piece of coursework conducted in Manchester, using a method of counter-mapping, thus, the coursework aims strongly influenced the processes involved, in form and in mapping. In line with this, we took a feminist approach of subversion; instead of a typical quantitative positivist mapping of women's fear, which we felt was inappropriate, we mapped women's fear using map art.

Art is unique in its ability to counteract and challenge, to subvert and provoke, whereas Katharine Harmon suggests that cartographers 'submit to a tacit agreement to obey certain mapping conventions, to speak in a malleable but standard visualized language… [whereas] artists are free to disobey these rules' (2009: 10). As such, there has been a growing trend among artists to use cartography to make a political statement (Watson, 2009). This then provides an alternative, and we argue more appropriate, method to traditional mapping, particularly when exploring the emotional and subjective notions of the perception of fear. The project attempts to subvert the spatial expression of patriarchy (Valentine, 1989), exploring how it is created and how map art is an appropriate technique for understanding its role in regulating the female body. The project had three aims:

- To look at women students' experiences of exclusion from 'public space' through fear;
- To give a political message about what it is to be excluded; and
- To challenge what is perceived to be a 'public space'.

Fear of Crime

Fear of crime is gendered. The threat of sexual violence, almost unfelt by men, causes women to distance themselves from potential attackers, both in space and time (Valentine, 1992; Stanko, 1995; Koskela, 1999; Madge, 1997; and Kosekela and Pain, 2000). Fear forces women to take precautions, which restricts their access to parts of the city (Koskela, 1999). This is often enacted in a spatial-temporal relationship; they do not enter certain places at night. Women who deviate from this norm are seen as responsible for any harassment they receive. The process of keeping the body safe becomes one of 'performative femininity' (Pain, 1999: 126) restricting their independent mobility as a form of self-preservation (Pain, 1997).

This 'fear' and 'blame' culture has resulted in a curfew mentality that represents a spatial expression of patriarchy (Valentine, 1989), which shapes life in cities (Pain, 2001). Effectively, this fear of violence regulates women's bodies (*ibid.*) limiting their access to 'public' space allowing control through 'patriarchal ideology and violence' (Dixon and Jones, 2006: 48). This project then, aims to challenge the notion of 'public space'. A space cannot be considered 'public' when it is unavailable to some of the people some of the time.

Figure 1 Mount Fear Manchester (Reynolds, 2009) (Used with the artist's permission)

In empirical studies of women's fear of crime, mapping has been utilized in order to understand the spatiality of this emotion (Pain, 1997). However, the statistical, empirical evidence of measured levels of fear is limited; data are unavailable and due to their emotive nature, sexual crimes are under-reported (Pain, 1997; Stanko, 1987; 1995; and NUS, 2010). One such example, attempting to visualize crime statistics is the series of Abigail Reynolds' works, entitled Mount Fear (MNtF). Her project drew on 2001–2002 police data to create space which seemed like imaginative fantasy, but challenged conventional mapping of crime by deploying height to represent the number of incidences of urban crimes, overlain on an A–Z map (see Figure 1). The project also covered Manchester, which, in line with our coursework aims, was the focus of our map. MNtF attempted to represent crime statistics in an extremely original way; however, through using statistically-collated data, it is positivist in its very nature. As such, it fails to acknowledge the crime that goes unreported and fails to look into people's feelings of fear.

The interpretation of much of this kind of data has also been misunderstood and misrepresented. In comparing the places of high incidence of assault and high fear, no correlation has been found, leading to the conclusions that such fear is 'irrational' (Pain, 1997; Hough and Mayhew, 1983). Moreover, women are seen to fear public rather than private space; whilst private space is, in reality more dangerous (Pain, 1999). However, often the objective data gathered does not actually answer the subjective question we were asking; why do women fear some places? It has been argued that the construction of a women's gender identity seems to include an idea of vulnerability in public space (Day, 2001). Thus, the perception of fear is a construction which regulates women's bodies and it is that which needs to be challenged. We wished to examine this socio-cultural perspective and the emotions behind it, rather than the objective reality offered in other maps.

Therefore, it can be argued that positivist methodology and quantitative analysis, traditionally used in mapping, are inappropriate when looking at human behaviour. Fear of crime is 'transitory and situational' (Fattah and Sacco, 1989: 211), multi-faceted and individual (Koskela and Pain, 2000). A qualitative approach ought to be taken which reflects the individual's experiences in time. Map art, which highlights experience and the aesthetic above scientific representation (Crampton, 2009), is therefore more appropriate.

Method

In producing a feminist map we needed our methodology to be underpinned with feminist ideology. As Moss (2002: 3) explains, this ideology should 'influence all aspects of the research process'. According to Kirsch (1999), this means ensuring 'research contributes toward enhancing – and not interfering with – the lives of others'; the best way to achieve this is through the collaboration of the researchers with the researched wherever possible. As women students we could blur the 'boundaries between researchers and participants' (*ibid.*: xiv) using ethnographic methods. However, in order not to assume women students are a homogeneous group, as some feminists are criticized for doing (see: The Combahee River Collective in McCann and Kim, 2003), we also drew upon the experiences of others. We felt the use of other people in our piece is justified, firstly as the aim of the map is to highlight their lived experiences and secondly, through sharing the finished piece with other women students at the 'Reclaim the Night' (RTN) march.

Where to Map

Pain (1997) suggests that women fear dark, lonely and unfamiliar places. This was confirmed by the data from the Women's Office (2010) on where 'students felt unsafe in Manchester'. Through considering their findings we looked at streets, alleyways, and shopping centres, before settling on parks. They were chosen in light of recent news reports in both Whitworth and Platt Fields Park detailing violent assaults against women at night (BBC, 2009).

While attacks elsewhere are also reported, aesthetics are a significant consideration for artistic mapping practice. This meant that our decision was based on more than just ethnography. The aesthetics of the parks and their association with nature allowed us to relate them to the female element with our project. Women are often viewed as being synonymous with nature (Ortner, 1974); as such, this stereotypical and rather loose concept was something we wished to play with through the idea of parks as inherently natural spaces that women (and other people) do not always use – or feel part of – and cannot be associated with. This subtle irony was something that finally swayed our decision.

We identified Platt Fields Park as the final location: it was most accessible, looked clear on Ordnance Survey mapping, was available through the Digimap service, and was well frequented by the team members. We chose to use Ordnance Survey mapping as it was easily accessible online, provided a clear, workable format, and used bold colours en bloc to highlight areas of the park with basic symbols, making it simple to copy, whilst quite importantly, still being recognizable when transferred onto the body.

We chose to create a highly performative and embodied piece of map art, as we wanted to create something which was dynamic and enacted (Perkins, 2009); speaking more of experience and emotion than of scientific representation. By 'branding' the woman with the 'scientific' map we were attempting to play with the notion of a patriarchal male gaze eroticizing and controlling the actions of a woman. By putting the model in the place we were mapping, she was walking a route she could not take at night; literally enacting the map which has been drawn on her.

Collecting Data

Kirsch (1999: xi) calls us to question 'whose words – whose reality – am I representing in my work?' asking us to look beyond ethnographic experiences to other women. This caused us to reflect on our own positionality within the work; we did not want to move away from the personal, but to simply analogue other people's experiences. With this in mind, alongside the idea that positionality is inherent in all research (Valentine, 2001) we collected data from a combination of ethnographic and open-ended questions that acknowledge and utilized our own experiences alongside the thoughts and feelings of others. This generated material that was 'rich, detailed and multi-layered' creating a picture of events and allowing individuals to share their own experiences (Silverman 1993: 15).

Using an informal survey, we discussed with women students the varying routes they would take during the day, and night. We struggled to find people at first, suggesting the park is less frequented than we had thought. But, of those we surveyed, we found most totally avoided the park at night due to fear of violence, or confrontation, even when it provided the quickest route. Of those that did use it at night most adjusted their route accordingly. This confirmed that the park, while labelled as a 'public' space, was actually not available to all of the people all of the time. We collated the data and identified three main routes through the park, which we then used to make one final route on which to base our map.

Making the Map

Using the data collated, we then set about making a route through the park which the woman would not use at night. We then identified three main locations at which we could pinpoint the woman in her journey, each one represented by one of the photos which made up our map. In the making of the map a number of considerations were discussed at the various points in the process (visualized in Figure 2). The initial considerations and preliminary photos gave us a chance to consider location, focus and positions. Eventually, we chose the neck, shoulders and lower back, in order to emphasize the femininity of the image but so as to not over-eroticize the body. These areas of the body were also able to show-off the OS-inspired map and the park well, as they covered a large enough portion of skin for the image to appear embodied. We wanted to dress the woman in a black evening dress, to show an ironic tension, as the evening gown represents the temporal nature of the relationship between women and 'public' space.

We also wanted the piece to represent 'natural' space, a place stereotypically defined as 'female' – to juxtapose with the reality of public space, which is dominated by men, for example, cities with urban industrial parks, are typically male – as implied and defined by male-orientated institutions. Alongside this, we had decided to use Ordnance Survey as the basis for a map which aspires to a (masculine) ideal of a 'scientific' and 'neutral' mapping service, however it makes a social claim that space is 'public' despite the social reality that access to this 'public' space is effectively restricted after dark. Through challenging this placement on the Ordnance Survey map, we wished to challenge the notion that space is public.

As feminists, we wanted our work to speak of the experiences of any and all women, which meant that we thought it was important not to show the woman's face, creating an anonymous image which highlights the facelessness of fear. The body then, becomes the contours of the map. Through this decision to keep the woman faceless and anonymous, obviously a woman and yet not overtly erotic, we were able to show that this is an experience any woman could face. However, we did find that keeping the image feminine whilst anonymous was a challenging task, especially as painting the map *in situ* in the park attracted attention from curious passers-by. Despite the pre-prepared templates, distraction from people in the shot and changes in lighting over the course of the day meant that our process took longer than estimated in the field. When moving locations during the shoot, we struggled to recreate our previous idea; the lighting had changed and the shadows were longer. Lots of adjustment was then necessary to ensure the natural lines of the neck and shoulder remained, otherwise the image had the potential to look disembodied.

Having selected our photos (see Figure 2 for the process) there was some debate over whether it was in line with feminist ideology to use the software 'Adobe Photoshop 7.0' to edit the images. We decided it would be, as long as the woman's body was not edited. The lighting in all three pictures was aligned and the background blurring adjusted so as the woman and the Digimap were prominent. The finished images were then printed and cut. The framing took into consideration the proportions and femininity of the body to background and the boldness of

Initial Considerations

Several things were considered before entering the field and further discussed on location in the park: Where and how to place the woman? Where to place the map on the body? Where to locate the shot?

Preliminary Photos

The preliminary photos were taken not only with the 'initial considerations' in mind but also several other technical and aesthetic points. In order to view these properly a range of pictures were taken in the field and then viewed on a larger screen before the final stills and locations could be selected. They took into consideration; angle, focus, light/shadow, people, and time.

Going out into the Field

At each of the three locations we:

drew around templates -> practised shot postions -> imprinted templates onto woman -> painted map onto woman -> took several photos

Choosing the Photos

All photos were then placed in separate documents. Together we cut pictures until we had collectively chosen three single pictures we thought appropriate. We selected them based on the following criteria: lighting, body shape/femininity and ratio of background to body.

Photoshopping

In keeping with our feminist ideology we did not 'enhance' the body, but we did photoshop other areas of our chosen pictures in order to bring the body forward and ensure all three images worked as a set. Using Photoshop we: blurred the contours of the body, blurred objects, sharpened colours, resized according to shape, increased focus on the body, and the painted map.

Mounting the Maps

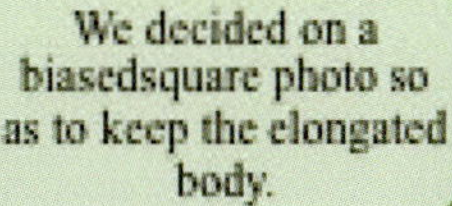

We decided on a biasedsquare photo so as to keep the elongated body.

Each map was cut, then mounted onto backing board before the frame was placed over.

Photos were secured with spray-on adhesive.

Figure 2
Diagram detailing the processes involved in making the counter-map

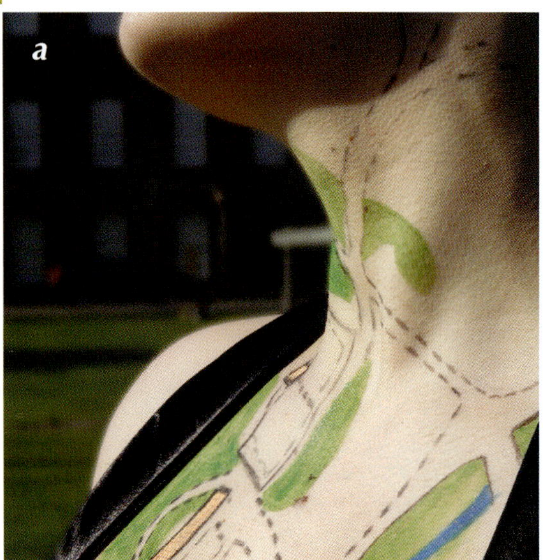

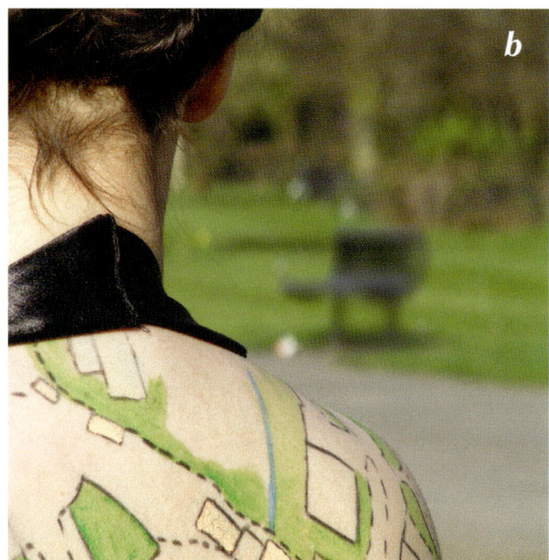

Figure 3a, b and c The individual elements of the artwork

the map. A triptych of square images (Figure 3) were then mounted alongside the legend, chosen in such a way as to tell the story of the woman walking along the path, through the park she wouldn't be able to access at night.

The Legend

Just as a placard in an art gallery guides you through a piece, so does a legend lead the reader through a map. Denis Wood (1992) has argued that legends are an important part of the map's authoritative knowledge; they do much more than simply guide, they actually help naturalize culture. In the same vein, we used our legend to further challenge traditional notions of mapping, particularly to highlight the ways in which most maps deny emotion and experience. In line with our earlier decisions to use Ordnance Survey Digimap data on the model, we chose to subvert an OS legend, as representative of a traditional (male) top-down, scientific mapping which ignores experience and emotion. We wanted to follow in the footsteps of other scholars (e.g. Kwan, 2002) and challenge traditionally masculine mapping which highlights the 'scientific' and 'natural' to reflect our own lived experience of the space through subverting the OS legend (Figure 4).

We edited an OS Explorer legend in Microsoft PowerPoint and emulated its proportions, font, layout and colouring wherever possible so that it appeared, at a glance, the same. The writing and symbology not only guide the reader through the piece but also challenge scientific claims of accuracy professed by Ordnance Survey, replacing them with feminist observations. The legend, alongside the framed map makes for a subversive journey through women's fear in 'public space'. Our images become the map itself.

Feedback and Political Realization

As part of our enactment of the mapping process we took our map to Manchester University Student's Union Reclaim the Night (RTN) movement, where we were able to perform the political message through both walking the very streets we were talking about and getting feedback. In this fashion, the feminist ideology could be realized. The RTN movement seeks to challenge the fear-driven 'curfew mentality' which controls the spatiality of women by advising them to stay indoors after dark (Chan, 2004). The Manchester student movement has recently focused on reclaiming public space including Platt Fields Park. Sharing our map with other women students, thus giving it a platform, allowed us to show our research for mutual benefit in line with our feminist ideals and methodology. We mounted it like a piece of art and requested feedback. By doing this we were able to see how our research reflected the real experiences of women and how they related to the map as both a political statement and a piece of art. Overall, the reaction was positive, with people being interested in the way we aimed to challenge the notion of 'public' space.

Later, we went on to join the RTN march, which went past the park where the original incident took place. By doing this we, the cartographers, became part of a political protest and engaged in fighting against the institutions of patriarchy we were attempting to make a statement about. Through RTN we were able to perform our map, by displaying, interacting and conversing with fellow feminists. At this stage in the process, the message ceased to be a theoretical

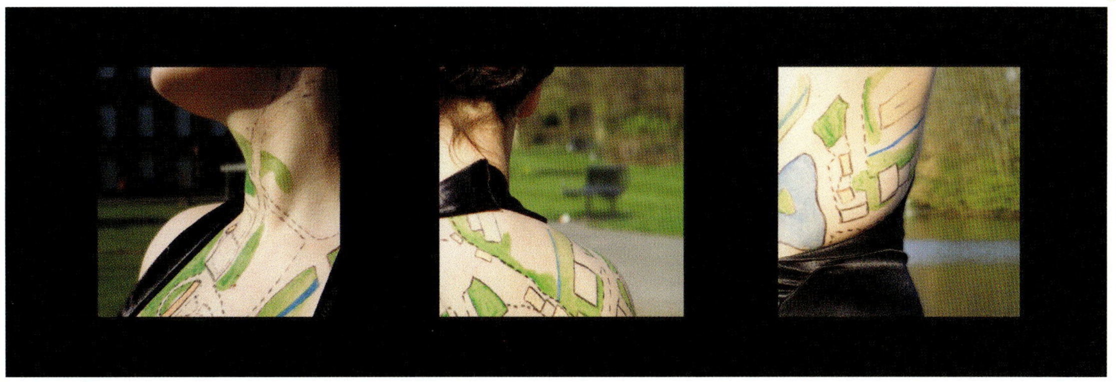

Figure 3d
The final piece

idea about 'performativity', and became a lived reality; the map was called into being through its performance. By literally performing our map art we were able to engage in a provocative part of protest; as art, as a map, and as a visualization of the fight against patriarchy itself.

Concluding Remarks

The project has shown the value of subjective methods and qualitative data in mapping experience. The piece of feminist map art not only provides a clear and

By Day but not By Night 001

Manchester

Platt Fields Park

Consumer Information

Revised with selected change 2010

We have endeavoured to ensure that the information on this product is subversive, but we cannot ensure that it is free from the politics of representation or the dominance of our patriarchal society, particularly with reference to the scientific map. Reproduction of the whole or part by any means is prohibited without the prior permission of the creators (unless it is for the purpose of bottom-up counter mapping or other enterprises that aim for equality and liberation of the genders).

General Information

NARRATING THE MAP

This map is to be read as a story not as scientific fact. It provides a narrative of the story of women's experience of the park. It combines both ethnographic information from the makers and information from other women students. Using these shared experiences it takes a popular day route through the park. However the symbols in the map places emphasis on the fact women do not use the park at night due to 'fear of violence'.

Symbology

Maps hide women's stories and through their 'scientific' appearing symbology they silence women's voice and women's experiences. They prioritise the scientific and intellectual over the experience and the emotional.

WOMEN'S BODY

The sexualised form

Faceless body/any woman

Evening Dress looks out of place in the day. Emphasises the limited access to the space

An embodiment of the space (see OS Map for more)

SPACE OF THE PARK

A seemingly natural space (traditionally associated with women), however in reality entirely 'man'-made.

Path that is only usable by day

'Public Space' with access restrictions to some people at some times of day.

OS MAP

The OS map has been literally branded on the woman, as she is subject to its dominant ideas that silence her experience of the space. However by placing the map on the body of a woman it takes the map away from its traditional place on paper - no longer a scientific representation, instead it becomes an emotional, woman's subversive map of space.

OS Map is a top-down, 'scientific', western, male gaze

The OS map suggests the park is 'public space' available for all, however due to 'fear of violence', women's access is limited.

The subversion of the map exists through its literal embodiment on the woman

Figure 4
Subversive legend detailing the journey, the space and the woman (image derived from OS mapping. Crown Copyright Ordnance Survey. All rights reserved.)

representative depiction of space but also challenges objective methods of mapping, with particular reference to fear. The politically-charged messages could be decoded through the use of the legend that in itself challenged patriarchal top-down mapping and was realized through its enactment, and interaction with RTN.

This project was ambitious, but effective, as can be seen through the feedback. It looked into and challenged ideas of 'public' space, giving women students a forum to express their experiences of exclusion due to fear. By using a feminist ethnographic methodology, and showing it at Reclaim the Night, we were able to perform our map art in such a way that it became a political statement about the way patriarchy controls space and the way in which women are forced into a role of performative femininity. We were also able to examine our own, and other students' exclusion from 'public space' through fear. Indeed, we were able to challenge what 'public space' means.

Art mapping enabled the performance, conveying this message, expressing emotion and subjective experience rather than an arguably far less appropriate quantitative approach to mapping fear. Ultimately, the map provided us with a powerful means of visualizing the relationship between patriarchy, power, exclusion, and space.

References

British Broadcasting Corporation (BBC) (2009) "Woman Dragged into Park and Raped" Available at: *http://news.bbc.co.uk/1/hi/england/manchester/8420360.stm* (accessed 03/05/10).

Chan, B. (2004) "Reclaim the Night" In Goodman, J. & Meekosha, H. (Eds) *Social Movements in Action: Conference Papers* (pp.20–23) Sydney: Research Initiative in International Activism.

Combahee River Collective (2003) "A Black Feminist Statement" In McCann, C.R. and Kim, S.K. (Eds) *Feminist Theory Reader: Local and Global Perspectives* (pp.164–171) New York: Routledge. (Original work published in 1977.)

Crampton, J.W. (2009) "Cartography: Performative, Participatory, Political" *Progress in Human Geography* 33 (6) pp.840–848.

Day, K. (2001) "Constructing Masculinity and Women's Fear in Public Space in Irvine, California" *Gender, Place & Culture* 8 (2) pp.109–127.

Dixon, D. and Jones, J. (2006) "Feminist Geographies of Difference, Relation, and Construction" In Aitken.S and Valentine G (Eds) *Approaches to Human Geography* (pp.42–57) London: SAGE Publications.

Hough, M. and Mayhew, P. (1983) *British Crime Survey: First Report* London: HMSO.

Fattah, E.A. and Sacco, V.F. (1989) *Crime and Victimization of the Elderly* New York: Springer.

Harmon, K. (2009) *The Map as Art* New York: Princeton Architectural Press.

Kirsch, G. (1999) *Ethical Dilemmas in Feminist Research: The Politics of Location, Interpretation, and Publication* New York: State University of New York Press.

Koskela, H. (1999) "'Gendered Exclusions': Women's Fear of Violence and Changing Relations to Space" *Geografiska Annaler (Series B: Human Geography)* 81 (2) pp.111–124.

Koskela, H. and Pain, R. (2000) "Revisiting Fear and Place: Women's Fear of Attack and the Built Environment" *Geoforum* 31 (2) pp.269–280.

Kwan, M-P. (2002) "Feminist Visualisation: Reenvisioning GIS as a Method in Feminist Geographic Research" *Annals of the Association of American Geographers* 92 (4) pp.645–661.

Madge, C. (1997) "Public Parks and the Geography of Fear" *Tijdschrift voor Economische en Sociale Geografie* 88 (3) pp.237–250.

McCann, C. and Kim, S. (2003) (Eds) *Feminist Local and Global Theory Perspectives Reader* New York: Routledge.

Moss, P. (2002) "Taking on, Thinking about, and Doing Feminist Research in Geography" In Moss. P (Ed.) *Feminist Geographies in Practice* (pp.1–20) Oxford: Blackwell Publishers.

National Union of Students (NUS) (2010) *Hidden Marks: A Study of Women Student's Experiences of Harassment, Stalking, Violence and Sexual Assault* NUS: London.

Ortner, S.B. (1974) "Is Female to Male as Nature is to Culture?" *Feminist Studies* 1 pp.5–31.

Pain, R. (1997) "Social Geographies of Women's Fear of Crime" *Transactions of the Institute of British Geographers* 22 (2) pp.231–244.

Pain, R. (1999) "Women's Experiences of Violence Over the Life-Course" In Teather, E.K. (Ed.) *Embodied Geographies: Spaces, Bodies and Rites of Passage* London: Routledge.

Pain, R. (2001) "Gender, Race, Age and Fear in the City" *Urban Studies* 38 (5, 6) pp.899–913.

Perkins, C. (2009) "Performative and Embodied Mapping" In Kitchin, R. and Thrift, N. (Eds) *International Encyclopaedia of Human Geography* (pp.126–132) Oxford: Elsevier.

Reynolds, A. (2010) "MNtF Manc" Available at: *http://www.abigailreynolds.com/mntF/mntFManc.html* (accessed 01/05/10).

Silverman, D. (1993) *Interpreting Qualitative Data: Methods for Analysing Talk, Text and Interaction* London: SAGE Publications.

Stanko, E.A. (1987) "Typical Violence, Normal Precaution: Men, Women and Interpersonal Violence in England, Wales, Scotland and the USA" In Hanmer, J. and Maynard, M. (Eds) *Women Violence and Social Control* (pp.122–134) London: Macmillan.

Stanko, E.A. (1995) "Women, Crime and Fear" *Annals of the American Academy of Political and Social Science* 539 Reactions to Crime and Violence pp.46–58.
Valentine, G. (1989) "The Geography of Women's Fear" *Area* 21 pp.385–390.
Valentine, G. (1992) "Images of Danger: Women's Sources of Information About the Spatial Distribution of Male Violence" *Area* 24 pp.22–29.
Valentine, G. (2001) *Social Geographies: Space and Society* London: Pearson Education.
Watson, R. (2009) "Mapping and Contemporary Art" *The Cartographic Journal* 46 (4) pp.293–307.
Women's Office (2010) Personal communication by letter, 12/04/10.
Wood, D. (1992) *The Power of Maps* New York: Guilford Press.

HELPING HAITI: SOME REFLECTIONS ON CONTRIBUTING TO A GLOBAL DISASTER REFLIEF EFFORT

Alexander J. Kent

Originally published in Volume 44, Nos 1 & 2, pp.39–45

One of the success stories to arise from the aftermath of the earthquake that struck Haiti on 12th January 2010 was the major role that volunteered geographical data played in assessing the damage and assisting many relief efforts. This paper provides some reflections on contributing to GEO-CAN (Global Earth Observation Catastrophe Assessment Network), a coordinated initiative in which over 600 experts and 131 private and academic institutions representing 23 countries took part to assess damaged buildings for the Haitian government using specially released satellite and aerial imagery. It aims to illustrate how, through effective coordination, integration and dissemination with other volunteered geographical data, the event became a milestone in disaster response and relief mitigation.

Introduction

At 4.53 pm local time on Tuesday, 12th January 2010, a magnitude 7.0 earthquake struck the island of Hispaniola in the Caribbean Sea, with an epicentre 25 km west of Port-au-Prince, Haiti (USGS, 2010). Official estimates maintain that around 230,000 people were killed (Associated Press, 2010), with a further 300,000 injured (Wooldridge, 2010). In less than a minute, the event levelled approximately 20 per cent of the buildings in greater Port-au-Prince and left a million homeless (Eguchi *et al.*, 2010).

When hearing news of a disaster striking another part of the globe, it is difficult to not feel a sense of helplessness. Work and home commitments can inhibit any direct action and physical access to the disaster zone – for those who are able to go – is often limited, dangerous, or both. One response might be to donate to those charities requesting money or material to send, either before or after a detailed assessment of the actual needs and aid requirements of the affected people has taken place. It may be assumed that states, working with the United Nations and other non-governmental organizations (NGOs), lead the relief effort and distribute aid as necessary, leaving any direct involvement by the individual (at least in the critical early stages) somewhat unlikely.

Geographical information plays a vital role in mitigating the effects of any disaster, from identifying and locating the most vulnerable to relocation and reconstruction efforts. As one of the world's poorest countries, Haiti lacked comprehensive databases of assets, infrastructure, population and location, and this situation was compounded by a dearth of up-to-date, accurate maps (even road maps and online maps), so that even the most fundamental geographical resources were not available (Zook *et al.*, 2010). The democratization of geospatial technology over the last few years, with, for example, the removal of Selective Availability (SA) signal degradation in GPS (global positioning system) in 2000 and the launch of Google Earth in 2005, has not only offered a greater sense of connection between people and place but has also opened up new ways of direct involvement in disaster relief – however remote that involvement may be.

Immediately following the Haitian earthquake, an unprecedented wave of volunteers from around the world rallied to fill the critical gaps in geographical data, working remotely to provide up-to-date maps and assess the damage. This paper focuses on the GEO-CAN (Global Earth Observation Catastrophe Assessment Network) initiative coordinated by ImageCat (based in Long Beach, California and Surrey, England) and EERI (Earthquake Engineering Resource Institute), in which a team of over 600 expert volunteers from around the world collaborated to assess damaged buildings. Specially released high-resolution satellite and aerial imagery were analysed and interpreted to create a GIS (geographical information system) database of damaged buildings in and around Port-au-Prince. The gathered information contributed to the Building Damage Assessment Report in support of the Post Disaster Needs Assessment (PDNA) and Recovery Framework, as submitted by the World Bank and the Global Facility for Disaster Reduction and Recovery (working with the United Nations and the European Commission), to the Haitian government on 3rd March 2010 (ImageCat and EERI, 2010).

Harnessing the Power of the Image and the Crowd

Remote sensing technology provides an opportunity to monitor wide geographical areas, to observe inaccessible places, and to reveal hidden characteristics by detecting reflected or emitted energy at wavelengths beyond normal human vision. These capabilities are particularly useful for assessing the impact of disasters, and the 'bigger picture' from satellite or aerial imagery can often provide the first holistic indication of physical conditions on the ground.

The strength of the GEO-CAN method lies in its coordinated analysis of remotely sensed imagery. Its approach is founded upon an earlier initiative that was

	**Grade 1: Negligible to slight damage (no structural damage, slight non-structural damage)** Hair-line cracks in very few walls. Fall of small pieces of plaster only. Fall of loose stones from upper parts of buildings in very few cases.
	**Grade 2: Moderate damage (slight structural damage, moderate non-structural damage)** Cracks in many walls. Fall of fairly large pieces of plaster. Partial collapse of chimneys.
	**Grade 3: Substantial to heavy damage (moderate structural damage, heavy non-structural damage)** Large and extensive cracks in most walls. Roof tiles detach. Chimneys fracture at the roof line; failure of individual non-structural elements (partitions, gable walls).
	**Grade 4: Very heavy damage (heavy structural damage, very heavy non-structural damage)** Serious failure of walls; partial structural failure of roofs and floors.
	**Grade 5: Destruction (very heavy structural damage)** Total or near total collapse.

Figure 1 *Classification of structural damage to masonry buildings from the European Macroseismic Scale 1998 (Grünthal, 1998)*

designed in response to an earthquake which struck Wenchuan, China in 2008. Using a prototype of Virtual Disaster Viewer (VDV), an online portal developed by ImageCat and supported by an international consortium of earthquake experts from Europe and the USA, the study area was divided into grid cells and allocated to remote volunteers, who analysed pre- and post-event Quickbird satellite imagery to classify damaged buildings according to the European Macroseismic Scale 1998, or EMS-98 (Figure 1) (Bevington *et al.*, 2010).

Following the 2010 Haitian earthquake, the VDV was customized to provide a grid of 2.5 km^2 cells, and a team of over 50 engineers, geoscientists and social scientists (comprising ImageCat employees and a network of volunteers familiar with the 2008 implementation) were assembled (Bevington *et al.*, 2010). Essentially, the process involved three successive phases of damage assessment, based on the interpretation of pre- and post-disaster imagery. The first of these initially focused on identifying collapsed buildings (Level 5 on the EMS-98) in an area covering 133.75 km^2 within the capital, Port-au-Prince (ImageCat and EERI, 2010). This used 50 cm GeoEye-1 satellite imagery, which was collected on 13th January and made publically available by Google, to identify over 5,000 buildings in 48 hours (Bevington *et al.*, 2010). Figure 2 shows pre- and post-event imagery and the

Figure 2 *Collapsed buildings identified and marked in Phase 1, over pre-event imagery from 27th July 2009 (left); and over specially released GeoEye-1 imagery from 13th January 2010 (right)*

pointers used to mark the damaged buildings as a KML (keyhole mark-up language) file in Google Earth, in preparation for Phase 2.

In the next phase, the study area was extended beyond Port-au-Prince (Figure 3) and a more detailed analysis of damaged buildings was undertaken. The network of volunteers from Phase 1 was expanded for Phase 2, and formally recognized as GEO-CAN, which eventually grew to include over 600 experts representing 131 private and academic institutions in 23 different countries, from Sudan to China and Germany to Costa Rica (Bevington *et al.*, 2010). The rapid growth of the network was largely due to its reliance on existing participants to enlist colleagues via social networking sites such as Twitter and Facebook (Eguchi *et al.*, 2010) and to contact via email institutions (such as universities) with an established record in remote sensing. Crucially, the network sought experts, for example, those holding at least a Master's degree in remote sensing/image processing plus 4–5 years' experience (Bevington, 2010). After screening by ImageCat, each new participant was then sent detailed instructions in PDF (portable document format).

Just over a week after the earthquake had struck Haiti, an email asking for volunteers to join the GEO-CAN initiative had been forwarded to my inbox at the University of Southampton. The timescale was short, as the work had to be completed over the weekend of 23rd–24th January. As well as the instructions, a set of clear guidelines were given on how to interpret the imagery for signs of debris – paying particular attention to the condition of neighbouring structures and examining shadows – to assess damaged buildings according to the EMS-98 scale. Participants were automatically assigned a grid square via the VDV to 'check-out' before opening a supplied set of KML files in Google Earth, through which to view some freshly released high-resolution 15 cm aerial imagery of the corresponding area and the marked buildings that were identified in Phase 1. Working from top left to bottom right in each cell, the footprint of every building deemed as falling into EMS-98 Level 4 (Very Heavy Damage) or 5 (Destruction or Collapsed) was digitized as a polygon (Figure 4). These polygons were accompanied by attribute data comprising the EMS-98 level, degree of confidence in the assessment (by assigning a value of 0–100), and a

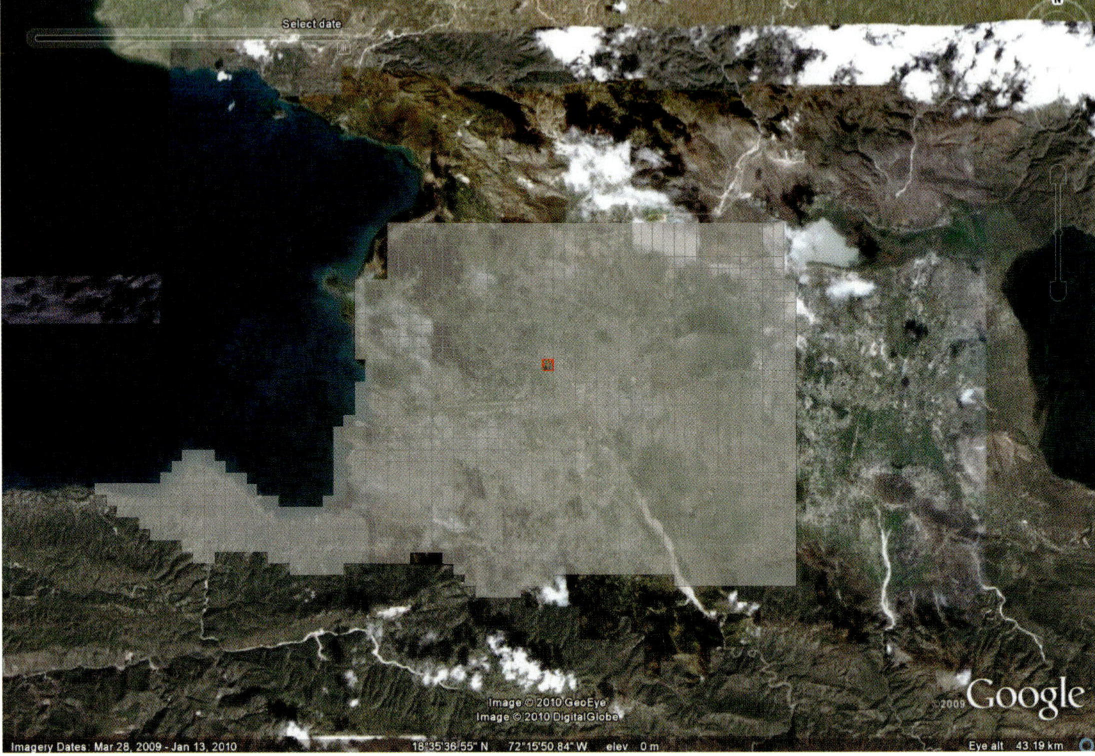

Figure 3 *The grid surrounding Port-au-Prince, which was extended to cover 346 km² for Phase 2*

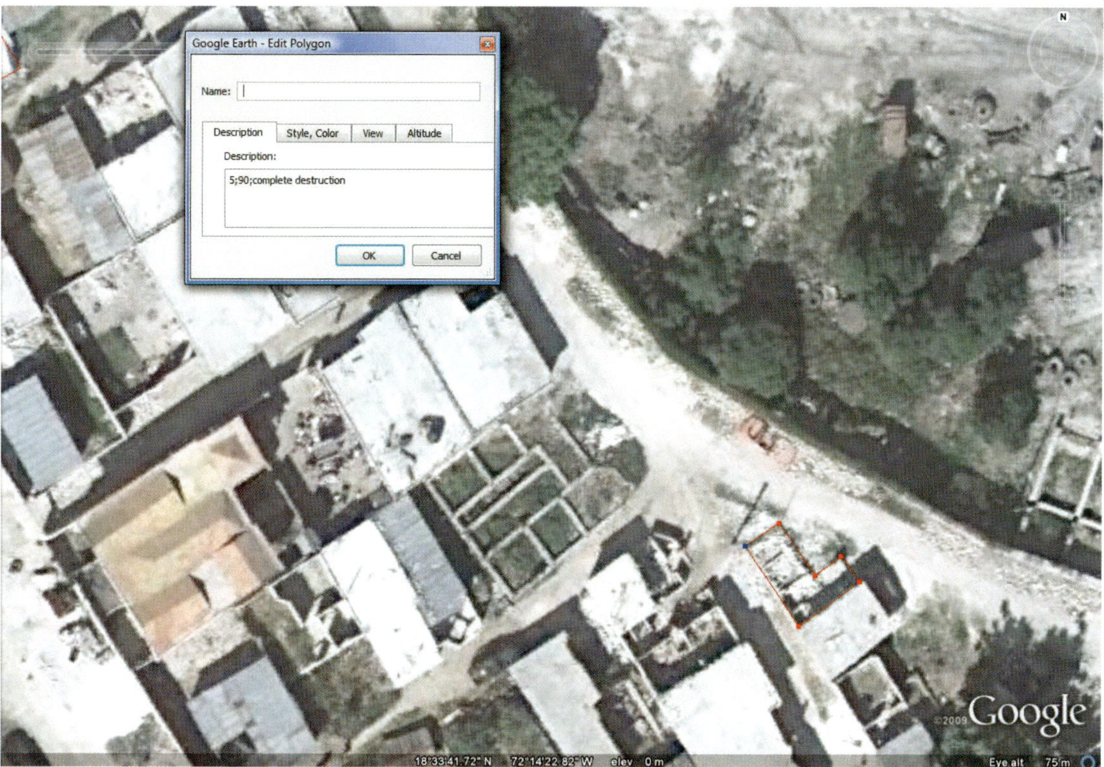

Figure 4 *Identifying damaged buildings and tracing their footprints in Google Earth as part of Phase 2, estimating the level of damage according to EMS-98 category (Level 4 or 5), rating confidence in the assessment (0–100), and providing a brief description of the damage*

short description of the damage. These data were configured as layers according to grid number, before being sent via email as zipped KML files (KMZ) to the central repository at ImageCat for inspection. The completed grid cell was then 'checked-in' via the VDV before the assignment of another. The whole process, of VDV 'check-out', interpret, digitize and add attribute data in Google Earth, send KMZ file, and VDV 'check-in' took some getting used to, while the most difficult stage in the process was perhaps knowing exactly how much confidence to place in your own assessment of the damage. That a recent ImageCat and EERI (2010) workshop has made recommendations such as including a menu of preset options and providing video guides for understanding the damage assessment process is encouraging.

Within a week of the earthquake, 30,000 buildings had been identified as heavily damaged or collapsed across the study area, using solely remotely sensed imagery as a data source (Eguchi *et al.*, 2010). The area covered by Phase 2 was expanded on 27th January to include 1024.75 km^2,

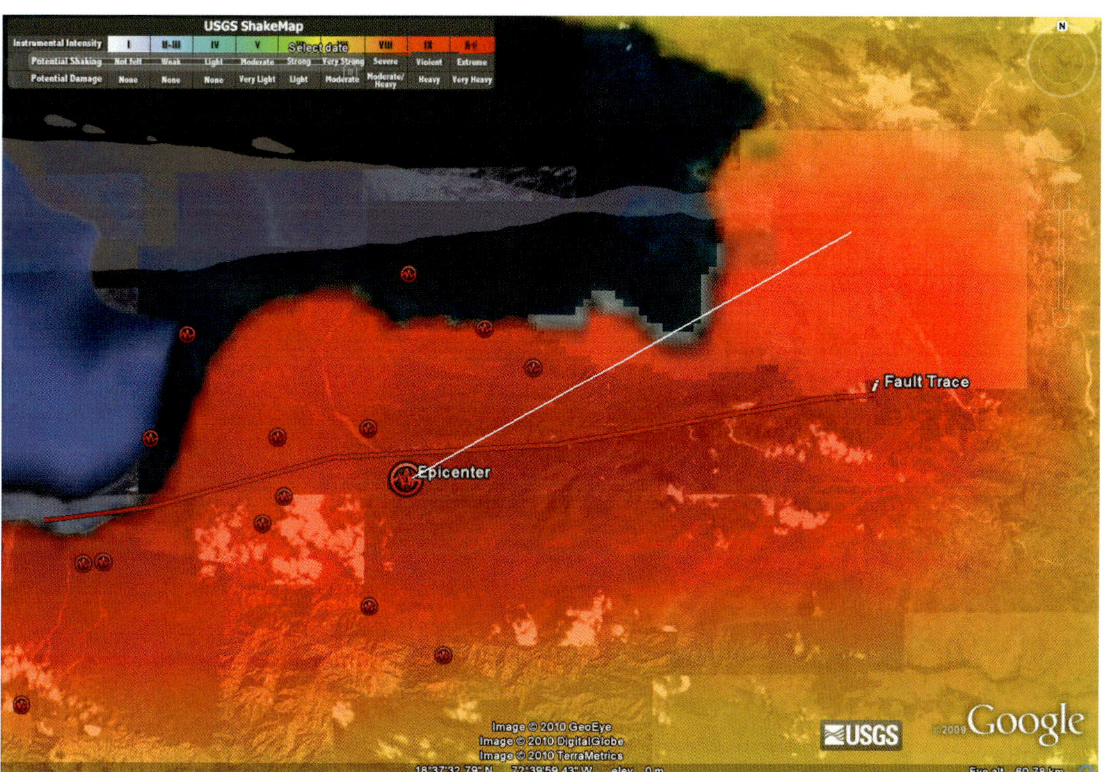

Figure 5 *The USGS 'shake map' provided to give a sense of the geography of the earthquake, allowing the Euclidean distance from grid square to epicentre to be calculated and help scrutinize damage (USGS, 2010)*

Figure 6 Pre- and post-event OSM coverage of Port-au-Prince, Haiti in December 2009 (left) and 14th January 2010 (right) (Maron, 2010)

which required a further 19 days to complete, during which a further 9,000 buildings were identified (ImageCat and EERI, 2010). The USGS released a 'shake map' so that the distance from the epicentre could be measured, if desired, to assist the interpretation process (Figure 5). A third phase was begun on 5th March 2010, using LIDAR (light detection and ranging) and thermal infrared imagery in an attempt to study the surface of the rupture, provide improved classifications, and conduct fire and thermal anomaly mapping (Eguchi and Adams, 2010).

Impact, Limitations and Legacy

As a result of the GEO-CAN initiative, over 90,000 buildings were identified as either destroyed or having sustained heavy damage (Eguchi *et al.*, 2010). The building footprint information gathered during Phase 2 was used to create damage maps and to calculate floor space, leading to an estimated cost of US$6 billion for the reconstruction and recovery of these buildings (Bevington *et al.*, 2010). However, while imagery is useful for exhibiting land cover and gaining an appreciation of the physical characteristics and condition of many features, it lacks the representation of land use and interpretation of the landscape that maps can provide, which is no less critical in supporting the relief effort.

The OpenStreetMap (OSM) community, in partnership with CrisisCommons, geared up in the days following the earthquake to update the basic open-source base maps of

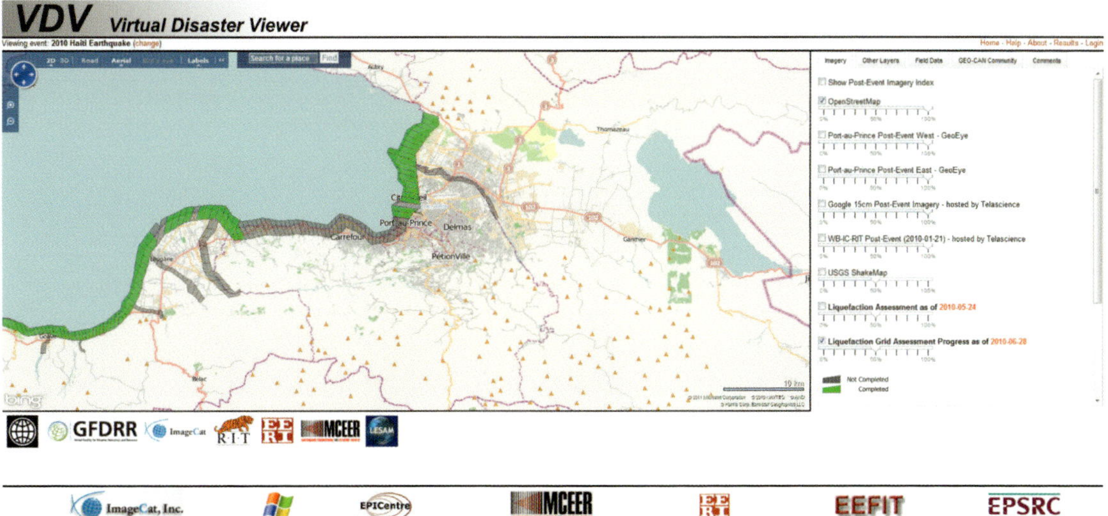

Figure 7 Screenshot from the Virtual Disaster Viewer, showing the integration of OSM data (www.virtualdisasterviewer.com)

Haiti, providing the most detailed mapping of locations of road networks and critical infrastructure (ImageCat and EERI, 2010). Google, DigitalGlobe, and GeoEye worked together to release high-quality satellite imagery within 24 hours of the disaster (Zook *et al.*, 2010) and OSM contributors began digitizing the imagery, as Keegan (2010) explains:

> When the earthquake happened it was a signal for OSM members around the globe to start downloading satellite images (either freely available or donated by Yahoo) and then to start tracing the outlines of streets on top so a map emerged. Volunteers on the ground in Haiti, often using Garmin GPS locators, added vital local information – such as which roads were passable, where the hospitals were situated, where refugee camps were, or walls, pharmacies, hedges and so forth – so rescue workers had an invaluable tool. The result is a new, detailed map that is updated frequently, unlike most commercial maps.

Figure 6 shows OSM coverage of Port-au-Prince before and after the earthquake, demonstrating the huge leap forward in understanding the region's infrastructure that this community provided.

Ironically, the huge response of remote volunteers to the Haiti earthquake and utilization of OSM and other crowd-sourced mapping initiatives such as Google Map Maker also instigated some problems. As Zook *et al.* (2010) explain, due to licensing issues, data were not portable between the two systems and efforts were undoubtedly duplicated, but more importantly, this incompatibility resulted in maps with varying degrees of coverage, depending upon the location within Haiti.

Inevitably, crowd-sourcing also raises concerns over accuracy. In requiring an 'expert crowd', the GEO-CAN initiative sought to maintain high standards of accuracy and reliability in its assessment. It was found, however, that some image interpreters were better than others, although an overall accuracy of 93 per cent is claimed (ImageCat and EERI, 2010). But imagery has inherent limitations, not least its plan perspective, which can hinder interpretation. Ground surveys played an important role in filling in a complete picture of the damage, particularly when the upper floors of buildings had collapsed onto lower floors (Eguchi *et al.*, 2010). Many of the organizations involved in GEO-CAN sent field reconnaissance teams to Haiti and were able to systematically survey damage in areas that were being analysed by the remote sensing scientists, allowing calibration and confirmation of the damage assessment (ImageCat and EERI, 2010). It is important to bear in mind that discrepancies will always exist and with enough people working together, any errors by one individual can be easily corrected by another (Zook *et al.*, 2010).

The development of the Virtual Disaster Viewer to incorporate OSM data (Figure 7) alongside high-resolution imagery and other field data has therefore been a major step forward. Its role as a centralized hub through which to coordinate and share crowd-sourced information, coupled with the capacity to harmonize the necessity for highly structured datasets with the organic nature of the crowd- sourcing, perhaps offers a future model. Indeed, the FAO (2010) uses OSM data in its Interactive Food Security Tool, an internet GIS. If the result would be to save valuable time and resources of those on the ground and maximize the collection of data while minimizing the duplication of effort, it would be a model worth pursuing.

Perhaps one of the most motivating experiences of my involvement with GEO-CAN was the sense that I was taking part in a global effort to directly help those in need. The ability to see the overall development of the project as time went on was rewarding and suggested that others around the world were working simultaneously with me to achieve something good. Mapping is not exclusively about possessing; it can also be about giving.

Conclusion

Through effective coordination, integration, and dissemination with other volunteered geographical information, the GEO-CAN initiative and Virtual Disaster Viewer have provided a wide-ranging and important set of tools for mitigating the effects of the 2010 Haiti earthquake. Crucially, the release of high-resolution satellite and aerial imagery soon after the disaster allowed a detailed assessment of the damage that was close to real-time. Although disaster relief will always require the presence of people on the ground, perhaps the real achievement of this initiative is how it was able to harness the expertise – and enthusiasm – of hundreds of individuals from around the globe so quickly. The empowering opportunity to help those in need directly, yet remotely, was a milestone in disaster mitigation and is a sure step towards directing relief efforts in future.

References

Associated Press (2010) "Haiti raises earthquake toll to 230,000" *The Washington Post* 10th February 2010 Available at: *http://www.washingtonpost.com/wp-dyn/content/article/2010/02/09/AR2010020904447.html* (accessed 31/12/10).

Bevington, J. (2010) "Operation GEO-CAN: Coordination of an unique global response community for disasters" *Paper presented at the AGI Northern Ireland event, Belfast, 19th May 2010 Available at: http://www.agi.org.uk/past-events/2010/5/19/agi-northern-ireland-event-presentations.html* (accessed 31/12/10).

Bevington, J., Adams, B.J. and Eguchi, R.T. (2010) "GEO-CAN Debuts to Map Haiti Damage" *Imaging Notes 25 (2)* Available at: *http://www.imagingnotes.com go/article.php?mp_id=208* (accessed 31/12/2010).

Eguchi, R.T. and Adams, B.J. (2010) "World Bank / GFDRR / ImageCat / RIT Remote Sensing and Damage Assessment Mission: Haiti, January 2010" (Project Sheet) Available at: *http://www.eqclearinghouse.org/20100112-haiti/wp-content/uploads/2010/02/ImageCat-Haiti-EQ-Pr oject-Sheet-EERI-20100209.pdf* (accessed 31/12/10).

Eguchi, R.T., Gill, S.P., Shubharoop, G., Svekla, W., Adams, B.J., Evans, G., Toro, J., Saito, K. and Spence, R. (2010) "The January 12, 2010 Haiti Earthquake: A Comprehensive Damage Assessment Using Very High Resolution Areal Imagery" *Paper presented at the 8th International Workshop on Remote Sensing for Disaster Management, Tokyo, 30th September–1st October 2010* Available at: *http://www.enveng.titech.ac.jp/midorikawa/rsdm2010_pdf/19_eguchi_paper.pdf* (accessed 31/12/10).

Food and Agriculture Organization of the United Nations (FAO) (2010) "Interactive Food Security Tool" (Internet GIS) Available at: *http://www.fao.org/haiti-earthquake/en* (accessed 17/01/11).

Grünthal, G. (Ed.) (1998) *European Macroseismic Scale 1998 (EMS-98)* Luxembourg: Centre Europèen de Géodynamique et de Séismologie.

ImageCat and Earthquake Engineering Resource Institute (EERI) (2010) "Remote Sensing and the GEO-CAN Community: Lessons from Haiti and Recommendations for the Future" (Workshop Report) Available at: *http://www.eqclearinghouse.org/20100112- haiti/published-reports* (accessed 31/12/2010).

Keegan, V. (2010) "Meet the Wikipedia of the mapping world" *Guardian Unlimited* 4th February 2010 Available at: *http://www.guardian.co.uk/technology/2010/feb/04/mapping-open-source-victor-keegan* (accessed 23/01/2011).

Maron, M. (2010) "Haiti OpenStreetMap Response" (Personal weblog) Available at: *http://brainoff.com/weblog/2010/01/14/1518* (accessed 31/12/10).

United States Geological Service (USGS) (2010) "Earthquake Hazards Program – Earthquake Details" Available at: *http://earthquake.usgs.gov/earthquakes/eqinthenews/2010/us2010rja6* (accessed 31/12/10).

Wooldridge, M. (2010) "Haiti will not die, President Rene Preval insists" (British Broadcasting Corporation [BBC] News website, 12th February 2010) Available at: *http://news.bbc.co.uk/1/hi/world/americas/8511997.stm* (accessed 23/01/11).

Zook, M., Graham, M., Shelton, T. and Gorman, S. (2010) "Volunteered Geographic Information and Crowdsourcing Disaster Relief: A Case Study of the Haitian Earthquake" *World Medical & Health Policy* 2 (2) pp.7–33.

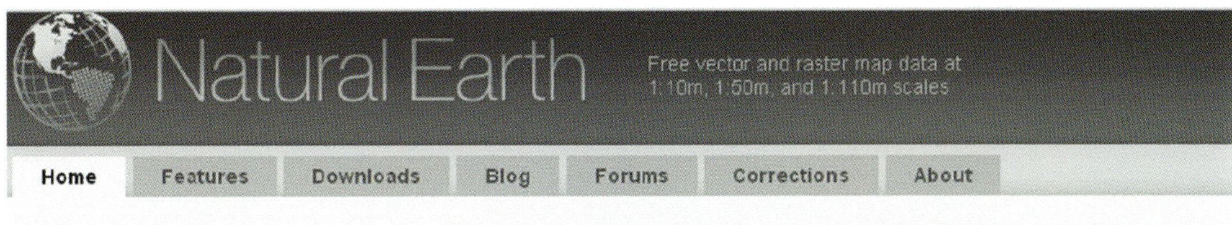

Natural Earth is a public domain map dataset available at 1:10m, 1:50m, and 1:110 million scales. Featuring tightly integrated vector and raster data, with Natural Earth you can make a variety of visually pleasing, well-crafted maps with cartography or GIS software.

Natural Earth was built through a collaboration of many volunteers and is supported by NACIS (North American Cartographic Information Society), and is free for use in any type of project (see our Terms of Use page for more information).

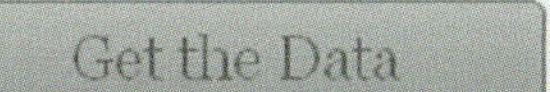

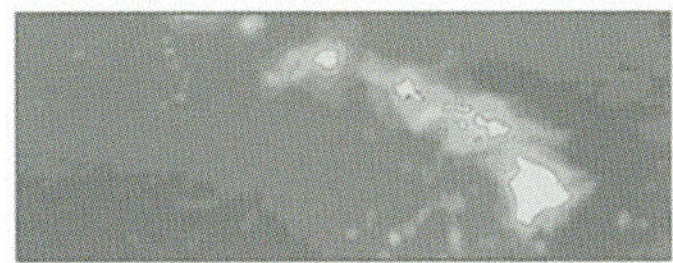

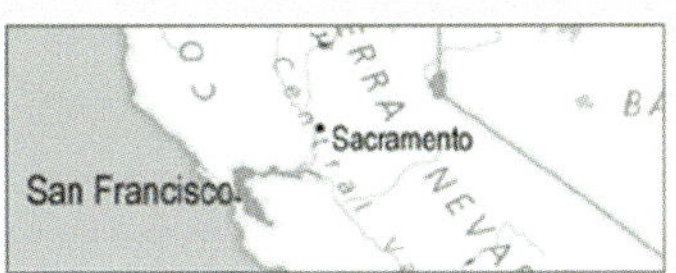

50m-admin-0-countries_area (242 areas selected

COUNTRYNAM	SCALERANK	FEATURECLA	SOVE
Afghanistan	1.00000000000	Countries	Afghanis
Aland	3.00000000000	Countries	Finland
Albania	1.00000000000	Countries	Albania
Algeria	1.00000000000	Countries	Algeria

OPENSTREETMAP IN GREAT BRITAIN: QUALITY AND COMPLETENESS

Oliver O'Brien

Originally published in Volume 44, No 1,2, pp.47–52

This paper looks at sources of data that have been used in creating the OpenStreetMap map in Great Britain. It considers the quality and completeness of these sources and how the map itself is becoming more complete, with improving accuracy and coverage as the project matures. Efforts on the ground by the volunteer community in certain areas result in a very rich map, while other parts of the country still have significant gaps in coverage.

1 The OpenStreetMap Project

OpenStreetMap (OSM) can be considered to be a 'Wikipedia of maps'. It is a crowdsourced global database of topological information. The data are made available free-of-charge under a flexible licence, allowing innovative use. A number of maps of selected features in the OSM database are available on websites, such as the default map at *http://www.openstreetmap.org* – its cartography being defined by a number of trusted contributors, including the Society of Cartographers' Chairman, Steve Chilton.

The project was initiated by Steve Coast in 2004, with most early contributions comprising features in Great Britain, particularly around London (Haklay, 2010). A number of data sources have been used to construct a comprehensive map and associated dataset of the country. These sources are often raster-based, requiring manual tracing and categorization in order to input the underlying topographical features to the OSM database. Each source has a differing contribution to the project, with varying levels of quality, completeness, accuracy and coverage.

2 Sources of Data for the Project in Great Britain

The project goes to considerable lengths to ensure that the data in the database do not have conditions which would not permit re-use under the project's standard licence.[1] As such, current products from Great Britain's national mapping agency, Ordnance Survey (OS), cannot be used to create the data, with the exception of products released under their new (April 2010) OpenData licence™ (Ordnance Survey, 2010). This is also true for Google Map imagery or mapping, which is often based on data supplied by commercial providers such as Tele Atlas or NAVTEQ. OpenStreetMap contributors in Britain have compensated for a lack of a standard base map by using a number of

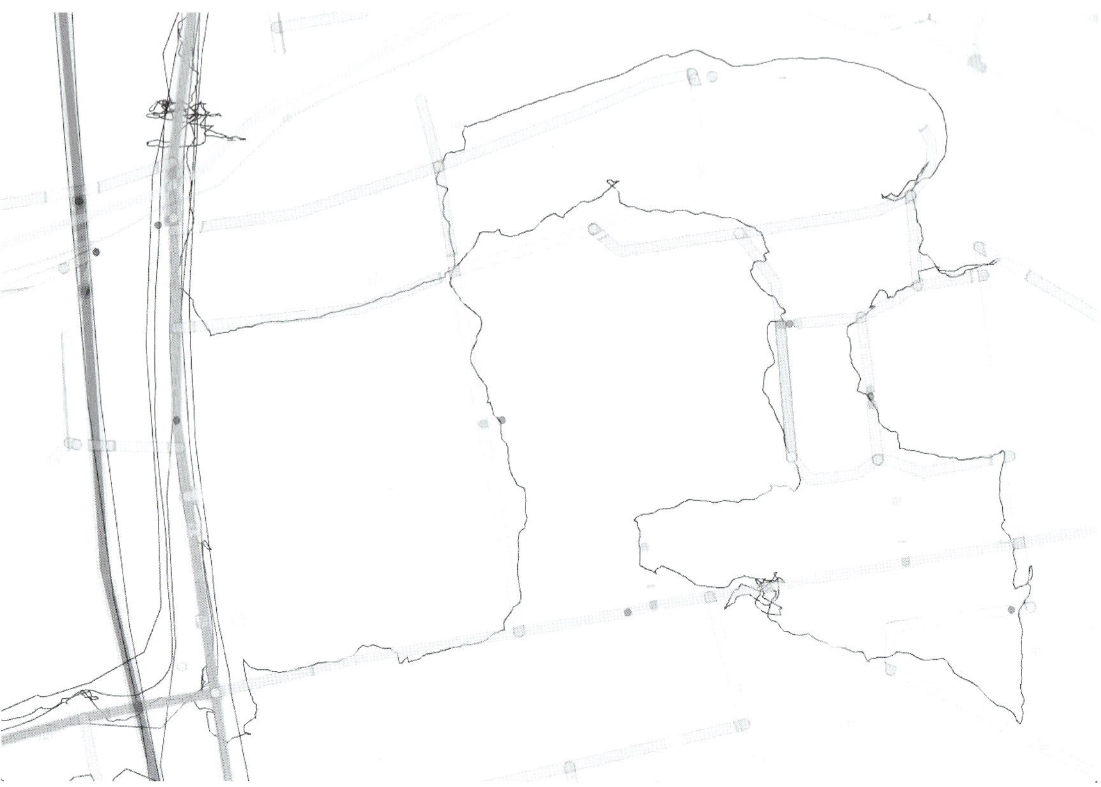

Figure 1 Screenshot of manually collected GPS traces superimposed on the corresponding OSM data in east London

different sources that are unencumbered under copyright or available under a sufficiently flexible licence.

2.1 GPS Receiver Data

Data from GPS (Global Positioning System) receivers, owned by a contributor to the project, can be considered to be the 'safest' form of data capture for the project in terms of copyright – the data being captured first-hand by the contributor, rather than second-hand from an existing map or imagery. When the project started, in 2004, consumer-grade standalone GPS receivers, with a storage and data-extraction capability, were becoming increasingly affordable and there were a large number of models available on the market. The data, in the form of 'track logs', can be uploaded to OSM, via its website and the resulting lines can then be viewed with other available tracks, traced over using OSM's online editor Potlatch, and metadata, such as road names, added.

The issues affecting the quality of this data are numerous. These include multipath interference from tall buildings, the fact that the contributors tend to be confined to the edges, rather than centrelines, of streets, and the variable accuracy and precision of the GPS units based on the chipset inside and the number and orientations of available satellites. The degree of error is hard to quantify – metadata describing any known error is typically not captured or made available. An additional source of GPS data, which made a significant contribution to the early central London road network map, was a donation from eCourier in December 2005. This consisted of a large number of GPS traces captured in their delivery vans as they drove around London. Quite quickly, the 'blank canvas' of central London began to take shape, providing a simple road network from which other features could be aligned and added iteratively.

2.2 Out-of-Copyright Maps

In Britain, Crown Copyright, which covers the Ordnance Survey, expires approximately 50 years after the map is first published.[2] A number of key British contributors to OpenStreetMap, including Andy Robinson, Richard Fairhurst and Steve Chilton, have arranged for their collections of such maps – typically Provisional/First Edition maps at 1:25,000 scale and Seventh Series and New Popular Edition maps at 1:63,360 scale – to be scanned and have allowed the resulting imagery to be used as base-maps for tracing features into the OpenStreetMap database (Chilton and Fairhurst, 2010).

The quality of the scans is variable. In many cases the base maps are low-resolution and have suffered from paper warping or otherwise deteriorated over the 50+ years. The detail on the maps, however, is still valuable. This is especially true for many rural areas because they typically lack significant development over the 50 years, have a

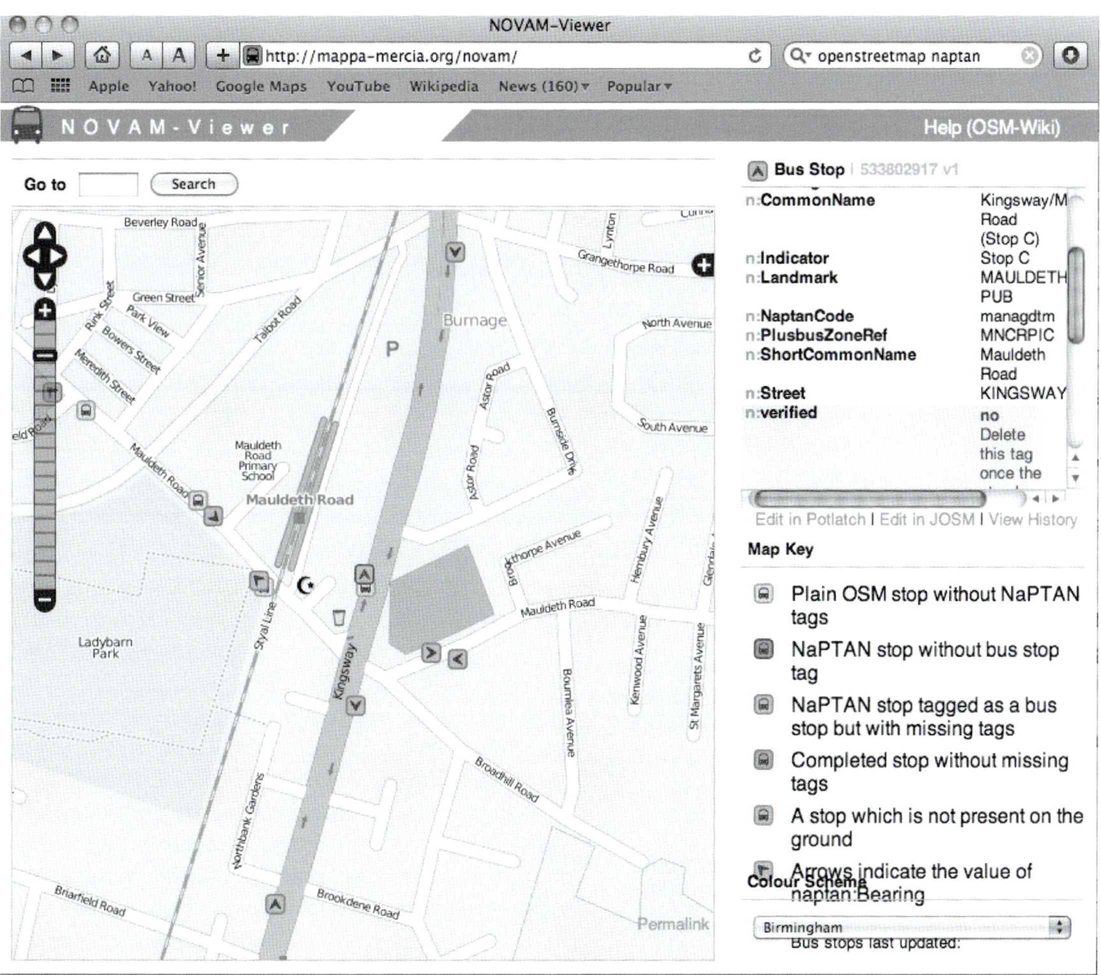

Figure 2
Screenshot of NOVAM-Viewer looking at NaPTAN bus stops in OpenStreetMap, south Manchester

Figure 3
Screenshot of part of the Isle of Sheppey in OSM, showing automatically created building objects from the OS Streetview raster imagery. The associated road network has yet to be added.

wealth of often difficult to measure natural features, and are unlikely to have active contributors on the ground. As each year passes, more maps from these series pass into the public domain and can start to be used as a basis for further OSM contributions.

2.3 Aerial Imagery

An agreement with Yahoo! in 2006 allowed their aerial imagery, as displayed on their own mapping service, to be used to trace features in many countries. In Britain, many large urban areas were covered, and significant time was spent by volunteers tracing over what appeared to be roads, onto the map, in order to quickly and effectively provide widespread, if poor quality, coverage. The approach was not without problems – roads in shadow could be missed, phantom ones could appear, and barriers or cycle-only gates were not normally visible, so the resulting data was not ideal for road routing. However it acted as a simple base map, from which contributors on the ground could improve the quality, such as adding in road names and other navigational features.

2.4 Data Imports

Bulk importing of existing datasets into the OpenStreetMap database has happened for a number of countries, notably in the U.S. (TIGER government data), Holland (from A.N.D., a digital mapping data company) and France (Corine land cover). Importing the data has the advantage of direct addition of data, without the accuracy loss inherent in any manual user tracing, although, at least in TIGER's case, the data quality was often quite low.

However, it is a contentious issue, particularly in Britain, where much manual work has already been undertaken before comprehensive datasets, e.g. the OS OpenData products, have become available for use. Issues regarding attribution of such datasets in the project have also not been fully resolved, particularly as transition to an alternative licence for OSM data is being considered.

One British dataset that has been gradually imported into OSM is the National Public Transport Access Nodes (NaPTAN) database from the Department of Transport. Railway station and port coverage was already near-complete, but few areas had bus stops comprehensively included, so these were imported in from NaPTAN for certain counties. A web tool, NOVAM-Viewer, was then developed to manage manual verification of the locations and metadata of the bus stops. Success has been mixed, perhaps as many would not consider checking bus stop information to be the most 'glamorous' of tasks, but nevertheless it has been a useful addition, both for building public transport accessibility into future routing products and also for detecting some missing key roads – a series of NaPTAN bus stops leading through a blank area on OpenStreetMap is indicative of a very incomplete area.

While the community has not yet used OS OpenData for significant bulk imports, and may never do (Amos, 2010), some experimentation has taken place with importing building outlines from the OS Streetview® product. This is a raster, but it is possible to write software to detect building shapes, which have a distinct colour, and turn these into objects for adding into the database. This has been carried out in a few places as a demonstration, including in the Isle of Sheppey. Additionally, the raster imagery has been re-projected and 'tiled', allowing its use as a base-map, similar to the out-of-copyright maps, for manual tracing.

OS OpenData's Meridian™ 2 product, a vector- based dataset allowing for direct import into OpenStreetMap, may also be useful to the project, although it is designed mainly for use at 1:50,000 and smaller scales, i.e. generally at a smaller scale for which OpenStreetMap is intended.

A third OS OpenData product, VectorMap™ District, appears to combine the best of both of the previous two – being a vector-based dataset but with a good resolution. Practical difficulties with the way the data are made available, and the ongoing debate on the use of OS OpenData vector data directly in OpenStreetMap, has prevented its widespread adoption in OSM so far.

2.5 Walking Papers

One final 'source' of data that is worthy of note is the Walking Papers website.[3] This has been designed to aid on-the-ground improvement of the map, particularly adding points of interest (POIs), by printing a specific rendering of the current OSM data, and even allowing placement of resulting annotations and scribbles as a base-map when adding the data into OSM using Potlatch. Ultimately, surveying on the ground, by OSM volunteers, results in the best and most up to date data for the project, and this website aims to make the process easier. It has been used at some of the recent 'Mapping Parties' in London which are social events aimed to target and 'complete' a specific area by the simultaneous deployment of a number of contributors.

3 Studies on Quality and Completeness

There have been a number of studies on quality and completeness of OpenStreetMap. Looking at data quality – specifically, how close the data is to what is on the ground, or to authoritative datasets – Haklay (2010) performed a 'comparative study' of OpenStreetMap and Ordnance Survey Meridian 2 datasets, focusing on motorway alignments. It was found that where the equivalent features existed in both sets, the OSM data was 'on average within about 6 m of the position recorded by the 80% overlap of motorway objects between the two datasets'. (The analysis was performed in 2008.) In a later study by Haklay and Ather (2009), OSM data was again compared, this time with Ordnance Survey MasterMap® Integrated Transport Network (ITN) data, an equal-area grid-based comparison of road lengths. It was discovered that 'when A-roads, B- roads and a motorway from ITN are compared to OSM data, the overlap can reach values that are over 95%' and that 'OSM is of better quality than Meridian 2'. The authors concluded that 'positional accuracy is satisfactory for many applications. Attribute accuracy is also satisfactory'.

Certainly the quality of OpenStreetMap data, in many areas of the country, is often as good as that seen on other consumer-grade online mapping websites. The quality in an area depends on the care and enthusiasm of the local 'champion' of that area, and so can vary widely.

When considering completeness of the map – the proportion of certain classes of features in real life that are included on the map – it is worth noting that it is nearly impossible for a map of the whole of Great Britain to ever be complete for popular features, simply because there will always be a time lag when new features are added to (or

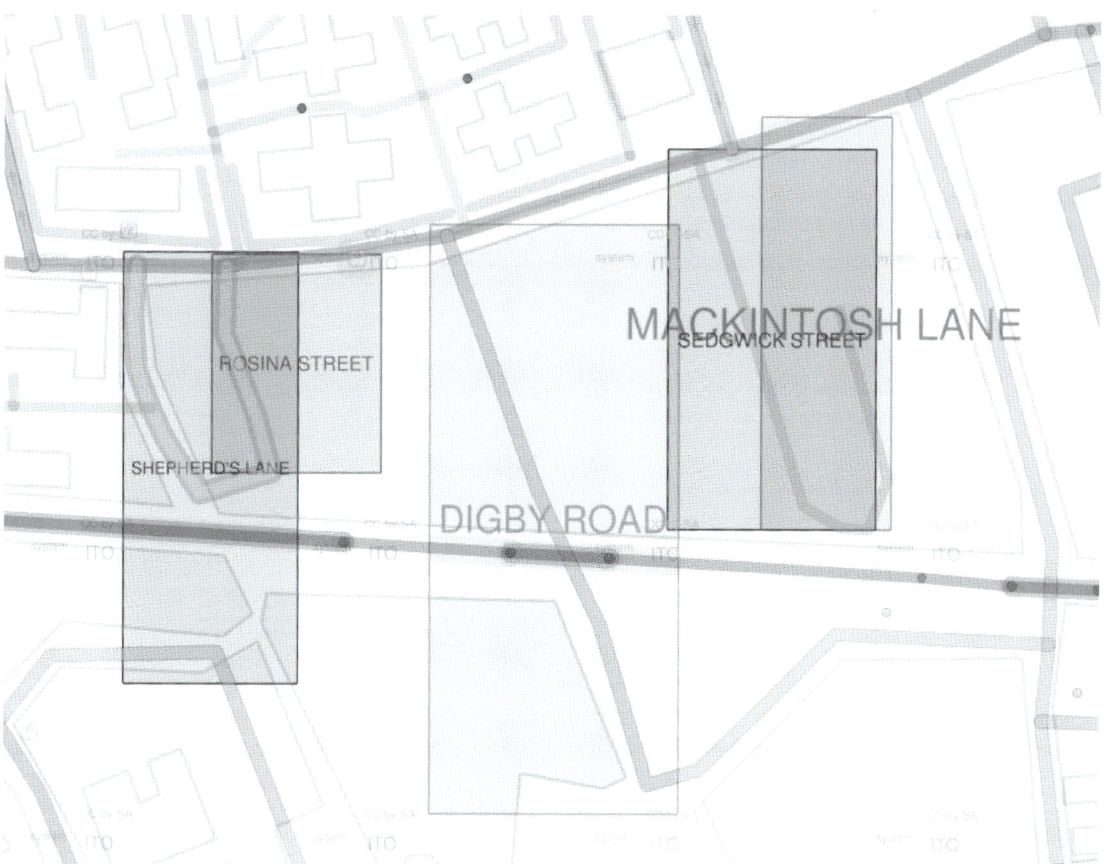

Figure 4 Screenshot of ITOWorld's OSM Analysis information, overlaid on a view of the OpenStreetMap data in East London and highlighting some roads with missing names and incorrect extents

old ones removed from) the ground, before any map can be updated to represent them. Completeness therefore again is normally a comparative measure against an authoritative source.

Haklay and Ellul (2010: 7–9) looked at the coverage of OpenStreetMap in England, including roads and street names. They found that in October 2009, 'OpenStreetMap already covers 65 per cent of the area of England, although when details such as street names are taken into consideration, the coverage is closer to 25 per cent. Significantly, this 25 per cent of England's area covers 45 per cent of its population'. They also noted a significant affluence bias, with the more wealthy south-east of England generally having a much higher completion, indicating a likely demographic bias of contributors to the project.

Peter Reed has performed a similar analysis in early 2010, comparing highway lengths, using information made available from the Department of Transport (DfT), with those in the OpenStreetMap database, on a county- by-county basis. Only a few areas of Great Britain were found to have a relative coverage of less than 50 per cent, and some areas, including a few London boroughs, were found to have a coverage exceeding 100 per cent – likely due to service roads being included in the analysis, that would not be included with the DfT data.

ITOWorld, a transport information company, has produced a tool, OSM Analysis, that allows graphic comparison of OpenStreetMap data with OS Locator™ data, another product available as part of OS OpenData. The tool also has a leaderboard,[4] showing the match between OSM and OS Locator roads, broken out by district or unitary authority. Because the match is performed on the name as well as the existence of a feature in the correct location, it is harder to achieve completeness, and the proportion cannot exceed 100 per cent – the OS Locator data is considered to be definitive and additional roads in OSM are ignored. In rare cases where the OS Locator data does not match with what is seen on the ground, special metadata tags are added in OSM to the affected roads. At the time of writing, four of 408 districts are shown to have 100 per cent coverage, and 117 have over 80 per cent coverage. The leaderboard ranks Derwentside as the district currently with the lowest match between OSM and OS Locator, with just 15 per cent of OS Locator roads in OSM. The statistics are updated daily.

The above studies all focus on roads as their criterion for completeness and accuracy – a not unreasonable emphasis, as OpenStreetMap is, as the name suggests, a map of streets – although in practice any geographical feature can be stored in the database.

4 Completing the Map – Tools and Efforts

The OSM Analysis tool, produced by ITO World, and mentioned above, is an effective way of spotting missing roads, incorrect road names and poor alignments with respect to the OS Locator equivalents. It provides a background map showing boxes representing each discrepancy. This can then be compared with aerial imagery or other sources, and the road corrected or added.

One effective way to build and reinforce a volunteer community around the process of completing the map in a particular area, is to hold a 'mapping party'. These are social events where volunteers agree to spend a few hours mapping a particular sector, or 'cake slice', of a town or city, on the same day. The mapping activity is then followed by a group editing session, where more experienced participants can provide simple training. Such mapping parties were key to starting the process of completing cities such as Manchester, where a 'Mapchester' mapping party was held in May 2006.[5] A similar event was held in London in January 2007, and several more occurred. The need for such broad-based events is perhaps diminished now with the more complete nature of the project in the country.

A variant of the mapping party is the 'mapping marathon' – in London, a series of bi-weekly events has taken place during the last few summers, typically taking place on a weekday evening but otherwise following the model of the mapping parties. The events have helped build the large and active London community. The summer series have progressively focused on roads, points of interest and, most recently, building outlines.

5 A Word on Metadata

OpenStreetMap has a loose taxonomy – that is, there are no rigid requirements on how a feature should be described, but guidelines do exist. The community proposes and documents the metadata associated with features, known as 'tags' which have keys and associated values, on the project's wiki, but does not enforce their use. However, in order for a user's features to be automatically picked up and appear on the 'standard' OpenStreetMap map and other popular map renderings of the OpenStreetMap data, the guidelines need to be followed. A strong incentive for many users is to see their work appear on the public map, so it is visible by the wider community (Coleman *et al.*, 2009), so the conventions are generally upheld. As well as metadata specifically entered by the user themselves, the user ID, date/time and editing application name are automatically added to all features.

6 In Conclusion

The OpenStreetMap project in Britain has had data input from a multitude of sources in the last six years. The data's accuracy has increased as new and improved sources have become available, mainly for use in tracing. The completeness of the map, in terms of the road network in the country, has also gradually improved although, for many areas, there is quite a way to go. Simultaneously, the volunteer community has continued to expand, the map benefiting from the continued enthusiasm and diligence of the contributors.

References

Amos, M. (2010) "Talk: Ordnance Survey Opendata" (OpenStreetMap Wiki) Available at: *http://wiki.openstreetmap.org/wiki/Talk:Ordnance_Survey_Opendata* (accessed 29/10/10).

Chilton, S. and Fairhurst, R. (2008) "Out-of-copyright maps" (OpenStreetMap Wiki) Available at: *http://wiki.openstreetmap.org/wiki/Out-of-copyright_maps* (accessed 29/10/10).

Coleman, D. J., Georgiadou, Y. and Labonte, J. (2009) "Volunteered Geographic Information: The nature and motivation of producers" *International Journal of Spatial Data Infrastructures Research* 4 pp.332–358.

Haklay, M. (2010) "How good is volunteered geographical information? A comparative study of OpenStreetMap and Ordnance Survey datasets" *Environment and Planning B: Planning and Design* 37 (4) pp.682–703.

Haklay, M. and Ather, A. (2009) "Beyond good enough? Spatial Data Quality and OpenStreetMap data" *Paper presented at State of the Map 2009, Amsterdam, The Netherlands, 10th–12th July.*

Haklay, M. and Ellul, C. (2010) "Completeness in volunteered geographical information – the evolution of OpenStreetMap coverage in England (2008–2009)" Under review in *Journal of Spatial Information Science as Article 35* Available at: *http://www.josis.org/index.php/josis/article/view/35* (accessed 29/10/10).

Ordnance Survey (2010) "OpenData Licence Terms and Conditions" Available at: *http://www.ordnancesurvey.co.uk/oswebsite/opendata/ licence/docs/licence.pdf* (accessed 29/10/10).

COMMUNITY MAPPING BY NON-LITERATE CITIZEN SCIENTISTS IN THE RAINFOREST

Michalis Vitos, Matthias Stevens, Jerome Lewis and Muki Haklay

Originally published in Volume 46, Nos 1 & 2, pp.3–11

Supporting local communities to share their environmental knowledge by utilizing scientifically accepted tools and methodologies can lead to improvement in environmental governance, environmental justice and management practices. Mbendjele hunter-gatherers in the rainforests of Congo are collaborating with the ExCiteS Research Group at University College London to record their local knowledge about illegal poaching activities, which will improve the control of commercial hunters and diminish the harassment they often experience at the hands of 'eco-guards' who enforce hunting regulations. Developing and deploying a system for non-literate users introduces a range of challenges that we have tried to tackle with our Anti-Poaching data collection platform.

1 Introduction

Sustainable natural resource management is one of the major development challenges facing humanity today. There is an urgent need for innovative solutions to enable scientifically informed, sustainable resource management of key environments such as the rainforests. The tendency to impose draconian floral and faunal management laws and practices designed by technocrats and politicians from outside the affected areas has disenfranchised local people from any say or involvement in the management of the areas their livelihoods depend upon. However, encouraging local people to share their environmental knowledge more effectively leads to improvements in environmental governance, environmental justice and management practices.

The Mbendjele are the indigenous people of northern Congo-Brazzaville. As expert hunters and gatherers of wild produce they move through huge areas of forest over the course of the year. The Mbendjele and other forest-dependent people are among the poorest citizens of Central Africa's countries and the groups most dependent on natural resources for their livelihoods, yet they are rarely consulted in decisions over the attribution, or involved in the management of, these areas. In 2005, the local logging company, Congolaise Industrielle des Bois (CIB), decided to seek certification by the Forest Stewardship Council (http://www.fsc.org/) as being environmentally and socially sustainable. Part of the requirements forced CIB to respect the rights and resources of indigenous and local forest people. A solution was developed by a consortium that introduced the Mbendjele to the use of rugged Personal Digital Assistance (PDA) devices, portable GPS (Global Positioning System) receivers, and bespoke software that allowed non-literate users to record observations using a pictorial decision tree (Hopkin, 2007; Lewis and Nelson, 2007; Lewis, 2012). In 2007 a similar initiative was set up in Cameroon (Lewis and Nkuintchua, 2012).

Nowadays, the Mbendjele are very concerned about over-hunting by commercial poachers in their traditional hunting areas. The traps such poachers leave concentrated in small areas ravage animals indiscriminately, and pose a danger to the hunter-gatherers and their children as they move throughout the forest. These poachers are dispersed in small forest camps and are typically armed with shotguns, Kalashnikovs and other rifles – posing a threat to locals, especially if they try to meddle in their activities. The poachers are known to bribe local law enforcers called 'eco-guards' and are often part of larger networks supported by local elites keen on profiting from their highly lucrative business. Hence they operate with relative impunity. Eco-guards looking for easier targets often visit Mbendjele and other local communities where they too often resort to violence and abuse. The Mbendjele experience this as unacceptable persecution for something that they see as their birthright – to live by hunting and gathering wild foods from the forest, as have their ancestors since time immemorial.

Building on their positive experience by mapping their resources to protect them from logging activities (Lewis, 2012; Lewis and Nkuintchua, 2012), some Mbendjele requested Lewis to design them a tool for recording their extensive knowledge of the whereabouts and habits of poachers. Together, they visited the offices of the Wildlife Conservation Society manager helping to organize the eco-guard patrols, to propose the idea and to discuss what issues they would like to monitor and, from the eco-guards' point of view, to identify the evidence they would need to record in order to effectively charge and arrest the poachers.

Our collaboration with the Mbendjele, as described in this article, is exemplary for the raison d'être of the ExCiteS group at University College London. Our mission is to co-develop information and communication technology (ICT) solutions with participating communities in order to enable them to capture their extensive environmental knowledge in ways that can be efficiently shared with outsiders and promote their control of their land and resources. In doing so, we seek to push beyond traditional citizen or community science projects

(Dickinson *et al.*, 2012; Haklay, 2013), which typically target educated people in affluent areas of the world. We call this approach Extreme Citizen Science (ExCiteS), and expect it to have transformative potential to deal with major sustainability challenges by making scientifically valid datasets available to a wide range of users in accessible formats – even if they are not literate.

2 Challenges

Deploying ICT systems in the rainforest and putting devices designed for educated, literate people in the hands of unschooled, forest people presents numerous foreseeable and unforeseeable difficulties. We tried to preempt the most obvious of these – the low or nonexistent literacy of the users of the system – in developing a suitable solution. The users have never had any formal education, and have only basic experience of technology through handheld devices that were used in the past. Illiteracy is a huge obstacle to using devices such as smartphones because virtually any standard user interface (UI) includes various textual and numerical elements. A related challenge is language: only a handful of Mbendjele speak an international language and their own language is spoken by few outsiders.

The overall technical challenge is providing the Mbendjele with a platform that allows them to report poaching-related activities in the forest and to feed this information into a central database in time for appropriate control activities to be organized successfully. We need a mobile device that can withstand the adverse conditions of the African rainforest: dust, mud, high humidity levels, and frequent rain. It also needs to be robust, such that it does not break when roughly handled for extended periods of time by people who have no experience in dealing with delicate electronic equipment. Furthermore, it should be equipped with a GPS receiver that is sensitive enough to get location fixes under dense forest canopy within a reasonable amount of time (e.g. within five minutes). Finally, it should be relatively cheap; affordable to non-governmental organizations and potentially indigenous communities themselves. In terms of software, although there are many existing frameworks and applications for mobile data collection, the nature of this project and its requirements made off-the-shelf solutions impractical.

There are also numerous practical challenges that need to be tackled. The lack of power facilities in the rainforest, especially when combined with the high power demands of smartphones or other handheld devices, poses a major challenge. Furthermore, although it is fast expanding, cellular network penetration and coverage remains limited in developing countries, especially in vast, sparsely inhabited areas (Parikh and Lazowska, 2006), which is problematic because we need data to be uploaded to, and synchronized with, a central database.

Finally, perhaps the most challenging problem is that of security; more specifically, personal safety. Given the nature of the data collection activity, the consequences of an Mbendjele member being caught in the act by poachers could be dramatic, possibly even fatal. Therefore the equipment should be discreet, easy to carry around, and, if necessary, easy to hide or discard. Moreover, the true purpose should be deniable – in part by restricting access to the data collection software.

3 Related Projects

Paper-based survey forms for collecting information in the field remain widely used across many scientific disciplines. However, advances in ICT have enabled data collection via digital forms and allow the data to be fed directly into an electronic database, making data collection more efficient, cheaper, and less error-prone when compared to traditional methods (Pundt, 2002). Before deciding to develop our own platform, we evaluated several existing platforms for mobile data collection. We found they all had limitations or restrictions which made them unsuitable for our goals.

The first mobile data collection platforms were designed for handheld computers or PDAs (Lane *et al.*, 2006). CyberTracker (*http://www.cybertracker.org/*) and EpiHandy (*http://www.epihandy.org/*) are examples of such early, PDA-based platforms. Nowadays both are outdated, primarily because they rely on expensive and equally outdated PDA devices that lack the processing power and built-in sensors of today's smartphones. The next generation of mobile data collection platforms targeted mobile phones and smartphones. EpiCollect (Aanensen *et al.*, 2009), an initiative of Imperial College London, and Open Data Kit (ODK) (*http://opendatakit.org/*), created by the University of Washington (Anokwa *et al.*, 2009; Hartung *et al.*, 2010), are examples of such modern platforms. Besides the data collection applications, both also offer online tools to design forms, visualize data (using Google Maps) and perform basic analysis. However, because EpiCollect is designed for literate scientists, it is heavily dependent upon textual interaction and does not support the use of pictorial icons and decision trees. On the other hand, ODK requires verbose and complex XForms structures to implement a decision tree consisting of pictorial icons. Moreover, due to textual information in the UI, the standard ODK Collect application would be too confusing for non-literate users. Still, because it is open source, we decided to use ODK Collect as the basis for our first prototype platform and modify it to suit our needs.

4 Platform Prototype

Currently, our Anti-Poaching data collection platform is in an advanced prototype state. Below we discuss the different hard- and software components.

4.1 Hardware

In order to withstand the harsh rainforest conditions and not-so-gentle treatment by the users, we looked for a rugged, water-resistant smartphone. While there is an increasing number of such devices on the market, most are designed for military or industrial clients and command correspondingly high prices. Fortunately some mainstream smartphone vendors have recently introduced much cheaper rugged devices such as the Samsung Galaxy Xcover (Samsung, 2011), which we eventually selected. This Android-based smartphone has a durable body, a scratch-resistant Gorilla Glass screen (*http:// corninggorillaglass.com/*), and is IP67-certified (International Electrotechnical Commission, 2012), which means it is dustproof and also waterproof (at depths of up to one metre).

We tested two possible solutions for charging the devices in the forest. The first was a combination of a roll-up solar panel and an external auxiliary battery that stores energy for later use. This solution was not expected to be sufficient for use in the rainforest because there is little direct sunlight reaching ground level due to the dense canopy. The second solution was the Hatsuden Nabe, a Japanese-produced customized cooking pan that converts thermal energy from a fire into electricity that can be used to charge electronic devices while cooking (TES NewEnergy Corporation, 2012). This solution is suitable for the lifestyle of the tribe, who always keep a fire going to cook food and keep animals at bay.

4.2 Software

Our Anti-Poaching application was built on top of a modified version of ODK Collect and to tackle the literacy issue we used a decision tree with pictorial icons as in an earlier project (Lewis, 2012; Lewis and Nkuintchua, 2012). The icons represent different signs of poaching activity (e.g. camps, footsteps, hidden weapons, traps, and so on), cases of abusive or corrupt behaviour by eco-guards, and sightings of live animals and other natural resources.

In order to make the icons comprehensible and relevant to the Mbendjele, they were designed according to their feedback in a clear, smooth style, using only black and white. To avoid confusing or distracting the users, we modified ODK to not display any textual information and to operate in full screen (to hide the Android status bar, shown at the top of the screen, which provides information on the time, battery status, and so on). What is left is a minimalistic, entirely graphical interface in which the icons are arranged in a grid, as shown in Figure 1.

In the current version the decision tree consists of 59 distinct icons spread across four levels. Every observation starts with the main menu (shown in Figure 1, left); the user can navigate to a lower level by touching one of the icons and is then presented with a new, more specific set of choices. At every level except the top level, a back button allows users to go back one level (shown in Figure 1, right). When the user has reached the bottom level of the tree this means a complete, specific observation was described, and at that point the observation can be optionally complemented with an audio recording, a photograph, or a video. Afterwards, the device will try to obtain geographical coordinates from the built-in GPS receiver to geo-tag the observation. When the coordinates are obtained, the observation (along with multimedia attachments and coordinates) is automatically saved to the memory card of the device and a sound (a beep) indicates that the process is complete. The application operates in a continuous loop, meaning that at the end of each complete observation it goes back to the main menu, ready for the user to make a new observation.

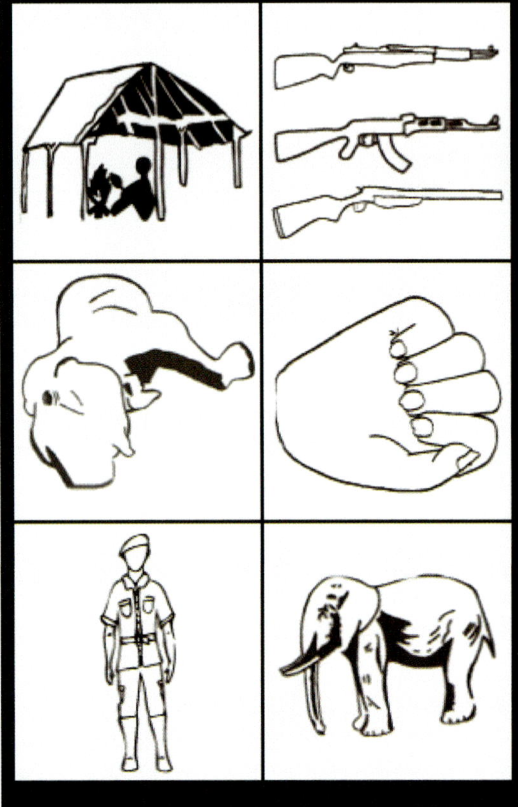

Figure 1 *Main menu interface (left) and lower level interface (right)*

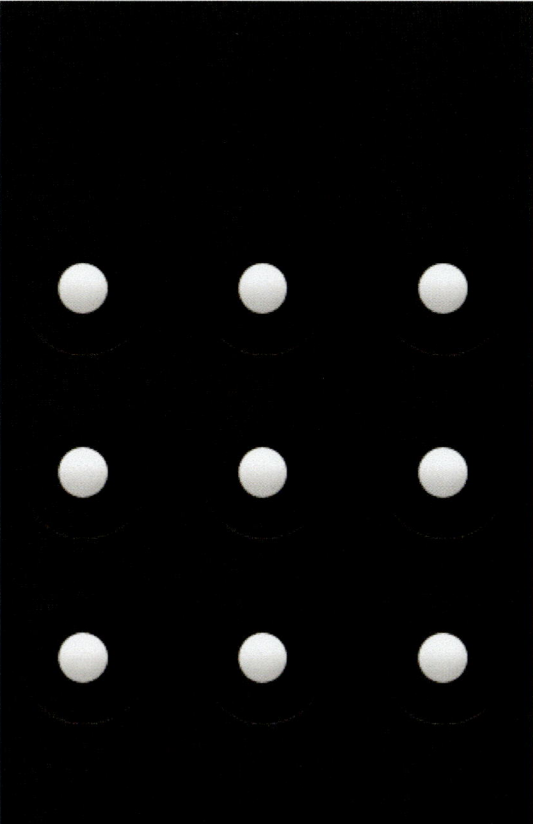

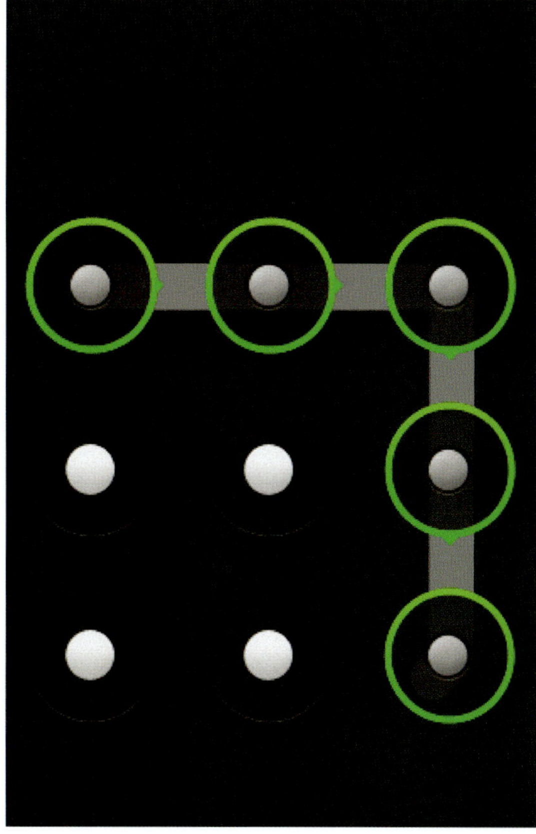

Figure 2
Opening screen, as locked (left) and unlocking pattern (right)

It is important to restrict access to the application, so that the true purpose of the device can be hidden or denied. Due to the illiteracy of the users, conventional authentication mechanisms such as passwords and PIN (Personal Identification Number) codes are not suitable. Instead, we use a pattern unlocking mechanism, which many Android users are familiar with. When the application is opened, the user is first presented with a screen as shown in Figure 2 (left) and to proceed the user must draw a previously agreed pattern by sliding a finger

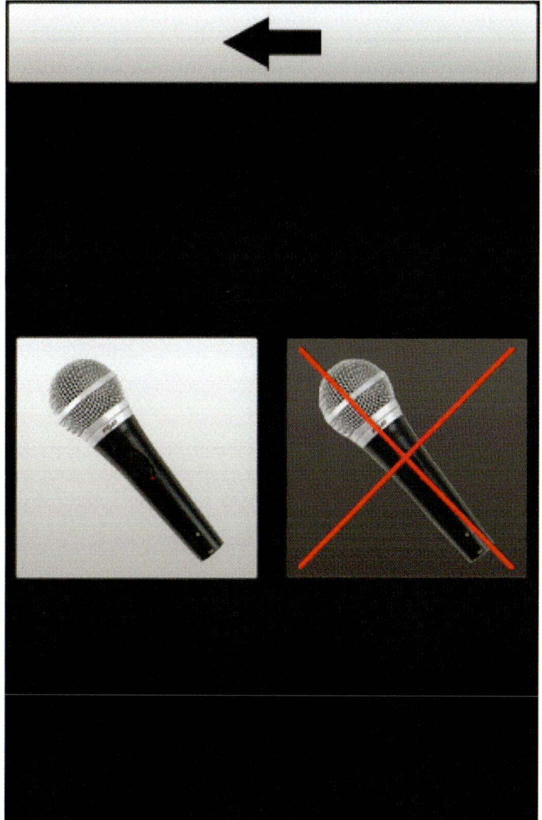

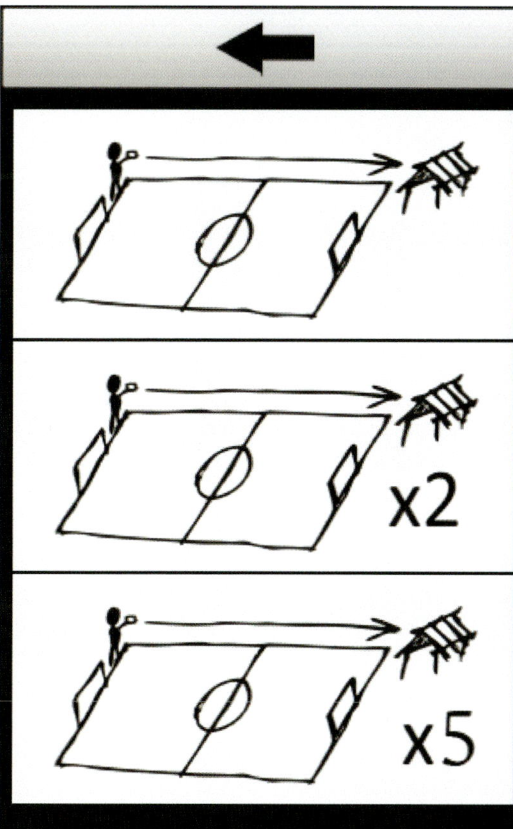

Figure 3
Audio recording interface (left) and distance estimation interface using football pitches (right)

Figure 4 Charging a smartphone with the Hatsuden Nabe HC-5

over the dots on the touchscreen, as shown (right). If the pattern is recognized, the user will be presented with the main menu. The mechanism is used to restrict access to our application, rather than to the device itself, so that other 'innocent' functions of the smartphone remain unobstructed.

By default, ODK relies on the standard Android applications for audio recording and the photo/video camera. These interfaces contain textual elements and a plethora of features and settings which are confusing and distracting for the intended, non-literate users. Hence, we have implemented a minimalistic audio recording interface, shown in Figure 3 (left), where recording is represented by a microphone (a familiar concept to some members of the tribe due to the local radio station interviews). We have not yet created a replacement for the standard photo/video camera application but plan to do so in the near future.

Making observations of (possibly active) poachers' camps presents the obvious risk of being spotted or caught. Thus we have implemented an innovative feature that allows users to point the device in the direction of the camp and provide an estimate of its distance away from the observer. The combination of the user's own position (obtained through GPS), the bearing (registered using the built-in compass), and the estimated distance, allows us to compute the approximate position of the camp. However, our users are unfamiliar with standardised distance units and have no (or limited) numeracy skills for expressing distance. Therefore, we allowed distance to be expressed in terms of football pitches; a familiar feature to the Mbendjele, who have seen them in logging towns. As illustrated in Figure 3 (right), the UI allows users to select a distance of one, two or five football pitches. Few Mbendjele can read numbers, but we know that the specific individuals who will do most of the data collection can recognize the numbers 2 and 5 from handling 2,000 and 5,000 CFA (Coopération financière en Afrique centrale) banknotes – something not all their peers have had the opportunity of doing very often. Although this method is not very accurate in terms of distance, it nevertheless allows us to record a reasonable indication of the position of potentially dangerous places, so that eco-guards can find them.

5 Evaluation

Over the course of April 2012, our prototype platform was tested by the Mbendjele (coordinated by Jerome Lewis).

5.1 Hardware

We provided the Mbendjele with a set of four Samsung Galaxy Xcover smartphones and we also supplied them with a Hatsuden Nabe unit, a roll-up solar panel, and an auxiliary battery.

Results from using the smartphones have been promising; the devices proved to be robust, they withstood

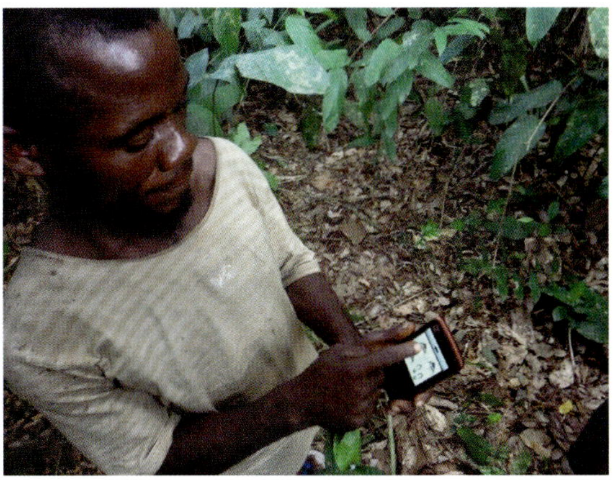

Figure 5
A user recording an observation

the dust and humidity of the forest as well as (mis)treatment by their users. Gorilla Glass screens proved to be as tough as promised, considering the smartphones were left without even the faintest scratch and were easy to operate, as the users quickly learned how to use the touchscreen. The built-in GPS receiver also proved to be adequate; in clear areas it took up to two minutes to obtain a first fix, and up to five minutes under forest cover. Obtaining subsequent fixes only took 10–25 seconds in most cases. Finally, we were impressed by the battery life of the smartphones; with ordinary use they could last for several days. However, people discovered they could use the smartphones to record their music and as torches during the night, causing dramatic battery drain.

The Hatsuden Nabe proved to be the most appropriate power source for recharging the smartphones, although it performed worse than the solar panel; it takes approximately 3–4 hours to fully recharge a smartphone, during which the water in the pan needs to be replaced up to six times (Figure 4 shows the pan in use). The tests were conducted with the Hatsuden Nabe HC-5, which has a diameter of 14 cm. In future, we may switch to one of the larger models to enable faster recharging. With the solar panel in direct sunlight it took about two hours to recharge two smartphones at once. As expected the solar panel turned out to be an excellent solution when sedentary in an open space, but is impractical while on the move or when under dense canopy. Some Mbendjele seemed to have no difficulty in executing tasks such as connecting the smartphones for charging or understanding the lack of power from a flat battery, while others required more coaching and repetition to become familiar with these tasks.

5.2 Software

Due to the secrecy of the project it was not possible to train as many users as we would have liked for software testing. Nonetheless, the Mbendjele collected a total of 427 observations, 151 photographs and 40 audio recordings during the training. Figure 5 shows one of the users recording an observation. Because this was the first evaluation of the platform, no risks were taken and hence the data does not reflect actual poaching activity. After the devices had been returned, we extracted the data and made an initial visualization using Google Earth. Figure 6 presents a sample of the observations.

The users who received training quickly grasped the concept of the decision tree, using the pictorial icons to represent observations, and overall, in working out what each icon represented. They seemed to like the beep sound that signalled the successful completion of an observation, although they would prefer a sound like a bird call or some other natural sound that would be less startling in the rainforest.

The users had little difficulty in learning how to use the pattern mechanism to unlock the smartphones, and while some found this confusing at first, after some guidance and

Figure 6
Initial visualization of observations in Google Earth

repetition they quickly learned to master this task.

The audio recording part of the application was very popular among the users and they worked out how to use this much easier than the camera. While audio recording was used extensively by the Mbendjele during an unsupervized 24-hour walkabout, the photographic application was difficult for those that were not used to two-dimensional representations. However, when using this was likened to operating a gun, i.e. 'aim, hold firm, and fire', their usage improved. Some common issues included putting fingers in front of the lens, not holding the device still, and not coping with the slight delay between pressing the shutter button capturing the actual picture. The Save button was the most difficult aspect to grasp and generally impossible for most, confirming the need to replace the standard camera interface with a simpler, text- free one. The quality of the camera itself is adequate: the pictures look clear and vibrant and are reasonably sharp, except when taken in low light conditions.

Finally, the users seemed to have no difficulty understanding the concept of pointing the smartphone in the direction of a dangerous place in order to record the position from a distance. Individual tribe members that will conduct most of the data collection were indeed able to understand the concept of one, two and five football pitches.

6 Conclusions and Future Projects

Our Anti-Poaching data collection platform takes proven concepts (Lewis, 2012; Lewis and Nkuintchua, 2012) and adds innovative features based on the functionality of today's smartphones. Moreover, the use of alternative power sources vastly increases the flexibility of the platform. The initial evaluation phase demonstrates that the prototype works as expected but it also resulted in helpful suggestions for further improvements, which we will implement.

Our goal for the future is to create a more generic tool that could be used by different communities who wish to address issues of concern using tools and methods of scientific research. Ideally, people with limited computing skills should be able to design and deploy their own decision trees, suited for specific projects beyond our own. Enabling such flexible re-use will be one of the main challenges in the near future. Part of the solution will be to create a Web-based tool for designing decision trees.

Another challenge will be to implement a flexible means with which to transmit data to a central server. We intend to use a background service running on the smartphones which regularly checks for network coverage. The Mbendjele regularly visit small towns, or get sufficiently near to them, and the software must use this opportunity to transmit accumulated data. Transmission should happen automatically via SMS (Short Message Service), Wi-Fi (Wireless Fidelity) or GPRS/3G (General Radio Packet Service/3rd Generation), depending on availability, without any user interaction. The transmitted data will be compressed and encrypted to offer short transfer times and security.

In this article we have not covered the visualization and analysis aspects of the ExCiteS vision. We are currently working on a concept we call 'Intelligent Maps': a novel, dynamic approach to presenting spatial data and informing about emerging trends, both in ways comprehensible to illiterate people. This approach will allow non-literate users to visualize and better understand the environmental data that they have collected, helping them to address and state their concerns.

References

Aanensen, D.M., Huntley, D.M., Feil., E.J., al-Own, F. and Spratt, B.G. (2009) "EpiCollect: Linking Smartphones to Web Applications for Epidemiology, Ecology and Community Data Collection" *PLoS ONE* 4 (9) e6968.

Anokwa, Y., Hartung, C., Brunette, W., Borriello, G. and Lerer, A. (2009) "Open Source Data Collection in the Developing World" *Computer* 42 (10) pp.97–99.

Dickinson, J.L.and Bonney, R. (Eds) (2012) *Citizen Science: Public Participation in Environmental Research* Ithaca: Cornell University Press.

Haklay, M. (2013) "Citizen Science and Volunteered Geographic Information: Overview and Typology of Participation" In Sui, D., Elwood, S. and Goodchild, M. (Eds) *Crowdsourcing Geographic Knowledge: Volunteered Geographic Information (VGI) in Theory and Practice* (pp.105–122) Dordrecht: Springer Netherlands.

Hartung, C., Anokwa, Y., Brunette, W., Lerer, A., Tseng, C. and Borriello, G. (2010) "Open Data Kit: Tools to Build Information Services for Developing Regions" *Proceedings of the 4th ACM/IEEE International Conference on Information and Communication Technologies and Development (London, 13th–16th December)* New York: ACM Article No.18.

Hopkin, M. (2007) "Conservation: Mark of respect" Nature 448 (7152) pp.402–403.

Imperial College London (2009) EpiCollect – Mobile/Web Application for Smartphone data collection Available at: *http://www.epicollect.net/* (accessed 09/11/12).

International Electrotechnical Commission (2012) IEC60529: *Degrees of Protection Provided by Enclosures (IP Code)* Geneva: International Electrotechnical Commission.

Lane, S.J., Heddle, N.M., Arnold, E. and Walker, I. (2006) "A review of randomized controlled trials comparing the effectiveness of hand held computers with paper methods for data collection" *BMC Medical Informatics and Decision Making* 6 (23) pp.1–10.

Lewis, J. (2012) "Technological Leap-frogging in the Congo Basin, Pygmies and Global Positioning Systems in Central Africa: What has happened and where is it going?" *African Study Monographs* Supplementary Issue 43 pp.15–44.
Lewis, J. and Nelson, J. (2007) "Logging in the Congo Basin: What hope for indigenous peoples' resources and their environments?" *Indigenous Affairs* 2006 (4) pp.8–15.
Lewis, J. and Nkuintchua, T. (2012) "Accessible technologies and FPIC: Independent monitoring with forest communities in Cameroon" *Participatory Learning and Action* 65 pp.151–165.
Parikh, T.S. and Lazowska, E.D. (2006) "Designing an architecture for delivering mobile information services to the rural developing world" *Proceedings of the 15th International Conference on World Wide Web (WWW '06), Edinburgh, 22nd–26th May* New York: ACM pp.791–800.
Pundt, H. (2002) "Field Data Collection with Mobile GIS: Dependencies Between Semantics and Data Quality" *GeoInformatica* 6 (4) pp.363–380.
Samsung (2011) Galaxy Xcover (GT-S5690) Available at: *http://www.samsung.com/uk/consumer/mobile-devices/smartphones/android/GT-S5690KOAXEU/* (accessed 09/11/12).
Stevens, M. (2012) "Community memories for sustainable societies: The case of environmental noise" *PhD thesis* Vrije Universiteit Brussel.

LONDON SUBTERRANEA

Stephen Walter

Originally published in Volume 46, Nos 1 & 2, pp.21–23

Beneath the ground lie many mysteries, secret burials, and hidden treasures.

London is one of the great palimpsests of our time. The land mass that it occupies is a document constantly being re-written and each time leaving a layer of evidence behind. Or more precisely 'below', for a new piece of the layer-cake is built on top of the old, leaving a piecemeal and multi-layered puzzle for us to unravel. Occasionally we dig far below any of our own makings and into a dense, dark and cold mass that tells of a far older story and fuels our imagination.

Figure 1 London Subterranea, 2012. 131 x 163 cm. Archival inkjet & screen-print with gold acrylic paint. Edition of 50. © Stephen Walter, 2012. Reproduced by kind permission of the artist and courtesy of TAG Fine Arts.

London Subterranea geographically tracks several things that exist under the surface of the city. It shines light on a clandestine world of utilitarian tunnels and passageways, and includes some folklore and legends attached to certain places. The map is a concise compendium of what I have learnt.

I gathered data from books, archives, websites, contacts, and by making occasional journeys below the ground. I tip my hat to MacDonald Gill's highly detailed works and Harry Beck's iconic utopian tube map. The underground lines are illustrated with their familiar colours.

Also included is Duncan Campbell's mapping of a 'secret' network of governmental tunnels under London published in 1980. A collection of archaeological discoveries from pre-history to the present, cemeteries, mass burial sites and places of worship are shown alongside well-known ley lines. The 'ghost' stations of the transport network, the Post Office railway, epithets to the 'underworld' of crime, and the scenes of notable murders are all pin-pointed.

I list past disasters and bombings; rename some stations to rekindle the area's past and anoint emotions to certain sections of the transport network. I also celebrate the site

Figure 2
London Subterranea (detail) © Stephen Walter, 2012. Reproduced by kind permission of the artist and courtesy of TAG Fine Arts.

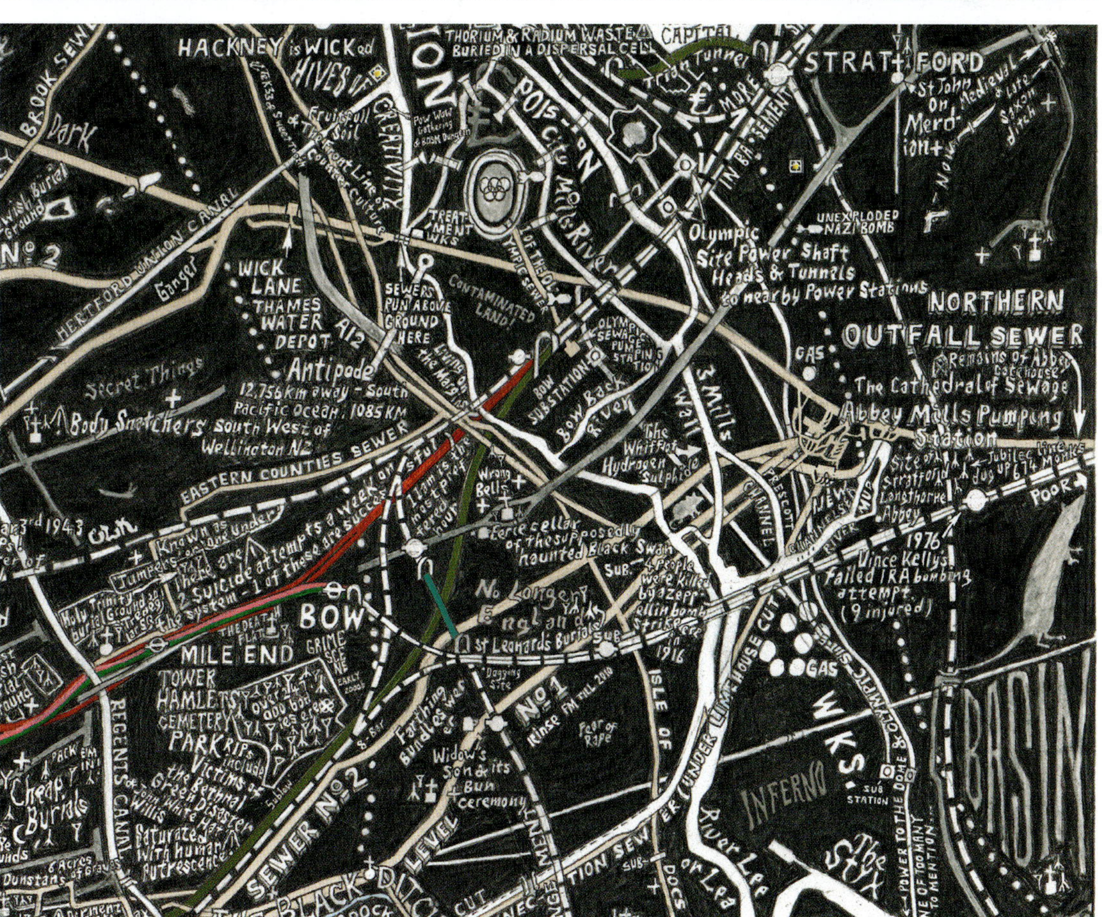

Figure 3
London Subterranea (detail) © Stephen Walter, 2012. Reproduced by kind permission of the artist and courtesy of TAG Fine Arts.

of the Tyburn Tree. It is said to be where a traffic island on Edgware Road meets Marble Arch, itself a Roman road, and where 40,000–60,000 corpses – after being hanged on the famous gallows – were buried in this vicinity, some of which have yet to be exhumed.

At first I assumed that the imaginative side of the project would come from rumours and perceptions of the 'underworld'. However, the lengths to which the authorities go in order to withhold information from the public have made the task of finding and mapping them an adventure in itself.

The bowels of London are its sewers. I celebrate these on my map along with its lost but not yet dead rivers.

It is a map that seeks to be a talisman or a fuel-base for those wishing to make further discoveries.

Society of
Cartographers